T0360092

NELSON

VICscience

Meredith McKague (Consultant)
Helene van Iersel (lead author)
Andrea Blunden
Michael Diamond
Kate Hutchison
Leigh Park
Adam Scanlon
Claire Wallace
Adina Wolters
Natasha Young

Contributing authors
Georgia Dawson
Brenda Dobia

psychology VCE UNITS 1 + 2

VICscience Psychology VCE Units 1 & 2 Student Book
4th Edition
Meredith McKague (series consultant)
Helene Van Iersel (lead author)
Andrea Blunden
Michael Diamond
Kate Hutchison
Leigh Park
Adam Scanlon
Claire Wallace
Adina Wolters
Natasha Young
Georgia Dawson
Brenda Dobia

Publisher: Eleanor Gregory
Content developer: Katherine Roan
Project editors: Felicity Clissold, Alex Chambers, Alana Faigen, Alan Stewart
Copyeditor: Jane Fitzpatrick
Proofreader: Nick Tapp
Indexer: Max McMaster
Series text design: Ruth Comey (Flint Design)
Chapter Maps: Leigh Ashforth (Watershed Design)
Series cover design: Emilie Pfitzner (Everyday Ambitions)
Series designer: Cengage Creative Studio
Permissions researcher: Liz McShane
Production controllers: Karen Young, Alex Chambers
Typeset by: Lumina Datamatics
Any URLs contained in this publication were checked for currency during the production process. Note, however, that the publisher cannot vouch for the ongoing currency of URLs.

Acknowledgements
Selected VCE Examination questions and extracts from the VCE Study Designs are copyright Victorian Curriculum and Assessment Authority (VCAA), reproduced by permission. VCE® is a registered trademark of the VCAA. The VCAA does not endorse this product and makes no warranties regarding the correctness or accuracy of this study resource. To the extent permitted by law, the VCAA excludes all liability for any loss or damage suffered or incurred as a result of accessing, using or relying on the content. Current VCE Study Designs, past VCE exams and related content can be accessed directly at www.vcaa.vic.edu.au .

For product information and technology assistance,
in Australia call **1300 790 853**;
in New Zealand call **0800 449 725**

For permission to use material from this text or product, please email **aust.permissions@cengage.com**

ISBN 978 0 17 046505 2

Cengage Learning Australia
Level 5, 80 Dorcas Street
Southbank VIC 3006 Australia

Cengage Learning New Zealand
Unit 4B Rosedale Office Park
331 Rosedale Road, Albany, North Shore 0632, NZ

For learning solutions, visit **cengage.com.au**

Printed in China by 1010 Printing International Limited.
1 2 3 4 5 6 7 26 25 24 23 22

Contents

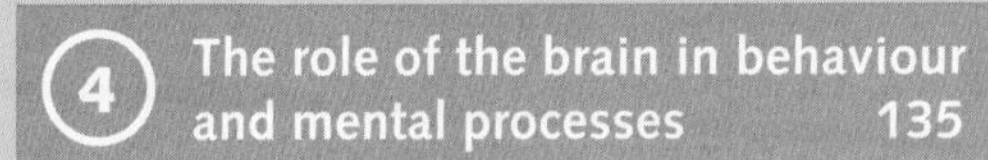

Area of study 2 How are mental processes and behaviour influenced by the brain?

Area of study 2 What influences a person's perception of the world?

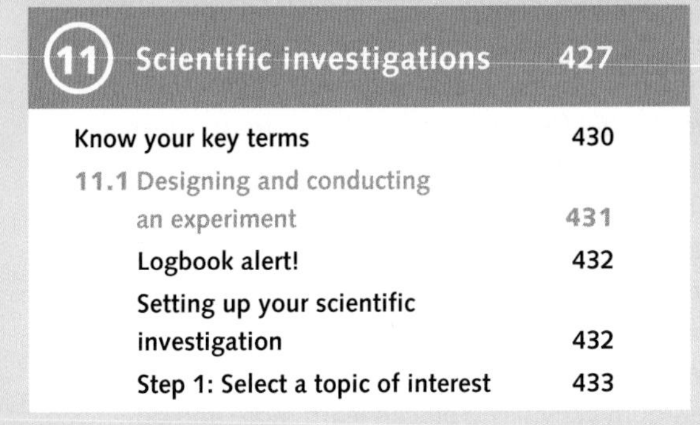
Area of study 3 How do scientific investigations develop understanding of influences on perception and behaviour?
11 Scientific investigations 427

Introduction

VICscience Psychology VCE Units 1 & 2 (fourth edition) has been written to meet the specifications of the *VCAA VCE Psychology Study Design 2023–2027.* Our author team has been chosen for their secondary or tertiary teaching experience and their comprehensive knowledge of the psychology discipline.

Our author team has produced a resource that is comprehensive, engaging and easily accessible to all students of VCE Psychology. The book map presented on pages xxii–xxiii provides a bird's eye view of the content covered in Units 1 & 2, and its interconnectedness. From this map, the content covered by each chapter is further expanded upon in the chapter maps that sit at the start of each chapter. Each chapter map is an easy-to-use visual navigational tool to gently guide students through the story and connections within each chapter.

VCE can be an exciting but sometimes overwhelming time. Student wellbeing has been a major priority in the construction of this resource. The role of colour and its ability to influence feelings, attention and behaviour when learning has been carefully considered in the designs of the cover and chapters. Authors have carefully constructed chapters taking into account choice of words, sentence length and use of diagrams to help explain concepts.

At the core of this approach is the inclusion of cognitive learning strategies within each chapter. These include pre-testing at the beginning of each chapter, promoting retrieval of prerequisite material from long-term memory. Key terms are front-loaded so that students can learn definitions early and retrieve them as they read through each chapter. Five evidence-based principles of effective learning drawn from research in cognitive and educational psychology have been embedded within the Key concept questions and higher-order thinking challenges that accompany the text (Weinstein et al., 2018). These include concrete examples **c**, retrieval practice **r**, interleaving of concepts **i**, elaboration **e** and dual-coding **d** of concepts through linking imagery with language-based definitions. Cognitive learning strategies promote the active construction of knowledge and assist students to make connections between concepts – the effective use of cognitive learning strategies deepens students' understanding and enhances their sense of agency and self-efficacy. The cognitive learning strategies can be easily extended to classroom activities. You can learn more and download freely available materials at http://www.learningscientists.org.

Activities in each chapter have been carefully designed to assist students in moving from lower-order-thinking in the *Try it* section to higher-order-thinking in the *Apply it* and *Exam ready* sections. Students can also practice and build upon their examination skills by attempting graded VCAA exam-style questions in the Area of Study reviews.

The study of Psychology is investigative. Key science skills associated with the scientific research that underpin the study of Psychology have been brought together in Chapter 1 Scientific research methods. These skills have been woven through all the chapters and students can practise applying them in the investigations and analysing research sections.

A dedicated chapter has been provided to assist students with successfully experiencing psychological research. Chapter 11 provides guidelines on how to complete Unit 2 Outcome 3.

VICscience Psychology VCE Units 1 & 2 (fourth edition) gives students a thorough grounding in VCE Psychology and sets them up for an enjoyable and successful year.

Reference: Weinstein, Y., Sumeracki, M., & Caviglioli, O. (2018). *Understanding how we learn: A visual guide.* Routledge.

The study of Psychology has been traditionally viewed from a Western science perspective, but study designers have become increasingly aware of the need to challenge assumptions and include diverse psychological knowledge and experience. Aboriginal and Torres Strait Islander knowledge and perspectives have been provided through engagement with highly qualified academic researchers, First Nations authors and reviewers. Special thanks go to:

Dr Brenda Dobia

Dr Dobia is a psychologist, educator and researcher and an Adjunct Fellow in the School of Education at Western Sydney University. She has a keen interest in issues of wellbeing and social justice that affect young people, particularly First Nations youth. Dr Dobia has led a number of research and educational initiatives focused on improving student wellbeing in schools. These include developing educational resources for the KidsMatter primary schools mental health initiative; reviewing the mental health and wellbeing needs of First Nations students; and evaluating community-based educational programs aimed at building respect and preventing violence.

Author team

Meredith McKague (author and consultant)

Meredith is a cognitive psychologist in the Melbourne School of Psychological Sciences at the University of Melbourne and was the Chief Assessor for the VCE Psychology exam between 2013 and 2020. Meredith is also a non-Indigenous member of the Reference Group and Community of Practice with the Australian Indigenous Psychology Education Project. Meredith conducts research and teaching in the areas of learning, memory and language.

Helene van iersel (lead author)

Helene taught in Victorian secondary schools for over 40 years and has been an assessor of VCAA Psychology for the past fifteen years. Her teaching experience ranged from Yr 7 through to Yr 12 but the bulk of her teaching has been at the senior level. Helene has taught classes in English and Humanities but for the past twenty years Psychology has been her main focus. She has always gained great satisfaction from seeing students deepen their understanding of human behaviour through their Psychology studies. Although Helene is now retired, she continues to tutor students in Psychology.

Andrea Blunden

Andrea has taught Psychology since its inception in the 1990s in both government and non-government schools throughout Victoria. She has also worked as a VCE Psychology exam marker. Andrea enjoys teaching Psychology as she believes it is a subject that is useful and relevant to everybody.

Georgia Dawson

Georgia works across private psychological practice, academic research and teaching at the Melbourne Graduate School of Education, University of Melbourne. She lectures in child development, exceptionalities and learning interventions for several postgraduate courses. As a registered psychologist, Georgia has experience in the primary and tertiary education sectors including individual counselling, psychological assessment of learning difficulties, and implementation of educational interventions. She has worked with a range of concerns related to mental health and educational.

Michael Diamond

Michael works at the intersection of psychology and data science. His areas of specialty within psychology are research methods, statistics, utilising technology, and creating teaching content at both the VCE and the tertiary level. Michael has worked as a Research Consultant for Unforgettable Research Services (a platform for the collection and analysis of data from smartphones, wearable devices, Internet of Things (IoT), and social media) and as the Hub Coordinator for the Complex Human Data Hub within the School of Psychological Sciences at the University of Melbourne. Michael is a recipient of the Australian Psychological Society APS prize, and holds a Bachelor of Psychological Sciences (Honours), a Graduate Diploma in Psychology, and a Bachelor of Medical and Health Science.

Kate Hutchison

Kate Hutchison completed her Science Degree majoring in Psychology at Melbourne University. It was her passion for the field and her love of learning that lead Kate to continue on to complete her Diploma of Education. She has over 10 years experience teaching both in Australia and overseas and has tought extensively across the year 10 to 12 Psychology curriculum. Kate is continually driven to find new and engaging ways for her students to learn and instilling a passion for the subject which has resulted in a number of her students continuing on to complete their own Psychology careers.

Leigh Park

Leigh has been involved in Psychology education for the past 30 years. He has presented at local, national and international conferences with a focus on the connection between Psychology and the outdoors. He is passionate about involving students in their learning to discover more about themselves, to explore the world around them and to develop an empathy for others.

Adam Scanlon

Adam Scanlon has worked as a VCE Psychology teacher for almost a decade in the state education system. He has also worked in school leadership and student management. He is extremely passionate about the subject and encourages students to think deeply about how the concepts of Psychology shape their experiences. In his work as a secondary teacher, he has supported students to achieve consistently high results, including perfect study scores, by focusing on writing quality short-answer responses.

Claire Wallace

Claire is an experienced science teacher as well as exam assessor. She has a degree in Psychology and Biochemistry from the University of Tasmania. Claire's favourite place is being in the classroom with her students, nurturing their interest in science with the belief that 'science is more than a body of knowledge, it is a way of thinking' (Carl Sagan).

Adina Wolters

Adina has a B.A, B.Sc., Dip. Ed. (Psychology and Maths) and is a current teacher of Psychology and Head of Psychology for Years 10–12 at Mount Scopus Memorial College. She has been teaching Psychology for 25 years with a particular passion for experiential learning. She is also a long-term presenter at Psychology teachers' conferences.

Natasha Young

Natasha Young completed a Bachelor of Applied Science (Psychology) in 2007 at RMIT University. Following a gap year, she went on to complete her teacher training in 2009 obtaining a Graduate Diploma in Secondary Education from Latrobe University. She started her teaching career in 2010 at Lakeside Secondary College teaching VCE Psychology. In 2012, she started working at Eltham High School and has been there every since. Throughout her time at Eltham she has worked as a classroom teacher predominantly teaching VCE Psychology and junior Humanities. She also worked as a Senior School coordinator for several years. Additionally, Natasha is trained in the area of Student Welfare and has had some experience working in student services at Eltham High School. This is the second edition of the Nelson Psychology text book she has authored. Natasha is passionate about teaching Psychology and loves to create engaging and practical teaching programs for her students.

Author acknowledgements

Leigh: I would like to thank Joanna, Jazmin and Brooke for their boundless support and fun.
Nicole: I would like to thank my friend and colleague Cheryl Watson for your advice. Thank you to Jim, Simone and Nina for all your belief and encouragement.
Adina: I would like to say thank you to my students for motivating me to continue finding new ways to engage with Psychology course content. You make the journey a fun adventure. Thank you to my husband Ian for his constant support and understanding when I spend countless hours making new activities and to my sons Dylan and Mitchell who let me test the activities out on them while they were studying Psychology.

Publisher acknowledgements

Eleanor Gregory sincerely thanks the following VCE Psychology teachers for their assistance in reviewing and commenting on the author's manuscripts during the development of this product:

- David Anderton, St Margaret's Berwick Grammar
- Elisa Baldwin, Wheelers Hill Secondary College
- Amelia Brear, Eltham High School
- Cheryl Watson, The Geelong College

And special thanks to:
Chalsea Chappel, Victoria University Secondary College – St Albans campus for compiling the Area of Study reviews.

To the student

The VCE Psychology course comprises both key knowledge and key science skills components, which will be assessed throughout your studies. We understand that undertaking VCE Psychology, especially at Units 1 & 2, can be an exciting but sometimes overwhelming time. You will learn a lot of content and develop scientific skills throughout very busy years which will culminate at the end of Units 3 & 4 in an external assessment. We have taken these stressors into account when designing the *VICscience Psychology* suite of products. You will not need to go beyond these learning materials to study VCE Psychology; they have been designed to work in unison so you can achieve at your very best level.

10 steps to study success

Ensure you take time to read the 10 ways we have organised your VCE Psychology journey. You will see that at various stages in your studies, different aspects of this textbook will be more useful. Whether you are learning new key terms and concepts for the first time, reviewing what you have learned or preparing for tests and exams, spending a little time now getting to know your textbook and what it offers will help you reach your learning potential for VCE Psychology.

1 Focus on the Study Design

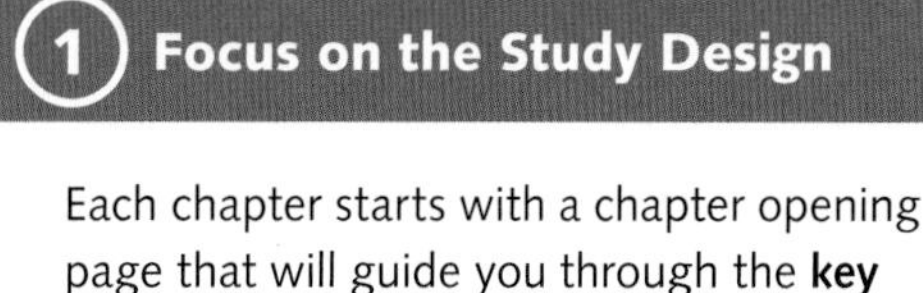

Each chapter starts with a chapter opening page that will guide you through the **key knowledge** and **key science skills** that are covered within the chapter.

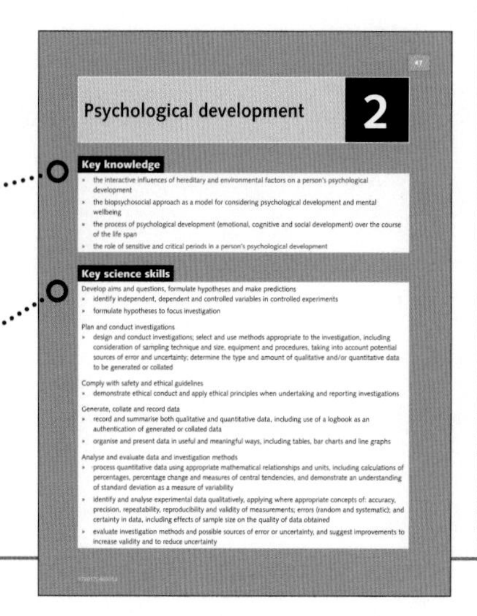

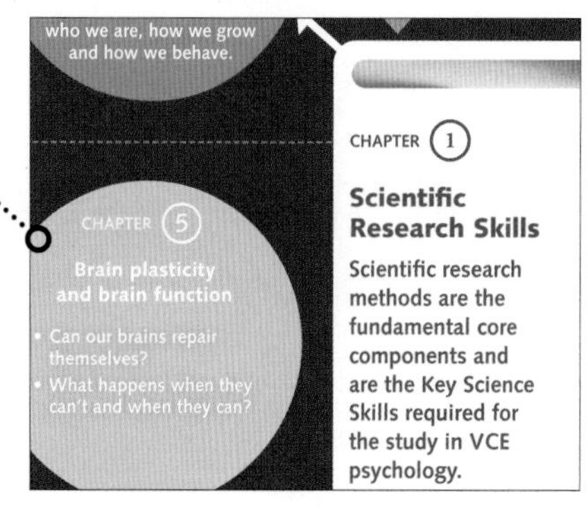

2 Overview of the VCE Psychology Study Design

You will find a book map on pages xxii–xxiii. The book map provides a bird's eye view of the content and how it is interconnected. From this map, the content covered by each chapter is further expanded upon in the chapter maps located at the start of each chapter. Each chapter map is an easy-to-use visual navigational tool to gently guide you through the story and connections within each chapter. Each chapter map:

- locates the chapter within the course
- enables you to see how the information in the chapter fits together
- is an easy-to-use navigational tool to guide you through each chapter
- offers a gentle entry into the more complex information.

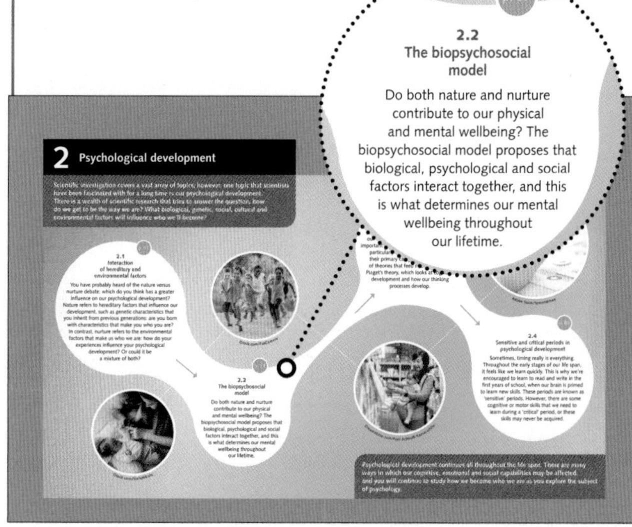

3 Remember, rehearse and retrieve key terms

We have frontloaded all the new key terms you will meet throughout the chapter at the beginning of each chapter in **Know your key terms**. Use the **flashcards** study tool to encode key terms along with their definitions. Then when you come across a key term in your reading you can retrieve the definition from your memory.

The definitions of these key terms are also found in the glossary at the back of the book.

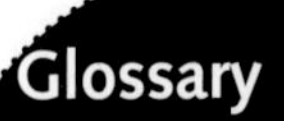

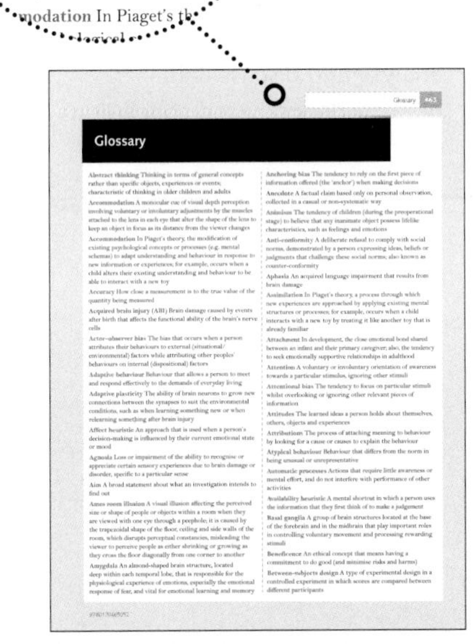

4 Test your retrieval

At the beginning of each chapter, use the **chapter pre-test** to assist you to retrieve from your memory previously learned concepts. You will assimilate these previously learned concepts into the new concepts as you progress through the chapter. Stronger foundations of knowledge make learning easier.

5 Develop your skills

Key science skills are an integral part of Psychology and they are examinable in the external assessment. **Chapter 1** focuses on all the key science skills so this is a good place to start.

Investigations within the textbook focus on developing and practising your key science skills.

Analysing research within chapters enables you to read a piece of psychological research and apply your skills to understand, analyse and evaluate it.

To further develop and refine the key science skills set out in the course, complete the activities in the accompanying *VICscience Psychology Skills Workbook*. Signposts to workbook activities are found throughout the textbook.

WB
2.1.3 EVALUATION OF RESEARCH

6 Understand the concepts

It is not only pictures that tell a thousand words, but tables and graphs as well. They are key to strengthening your understanding. Ensure you look carefully at each **figure** and **table** and read the labels and captions so that you can understand what they are telling you.

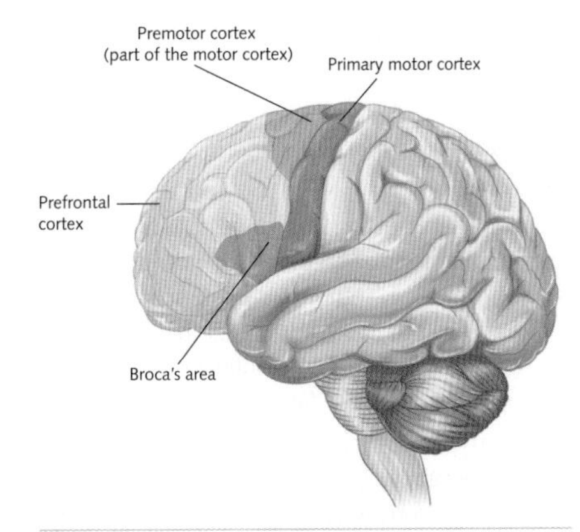

Figure 4.17 The structure of the frontal lobes

Important ideas, concepts and theories are summarised in **Key concept boxes**. **Concept questions** follow each Key concept box. These questions will help you to determine whether you have fully understood the content before you progress further in the chapter.

KEY CONCEPTS 5.3

- All neurological disorders are the result of damage to the nervous system.
- Neurological disorders affect the brain, the spinal cord, and the nerves that connect them.
- Neurological disorders can affect biological, psychological and social functioning.
- Research may be in vitro or in vivo.
- Researchers often use animal models to investigate the causes, progress and treatment of neurological disorders in humans.
- Neuroimaging technologies such as CT scans, MRI scans and PET scans provide images of the physiological impacts of neurological disorders.
- New neuroimaging technologies such as DTI, MRS, the trimodal model and MEG can provide even more detailed images of the workings of the brain.

Concept questions 5.3

Remembering

1 What is a neurological disorder? r
2 List three examples of a neurological disorder. r

Applying

7 A neuropsychologist was presented with a patient who had a suspected ABI. Explain how

Concept questions have been carefully structured using cognitive learning strategies to use the power of your brain to maximise your learning efficiency. Five key learning strategies have been signposted:

Retrieval practice **r**
Elaboration **e**
Concrete examples **c**
Dual coding **d**
Interleaving **i**

You can find out more about these strategies at https://www.learningscientists.org/posters.

If you are feeling confident with the concepts you can give the **HOT Challenges** a go! You can find these at the end of Concept questions. These questions require higher cognitive application and may need further research. Try to give them a go as they will extend your understanding to a higher level.

7 Explore and learn

You will collaborate, explore and discover the psychological world through **investigations** and also come to appreciate the collegial nature of VCE Psychology.

Explore key knowledge and develop, use and demonstrate the key science skills through the Investigations. Investigations provide an opportunity to:

- explore the different methodologies
- use the logbook template provided to build up your logbook
- analyse secondary data
- collect and display your own primary data
- analyse and discuss data
- draw evidence-based conclusions
- discover psychology for yourself.

INVESTIGATION 8.1
Conformity at school
Scientific investigation m
Fieldwork
Aim
To investigate examples of
roduction

y a need to fit i
nformity include norma
information about these fac
Procedure
In this investigation you will d
environment. You are to work
examples of when they have
different year levels to iden
lyse the questionnai

ment. You are
mples of when they ha
ifferent year levels to ident
analyse the questionnaire resp
Results
Record, organise and present d
appropriate mathematical rela
Discussion
What was the size an

INVESTIGATION 8.1

Conformity at school

Scientific investigation methodology

Fieldwork

Aim

To investigate examples of conformity within the school environment

Introduction

Conformity is adjusting one's beliefs, judgements or actions to bring them into agreement with those of a social group or situation. Conformity occurs when individuals change their behaviour as the result of real or implied pressure from others in a range of social settings including schools and professional settings. Research suggests that conformity is influenced both by a need to fit in and a belief that other people are smarter or better informed than you are. Factors that affect conformity include normative influence, informational influence, unanimity, deindividuation and social loafing. Revise information about these factors using your textbook or other resources.

Procedure

In this investigation you will develop, conduct and analyse a questionnaire that investigates conformity in your school environment. You are to work with your classmates to develop a questionnaire that helps fellow students to identify examples of when they have conformed and why. You may choose to conduct the survey on a range of students across different year levels to identify differences in conformity across age groups. Once you have carried out the survey, analyse the questionnaire responses to identify the range of factors affecting conformity.

Results

Record, organise and present data using an appropriate table and/or graph. Process any quantitative data using appropriate mathematical relationships and units.

8 Complete your outcome

A **dedicated chapter** within your textbook will guide you through the research and investigation outcomes. **Chapter 11** provides you with important advice and steps you through Unit 2 Outcome 3. It has been written to help you, so make full use of it.

9 Prepare for tests and exams

Activities appear within each chapter. Each activity is divided into three sections. The first section, *Try it*, challenges you to do something at a low cognitive level. The second section, *Apply it*, asks you about what you have just done at a higher cognitive level, and the third section, *Exam ready*, gets you to use what you have learned so far in the activity to answer an exam-style question.

You can then take the skills that you learn from completing the activities into the **Area of Study reviews** at the end of each Area of Study. These allow you to check your knowledge by completing exam-style questions that are graded for difficulty.

You will find the answers to the activities and Area of Study reviews at the back of the book.

10 Consolidate your learning

At the end of every chapter, you can consolidate your knowledge using:

- **chapter summary of key concepts** that you have met throughout the chapter. You can download a copy of the concepts by accessing the MindTap icon. Use this to assist you in revising and studying for internal and external assessments
- **end-of-chapter exam** will help you to retrieve, revise, understand and apply the concepts from the chapter.

A **glossary** of all the key terms plus their definitions can be found at the end of the book.

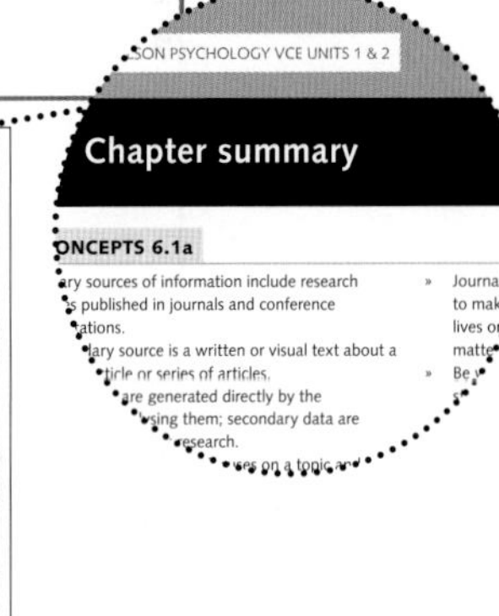

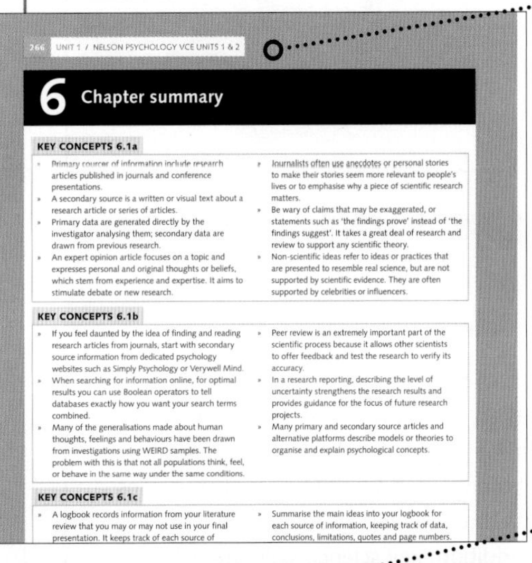

266 UNIT 1 / NELSON PSYCHOLOGY VCE UNITS 1 & 2

6 Chapter summary

KEY CONCEPTS 6.1a

- Primary sources of information include research articles published in journals and conference presentations.
- A secondary source is a written or visual text about a research article or series of articles.
- Primary data are generated directly by the investigator analysing them; secondary data are drawn from previous research.
- An expert opinion article focuses on a topic and expresses personal and original thoughts or beliefs, which stem from experience and expertise. It aims to stimulate debate or new research.
- Journalists often use anecdotes or personal stories to make their stories seem more relevant to people's lives or to emphasise why a piece of scientific research matters.
- Be wary of claims that may be exaggerated, or statements such as 'the findings prove' instead of 'the findings suggest'. It takes a great deal of research and review to support any scientific theory.
- Non-scientific ideas refer to ideas or practices that are presented to resemble real science, but are not supported by scientific evidence. They are often supported by celebrities or influencers.

KEY CONCEPTS 6.1b

- If you feel daunted by the idea of finding and reading research articles from journals, start with secondary source information from dedicated psychology websites such as Simply Psychology or Verywell Mind.
- When searching for information online, for optimal results you can use Boolean operators to tell databases exactly how you want your search terms combined.
- Many of the generalisations made about human thoughts, feelings and behaviours have been drawn from investigations using WEIRD samples. The problem with this is that not all populations think, feel, or behave in the same way under the same conditions.
- Peer review is an extremely important part of the scientific process because it allows other scientists to offer feedback and test the research to verify its accuracy.
- In a research reporting, describing the level of uncertainty strengthens the research results and provides guidance for the focus of future research projects.
- Many primary and secondary source articles and alternative platforms describe models or theories to organise and explain psychological concepts.

KEY CONCEPTS 6.1c

- A logbook records information from your literature review that you may or may not use in your final presentation. It keeps track of each source of
- Summarise the main ideas into your logbook for each source of information, keeping track of data, conclusions, limitations, quotes and page numbers.

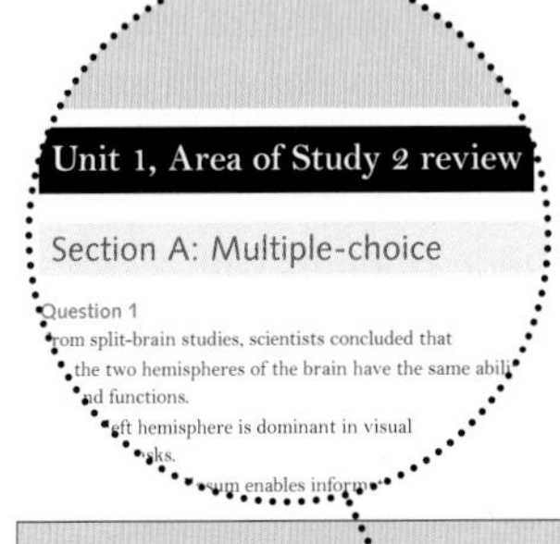

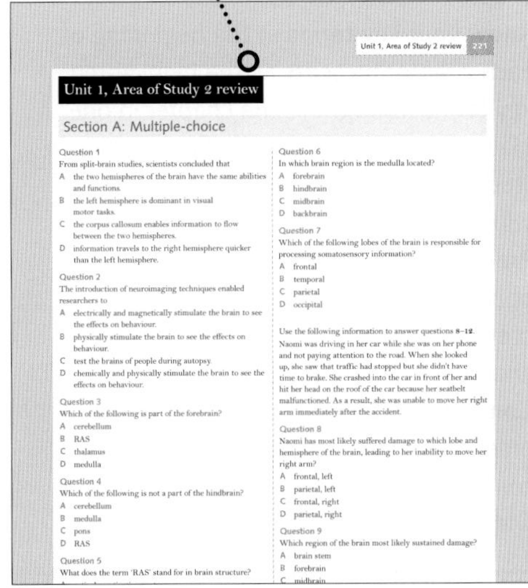

Unit 1, Area of Study 2 review 221

Unit 1, Area of Study 2 review

Section A: Multiple-choice

Question 1
From split-brain studies, scientists concluded that
A the two hemispheres of the brain have the same abilities and functions.
B the left hemisphere is dominant in visual motor tasks.
C the corpus callosum enables information to flow between the two hemispheres.
D information travels to the right hemisphere quicker than the left hemisphere.

Question 2
The introduction of neuroimaging techniques enabled researchers to
A electrically and magnetically stimulate the brain to see the effects on behaviour.
B physically stimulate the brain to see the effects on behaviour.
C test the brains of people during autopsy.
D chemically and physically stimulate the brain to see the effects on behaviour.

Question 3
Which of the following is part of the forebrain?
A cerebellum
B RAS
C thalamus
D medulla

Question 4
Which of the following is not a part of the hindbrain?
A cerebellum
B medulla
C pons
D RAS

Question 5
What does the term 'RAS' stand for in brain structure?

Question 6
In which brain region is the medulla located?
A forebrain
B hindbrain
C midbrain
D backbrain

Question 7
Which of the following lobes of the brain is responsible for processing somatosensory information?
A frontal
B temporal
C parietal
D occipital

Use the following information to answer questions 8–12.
Naomi was driving in her car while she was on her phone and not paying attention to the road. When she looked up, she saw that traffic had stopped but she didn't have time to brake. She crashed into the car in front of her and hit her head on the roof of the car because her seatbelt malfunctioned. As a result, she was unable to move her right arm immediately after the accident.

Question 8
Naomi has most likely suffered damage to which lobe and hemisphere of the brain, leading to her inability to move her right arm?
A frontal, left
B parietal, left
C frontal, right
D parietal, right

Question 9
Which region of the brain most likely sustained damage?
A brain stem
B forebrain
C midbrain

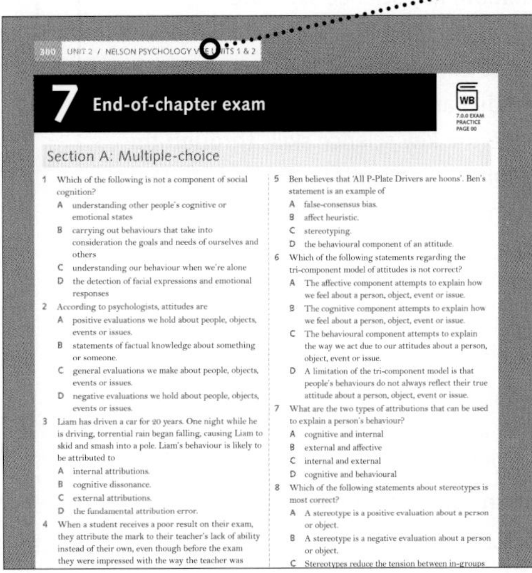

300 UNIT 2 / NELSON PSYCHOLOGY VCE UNITS 1 & 2

7 End-of-chapter exam

WB 7.0.0 EXAM PRACTICE PAGE 00

Section A: Multiple-choice

1 Which of the following is not a component of social cognition?
A understanding other people's cognitive or emotional states
B carrying out behaviours that take into consideration the goals and needs of ourselves and others
C understanding our behaviour when we're alone
D the detection of facial expressions and emotional responses

2 According to psychologists, attitudes are
A positive evaluations we hold about people, objects, events or issues.
B statements of factual knowledge about something or someone.
C general evaluations we make about people, objects, events or issues.
D negative evaluations we hold about people, objects, events or issues.

3 Liam has driven a car for 20 years. One night while he is driving, torrential rain began falling, causing Liam to skid and smash into a pole. Liam's behaviour is likely to be attributed to
A internal attributions.
B cognitive dissonance.
C external attributions.
D the fundamental attribution error.

4 When a student receives a poor result on their exam, they attribute the mark to their teacher's lack of ability instead of their own, even though before the exam they were impressed with the way the teacher was

5 Ben believes that 'All P-Plate Drivers are hoons'. Ben's statement is an example of
A false-consensus bias.
B affect heuristic.
C stereotyping.
D the behavioural component of an attitude.

6 Which of the following statements regarding the tri-component model of attitudes is not correct?
A The affective component attempts to explain how we feel about a person, object, event or issue.
B The cognitive component attempts to explain how we feel about a person, object, event or issue.
C The behavioural component attempts to explain the way we act due to our attitudes about a person, object, event or issue.
D A limitation of the tri-component model is that people's behaviours do not always reflect their true attitude about a person, object, event or issue.

7 What are the two types of attributions that can be used to explain a person's behaviour?
A cognitive and internal
B external and affective
C internal and external
D cognitive and behavioural

8 Which of the following statements about stereotypes is most correct?
A A stereotype is a positive evaluation about a person or object.
B A stereotype is a negative evaluation about a person or object.
C Stereotypes reduce the tension between in-groups

To the teacher

The VCE Psychology course comprises both key knowledge and key science skills. The *VICscience Psychology* suite of products provides you with the perfect resource to teach all the key knowledge and key science skills in an integrated and engaging way and to prepare your students thoroughly for the school-based and external assessments.

Follows the Study Design

This textbook has been written so that all of its content closely aligns with the *VCAA VCE Psychology Study Design* (2023–2027). It has been authored and reviewed by experienced Psychology teachers, academics, researchers and First Nations consultants to ensure up-to-date scientific and accurate content for students.

Integrates key science skills into the course

Chapter 1 *Scientific research methods* provides an overview of the **key science skills** in the VCE Psychology Study Design. This is built upon in Chapter 11, which guides students through Unit 2 Outcome 3.

In addition, key science skills are integrated into each chapter.

Students get to explore relevant key knowledge and to develop, use and demonstrate the key science skills through the **Investigations**. Investigations provide students an opportunity to:

- explore the different methodologies
- use the logbook template provided to build up their logbook
- analyse secondary data
- collect and display their own primary data
- analyse and discuss data
- draw evidence-based conclusions
- communicate their findings.

Analysing research sections provide students with the opportunity to read authentic psychological research and then apply their understanding of the key science skills by answering the questions that follow.

The *VICscience Psychology Skills Workbook* provides a great source of skill-based activities. Each activity is aligned to the key knowledge and enables students to develop and refine their key science skills. Signposts to the workbook activities are found throughout the textbook. See the chapter teaching plans to see how the textbook and workbook work together.

3 Gives you access to differentiated material

Differentiation is built into each chapter to assist you in helping those students that may struggle with content or skill development and extending those students who want to achieve at a higher level.

- **Chapter maps** provide students with a gentle and visual introduction to each chapter, enabling students to engage with the chapter content prior to entering the chapter.
- **Know your key terms** at the beginning of each chapter presents all the highlighted key terms throughout the chapter in one place. Students can use the flashcards study tool to learn and review key terms with their definitions.
- **Pre-tests** provide students the opportunity to retrieve concepts previously learned that will be revisited and built upon during the chapter.
- **Concept questions** are pitched to be lower-order questions to assist with learning consolidation, but each question set ends with HOT Challenge questions for those

students who would benefit from answering higher-order thinking questions.

- **Worksheets** with associated figures and tables allow students to analyse the information in more depth.
- **Activities** are designed to shift students from lower order thinking in *Try it*, to higher order thinking in *Apply it*. Exam-style questions are provided in *Exam ready* to illustrate how this concept may be examined.
- **Investigations** provide students with the opportunity to experience psychology research.
- **Weblinks** to external, vetted websites provide extra information.
- **End-of-chapter exams** and **Area of Study reviews** provide students with experience at answering exam-style questions.

4 Prepare for the exam

Students of VCE Psychology are working toward external assessment at the end of Units 3 & 4. To fully prepare for this exam, students require access to a large number of quality exam-style questions with answers. *VICscience Psychology* gives you the full complement including the following.

- **End-of-chapter exams** consist of 20 difficulty-graded multiple-choice and three difficulty-graded short-answer questions written in VCAA style.
- **Area of study reviews** at the end of each Area of Study provide students with 30 difficulty-graded multiple-choice and five short-answer questions that have been adapted from VCAA exam questions, with answers provided at the back of the book.
- *VICscience Psychology VCE Units 1 & 2 Skills Workbook* develops the skills required by students to confidently answer scientific research questions.
- **examplus (Units 3 & 4 only)** simulates real exam practice and comprises thousands of unseen exam-style and past VCAA exam questions with answers to use in your teaching. Simply select your questions for a quiz, topic test or practice exam and examplus generates a practice test or exam.

- Consider bundling **A+ Study Notes** and **Practice Exams** with your *VICscience Psychology* booklist for the most economical solution for students' exam preparation and readiness.

5 Support for the teacher

There is a wealth of teacher support materials on the MindTap Schools site that accompanies this product. These include:

- **answers** to all textbook questions, activities, case studies, analysing research, investigations, end-of-chapter exams and Area of study reviews.
- **sample SACs** with suggested marking schemes (Units 3 & 4 only)
- **unit practice exams** with exemplar answers
- **slideshow summaries** to use in a flipped classroom
- **teaching plans** for every chapter showing how all the components of the *VICscience Psychology* suite are integrated to provide your students with a thorough and complete learning experience designed to prepare them for internal and external assessment.

UNIT 1 How are behaviour and mental processes shaped?

AREA OF STUDY 1

CHAPTER 3

Defining and supporting psychological development

- How do influencing factors interact, or are they independent?
- How can we measure and study this complexity?

CHAPTER 2

Psychological development

Research has shown that the complexity of psychological development cannot be underestimated. What internal and external factors influence who we are, how we grow and how we behave?

CHAPTER 1

Scientific research methods

Scientific research methods are the fundamental core components and are the Key Science Skills required for study in VCE psychology.

AREA OF STUDY 2

CHAPTER 4

Role of the brain in behaviour and mental processes

Our brain is made up of different parts that are responsible for how we behave, how we think and how we feel.

CHAPTER 5

Brain plasticity and brain injury

- Can our brains repair themselves?
- What happens when they can't and when they can?

AREA OF STUDY 3

CHAPTER 6

Contemporary psychological research

- Can we really study our thoughts, feelings and behaviours?
- How can we research and study such complex humans? Surely there are limitations?

UNIT 2 How do internal and external factors influence behaviour and mental processes?

AREA OF STUDY 1

CHAPTER  7

Social cognition

- Is my truth the same as your truth?
- Do we all experience the world in the same way?
- Can people influence and change the way we live in the world?

CHAPTER  8

Factors that influence individual and group behaviour

- How do we develop our values and rules for living?
- Do we behave differently on our own or in a group?
- Is conformity inherent or are we influenced by those around us?

AREA OF STUDY 2

CHAPTER  9

Perception

- We do we make sense of a world exploding with sights, smells, noise, tastes and textures.
- How do we notice it all?
- What information about the world is important for us to function?

CHAPTER  10

Distortions of perception

- How good is our brain with information?
- What does it know and what does it infer?
- Do you see what I see?
- What is real?

AREA OF STUDY 3

CHAPTER  11

Scientific investigations

- How can we study how we live in our multisensory world?
- What methods do we use and how do we communicate them?

Nelson MindTap

An online learning space that provides students with tailored learning experiences.

- Access tools and content that make learning simpler yet smarter to help you achieve mastery.
- Includes an eText with integrated interactives and online assessment.
- Margin links in the student book signpost multimedia student resources found on MindTap.

For students:

Nelson MindTap provides you with material that will help you understand, explore, engage and organise the key knowledge and key science skills you have learned about in your textbook. On MindTap, you will find chapter resources such as:

- Interactive ebook
- Key term flashcards
- Chapter pre-test
- Interactive learning activities
- Worksheets
- Slide show of key concepts for each chapter
- Weblinks to online videos and further information
- Templates to assist you in completing statistical analysis of your data
- Downloadable chapter maps and Key concept checklists

For teachers*:

Nelson MindTap allows you to teach in a way that caters to the needs of your students. Monitor student progress and customize your course in a way that makes sense for your classroom. In addition to the resources found on students' MindTap, you will also find teacher support materials such as:

- Teaching plans
- Logbook templates

* Complimentary access to these resources is only available to teachers who use this book as part of a class set, book hire or booklist. Contact your Cengage Education Consultant for information about access and conditions

Content warning This resource includes discussion of topics that may cause distress to Aboriginal and Torres Strait Islander people, including the discussion of racist policies and trauma associated with events such as massacres and forced removal of children. The content may also include reference to the names, images and voices of people who have died.

1 Scientific research methods

Key science skills

Develop aims and questions, formulate hypotheses and make predictions

» identify, research and construct aims and questions for investigation

» identify independent, dependent and controlled variables in controlled experiments

» formulate hypotheses to focus investigation

» predict possible outcomes of investigations

Plan and conduct investigations

» determine appropriate investigation methodology: case study; classification and identification; controlled experiment (within subjects, between subjects, mixed design); correlational study; fieldwork; literature review; modelling; product, process or system development; simulation

» design and conduct investigations; select and use methods appropriate to the investigation, including consideration of sampling technique (random and stratified) and size to achieve representativeness, equipment and procedures, taking into account potential sources of error and uncertainty; determine the type and amount of qualitative and/or quantitative data to be generated or collated

» work independently and collaboratively as appropriate and within identified research constraints, adapting or extending processes as required and recording such modifications

Comply with safety and ethical guidelines

» demonstrate ethical conduct and apply ethical guidelines when undertaking and reporting investigations

» demonstrate safe laboratory practices when planning and conducting investigations by using risk assessments that are informed by safety data sheets (SDS) and accounting for risks

» apply relevant occupational health and safety guidelines while undertaking practical investigations

Generate, collate and record data

» systematically generate and record primary data and collate secondary data appropriate to the investigation

» record and summarise both qualitative and quantitative data, including use of a logbook as an authentication of generated or collated data

» organise and present data in useful and meaningful ways, including tables, bar charts and line graphs

Analyse and evaluate data and investigation methods

» process quantitative data using appropriate mathematical relationships and units, including calculations of percentages, percentage change and measures of central tendencies (mean, median, mode) and demonstrate an understanding of standard deviation as a measure of variability

» identify and analyse experimental data qualitatively, applying where appropriate concepts of: accuracy, precision, repeatability, reproducibility and validity; errors; and certainty in data, including effects of sample size on the quality of data obtained

- identify outliers and contradictory or incomplete data
- repeat experiments to ensure findings are robust
- evaluate investigation methods and possible sources of error or uncertainty and suggest improvements to increase validity and to reduce uncertainty

Construct evidence-based arguments and draw conclusions

- distinguish between opinion, anecdote and evidence and scientific and non-scientific ideas
- evaluate data to determine the degree to which the evidence supports the aim of the investigation and make recommendations, as appropriate, for modifying or extending the investigation
- evaluate data to determine the degree to which the evidence supports or refutes the initial prediction or hypothesis
- use reasoning to construct scientific arguments and to draw and justify conclusions consistent with evidence base and relevant to the question under investigation
- identify, describe and explain the limitations of conclusions, including identification of further evidence required
- discuss the implications of research findings and proposals, including appropriateness and application of data to different cultural groups and cultural biases in data and conclusions

Analyse, evaluate and communicate scientific ideas

- use appropriate psychological terminology, representations and conventions, including standard abbreviations, graphing conventions and units of measurement
- discuss relevant psychological information, ideas, concepts, theories and models and the connections between them
- analyse and explain how models and theories are used to organise and understand observed phenomena and concepts related to psychology, identifying limitations of selected models/theories
- critically evaluate and interpret a range of scientific and media texts (including journal articles, mass media communications, opinions, policy documents and reports in the public domain), processes, claims and conclusions related to psychology by considering the quality of available evidence
- analyse and evaluate psychological issues using relevant ethical concepts and principles, including the influence of social, economic, legal and political factors relevant to the selected issue
- use clear, coherent and concise expression to communicate to specific audiences and for specific purposes in appropriate scientific genres, including scientific reports and posters
- acknowledge sources of information and assistance and use standard scientific referencing conventions

Source: VCE Psychology Study Design (2023–2027), pp. 12–13

1 Scientific research methods

You have no doubt been learning about the principles of scientific research during your science classes. These principles underpin all science investigations including all psychological investigations. Scientists always follow the same systematic steps: proposing and investigating hypotheses, collecting and analysing data and drawing evidence-based conclusions from this data.

p. 8

1.1
The process of psychological research investigations

All scientific research starts by asking a question: What? Who? Why? When? Where? Once you have defined your research question, there is a particular process that you can use to design a study to answer the question. Just make sure you record everything you do in your logbook!

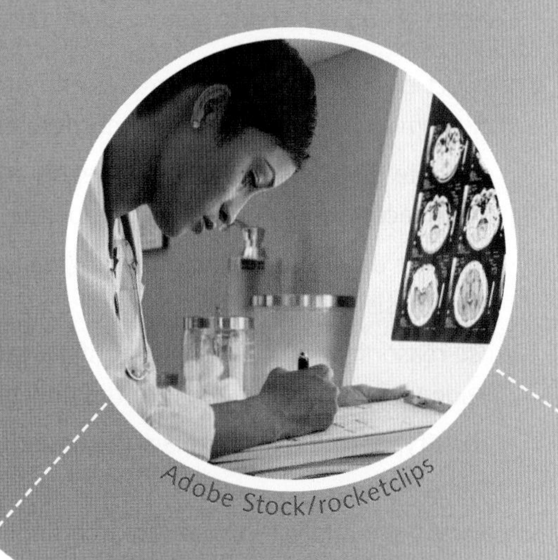

Adobe Stock/rocketclips

p. 14

1.2
Scientific investigation methodologies

You are probably most familiar with the controlled experiment; however, there are many other types of scientific investigation methodologies out there. Each method has its strengths and weaknesses, and is more suited to certain types of research questions than others.

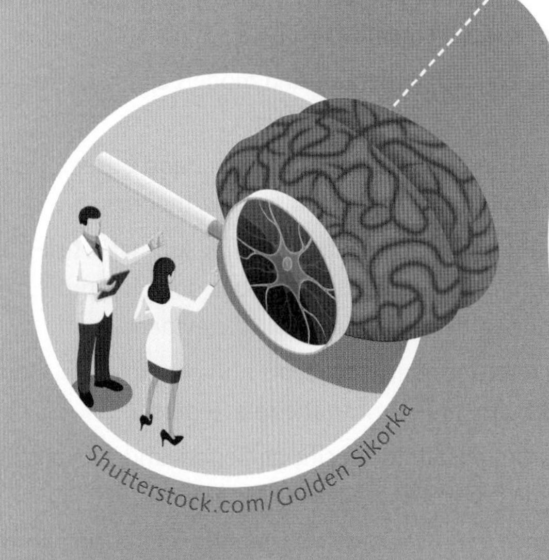

Shutterstock.com/Golden Sikorka

p. 21

1.3 The controlled experiment in detail

In a controlled experiment, it is important to properly identify the different types of variables involved. Systematically manipulating a particular variable in a controlled experiment can help us find a cause-and-effect relationships between two variables (if there is one). Be careful when you are deciding what type of experimental design to use, because it will affect how you select the participants in your study.

iStock.com/SolStock

p. 29

1.4 Analysing and evaluating research

Now it is time to work out what the data are telling you. Researchers use descriptive statistics to analyse data, and often organise data into tables and graphs to make it easier to see trends (if there are any). Not only do we have to analyse the data, but we also need to check the quality of the data to make sure that any conclusions drawn are valid.

Shutterstock.com/Thanakorn.P

Scientific investigation permeates all aspects of science. Only through investigation do we find out new information, but for this information to be valued by the scientific community it must be carried out in a systematic and accepted form. As you continue with your studies of psychology, always ask questions. You never know, you might just come up with a new area of research!

Test
Chapter 1 pre-test

Flashcards
Chapter 1 flashcards

Slideshow
Chapter 1 Slideshow

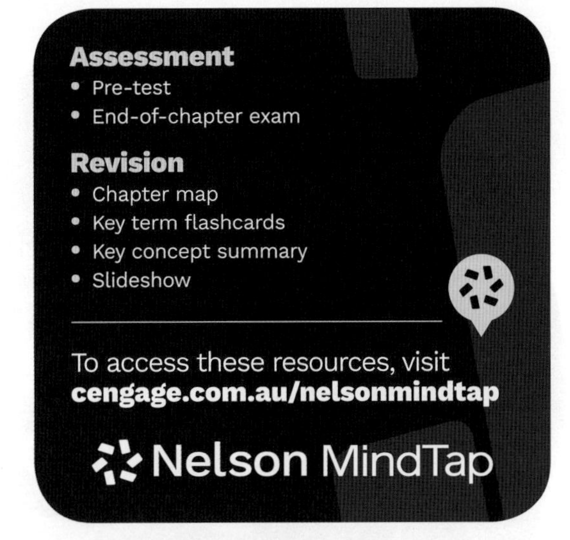

Know your key terms

Accuracy
Aim
Beneficence
Between-subjects design
Case study
Conclusion
Confidentiality
Confounding variable
Control group
Controlled experiment
Controlled variable
Convenience sampling
Correlational study
Data
Debriefing
Deception
Dependent variable (DV)
Descriptive statistics
Ethical concepts
Ethical guidelines
Experimental group
External validity
Extraneous variable (EV)
Fieldwork
Hypothesis
Independent variable (IV)
Informed consent
Integrity
Internal validity
Justice
Limitations literature review
Logbook
Mean
Measure of central tendency
Median
Mixed design
Mode
Non-maleficence
Order effect
Outlier
Population
Precision
Primary data
Psychological construct
Psychological model
Psychological theory
Qualitative data
Quantitative data
Random allocation
Random error
Random sampling
Repeatability
Representative sample
Reproducibility
Research question
Respect
Risk assessment
Sample
Sample size
Sampling
Secondary data
Standard deviation (SD)
Stratified sampling
Systematic error
True value
Uncertainty
Validity
Variability
Variable
Voluntary participation
Withdrawal rights
Within-subjects design

Psychological research aims to understand aspects of the mind, brain and behaviour. This is no easy task, because human thoughts, feelings and behaviours are complicated and cannot always be observed directly. Researchers use the scientific method to discover new information about psychological processes. The scientific method is a well-defined, step-by-step process that allows us to generate new knowledge. It is used in all subfields of psychology, including clinical psychology, biological psychology, developmental psychology, cognitive psychology, personality psychology and social psychology. The scientific method is also the process used in other sciences, such as physics, chemistry, biology and environmental science.

Research psychologists develop models and theories to organise and explain the psychological concepts and processes they are interested in. In psychological science, psychological concepts are expressed and defined as psychological constructs. **Psychological constructs** are terms used in psychology that define specific psychological structures, mechanisms and processes that are thought to be the basis of behaviour and mental experiences. For example, the word 'attention' has an everyday sense that people use and understand. However, in psychology the term attention is a psychological construct that has a much more precise definition than its everyday meaning. Unit 4 of the VCE Psychology Study Design asks you to consider 'sleep as a psychological construct'. This means that you need to study sleep as a concept that is precisely defined by a set of psychological, physiological and neurological properties and processes. Psychological science aims to reduce uncertainty about the true nature of the mind, brain and behaviour by developing our understanding of psychological constructs and how they relate in ever more precise ways.

A **psychological theory** is an organised set of interrelated psychological constructs that describes and/or explains a psychological system, process or experience. For example, a psychological theory of the memory system is made up of a set of constructs, mechanisms and processes that are involved in remembering. Similarly, a psychological theory about racism comprises a set of ideas about the beliefs, attitudes and unconscious biases that drive racist behaviours.

Psychological models are used to make the ideas within a theory more concrete. Models come in a variety of forms. Some examples of models include Ivan Pavlov's model of classical conditioning, Richard Lazarus and Susan Folkman's transactional model of stress and coping, and Richard Atkinson and Richard Shiffrin's multi-store model of memory. Models and theories are built from our current understanding, so they can have limitations, including incorrect assumptions or oversimplifications. When there are inconsistencies between the predictions of a model or a theory and the results gathered from investigations, this indicates that the model or theory may be incomplete or wrong. With new evidence, researchers update theories and models to describe processes more accurately.

Psychology researchers conduct research investigations to gather evidence to better understand psychological processes. When designing a study, the researcher must choose a **population** of interest that they wish to investigate. This population of interest will relate to the construct they are investigating. For example, the population of interest could be:

» students
» teenagers
» women
» people diagnosed with anxiety.

Often in psychology, the phenomenon being investigated applies to all humans (for example, memory processes), so the population of interest can also be as broad as the entire human population. It isn't practical to include every person from the population of interest in a research investigation, so researchers will recruit a smaller sample of participants. A **sample** is a group of people who are recruited from a larger population of interest for the research. As we will explore later, the validity of research findings for the population of interest depends on how well the sample represents that population.

To generate new knowledge from a study, researchers break down the processes, systems or constructs they are investigating into variables that can be measured. A **variable** is any factor in a study that can vary in its score, amount or type and that can be measured, recorded or manipulated. In psychological science, variables can be factors such as:

» individual characteristics (for example, a personality trait such as extraversion)
» properties of a stimulus (for example, brightness of light)
» behaviours (for example, responses on a memory test)
» processes (for example, excitation of neurons).

Some other examples of variables include height, gender, ethnicity, employment status, reaction time on a hazard perception test and happiness scores on a survey. You will notice that all of these variables are phenomena (that is, things or events) that can:

» be measured
» take a range of values
» change over time
» be different between people or between groups.

Defining and measuring variables allows researchers to investigate how factors or psychological constructs are related to each other. Discovering relationships between variables allows researchers to draw conclusions about psychological processes.

When we interpret the conclusions of scientific research, we must consider whether any claims made are valid. **Validity** takes two forms:

internal validity and external validity. **Internal validity** relates to how effective the design and measures used in a study are for understanding the research question. **External validity** indicates how well the results of a study can be applied meaningfully to real-world contexts, situations and behaviours. Researchers must make careful decisions when designing a research investigation because their choices affect the validity of the study's conclusions.

This chapter introduces you to the scientific research methods used in psychology. This information will allow you to understand different methods of scientific investigation, develop key science skills and integrate the links between knowledge, theory and practice. Most importantly, it will deepen your understanding of the tools that psychology researchers use to generate new knowledge about the mind, brain and behaviour. We hope that it will spark your interest in this marvellous discipline. In the words of Ivan Pavlov:

"Do not become a mere recorder of facts, but try to penetrate the mystery of their origin."

1.1 The process of psychological research investigations

How do I use the research process to investigate a psychological phenomenon that interests me? In this section, we explore the step-by-step process that researchers use to design a study to answer a question about psychology. These steps are shown in Figure 1.1.

The steps in Figure 1.1 are:

1 decide on a research question that you will attempt to answer
2 choose an aim to focus your research
3 create a broad hypothesis that you wish to test
4 design a study to test the hypothesis
5 generate a specific hypothesis that predicts the results/outcomes of the study
6 conduct the study and collect data
7 analyse data and report results
8 draw evidence-based conclusions about whether the results support the hypothesis
9 identify limitations and make recommendations for modifying or extending the investigation.

Each of these steps is described in detail shortly. However, first it is important to understand the role of your logbook in this process.

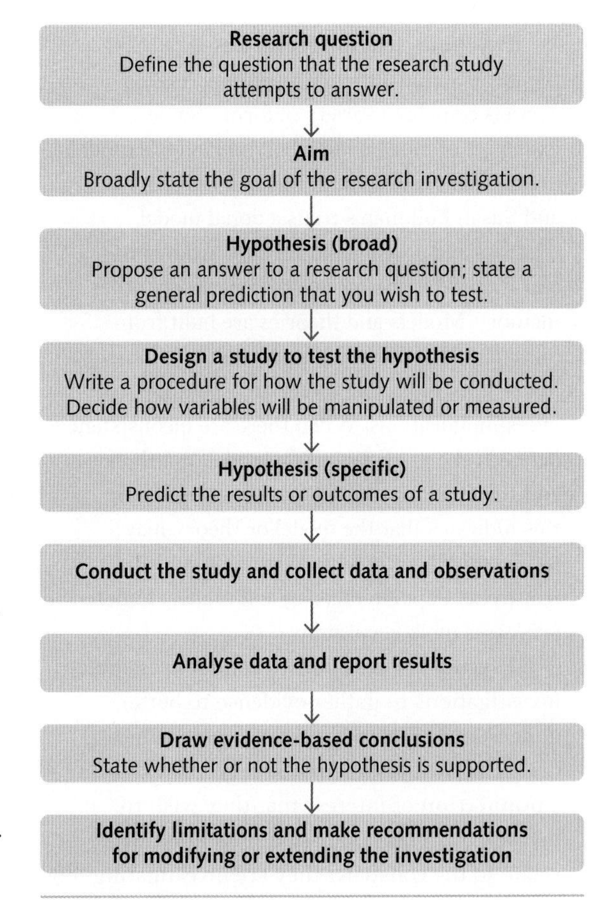

FIGURE 1.1 The research process, step-by-step

The logbook

Whenever you conduct a research investigation it is crucial that you record each of the steps you take in a **logbook**. Your logbook is a document that you will use throughout the year to record all of your practical work so that it can be assessed. It is particularly important to use your logbook to describe and record the steps involved in developing your student-designed scientific investigation (Outcome 3 of Units 1 and 2). You will use your logbook to:

» describe the process you used to develop your research question, aim and hypothesis
» outline the design of your study and the methods you used
» record the data you collect
» describe your data analysis method and results
» explain the conclusions you have drawn.

Developing your research question, aim and hypothesis

Imagine you are a researcher interested in how meditation training affects wellbeing. Before you begin to think about how you can test your ideas about meditation, you need to find out what is already known about this topic. You can do this by conducting a **literature review**. This would involve reading about meditation and its relationship to wellbeing in scientific journals, learning about any existing theories or models that are relevant, and determining what questions need to be addressed (see Chapter 6, page 229). During your literature review, you may discover something you are interested in that is not fully understood (that is, where there is some uncertainty around the issue or around a part of how the process works). For example, you may find some research articles suggesting that meditation training may improve mental wellbeing.

Research question

Once you have surveyed the literature to see what questions may be worthwhile to investigate, you can define your research question more precisely. A **research question** should express the exact question you are trying to answer with your research, including the population of interest (which could be a particular subpopulation such as 'people with anxiety' or 'teenagers', or could be broad and relate to people in general). An example of a research question is 'Does meditation training improve people's mental wellbeing?'. When choosing a research question, you should consider the unanswered questions that arose during your focused reading or literature review. You will try to answer your research question with your research study, or more specifically you will try to reduce the uncertainty around this question.

Aim

Next, you should create your research **aim**. Your aim is a broad statement about what you intend to investigate. You can think about the process of developing your aim as turning your research question into a statement of your research goal. For example, your aim could be 'To determine whether meditation training improves people's mental wellbeing'.

Hypothesis

While reviewing the literature to develop your research question and aim you may have come across theories or models that relate to your research question. You may also have developed your own ideas about possible answers to your research question. These ideas about possible answers to your research question are called hypotheses. A **hypothesis** is a statement that expresses a possible (that is, hypothetical) answer to a research question. A hypothesis describes the expected relationship between the variables of interest and should specify the predicted direction of the relationship between them.

A hypothesis can be expressed at two levels:

» a broad hypothesis that makes a general prediction about a relationship between variables in the *population* of interest (sometimes called a research hypothesis)
» a specific hypothesis that makes a prediction about the expected results of a study that relates to a *sample* of participants.

Researchers often begin with a broad hypothesis before they design a study. As they refine their study design, they create a specific hypothesis that can be tested by the study (sometimes called a prediction or operationalised hypothesis). Figure 1.1 (page 8) shows the statement of your broad hypothesis as Step 3 of the research process and the statement of your specific hypothesis as Step 5.

A broad hypothesis can be thought of as a clearly expressed statement that proposes a possible answer to your research question. For example, for our research question 'Does meditation training improve people's mental wellbeing?', the broad hypothesis could be 'Meditation training improves people's mental wellbeing'. Notice how this broad hypothesis refers generally to meditation training (not to a specific implementation of meditation) and to mental wellbeing (not a specific measure of mental wellbeing). It also makes a broad prediction that mediation will improve mental wellbeing and refers to the population of interest (people).

The specific hypothesis is a precise statement that makes a prediction about the expected results of your study. For example, in our meditation and mental wellbeing study, imagine you have chosen to:

- define meditation as a daily 5-minute mindfulness session using an online app over one week
- expose one group of participants to the meditation training and to have another group that does not do the training
- measure people's mental wellbeing using a common wellbeing questionnaire called the Positive and Negative Affect Schedule (PANAS; Watson et al., 1988). Higher scores on the PANAS indicate higher levels of mental wellbeing.

In this scenario, your specific hypothesis will state a prediction for how daily use of the meditation app over one week will affect people's scores on the PANAS. For example, your specific hypothesis might predict that 'Daily use of the meditation app for one week will result in higher scores on the PANAS compared to people who do not use the app'. In this example of a specific hypothesis, notice that we have referred to the variables of interest (meditation and mental wellbeing) in terms of how they were implemented or measured in the study. The specific hypothesis also provides a prediction about how mental wellbeing scores will differ between the groups and it makes reference to the sample (people in the two groups).

1.1.1 FORMULATING HYPOTHESES

Notice that although the broad hypothesis and the specific hypothesis have different levels of detail, they both include three core elements; when you are asked to write a hypothesis you must include these three elements, as follows.

- You must refer to the *variables of interest.* (As we discuss later in this chapter, if your study is a controlled experiment, then these will be the independent variable/s (IV) and the dependent variable/s (DV). However, if your study is a correlational study or some other kind of non-experimental study, then the variables of interest are the two (or more) factors that are thought to be related to each other. Only controlled experiments have variables called IVs and DVs. (The different kinds of research investigations and variables are described in detail in section 1.2 Scientific investigation methodologies, page 14.)
- You must state the *direction* of the predicted relationship between your variables. For example, the hypothesis could predict that scores for an outcome variable will increase for one group more than another. By including the predicted direction of the outcome, your hypothesis indicates not only that the scores on a variable will change, but also how they will change.
- You must refer to either the *population of interest* or to the *sample*, depending on whether you are stating a broad or a specific hypothesis, respectively.

The VCE Psychology Study Design does not distinguish between a broad hypothesis and a specific hypothesis. We have defined these two kinds of hypotheses to show how a hypothesis can be expressed with different levels of precision at different stages of the research design. A broadly expressed hypothesis is a proposed answer to the research question expressed at the population level. A specific hypothesis expresses a prediction for the outcome of a particular study with reference to the sample used in your investigation. If an exam question asks you to write a hypothesis for a particular study, you could express your hypothesis at either of these levels and be marked correct.

Designing a research investigation

Once you have developed a research question, an aim and a broad hypothesis, you are ready to design a research investigation (that is, a study) that you can use to test this hypothesis. The results of your investigation will either provide evidence to support your hypothesis or provide evidence against it. Of course, we can never prove that a hypothesis is true, but a carefully designed research investigation can reduce uncertainty about it.

There are many kinds of research designs that you could choose from to test the hypothesis 'Meditation training improves people's mental wellbeing'. The purpose of a research design is to create a method that allows you to observe and measure the psychological constructs (in this case 'meditation' and 'mental wellbeing'), in order to investigate how they are related.

For example, when designing your investigation of the effect of meditation on mental wellbeing, you need to make concrete decisions about things such as what kind of meditation training will be used, how it will be taught (in-person or online?), and how long people will use it (just 1 hour or several sessions?). You also need to decide exactly how you will define and measure mental wellbeing (which scale or questionnaire will you use to measure this construct?). Will you compare two different groups to each other (one who practises meditation and one who does not), or will you measure the *amount* of meditation people engage in and see if this is related to a measure of mental wellbeing? The kind of design and the kinds of measures you choose will determine the kind of data that you generate. We consider the different kinds of research designs in detail later in this chapter.

Collecting data

Once you have designed your study, you are ready to collect and record your data. To collect data, you need to understand the different kinds of data you could collect and how to select a sample of people to generate the data. You also need to think about how you will record your data so that you can analyse it later.

Kinds of data

Data is the term we give to any information that is collected or used in a scientific investigation. **Primary data** are data that we collect ourselves from a study that we have designed. **Secondary data** are data that have been collected by someone else that we use when conducting a literature review of the existing knowledge on a research topic. In VCE Psychology Units 1 and 2, you will have the opportunity to collect your own primary data when you complete Area of Study 3, Outcome 3 in Unit 2, and you will use secondary data when you conduct a literature review for Area of Study 3, Outcome 3 in Unit 1.

Primary and secondary data can take different forms, depending on whether or not the information is numerical. **Quantitative data** are numerical. They are recorded in the form of numbers (for example, a response-time measure or a score on a test). They must be collected through systematic and controlled procedures to ensure that the measurements are accurate and precise across people or trials. **Qualitative data** are non-numerical. They are descriptions of states or qualities that are often organised into themes. Examples of qualitative data include descriptions or personal accounts of feelings, attitudes, experiences or behaviours, or descriptions of changes to the quality of a variable such as changes in the experienced intensity of an emotion or vividness of a mental image. In psychological research qualitative data are often collected through questionnaires or interviews.

Sampling

When a psychology researcher wants to collect data, they need to recruit a sample of participants to take part in their study. The sample will be determined by the population of interest. For example, if a researcher is interested in how meditation affects the mental wellbeing of VCE students, then their population of interest is VCE students. It is not feasible to collect data from all VCE students, so the researcher must select a sample from this population. A sample is a group of people who are recruited from a larger population of interest (Figure 1.2a).

The term 'generalise' is not mentioned in the VCE Psychology Study Design but we use it to help you understand the relationship between a sample and a population. It will also be helpful for understanding external validity.

A researcher will conduct a study using a sample of participants to collect data and determine whether they can find an effect or a relationship in this data. However, researchers are not just interested in whether these effects exist in their sample; they are interested in whether an effect or relationship exists in the population from which the sample was drawn. That is, researchers want to be able to *generalise* the results they get from their sample to the entire population of interest. Generalisability is the extent to which research findings (found using a sample) can be applied to the population of interest (Figure 1.2b).

"Data" is a plural noun, so we write "data are". Datum is the singular noun.

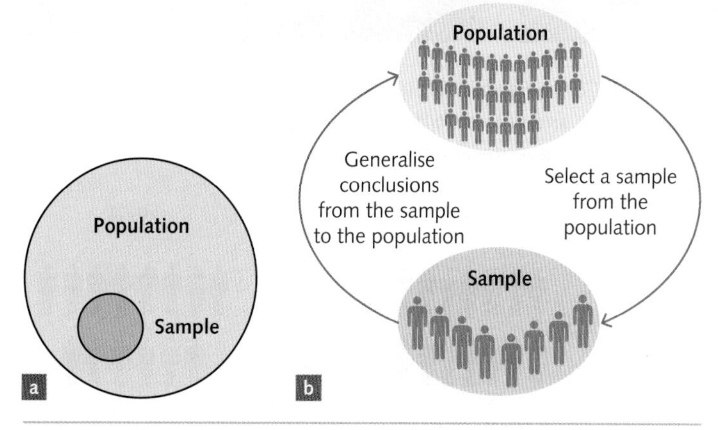

FIGURE 1.2 a A sample is a subset of a population. b A researcher can generalise from the sample to the population if the sample is representative of the population

1.1.2 SAMPLING PROCEDURES

For the findings of a study to be generalisable to the population, the sample must be **representative** of the population. **Sampling** is the process of selecting participants from a population of interest to participate in a research investigation. Researchers must consider sampling techniques to recruit an appropriate sample of participants that is representative of the population (a **representative sample**) of interest. Three important sampling methods are convenience sampling, random sampling and stratified sampling.

WB 1.1.3 APPLYING RANDOM SAMPLING

Convenience sampling is a method of sampling in which the researcher recruits a sample of participants that is convenient to recruit (for example, friends and family members of the researcher, or a class of first-year psychology university students; see Figure 1.3). While it is relatively easy to recruit a convenience sample, it is unlikely that this sample will be an accurate representation of the general population. For example, imagine a researcher recruits a sample of first-year psychology university students for a study where the population of interest is all people. A sample made up of these participants may have large differences (for example, in age, gender, education etc.) from the average person in the population of interest.

Alamy Stock Photo/Jeffrey Isaac Greenberg 7+

FIGURE 1.3 Convenience samples consist of participants who are readily available and where no mechanism has been used to ensure that participants are representative of the population

Random sampling is a sampling technique that uses a chance process to ensure that every member of the population of interest has an equal chance of being selected for the sample. For example, the researcher could assign each member of the population a number and select which participants will participate using a random number generator (a computer program) to select people to be approached for the study. A simple way to think about the random selection process is that it is equivalent to drawing names from a hat.

Stratified sampling is a sampling technique used to ensure that a sample contains the same proportions of participants from each social group (that is, strata or subgroup) present in the population of interest. First, strata are identified within the population of interest. Then the demographics of the population of interest are assessed to determine the number (or the proportion) of people in each stratum. Finally, participants are recruited to the sample from each of the stratum in the same proportions as their numbers in the population (that is, the number of participants recruited from each stratum for the research sample is proportionate to their numbers in the population) (Figure 1.4).

As an example of stratified sampling, consider a researcher studying wellbeing in the general population of Australia who plans to use a stratified sampling method to ensure the age of the sample is representative of the Australian population. The researcher would decide on the strata (how they want to divide up the age groups) and would look up the percentage breakdown of each age group in Australia. The number of participants recruited from each age group would then be based on that age group's proportion in the larger population.

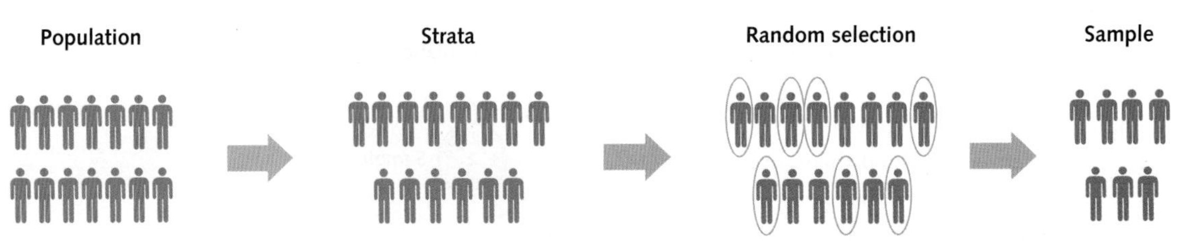

FIGURE 1.4 Stratified sampling is used to ensure that a sample contains the same proportions of participants from each stratum

Sample size

Another important aspect of sampling is obtaining a large enough **sample size**. This means recruiting enough participants for the study. Larger sample sizes are more likely to be representative samples, which means the sample is more representative of the population of interest, and this improves generalisability. With a larger sample, it is more likely a researcher will find any effect or relationship that exists. Samples that are too small also affect the accuracy and precision of any data collected.

Historically, most psychological research has been conducted using samples selected from what the psychology research community terms 'WEIRD' societies. WEIRD stands for Western, educated, industrialised, rich and democratic. Of all published studies in behavioural science, 96 per cent rely on participants recruited from populations that meet the WEIRD criteria (Henrich et al., 2010). A large proportion of psychology research takes place in the US, Europe and Australia, where WEIRD samples are more readily accessible. These samples under-represent a large proportion of the global population. Studies that recruit samples from WEIRD populations may not produce findings that can be generalised to people from other cultures.

Recording your data

The quality of the data you collect is determined by the accuracy and precision of your measurements and how carefully you record your measurements and observations. Be careful to ensure that data are labelled accurately so that they can be analysed later. Your logbook and tools such as spreadsheets are crucial for making accurate and organised records of your observations and measurements, and using them helps you avoid errors.

Analysing data

Data analysis means taking the raw data that you have collected and processing it so that you can report your findings. This step requires high attention to detail to ensure that your data are analysed accurately. For example, imagine in your study of the effects of meditation on mental wellbeing that you have collected:

» information about whether (or to what extent) participants engaged in meditation
» sets of responses from participants about their experience of mental wellbeing.

Analysing these data involves matching participants' mental wellbeing scores with the information about their meditation practice so you can describe the relationships you have observed between these variables. This might involve things like calculating the average level of mental wellbeing reported by the group of participants who participated in meditation and comparing this with the average level of mental wellbeing reported by the group of participants who did not participate in meditation. We further discuss specific methods for analysing data later in this chapter (page 34).

1.1.4 SAMPLING FLOWCHART

Drawing conclusions

After collecting and analysing your data, it is time to interpret your results so that you can determine what conclusions can be drawn. Drawing **conclusions** involves describing and explaining the results of your study and discussing how your findings relate to the aim and the hypothesis of your study. Your conclusion will involve making and justifying claims about whether or not the results support your hypothesis. You must also carefully consider to what extent your results can be generalised to the population of interest. Drawing conclusions involves a careful process of evaluating your claims against the evidence provided by your study. We explore the process of evaluating evidence from research investigations in greater detail in Evaluating data and investigation methods on page 38.

1.1.5 WAYS TO OBTAIN DATA

Limitations and recommendations

It is important to identify any **limitations** of your study. This means considering issues and unexpected problems that may have compromised the internal and external validity of the investigation. You should also make recommendations for modifying or extending the investigation in future studies and discuss what further evidence may be required to make conclusions. This may help to enhance the validity of future studies and help other researchers to avoid the pitfalls that you encountered.

We have now covered the nine steps of the research investigation process (Figure 1.1, page 8). The final step (of identifying limitations and making recommendations for modifying or

1.1.6 FEATURES OF A SCIENTIFIC RESEARCH REPORT

extending the investigation) sets off the whole research process again. Future studies will build on the recommendations, and try to correct and improve on the limitations of existing studies. As the research community continues to generate new knowledge, the cycle continues and the scientific process repeats, with the next study starting again at Step 1 of the process.

One study at a time, researchers improve the collective scientific understanding. We should note that all stages of the research process must be guided by ethical considerations (see page 39).

KEY CONCEPTS 1.1

» A research question defines the question that a research investigation tries to answer.
» An aim is a broad statement about the goal of a research investigation.
» A hypothesis is a proposed answer to a research question (made before the investigation is conducted). It is usually a directional statement about the relationship between the variables and states the expected results of a research investigation.
» A research investigation is used to test a hypothesis.
» Data are any information collected in scientific investigations.
» Primary data are data we collect ourselves from a study.
» Secondary data are data collected by someone else that we use when conducting a literature review.
» Quantitative data are numerical (recorded in the form of numbers).
» Qualitative data are non-numerical (descriptive).
» Sampling is the selection of participants from a population to participate in a research investigation.
» Convenience sampling means recruiting participants that are readily available.
» WEIRD samples (Western, educated, industrialised, rich and democratic) under-represent a large proportion of the population.
» Random sampling gives every member of the population an equal chance of being recruited.
» Stratified sampling ensures that the sample contains the same proportions of participants from each stratum as the proportions in the population of interest.
» Using larger samples can improve representativeness.

1.2 Scientific investigation methodologies

There are many different types of scientific investigation methodologies that researchers can choose from. The ideal scientific investigation is the controlled experiment, but not all research questions can be studied this way. Other kinds of investigations include:

» correlational studies
» case studies
» classification and identification
» fieldwork
» modelling
» simulation
» product, process or system development.
» literature reviews

In this section we begin with a brief outline of controlled experiments and then explore the other kinds of research investigations to understand what makes them different from experiments, and what kinds of research questions they can address. We then explore the controlled experiment in detail in section 1.3 (page 22).

Controlled experiments in brief

A **controlled experiment** is a methodology used to test a hypothesis in which the researcher systematically manipulates (changes) one or more variables to investigate what effect these manipulations have on another variable. Unlike other methodologies, controlled experiments are designed to enable the researcher to draw conclusions about the *causes* of the phenomena or process they are investigating. In psychological science, controlled experiments allow us to determine the causes of behaviours, mental states and psychological processes.

Correlational studies

A **correlational study** is a non-experimental study where the researcher investigates relationships between variables. Unlike in a controlled experiment, the researcher does not try to control or change any of the variables. Instead, they observe and measure the variables as they naturally occur. To conduct a correlational study, the researcher chooses two or more variables they wish to investigate. They then recruit a sample of participants and measure these variables in their sample. The analysis involves describing how (or whether) the two variables are related to each other.

The correlational study is one of the most commonly used research designs in psychology. This is because it is not always possible, desirable or appropriate to investigate some questions with an experiment. Because no manipulation is required in a correlational study, it can be less invasive than a controlled experiment that investigates the same research question. A correlational study can also be useful to inform future work, because it can be used to identify which variables may be more important to study further.

An example of a correlational study in psychology could be an investigation of the relationship between levels of wellbeing and amount of sleep. Researchers could recruit a sample of research participants to report their average sleep duration and complete the PANAS wellbeing survey. Each participant's score on the PANAS (variable 1) and their average sleep duration (variable 2) could then be graphed to observe the relationship between sleep duration and wellbeing. A correlation analysis could also be performed to assess the strength and direction of this relationship mathematically.

What is a correlation?

A correlation is a relationship or association between two variables in a data set. It is a measure of the strength and direction of the relationship between the two variables. Some examples of correlations are:

- people with high values on one variable (for example, longer time spend studying) often having high values on another variable (for example, exam score)
- people with high values on one variable (for example, alcohol consumption) often having low values on another variable (for example, driving accuracy).

Correlations can be visually represented by plotting the values of two variables on a type of graph called a scatterplot. Each point on a scatterplot represents one participant and the position of the data point on the plot provides information about that person's scores on both variables. Specifically, scores for variable 1 are represented on the x-axis (horizontal) and scores for variable 2 are represented on the y-axis (vertical). The data point that represents the relationship between the two variables for each participant is shown at the point on the scatterplot for which these values intersect.

Figure 1.5 shows a single data point on a scatterplot that represents one participant's scores on two variables. For example, variable 1 could be their score on a sleep quality scale, represented on the horizontal axis with a score of 8, and variable 2 could be their score on the wellbeing scale, represented on the vertical axis with a score of 4.

In a correlational study, we can record the scores of two variables for a whole sample of participants

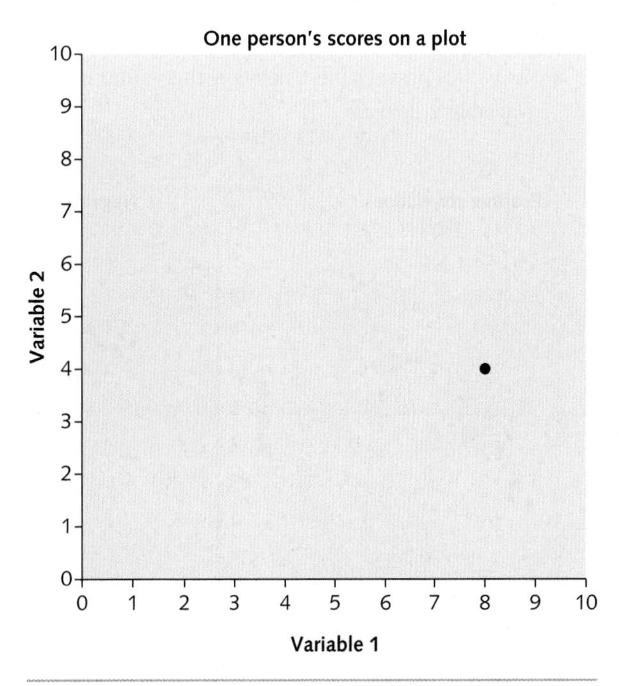

FIGURE 1.5 A scatterplot showing one person's scores on two variables. For example, variable 1 could be sleep quality and variable 2 could be wellbeing

and display the results on a scatterplot. We can then look for patterns in the plot to see if the sample shows a relationship between the two variables.

Direction of correlation

First, let's explore the direction of a correlation: positive or negative. Figure 1.6 shows three possible types of patterns that could emerge in a data set:

- positive correlation
- no correlation
- negative correlation.

A positive correlation is where the pattern of data slopes up diagonally from left to right on a graph (Figure 1.6a). Positive correlation means that:

- as values on variable 1 *increase*, the values on variable 2 also *increase*
- people who score *high* on one variable are likely to score *high* on the other variable
- people who score *low* on one variable are likely to score *low* on the other variable.

As discussed earlier, a good example of a positive correlation is time spent studying and final exam grade. There is a positive relationship between these variables because students who study more generally score higher in exams than students who study less.

A negative correlation is where the pattern of data slopes downward from left to right on a graph (Figure 1.6b). Negative correlation means that:

- as values on variable 1 *increase*, the values on variable 2 *decrease*
- people who score *high* on variable 1 are likely to score *low* on variable 2
- people who score *low* on variable 1 are likely to score *high* on variable 2.

As discussed earlier, a good example of a negative correlation is alcohol consumption and accuracy on a driving simulation test: As consumption increases, accuracy decreases.

If there is no correlation (Figure 1.6c):

- scores on variable 1 are not related to scores on variable 2
- data points appear on a scatterplot as a random cloud with no particular pattern.

Strength of correlation

The strength of a correlation is an indication of how strongly the variables are related. Figure 1.7 displays scatterplots of positive correlations of different strengths, ranging from strong to weak. Stronger correlations between variables produce data points that are more tightly clustered along the diagonal of the scatterplot.

Correlation vs causation

It is important to know the difference between correlation and causation. If two variables are correlated, there is a relationship between them, but this does not mean that changes in the value of one variable causes changes in the value of the other variable. For example, you could conduct a study that finds a positive correlation between sleep duration and mood. This finding does not

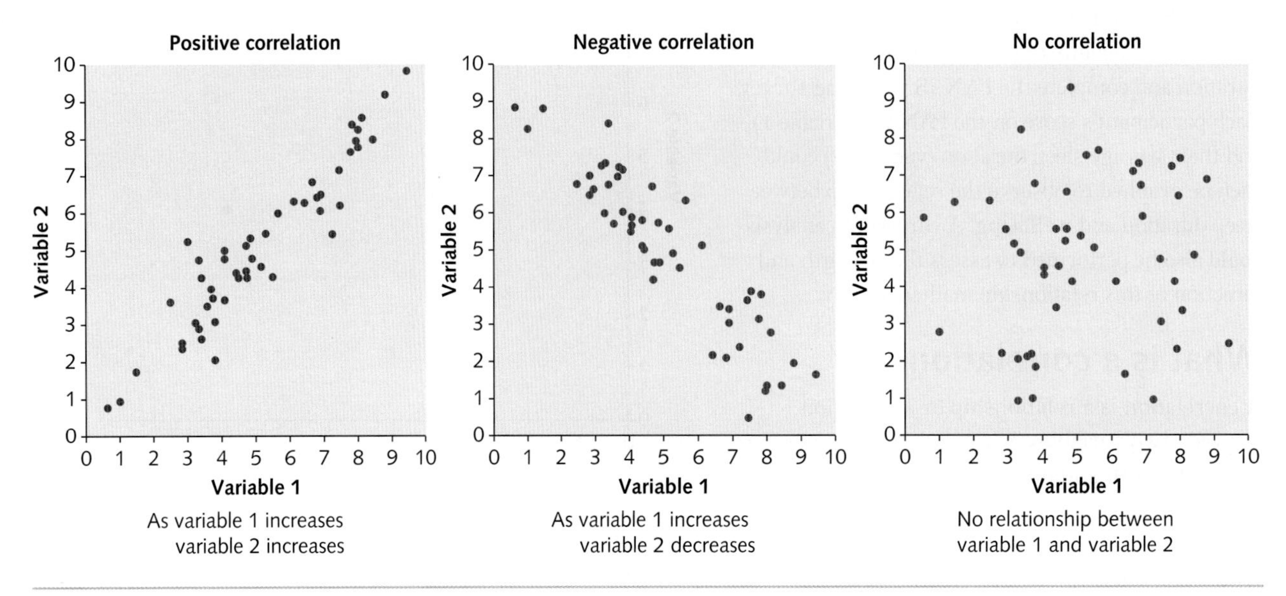

FIGURE 1.6 Types of correlation

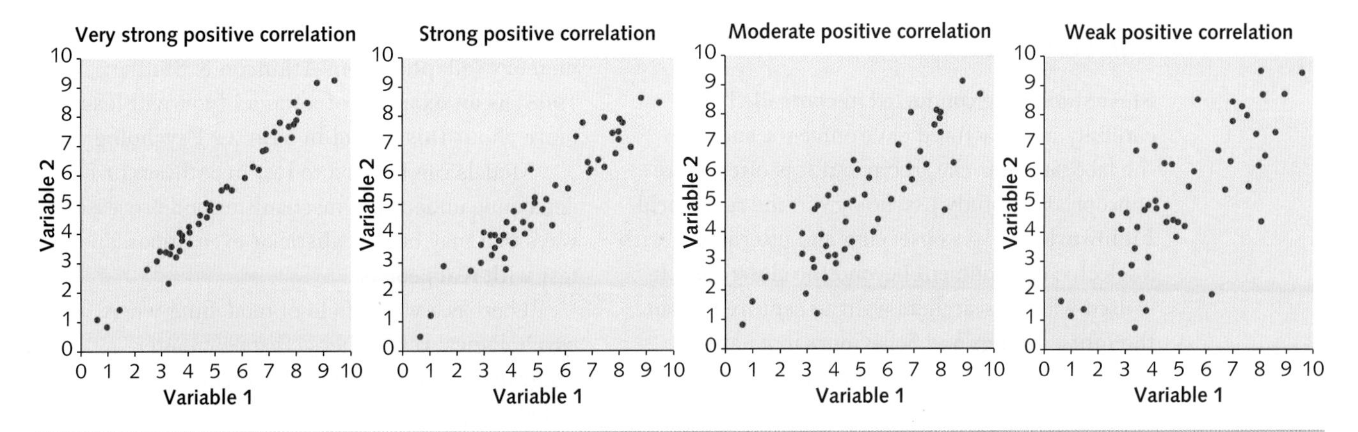

FIGURE 1.7 Positive correlations of various strengths: very strong, strong, moderate and weak

necessarily mean that longer sleep causes better wellbeing. Instead, it could be that people who get more sleep also have more structured routines in general and this could be the factor responsible for the correlation, rather than the sleep variable measured.

A common phrase in science communication that summarises this point is: 'Correlation does not equal causation'. This is a valuable phrase to remember, both for scientific research and for life in general. Many logical errors arise when evidence of a correlation is confused with causation.

To identify a causal relationship, a controlled experiment is needed.

Case studies

A **case study** is a detailed investigation of one instance of a broader phenomenon. A case study focuses on a particular person, activity, behaviour, event or problem in a real or hypothetical situation.

Case studies can take various forms. For example, a case study may involve:

- direct observation of a situation
- analysis of historical information, which could include a discussion of knowledge learned from the situation or exploration of the causes and consequences of an event
- a real situation or a role-play of an imagined situation (in particular, if recommendations are to be made)
- problem-solving (in particular, if there is a need to develop a new design or methodology).

Case studies are very common in clinical psychology because the investigation of mental disorders often occurs at the level of the individual person. Such case studies may include detailed descriptions of the person, their symptoms and the factors that make up their experience. In social psychology, case studies can be used to examine how groups behave under certain conditions.

An advantage of the case study methodology is that the data are rich and highly detailed. A case study can include the complexities that are encountered in the real world (outside of the laboratory). A disadvantage is that the information is specific to one particular case, so many of the details may be specific to that single case and may not apply to the wider population or to other situations of interest.

Classification and identification

Classification is a scientific activity that seeks to systematically organise phenomena, objects or events into manageable sets. Identification is the process of recognising phenomena as belonging to particular sets or possibly being part of a new or unique set. The *Diagnostic and Statistical Manual of Mental Disorders* (*DSM-5-TR*™) produced by the American Psychological Association is the most common diagnostic classification system for mental disorders. This manual classifies mental health disorders into categories based on the presentation of symptoms. When a clinical psychologist uses the *DSM-5-TR*™ criteria to diagnose a patient, they are engaging in the process of identification.

Fieldwork

Most studies are conducted in controlled and carefully manufactured environments, such as the laboratory or classroom, but it is often more appropriate to study psychology in the real world. **Fieldwork** involves observing and interacting with a selected environment beyond the classroom. It is used when researchers want to capture human thoughts, feelings and behaviours in a natural setting. Fieldwork can be conducted through direct observation of behaviour (for example, participant observation) or by using sampling methods to gather a group of people in a natural environment. This observation process can be qualitative (for example, describing how people act or identifying themes from interview questions) and/or quantitative (for example, collecting numerical information, such as counting how many times a person engages in a particular action). Fieldwork could also be asking people about their opinions or behaviours through qualitative interviews, questionnaires, focus groups or yarning circles. Fieldwork usually aims to determine correlation rather than a causal relationship.

1.2.1 RESEARCH METHODOLOGIES AND INVESTIGATION AIMS

Modelling and simulation

In psychological research methods, modelling means creating a conceptual, mathematical or physical representation (that is, a model) of a system of concepts, events or processes.

Figure 1.8 shows the multi-store model of memory (adapted from Atkinson & Shiffrin, 1968) as an example of a model (you will learn more about this model in Year 12 Psychology).

Models can be used to test hypotheses or to determine underlying mechanisms and processes in ways that may be unrealistic or even impossible to test with real people.

There is a whole field of modelling where psychological researchers create computer programs to simulate mental processes. This is called computational cognitive modelling. The data generated from these computational models can be compared to data collected in experiments and, if the patterns of data match, this is evidence supporting the computational model.

Product, process or system development

Product, process or system development is the design or evaluation of a process, system or artefact to meet a human need. This may involve technological applications as well as scientific knowledge and procedures.

Psychology-focused smartphone applications (apps) are one area where this methodology is flourishing. Smartphone apps are now providing people with information, structures and techniques that can help to improve mental wellbeing or modify behaviour. Some examples are:

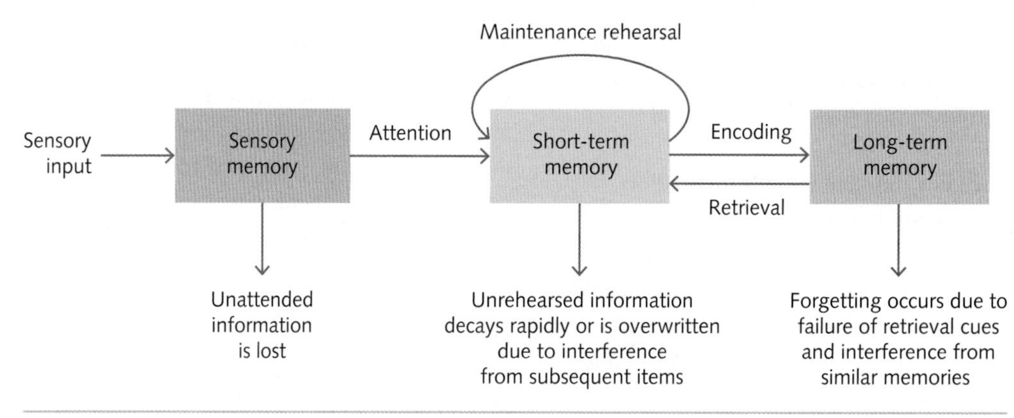

FIGURE 1.8 **The multi-store model of memory (adapted from Atkinson & Shiffrin, 1968) is an example of a model. Models allow psychologists to represent theories or concepts more concretely**

- meditation apps such as the headspace app, which aims to help people reduce anxiety and improve their mental health
- habit tracker apps such as the Habitica app, which helps people to track and intentionally modify their habitual behaviours.

Smartphone apps are also being used in psychological research. For example, apps have been developed that deliver short surveys to people on smartphones at regular intervals. This enables experiences or emotions to be measured in real time while the participant is going about their day, and allows surveys to be delivered more frequently. Some apps collect information for researchers, so they facilitate access to more representative samples. Case study 1.1 explores an example of one of these applications.

CASE STUDY 1.1

SEMA3: an example of a product, process or system development

SEMA3 (Smartphone Ecological Momentary Assessment Version 3) is a smartphone app developed by psychology researchers in Melbourne (Figure 1.9). It is an ideal case study to explore product, process or system development in Psychology while also highlighting other relevant concepts that are important for VCE Psychology. SEMA3 allows researchers to set up short, automated surveys that are sent to participants' smartphones at regular intervals (for example, multiple times per day). Researchers collect information from people using the app, and the repetition of surveys allows them to explore the relationships between responses (for example, relationships between emotions) and changes over time.

An advantage of this methodology is that thoughts, feelings and behaviours can be captured in real time in the real world, instead of in a laboratory. This can improve both the internal validity and external validity of a study.

Being able to get survey responses from participants while they are going about their regular life allows researchers to gain a more realistic assessment of a person's experience. It also avoids some drawbacks of laboratory studies, such as participants being unable to recall information when they are asked questions about the past. These features can help surveys to measure what they intend to measure, thus improving the internal validity of the study.

Many people own smartphones throughout the world, in both advanced and emerging economies (Figure 1.10). By surveying people via their smartphone, researchers are able to reach a more diverse sample of participants for their research studies, including people who they may not have previously had access to. This can improve external validity by allowing researchers to use a sample that more closely resembles the overall human population. Technologies of this kind could help researchers address the common problem that most research is conducted with Western, educated, industrialised, rich and democratic (WEIRD) samples that under-represent a large proportion of the overall population.

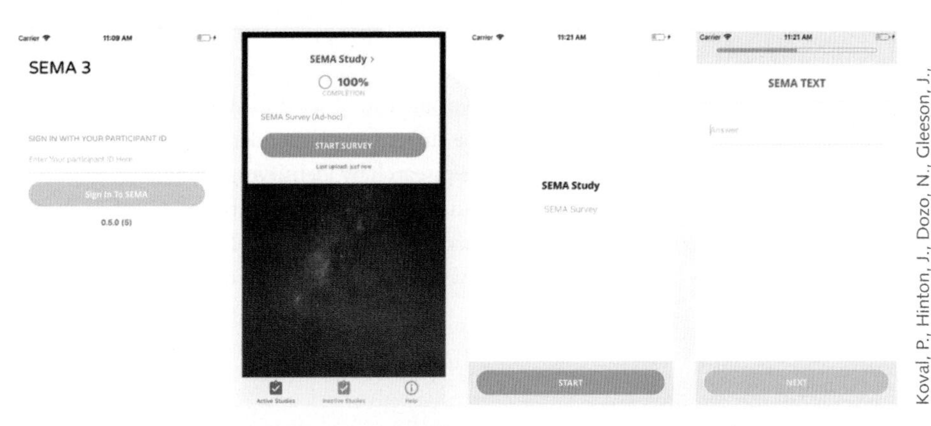

Koval, P., Hinton, J., Dozo, N., Gleeson, J., Alvarez, M., Harrison, A., Vu, D., Susanto, R., Jayaputera, G., & Sinnott, R. (2019). SEMA3: Smartphone Ecological Momentary Assessment, Version 3. [Computer software]. Retrieved from http://www.sema3.com

FIGURE 1.9 Three screens from the SEMA3 app

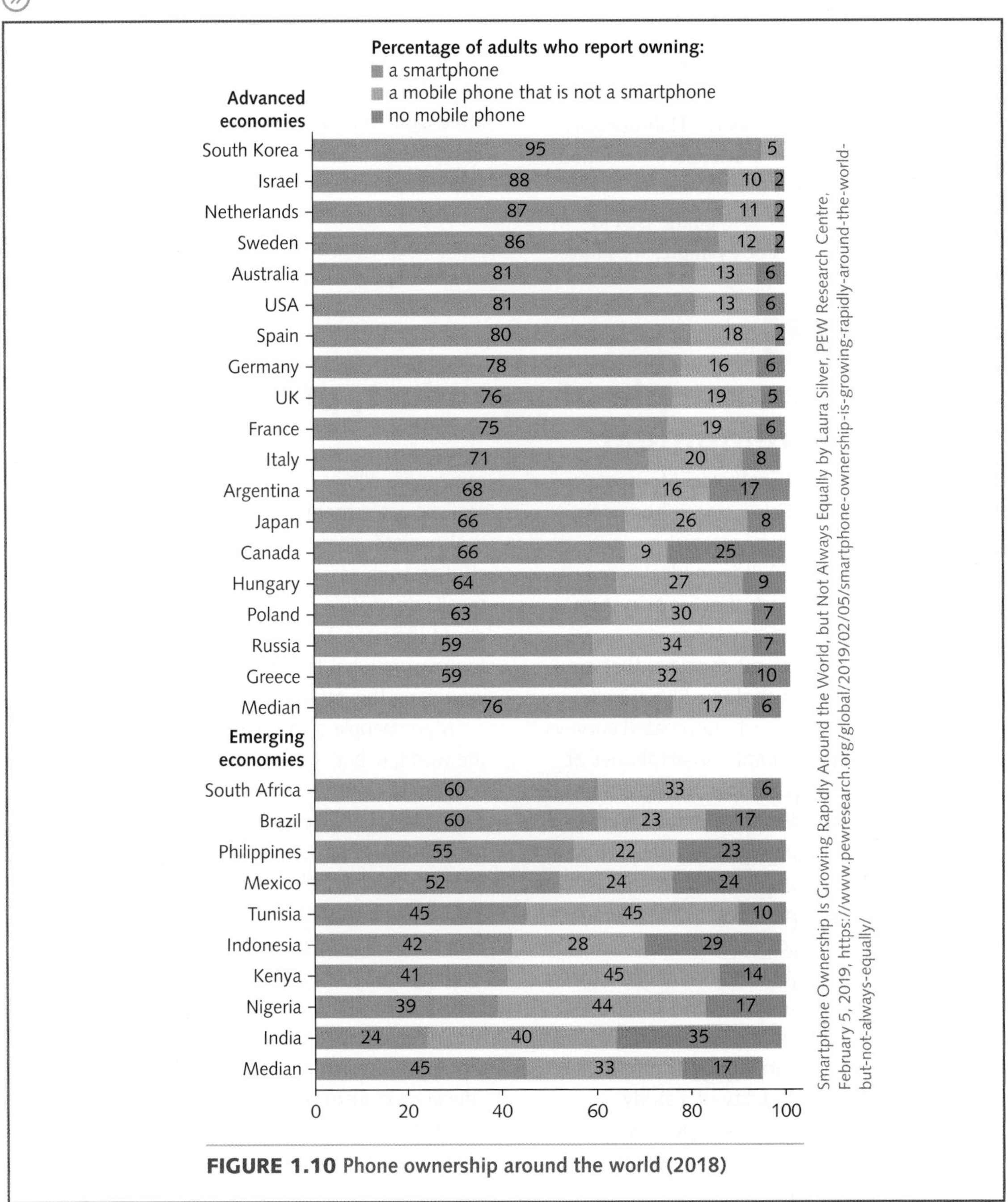

FIGURE 1.10 Phone ownership around the world (2018)

Literature reviews

A literature review is a report produced by reading scientific research on a particular area and writing a summary. Literature reviews play an important role in analysis because they organise what is already known, and can be used to synthesise new ideas based on the current level of understanding. They also guide research by identifying knowledge gaps in the literature. This allows researchers to refine current research questions and develop new ones. (Literature reviews will be discussed in detail in Chapter 6.)

KEY CONCEPTS 1.2

- » Each type of research design has specific benefits and limitations.
- » A controlled experiment involves experimental manipulation of a variable to determine the effect on an outcome(s) of interest.
- » A correlational study is a non-experimental investigation of the relationship between variables.
- » Correlations can be visually represented by plotting two variables on a graph.
- » Correlation does not equal causation.
- » A positive correlation is where high scores for one variable occur with high scores for another variable.
- » A negative correlation is where high scores for one variable occur with low scores for another variable.
- » A case study is an investigation of one particular example of an activity, behaviour, event or problem, to acquire knowledge about the process as a whole.
- » Classification and identification are processes used to organise phenomena into categories and identify examples of that categorisation.
- » Fieldwork involves observing and interacting with an environment in the real world.
- » Modelling means creating a representation of an event, process or system of concepts.
- » Product, process or system development is the design or evaluation of a process, system or artefact to meet a human need.
- » A literature review is a report produced by reading scientific research on a particular area and summarising it.

1.3 The controlled experiment in detail

Controlled experiments use a research methodology in which the researcher systematically manipulates one or more variables to investigate how this affects an outcome variable/s of interest (the dependent variable(s)). The systematic manipulation of variables gives the controlled experiment the unique ability to find cause-and-effect relationships between variables.

The simplest form of controlled experiment consists of two conditions: the **experimental condition** (sometimes called the treatment condition) and the **control condition**.

A randomised controlled trial (RCT) is an example of a controlled experiment design in which one variable is manipulated to find its effect on another variable. RCTs are used by psychologists when they want to determine whether a treatment is effective.

For example, to determine whether practising meditation causes improvements in people's mental wellbeing, we could 'manipulate' people's exposure to meditation to see if this affects their reported level of mental wellbeing. The simplest experimental manipulation would be to create one group of participants who experience meditation and another group of participants that do not. The group that experiences meditation is called the experimental group (or **treatment group**) and the group that does not is called the **control group**. The experimental group experiences the experimental (or treatment) condition and the control group experiences the control condition. This design allows us to compare self-reported levels of mental wellbeing between the experimental and control conditions to determine whether meditation improves mental wellbeing. Of course, drawing a conclusion about the effectiveness of meditation based on this study will depend on whether the two groups differed from each other *only* in the variable of interest (exposure to meditation) and not in any other variable(s) that might also affect mental wellbeing. This brings us to consider the kinds of variables that influence experiment outcomes.

Variables in controlled experiments

A variable is any condition (for example, stimulus, event, quality, trait or characteristic) that can take a range of values and that can be measured or manipulated in a scientific investigation. You need to understand the different types of variables and be able to distinguish between them to understand how controlled experiments work.

Independent and dependent variables

The independent variable and the dependent variable are the two fundamental variables used in controlled experiments. The **independent variable (IV)** is the variable that the researcher manipulates. The **dependent variable (DV)** is the outcome variable that the researcher measures to determine whether manipulating the independent variable had an effect. In our example of an RCT study to determine the effect of meditation on levels of mental wellbeing, the meditation condition (meditation or no meditation) is the IV, and people's scores on a mental wellbeing scale is the DV.

Knowing where these terms come from may help you to remember them. If these variables are causally related (which is what the controlled experiment seeks to determine), then the value of the DV *depends on* the value of the independent variable. In contrast, the IV is not dependent on the DV (it is *independent* of the DV).

In a controlled experiment, participants will be exposed to different conditions and each condition is called a 'level' of the IV. For example, in an experiment where the IV is participation in a mindfulness meditation training program, the two levels of the IV could be:

1 meditation training (experimental condition)
2 no meditation training (control condition).

In an experiment where the IV is caffeine consumption, the two levels of the IV could be:

1 caffeinated coffee (experimental condition)
2 decaffeinated coffee (control condition).

In an experiment where the IV is sleep duration, the two levels of the IV could be:

1 restricted sleep (experimental condition)
2 normal sleep (control condition).

Each of these examples has just one independent variable that is manipulated so that it has two levels (conditions). More complex experiments can have the IV at more than two levels. For example, you might want to compare two different kinds of meditation to a no-meditation control condition. The IV then has three levels, one for each kind of meditation and one for the control group. Some experiments have more than one IV, each with two or more levels (for example, the mixed design we consider on page 28).

How do we know which variable in our experiment should be the IV and which variable should be the DV? To answer this, we need to think about which variable *affects* which. Imagine we hypothesise that warm temperatures make people happier. To test this hypothesis in a controlled experiment, temperature would be the IV and score on a happiness survey would be the DV. We are testing whether happiness *depends* on the temperature.

More examples of IVs and DVs

Example 1

Hypothesis (broad): Mindfulness meditation improves people's mental wellbeing.
Hypothesis (specific): Teenagers who participate in mindfulness meditation training will score higher on the PANAS wellbeing survey than participants who do not participate in the training.
IV: Engagement in mindfulness meditation training with two levels: training or no training
DV: Score on the PANAS wellbeing survey

Example 2

Hypothesis (broad): Caffeine improves students' short-term memory.
Hypothesis (specific): Students who consume a caffeinated coffee score better on the Digit Span Memory Test than participants who consume a decaffeinated coffee.

IV: Caffeine consumption with two levels: Caffeinated coffee or decaffeinated coffee.
DV: Score on the Digit Span Memory Test

Further points about IVs and DVs

The terms 'independent variable' and 'dependent variable' are only relevant to controlled experiments. You would not use these terms to describe the variables in a correlational study, because correlational studies investigate the relationship between variables, but not the *causal relationship* between variables. In a correlational study we do not assume that one variable depends on another variable and there is no manipulation of an IV.

You may be asking what is meant when we say the researcher 'manipulates' the IV. This depends on the type of controlled experiment, but it can mean changing, selecting or controlling the variable.

A common error is to say 'we gave the experimental group the IV'. In our caffeine example, this would incorrectly imply that the coffee itself is the IV. The IV is whether or not people consumed caffeine and the two levels of the IV would be the caffeinated coffee condition and decaffeinated coffee condition.

On a graph, the IV is always placed on the horizontal axis (that is, the x-axis) and the DV is placed on the vertical axis (that is, the y-axis).

Extraneous variables

An **extraneous variable (EV)** is any variable other than the IV that *may* affect the DV. The presence of EVs in an experiment is a problem because they make it difficult to be sure that the IV was responsible for any observed change in the DV.

For example, in a study investigating the effect of meditation training on wellbeing, different participants may:

» experience varying amounts of daily stress in their lives outside of the experiment

» have varying amounts of past experience with different forms of meditation.

These are both examples of EVs. They are factors other than the meditation training intervention that *may* affect people's levels of mental wellbeing (the DV).

Note that EVs become a major problem if their effects are distributed differently between the two conditions. When this occurs, an EV becomes a confounding variable.

Confounding variables

A **confounding variable** is a variable other than the IV that *has* systematically affected the DV because its influence is not evenly distributed across the levels of the IV. Confounding variables are a major problem for the internal validity of an experiment because they provide an alternative explanation for the results. (See pages 35–6 for more detail on internal validity.)

For example, in our meditation study, the EV of daily stress level would become a confounding variable if stress levels were systematically higher in one of the groups than in the other. This could happen if most people allocated to the meditation group happened to experience high levels of daily stress, but most people in the control group did not. If this were the case and our study finds that there is no beneficial effect of meditation on mental wellbeing, then our results are said to be 'confounded'. This is because we cannot determine whether our failure to find an effect of meditation on mental wellbeing was because meditation is ineffective or because the meditation group also happened to experience much higher stress throughout the experiment than the control group.

Let us consider another example of an EV that becomes a confounding variable. In a study designed to investigate the effect of caffeine on the rate of learning, participants could be divided into two groups:

1.3.1 EXTRANEOUS VARIABLES

1 an experimental group that receives caffeinated coffee
2 a control group that receives decaffeinated coffee.

The time at which participants are tested could be a confounding variable if the experimenter tested the learning rate of the experimental group in the morning and the control group in the late afternoon. In this example, the effect of caffeine on learning rate is confounded with participants' levels of alertness due to the time of day, because both caffeine and time of day affect levels of alertness and the two groups were systematically tested at different times of the day. If the results were that the experimental group learned faster than the control group, the researcher cannot be sure whether participants in the experimental group learned faster due to caffeine, or because they were more alert from being tested earlier in the day (or whether both caffeine and the time of testing influenced the results). Time of day would *not* be a confounding variable if both groups were tested at a similar time of day. In this case, time of day is still an EV that affects alertness, but it is not confounded with one of the experimental conditions.

Controlled variables

Controlled variables are variables that the researcher holds constant (controls) in an investigation. If an EV is identified before the study is conducted and the experimenter holds its effect constant throughout the experiment, it becomes a controlled variable. It is controlled because its influence has been managed so that it cannot bias the results of the experiment one way or the other. Controlled variables are kept constant to ensure that changes in the DV are caused by the manipulation of the IV, rather than by variation in other variables that are not of interest to the study. Researchers control variables to eliminate or neutralise their potential effects on the results.

A controlled variable is *not* the same as a control group. Be careful not to confuse these terms.

A researcher can control variables by trying to keep all aspects of an experiment (except the IV) identical for each condition of the experiment. For example, in the example of analysing the effect of caffeine on the rate of learning, prior caffeine consumption could be a controlled variable; the researcher may instruct participants to abstain from drinking coffee or energy drinks on the day of the experiment. The testing location could be a controlled variable; the researcher may test each participant in the same place to control for variation in the environment, such as noise and distractions. The time of day the participants are tested could also be a controlled variable, to control for level of alertness.

Controlled variables are not variables that are part of an investigation itself. The researcher wants to control variables (keep them constant) so that they do not affect the investigation. Controlled variables should not be confused with the variables of interest in the experiment (that is, the IV and the DV), because *controlling* a variable is not the same as *manipulating* a variable: the aim of controlling a variable is to neutralise its effect, whereas the aim of manipulating a variable is to purposefully make changes to create different conditions (that is, different levels of the IV).

Group allocation and variables

An important way that researchers control the effects of EVs is by managing how participants are allocated to conditions. Random allocation of participants to groups is used to increase the likelihood that the broad range of potential EVs (such as age, gender, and amount of sleep) will be equally distributed between groups. Alternatively, suppose there is a particular EV that the researcher wants to control. In that case, the researcher may seek to match each of the groups on this EV by ensuring that equal numbers of people with the specific feature (e.g., age group) are in each group. However, even when participants are randomly allocated to groups, there may still be underlying differences between the groups due to chance.

A researcher can match groups on variables such as age or gender to control (that is, neutralise) the effect of the variable on the outcome of the study. For example, to match groups by gender would be to allocate an equal number of males and females to each group. We explore the importance of allocation to groups in experimental designs in detail later in this chapter.

Variables summary

In summary, there are five kinds of variables that are relevant to controlled experiments (Figure 1.11). The independent variable (IV) and dependent variable (DV) are the variables of interest in a controlled experiment. The IV is the variable that the researcher manipulates. The DV is the outcome variable that the researcher measures to determine the effect of the IV. An extraneous variable (EV) is a variable that *may* have an unwanted effect on the DV. An EV becomes a confounding variable when its effect on the DV differs across the levels of the IV. An EV becomes a controlled variable when its effect is held constant by the researcher to prevent it from affecting the DV.

Understanding the variables defined in this section gives us a more comprehensive definition of the controlled experiment: a controlled experiment is a scientific investigation where the researcher systematically manipulates one or more IVs to determine the effect on one or more DVs, while attempting to control (eliminate or neutralise) the influence of EVs. At a minimum, a controlled experiment involves a comparison of outcomes between at least one experimental condition and one control condition. This design allows the researcher

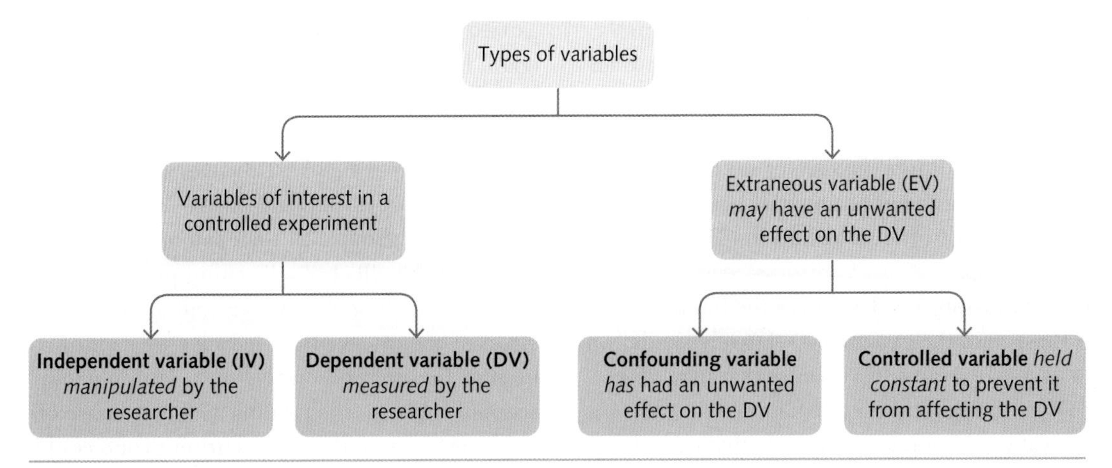

FIGURE 1.11 **Important types of variables: the independent variable, dependent variable, extraneous variable, controlled variable and confounding variable**

to determine whether a hypothesised causal relationship exists between the IV and the DV.

Kinds of designs for controlled experiments

There are three different designs for controlled experiments: the between subjects design, the within subjects design, and the mixed design. Before we explore each of these designs in detail, it may help to know where these terms come from. In the past, researchers would refer to a person who takes part in a research study as a 'subject' (we now call them 'participants'). In a **between-subjects design**, scores are compared *between* different participants. In a **within-subjects design**, scores are compared *within* the same participants (that is, each person's score is compared with their own score at a different time). In a **mixed design**, a mix of both types of design is used, with both between-subjects and within-subjects comparisons.

Between-subjects design

The between subjects design is the kind of experimental design we have been using as examples so far in this chapter. In a between-subjects controlled experiment, participants are allocated to different groups, each exposed to a different condition. For example, consider a study of the effect of a treatment intervention. The two groups are:

1 the experimental group, which is exposed to the intervention
2 the control group, which is not exposed to the intervention.

In other words, the researcher manipulates the IV by varying the conditions experienced by participants in each group. The researcher measures the DV for each participant in each group and then compares the scores obtained for each group to the other group/s. If the DV scores are significantly different between the groups, this supports the hypothesis that the IV affects the DV. Figure 1.12 shows a flowchart of the between-subjects design.

Let us consider an example of a between-subjects research design to test the effect of a learning program on reading ability.

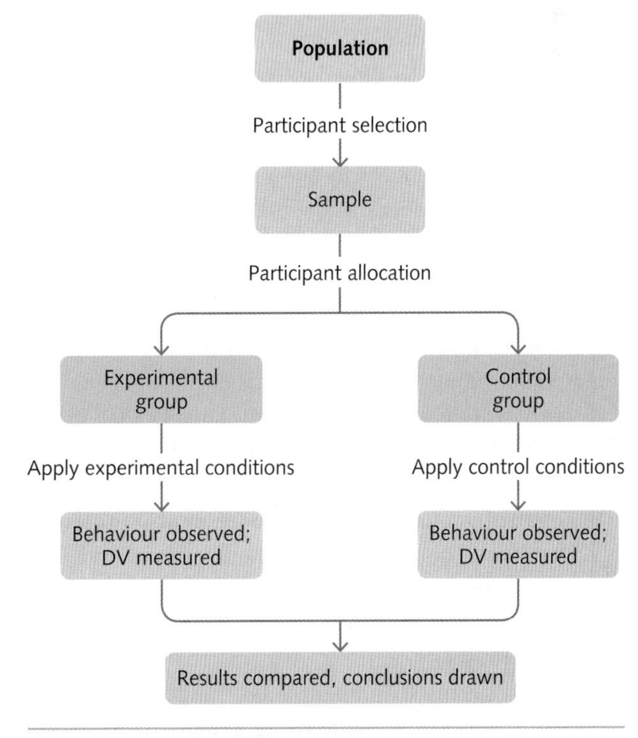

FIGURE 1.12 **A simple between-subjects controlled experiment**

The randomised controlled trial described at the start of this section (page 22) is an example of a between-subjects controlled experiment.

In this experiment, the IV is participation in the learning program. Students are split into an experimental group and a control group. The DV in this study could be the difference in the accuracy of reading a passage of text. The scores on the DV from the experimental group are then compared to the scores on the DV for the control group.

Between-subjects design has the limitation that it assumes the groups are relatively similar on the range of extraneous variables that may affect the DV. If the groups are substantially different this creates a confounding variable that can affect the results. For the learning program example, groups could differ substantially if, for example, a Year 6 class was used as the experimental group and a Year 5 class was used as the control group. This potential limitation of differences between groups can be addressed by appropriately assigning participants to groups.

An advantage of the between-subjects design is that there are no **order effects** (see within-subjects design, on the next page).

Allocation to groups

The allocation of participants to groups is important in a between-subjects controlled experiment. One method is random allocation. **Random allocation** to groups uses chance to determine how participants are assigned to groups. For example, the researcher may use a random number generator to generate a number for each participant and then put participants with even numbers in one group and participants with odd numbers in the other. This gives every participant an equal chance of being assigned to each group. Random allocation minimises the likelihood of extraneous participant variables (for example, age or level of stress) becoming confounding variables. This is because when group allocation is random, it is likely that any EVs that may affect the DV will be evenly distributed between the experimental and control groups. Random allocation also protects against unconscious experimenter biases in which they may unintentionally influence group allocation in a way that is advantageous to their hypothesis.

If group assignment is not randomised, but instead relies on existing groups such as two existing classrooms at a school, there could be inherent differences between groups that confound the results. For instance, if the experimental group was a Year 6 class and the control group was a Year 5 class, age could be a confounding variable because it is different between the groups. In the example of a study assessing the effect of a learning program on students' reading ability, age confounds the effect of the intervention because we cannot be sure that the reason the intervention group scores higher on the reading test is because of the intervention; the intervention group could have done better simply due to their age. This issue may be avoided by randomly allocating participants to groups, because randomisation makes it more likely that an approximately equal number of Year 5 and Year 6 students will be in each group. Although randomisation does not guarantee that age distribution will be the same in both groups, it reduces the probability that differences are present.

Exam tip: Random allocation to groups is not the same as random sampling.

- ✓ **Random allocation** *is a method used to minimise bias in assigning participants to groups for between-subjects experiments.*
- ✓ **Random sampling** *is a method used to minimise bias in selecting a sample from the population.*

As an alternative to random allocation, a researcher may match groups on variables such as age or gender to control (that is, neutralise) the effect of this variable on the outcome of the study. This is called a **matched- groups** or **matched-participants design**. For example, to match groups by age would be to intentionally allocate an equal number of Year 5 and Year 6 children to each group.

Random allocation to groups is not possible when the experimenter is interested in the effect of a characteristic that defines different groups of people. For example, if the researcher is studying the effect of meditation for people who experience anxiety compared to non-anxious people, then group allocation is defined by a person's level of anxiety and cannot be random. We call these kinds of experiments **quasi-experimental designs** because they don't quite meet the standards of a controlled experiment. In quasi-experimental designs, the researcher may choose a matched-participant design to try to control the EVs that may differ between anxious and non-anxious people. In our example, the experimenter may match participants in the non-anxious group to people in the anxious group on factors such as age, gender and prior exposure to meditation.

Within- subjects design

In a within-subjects controlled experiment, each participant is exposed to both the experimental condition and the control condition. The within-subjects design is also known as a **repeated-measures design** because each participant has to repeat the experiment in order to collect data for both conditions. Each participant can then act as their own control. That is, each participant has one score from their experimental condition to test the intervention or manipulation and another score from their control condition.

For example, a within- subjects research design to test the effect of a learning program on reading ability could test each participant's reading ability before and after they participate in a learning program. If test scores are higher after the reading program than before the reading program, this can be seen as evidence that the reading program improves students' reading ability.

A benefit of the within- subjects design is that individual differences between people do not influence the results because each participant is compared to themselves (instead of being compared to other participants). People can be very different from each other and the within-subjects design overcomes the variation between participants. This variation could include attributes such as gender, ethnicity, personality, ability, education, socioeconomic status, memory, motivation and mood.

A limitation of the within-subjects design is that it susceptible to the order effect (also known as the practice effect). As each participant has to participate in both conditions, they engage in the testing procedure twice. In the example of a reading test, there could be a systematic improvement to their reading score simply because they have already taken a similar test; they have practised the test once and this makes them better the second time they do it. Another example of the order effect is if somebody had to do a test of reaction time twice in a row and their score in the second test was lower because of fatigue.

The order effect can sometimes be overcome by counterbalancing. Counterbalancing is where the order of the conditions is split, so not everybody completes the same conditions in the same order. For example, half of the participants could undertake the control condition first, followed by the experimental condition, and the other half undertake the experimental condition first, followed by the control condition (Figure 1.13).

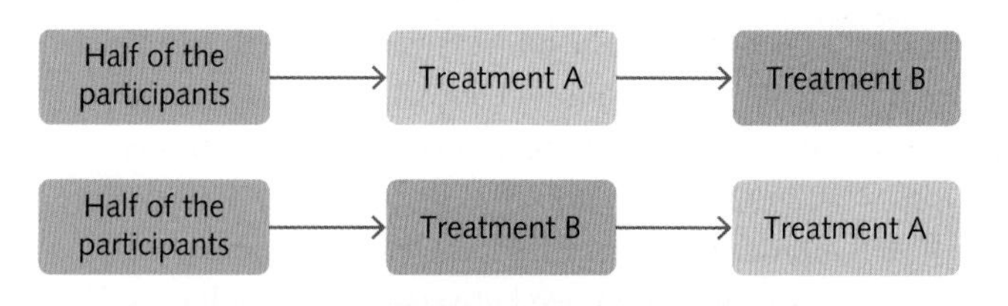

FIGURE 1.13 In a within-subjects controlled experiment, counterbalancing can be used to minimise order effects

Counterbalancing averages out any potential order effects across both conditions.

However, counterbalancing is not always possible due to the nature of particular studies. For instance, it is impossible to counterbalance the study examining the effect of the learning program on reading ability because the same students need to be tested before and after the learning program, and the researcher can not reverse the order of these events.

Some examples where counterbalancing is possible in within subjects designs are the investigation of restricted sleep on wellbeing and the investigation of caffeine consumption on memory. In these studies, the experimenter has control of the order in which participants experience each condition: in the sleep and wellbeing study, half of the participants could experience the restricted sleep condition before the long sleep condition and the other half could experience each condition in the opposite order, and in the caffeine and memory study, half of the participants could experience the caffeinated coffee condition before the decaffeinated coffee condition and the other half could experience these conditions in the opposite order. Of course, the researcher must allow enough time between each condition so that one condition's effects have worn off before testing the following condition.

Mixed design

1.3.2 RESEARCH DESIGNS

A mixed design has the elements of both a between-subjects design and a within-subjects design. In its most simple form, a mixed design has two IVs, and each IV has two levels. One of the IVs is a between subjects variable, and the other IV is a within subjects variable.

For example, when studying the effectiveness of the reading program, the ideal design is to compare two groups of students: one that experiences the reading program and one that experiences a control condition in which they continue learning as usual.

This is a between subjects variable with two levels (reading program vs. control). The researcher can measure reading accuracy for both groups before and after the reading program. This design allows the researcher to be more confident that any change they see between groups is due to the treatment itself and not simply to improvements that would have occurred during normal learning during that time. It combines the advantage of the within subject measure of reading performance with the advantage of a control group design. It has the disadvantages of being more difficult to carry out and producing results that can be more difficult to analyse.

TABLE 1.1 A mixed design controlled experiment to study a learning program intervention. The within-subjects factor (time) are the vertical columns of the table and the between-subjects factor (learning group) are the horizontal rows of the table.

		Within-subjects factor (Time)	
		Before the learning program	**After the learning program**
Between-subjects factor (Gender)	**Girls**	Girls' results before the learning program	Girls' results after the learning program
	Boys	Boys' results before the learning program	Boys' results after the learning program

KEY CONCEPTS 1.3

- Manipulation of a variable gives the controlled experiment the unique ability to find cause-and-effect relationships between variables.
- The independent variable (IV) is the variable that the researcher manipulates.
- The dependent variable (DV) is the variable that the researcher records to see whether it has been affected by a change in the IV.
- IVs and DVs are only relevant to controlled experiments.
- When plotted on a graph, the IV goes on the horizontal axis and the DV goes on the vertical axis.
- An extraneous variable (EV) may affect the DV and influence the results in an unwanted way.
- A confounding variable does affect the DV and influences the results in an unwanted way.
- A controlled variable is held constant in an investigation.
- A controlled variable is not the same as a control group.
- A between-subjects design is a controlled experiment where participants are allocated to different groups (for example, the experimental group or the control group).
- Between-subjects designs have no order effects.
- Random allocation of participants to groups helps to reduce confounding variable bias in between-subjects designs.
- A within-subjects design is a controlled experiment where each participant is exposed to both the experimental and the control conditions.
- In within-subjects designs, individual differences between people do not influence the results.
- Within-subjects designs can suffer from the order effect.
- Counterbalancing can overcome the order effect.
- A mixed design combines the between-subjects and within-subjects design, with multiple groups and data recorded at multiple times.

Interim summary

So far, this chapter has explored how to precisely create a hypothesis and test it with a study, and the research designs used for research investigations. We focused much of our attention on the controlled experiment, but we also explored correlational studies and other methodologies and showed why it is not always possible, desirable or appropriate to investigate some questions with an experiment.

Next we will explore some key science skills relevant to analysing and interpreting data collected in research investigations. We will also consider some theoretical concepts relevant to understanding research investigations and data analysis and ethical concepts and guidelines. These are all important foundations for understanding the research process and for ensuring that research investigations validly and ethically assess what they intend to investigate.

1.4 Analysing and evaluating research

Evaluating research includes closely considering different elements of a study and how any data it produces has been processed and analysed. In this section we will first describe how quantitative data can be processed and presented. We will then introduce the important concepts that underlie data analysis, including accuracy, precision, repeatability, reproducibility and validity. Next, we will consider how to evaluate whether findings support a hypothesis, how to identify limitations of a study, and how to understand and evaluate possible sources of error and uncertainty. Finally, we look at both the ethical concepts and guidelines and the health and safety considerations that must apply to all research.

Processing quantitative data

Whether you are deciding how to present your own data or analysing a published research investigation, it's important to understand how data can be processed and presented to best communicate the results.

Displaying data in tables, bar charts and line graphs

Data can be displayed in tables, charts or graphs. These can be used to organise data, compare variables and visualise relationships between them. Figure 1.14 provides an example of a well-constructed table, bar chart and line graph. A table is a grid with horizontal rows and vertical columns that is used to record and organise data. Statistics that summarise data can also be included in tables.

The terms 'chart' and 'graph' are often used interchangeably but, technically speaking, a chart is a visual representation of data that can take many forms and a graph is a specific type of chart that has two axes representing two variables. A bar chart is a graph that shows data using separated rectangular columns or 'bars' to represent the total number (or other measures such as the mean) for distinct categories of data. It shows how frequently a particular group or score occurs in a data set. A line graph represents the relationship between two variables by a line that connects each data point so that the reader can see the change from point to point and the overall trend. When showing the variables of a controlled experiment on a graph, the IV is plotted on the horizontal axis and the DV is plotted on the vertical axis.

Distribution of data

The distribution of data can be thought of as the shape and symmetry of data when it is plotted on a histogram. A **histogram** is a bar chart that graphs just one variable, where the horizontal axis represents the variable (the score or thing being measured) and the vertical axis lists the frequency at which that score is found in the data set (for example, the number of participants that have that score on that variable).

A data set is described as *normally distributed* when its curve has the shape of a bell; it is highest around the centre (indicating that there is a typical score for most participants) and gradually decreases in a relatively symmetrical pattern as scores move away from the centre (indicating that there are fewer participants who score towards the extremes).

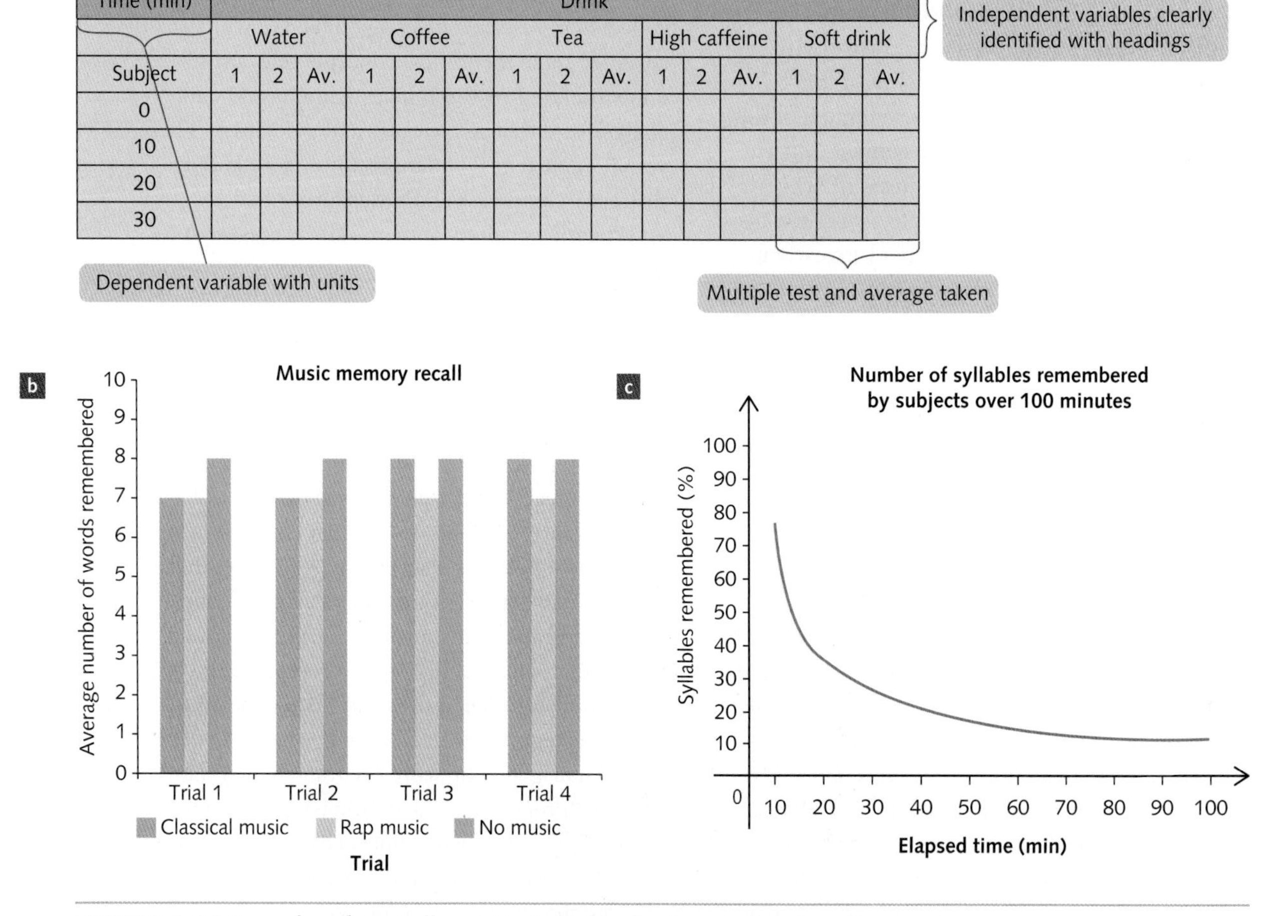
a Number of words recalled at 10-minute intervals after consuming a particular drink

Time (min)	Drink														
	Water			Coffee			Tea			High caffeine			Soft drink		
Subject	1	2	Av.	1	2	Av.	1	2	Av.	1	2	Av.	1	2	Av.
0															
10															
20															
30															

FIGURE 1.14 Examples of a a well-constructed table, b a bar chart and c a line graph

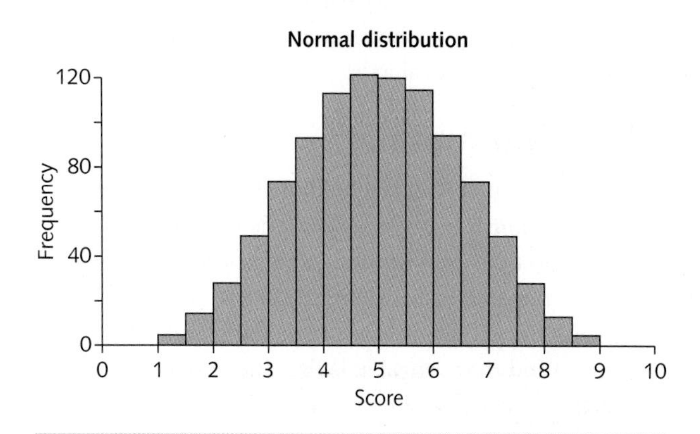

FIGURE 1.15 A histogram of normally distributed data

An example of normal distribution is the height of a randomly selected sample of people from the population. Figure 1.15 shows a frequency distribution of normally distributed data.

A data set is skewed when the curve is not symmetrical (that is, uneven on one side of the curve). An example of something that follows a skewed distribution is income. Most people lie relatively close to the mean, with some variation on either side, but some very wealthy people earn far more than the average person. Figure 1.16 shows two examples of histograms showing data that has a skewed distribution.

The terms 'normally distributed' and 'skewed' can describe one specific data set or they can describe the distribution of the population from which the sample was recruited. For example, imagine we have a data set with one value that is very different from the rest. This unusual value is called an outlier. We may say that the outlier in our data set (obtained from measurements of our sample) could reflect the underlying distribution (the distribution of the data of the whole population) being a skewed distribution, or from a different underlying distribution.

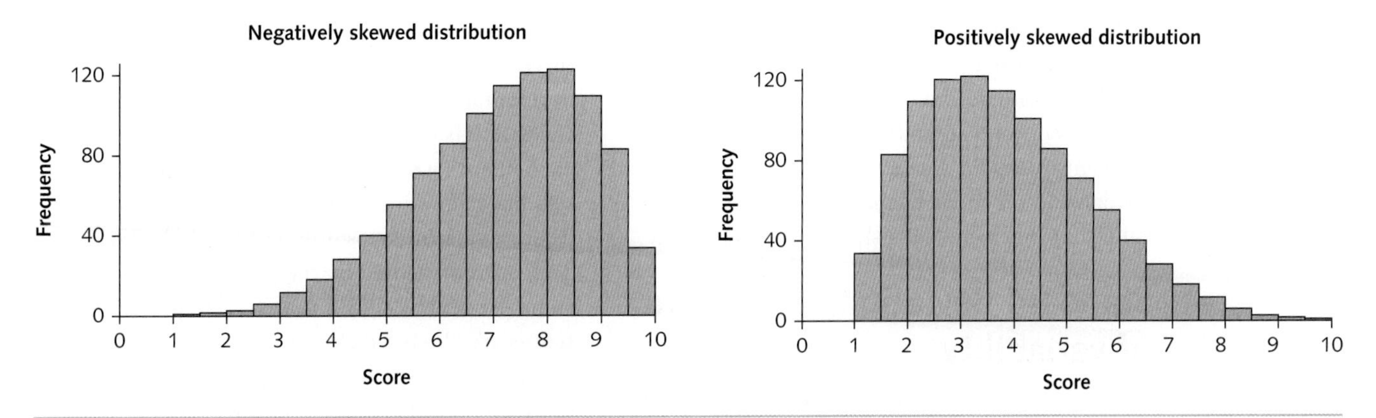

FIGURE 1.16 Two histograms showing examples of data following a skewed distribution

Descriptive statistics

Descriptive statistics summarise the main features of an overall data set. **Descriptive statistics** *describe* the data set by condensing a set of values down to a single numerical value. They can also be visualised in charts, tables and graphs. Descriptive statistics include measures of central tendency (such as the mean, mode, and median values) and measures of the range and spread of the data (such as standard deviation).

Measures of central tendency provide a number that describes a 'typical' score around which other scores lie. Central tendency is commonly measured by using three descriptive statistics: the mean, median and mode.

» The **mean** is a measure of central tendency that gives the numerical average of a set of scores, calculated by adding all the scores in a data set and then dividing the total by the number of scores in the set.
» The **median** is a measure of the middle score in a data set. It is calculated by arranging scores in a data set from the highest to the lowest and selecting the middle score. If there are two middle scores (if the data set contains an even number of scores), take the mean of these.
» The **mode** is a measure of central tendency found by selecting the most frequently occurring score in a set of scores.

1.4.1 COLLECTING AND COLLATING DATA

Figure 1.17 shows the three measures of central tendency (the mean, median and mode)

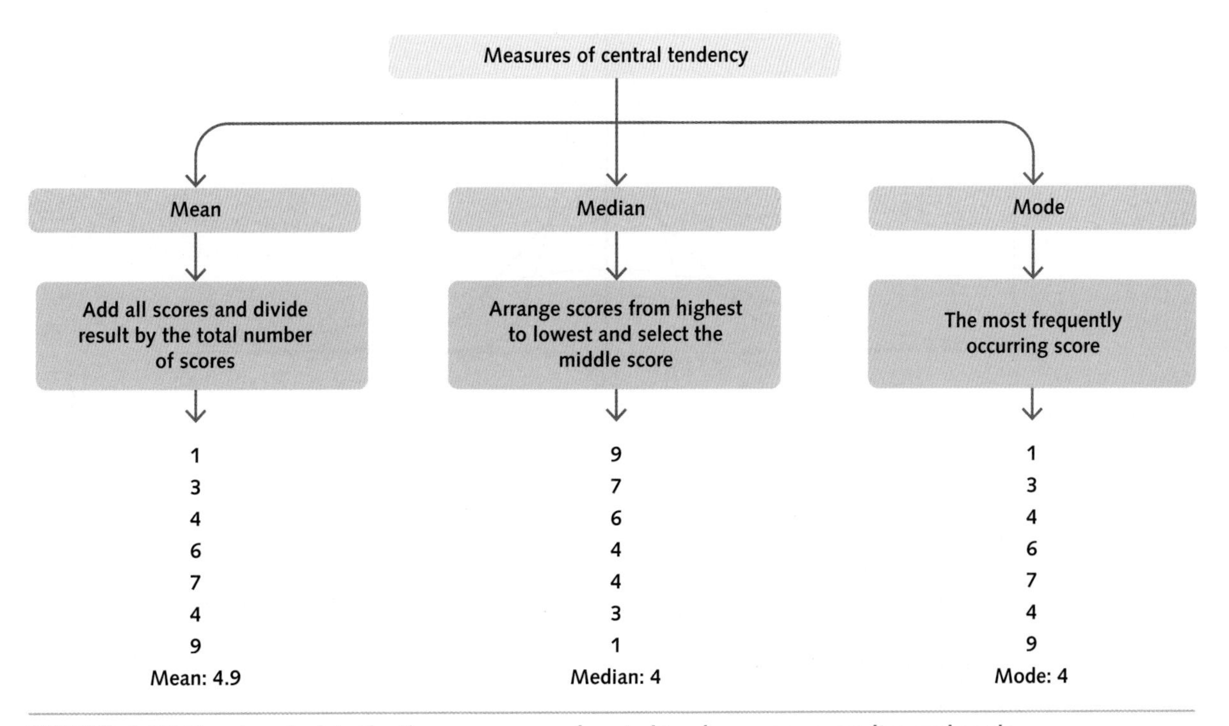

FIGURE 1.17 How to calculate the three measures of central tendency: mean, median and mode

and demonstrates how to calculate these, using an example data set.

This example shows that the different measures of central tendency can be the same or different values for a data set: in our example, the median and the mode happened to be the same value, but the mean had a slightly different value.

Standard deviation: a measure of variability

While measures of central tendency provide valuable information about the central or typical value in a data set, they provide no information about the *variability* of scores in the data set. **Variability** means how spread out or clustered together the scores are. **Standard deviation (SD)** is a measure of variability that the average deviation (or distance) of a set of scores from the mean (that is, the average distance that a set of data points are from the mean). Standard deviation uses a mathematical formula to provide a standardised measure of the average deviation of scores from the mean.

A standard deviation that is a *large* number demonstrates *high variability*, as it indicates that scores have a *large* average distance from the mean. A standard deviation that is a *small* number demonstrates *low variability*, as this indicates that scores have a *small* average distance from the mean.

Figure 1.18 shows the distribution of three data sets plotted on the same graph. The horizontal axis represents the score and the vertical axis can be thought of as the number of participants who have that score. All three data sets have a mean of 50, but the spread of scores (that is, variability) is different for each. The data set represented by the purple line has a low standard deviation (SD = 5), which means most scores are clustered close to the mean (you can see that the distribution is high at the centre and there are not many scores at the extreme values because the distribution does not cover a wide area). The data set represented by the green line has a high standard deviation (SD = 20) and so this data set has scores that are much more spread out. Notice how the green line does not rise very much at the centre, because there are fewer data points around the central value, but the distribution covers a much wider area, meaning there are more scores at the extreme values. The data set represented by the pink line has a standard deviation in between the other two data sets (SD = 10).

Standard deviation example

Imagine a study has investigated two forms of psychotherapy to treat anxiety. The first form of psychotherapy is a traditional and

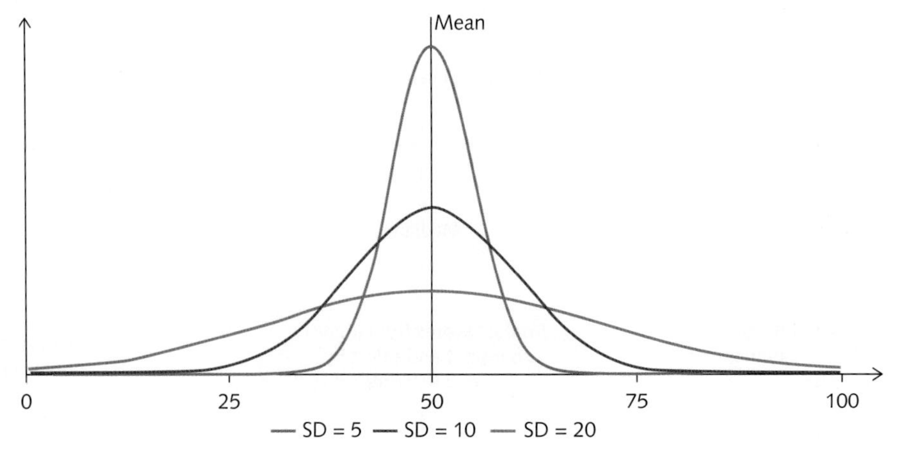

Figure 1.18 Three data sets with different standard deviations. A low standard deviation means that data are clustered close to the mean. As standard deviation increases, so does the spread of data. On this graph, the horizontal *x*-axis represents a variable (for example, score) and the vertical *y*-axis represents the frequency with which each score was found.

well-established form of therapy that is found to decrease anxiety by a relatively consistent amount for all research participants. The second form is a new form of psychotherapy that was found to reduce anxiety by a large amount in some participants and a small amount in other participants, but it increases anxiety in some participants. Both forms of psychotherapy could have the same mean, but the variability measured as standard deviation would be much larger for the new form of psychotherapy.

Percentage and percentage change

To calculate a percentage, you divide the number of the subset by the total number and then multiply the result by 100. Percentage change is the difference between two percentages.

Let us consider the earlier example of an experiment testing whether a new learning program improves students' reading ability. In this example, we will only consider the experimental group (that is, those who participated in the learning program). To pass the reading test, the student must read the passage aloud with no errors. Before the program, we ask 40 students to read a passage aloud and 12 students pass the test. To calculate the percentage who passed the test, we divide the number of students who passed the test (12) by the total number of students (40) and then multiply it by 100.

$$\frac{12}{40} \times 100 = 30\%$$

After the reading program, 28 students pass the reading test. Once again, we divide the number of students who passed the test (28) by the total number of students (40) and then multiply it by 100.

$$\frac{28}{40} \times 100 = 70\%$$

The percentage change is the difference between these two percentages, so we need to subtract the initial percentage value from the final percentage value.

$$70\% - 30\% = 40\%$$

Here the percentage change is 40%, which means that there was a 40% change in the number of students who passed the test after the intervention of the learning program.

Data analysis concepts

Analysing data requires us to look behind the actual data itself. It requires us to also to make a judgement on the *quality* of the data collected.

True value

The concept of a true value is something many other concepts build upon. The **true value** can be defined as the value or range of values that would be found if a quantity could be measured perfectly. However, this idea is somewhat more complex to consider in psychology than in other sciences. In psychology, we investigate concepts that are difficult to measure. Often concepts of interest cannot be directly measured, but instead we have to find a way to score the concept by measuring things that reflect the concept. Nevertheless, we can make use of the idea of true value when we analyse numerical data. To make use of the true value idea, we treat our constructs as if they would have a true value if it were possible to measure them perfectly.

Accuracy and precision

In science, the **accuracy** of a measurement means how close it is to the true value of the quantity being measured. While accuracy is not quantifiable, it can be used to describe measurement values as more accurate or less accurate. That means accuracy is a *relative* description, used to compare the accuracy of different values with each other. Researchers aim to make improvements in their research design to make their measures more accurate (that is, closer approximations of the true value).

Precision refers to how close a set of measurement values are to one another. Precision is determined by the repeatability and/or the reproducibility of the measurements obtained using a particular measurement instrument and procedure (for example, equipment for measuring participant response times, a psychological questionnaire or a brain imaging technique). Precision gives no indication of how close the measurements are to the true value and is therefore a separate consideration from accuracy. It is possible for a measurement instrument to produce precise (repeatable and reproducible) measurements, but those measurements may not be accurate. It is also possible to get an accurate measurement of the true value by averaging a set of highly variable (imprecise) measurements.

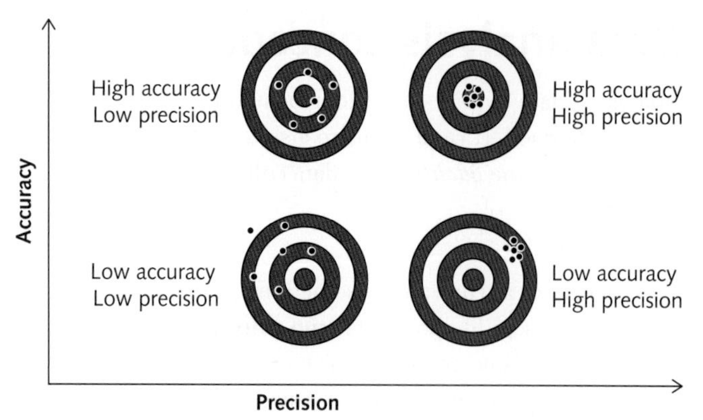

FIGURE 1.19 A representation of the precision and accuracy of measurements, where the bullseye of the dartboard represents the true value

Figure 1.19 shows a representation of precision and accuracy on a dart board, where the bullseye in the centre represents the true value and the darts (the black dots) represent individual measurements. This shows measurements can be:

» precise and accurate
» accurate but not precise
» precise but not accurate
» not accurate and not precise.

Another word for 'reproducible' is 'replicable'. The VCE Psychology Study Design uses 'reproducibility' as the key term, but interchangeably uses the term 'replicable' (and its opposite 'irreplicable', which means not replicable/ reproducible).

Repeatability and reproducibility

Figure 1.20 shows examples of repeatability and reproducibility. The VCE Psychology Study Design defines them as follows.

» **Repeatability** is the closeness of the agreement between the results of successive measurements of the same quantity being measured, carried out under *the same* conditions of measurement. These conditions include the same measurement procedure, the same observer, the same measuring instrument used under the same conditions, the same location, and repetition over a short period of time one researcher repeating a measurement of their own to see if they get the similar results each time.
» **Reproducibility** is the closeness of the agreement between the results of measurements of the same quantity being measured, carried out under *changed* conditions of measurement. These different conditions include a different method of measurement, different observer, different measuring instrument, different location, different conditions of use, different time and/or different culture(s).

1.4.2 REPRODUCIBILITY IN RESEARCH

Figure 1.20 Repeatability is the agreement of results or measurements under *the same* conditions of measurement, and reproducibility is the agreement of results or measurements under *changed* conditions of measurement.

Repeatability and reproducibility both refer to how close successive measurements of the same quantity are. The closer the results are to each other, the higher the repeatability or reproducibility. However, the terms differ in that:

» repeatability is the agreement of results carried out under *the same* conditions of measurement
» reproducibility is the agreement of results carried out under *changed* (different) conditions of measurement.

The conditions of measurement include the measurement procedure, instruments, observers (that is, the people recording the measurement), culture of observers, location of study and timespan between measurements (that is, repetition over a short or long period of time).

Repeating experiments can help to determine whether substantial errors exist, which is important because there will always be a level of error with any tools. Research findings are said to be robust if they are not easily affected by the presence of some errors in the data. Therefore, researchers should repeat their experiments to ensure the findings are robust.

Scientific findings can be considered as reproducible or not reproducible. If results are reproducible, other researchers find the same results when they repeat the experiment. If results are not reproducible, other researchers get different results when they repeat the experiment. Scientific findings that are not reproducible (replicable) may lack credibility, because this suggests that the initial results may not reflect a true, measurable relationship, but may rather have been a result of error or chance.

Validity

Validity means how well the design of a scientific investigation and its measurements provide meaningful and generalisable information about the psychological constructs of interest. The validity of a psychological investigation tells us how well the results from the study participants represent true findings among the population of interest outside of the study. There are two types of validity: internal validity and external validity.

A psychological investigation has internal validity if it investigates what it sets out to investigate. The internal validity of an investigation depends on:

- how appropriate the investigation design is
- the sampling and participant allocation techniques used
- whether there are extraneous and confounding variables that affect the results.

A lack of internal validity means that the study results deviate from the truth and therefore no conclusions can be drawn.

A psychological investigation has external validity if the results of the research can be applied to similar individuals in a different setting (outside the research context/in the real world). A lack of external validity means that the research results may not apply to individuals or situations outside the study population. External validity can be increased by using sampling techniques that ensure the study population more closely resembles the general population. External validity is also increased by using measures that reflect the way psychological processes operate in natural contexts. For example, measuring memory using lists of words in a lab may not have very high external validity with memory for events in real-world contexts.

If a study is not internally valid, the concept of external validity is irrelevant. If a study has not measured what it intended to measure, it is not possible to generalise its results to the population or make inferences about whether similar results would occur in a real-world context.

Errors

1.4.3
ERRORS

Measurement error is the difference between the measured value and the true value of what is being measured. Two types of measurement errors should be considered when you evaluate the quality of data: random errors and systematic errors. Figure 1.21 shows how errors are categorised.

Random errors

Random errors are unpredictable variations that can occur during measurement. When you take multiple readings of the same thing, random measurement errors cause small variations so that you end up recording a spread of readings. Such errors affect the precision of a measurement (that is, how close a set of measurement values are to one another). Random errors can be caused by limitations of instruments, environmental factors (such as sudden noises or interruptions) and slight variations in procedures.

Remember: ***Random errors*** *affect the* ***precision*** *of a measurement.* ***Systematic errors*** *affect the* ***accuracy*** *of a measurement.*

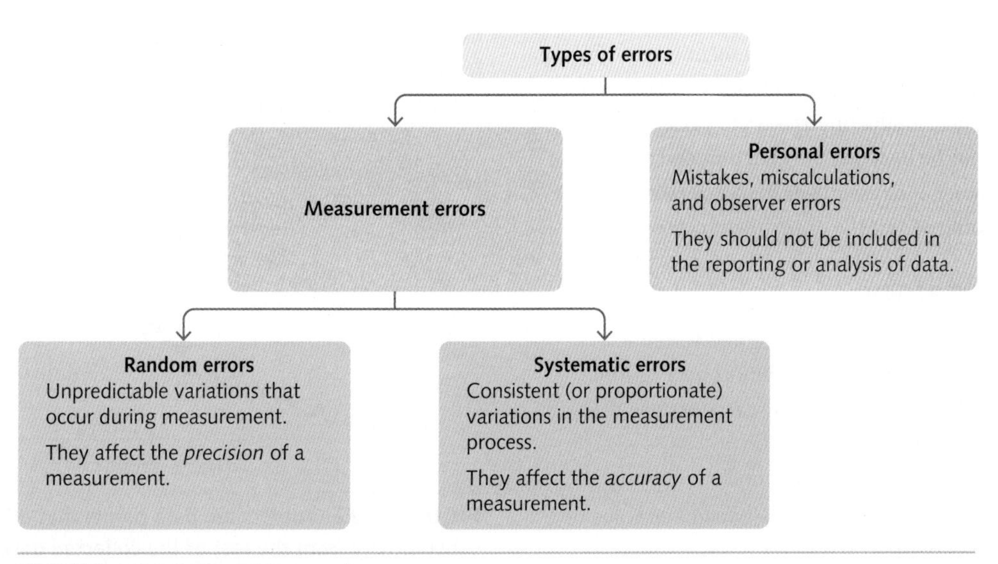

FIGURE 1.21 Different types of errors

The effect of random errors can be reduced by making more or repeated measurements and calculating a mean. The average value of repeated measurements can be used as the final value. The resulting reduction in random errors should mean that the final value is a closer approximation of the true value. The effect of random errors can also be reduced by increasing the sample size and/or by refining the measurement method or technique.

Systematic errors

Systematic errors affect the accuracy of a measurement by causing all of the readings to differ from the true value. They affect the reading by a consistent amount (or by the same proportion) each time a measurement is made. This causes all of the readings to be shifted away from the true value in the same direction.

Systematic errors may be caused by measuring instruments not being correctly calibrated or by environmental interference. Systematic errors can also be caused by observational error if there is a consistent distortion in the way we view things that causes errors that are the same every time (for example, a tall person reading a thermometer from a higher viewpoint and recording a lower measure than the true value every time). Most systematic errors can be reduced by knowing how to use the measurement instrument correctly and being familiar with its limitations. You cannot improve the accuracy of measurements that have systematic errors by repeating those measurements, because systematic errors will be present in all the measurements you make.

Don't confuse measurement errors with personal errors.

Personal errors

Personal errors are not measurement errors. Personal errors are mistakes, miscalculations and observer errors when conducting research. Personal errors should not be included in reporting and analysis of data. Rather, if a researcher makes personal errors, they should repeat the experiment correctly.

Don't confuse the terms 'error' and 'uncertainty'. They are not synonyms.

Uncertainty

The scientific method is designed to reduce the degree of uncertainty about observations, relationships and causes. Researchers use it to reduce uncertainty:

» in the observations they make
» about the relationships between variables of interest
» about the causes of the relationship they are interested in.

The **uncertainty** of a measurement reflects the lack of exact knowledge of the true value of the measurement. All measurements are subject to uncertainty because all measurements have many possible sources of variation. Because all inferences and conclusions in research depend on uncertain measurements, this uncertainty extends to all inferences and conclusions.

Psychological measurement tools (for example, surveys) are used to measure psychological constructs, which are, by nature, not directly measurable. Therefore, there is *always* a degree of uncertainty associated with measurements of psychological constructs.

The concept of uncertainty is especially relevant when evaluating data. You should always be vigilant for things that increase uncertainty, such as possible sources of bias, contradictory data and incomplete data. Incomplete data are data that are missing, such as questions without answers or variables without observations. Contradictory data are *incorrect* data: for example, a 5-point scale intended to be scored from 1 'strongly disagree' to 5 'strongly agree', that may have been miss-scored so that 1 represents 'strongly agree' and 5 represents 'strongly disagree', thus reversing the interpretation of the findings. Make sure you carefully consider sources of uncertainty both in data you have collected yourself and in data provided by others.

Outliers

Outliers are data points that differ substantially from the rest of the collected data. Figure 1.22 shows a plot of results where an outlier is present.

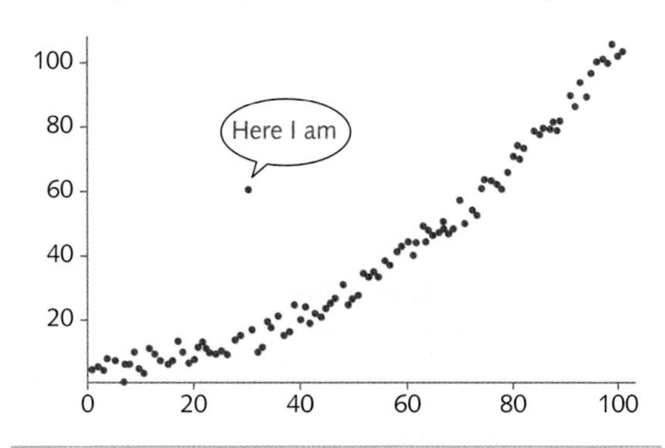

FIGURE 1.22 Outliers are data points that differ substantially from the rest of the collected data.

An outlier can be caused by a measurement or recording error or it can simply be an uncommon value. Uncommon values can appear in a data set by chance and so are more likely to be found in a larger sample. Therefore, you would expect to find a small number of outliers in a data set of a large sample. An outlier from a measurement or recording error can have a significant impact on the results and can affect the validity of the research.

An outlier can also be a valid measurement that stands out from the rest (that is, a true reading of a rare data point). Such outliers can occur in a data set due to natural variation. An outlier score can be a valid data point from an underlying skewed distribution or other non-normal distribution. In a skewed distribution, the data set may include very high or very low data points that are valid. For example, in a study of income (which follows a skewed distribution in the population), having a very wealthy person in a sample would give a valid outlier score.

If you obtain outliers in research results, you must distinguish between those that are valid measurements and those that result from measurement error. One way to investigate whether an outlier is an error or an uncommon value is by repeating measurements. If the outlier is a valid value, it should remain in the data set on the second repetition of data collection, whereas if the outlier is due to a measurement error, it would be unlikely that the same measurement error would appear again.

When you analyse a study, make sure you reflect on how different types of outliers would affect the testing efforts and validity of the research. If an outlier is just an uncommon value, it can still be considered a valid data point. However, if an outlier is an error, it can compromise the validity of the research.

Sample size and the quality of data

Sample size has a significant impact on the quality of data. Using a larger sample size for a study makes it more likely you will find an effect or a relationship, if one exists. Larger sample sizes improve data quality for several reasons.

Outliers and errors have less impact on the results you get from a larger sample. In small samples, natural variation can significantly influence the results, whereas natural variation is usually averaged out in larger samples. If an error or an outlier is present in a data set from a sample size of 100, it will have less effect than an error or outlier present in a data set from a sample size of 10. The conclusions from the data collected from the larger sample will therefore be less affected by this error, which means that increasing the sample size can improve the quality of the data.

A larger sample size also means that the sample is more likely to be representative of the population because there is more chance that it will include any natural variation and less chance that it will be a unique sample that does not reflect the general population. For these reasons, a larger sample size improves the study's validity and enhances certainty in the data. It follows that a large sample size makes a study more repeatable and reproducible, because it makes it more likely that findings reflect effects present in the real world.

The issue of small sample sizes is a serious problem for psychological research and for scientific research in general. Many historically important scientific studies were conducted with small samples. When other researchers attempted to reproduce these studies to confirm the findings, they got different results. This problem has contributed to what is called the 'the replication crisis' (or the reproducibility crisis) and has received significant attention in the past decade. There is a movement in the psychological research community to increase the sample sizes used in studies to address this issue and improve the quality of research findings.

1.4.4 ANALYSING DATA

Don't confuse errors and outliers. Remember that outliers must be further analysed and accounted for, rather than being automatically dismissed.

Evaluating data and investigation methods

When you evaluate an investigation, you need to consider both the data and investigation method itself. To do so, you need to apply the concepts that you have learned in the previous sections.

1.4.5 EVALUATION OF RESEARCH

Evaluating a research design

Students are often asked to *evaluate* a research design for the extended response question on the VCE exam. This type of question often provides an example of a study that you will need to read and evaluate. The question may tell you which concepts to write about or you may need to determine which concepts are relevant. The kinds of things the examiner may be looking for can be found in the key science skills section of the VCE Psychology Study Design; in particular, in the dot points that use the word 'evaluate'. The concepts that you may need to draw on to answer this question include:

- you may be asked to evaluate data to determine the degree to which the evidence supports or refutes the initial prediction or hypothesis, or supports the aim of the investigation
- you may need to evaluate investigation methods and possible sources of error or uncertainty
- you may need to suggest improvements to increase validity and to reduce uncertainty
- when evaluating data, you may need to identify outliers and contradictory or incomplete data; or comment on certainty in data, which includes the effects of sample size on the quality of data
- you may need to apply the science skills and concepts of accuracy, precision, repeatability, reproducibility, validity and errors
- you may also need to consider cultural biases that could affect the results (for example, WEIRD samples) or consider ethical concepts and guidelines.

Does the evidence support the aims?

When you interpret the results of a research investigation, it is important to reflect on and critique whether the evidence supports the aims. Ask yourself, 'Has the study answered the question it set out to answer?' Limitations of the study can be difficult to anticipate and are sometimes only identified during or after the research has taken place. For example:

- after recruitment, the researcher may identify that the sample is not representative of the population; this would affect external validity because the findings are unlikely to generalise to the broader population
- during the investigation, the researcher may identify unexpected factors that compromise the internal or external validity of the study
- during data analysis, other problems, such as measurement errors, may be identified.

Identify limitations of conclusions

All studies have limitations and all conclusions drawn from studies are inherently limited. The evidence may support the conclusions being drawn but there is always a level of uncertainty and room for improvements in the study design. Limitations are often unavoidable and may be caused by design constraints such as difficulties recruiting participants, small sample size, time constraints or inability to control EVs.

When considering the potential limitations of findings, researchers should consider whether their interpretations of the results would apply in different cultural contexts. Researchers should also consider whether cultural biases are present in their testing methods. For example, people may perform poorly on a test that has been developed and standardised for a different cultural group. The test might assume participants have background knowledge or particular ways of interpreting material; however, some cultural groups may not have been exposed to the testing methods or may interpret them differently through their own cultural context. Also, what is considered normal behavioural, cognitive and socio-emotional functioning can be influenced by cultural contexts, such as the traditional customs and beliefs of a particular culture or society.

Ethics

Ethics is a guiding framework that all research investigations must follow. Because psychological research involves human (and sometimes animal) participants, research investigations should only proceed if they can be carried out ethically. Ethical concepts and guidelines provide moral guidance for making decisions about the design and implementation of a research investigation. You must apply your ethical understanding across all units of VCE Psychology, particularly when you design and conduct your own research investigation.

VCE Psychology requires students to know:

- five **ethical concepts**, which you can use to explore the conduct of psychological investigations and to determine whether a research investigation is ethically acceptable
- six **ethical guidelines**, which you should consider when you are conducting and evaluating psychological investigations.

The five **ethical concepts** are: beneficence, integrity, justice, non-maleficence and respect (Table 1.2).

The six **ethical guidelines** that underpin psychological research are confidentiality, debriefing, informed consent procedures, use of deception in research, voluntary participation and withdrawal rights (Table 1.3).

The full definitions of the ethical concepts and guidelines as listed in the VCE Psychology Study Design are provided below.

The ethical concepts are:

- **Beneficence**: The commitment to maximising benefits and minimising the risks and harms involved in taking a particular position or course of action.
- **Integrity**: The commitment to searching for knowledge and understanding and the honest reporting of all sources of information and results, whether favourable or unfavourable, in ways that permit scrutiny and contribute to public knowledge and understanding.
- **Justice**: The moral obligation to ensure that there is fair consideration of competing claims; that there is no unfair burden on a particular group from an action; and that there is fair distribution of and access to the benefits of an action.
- **Non-maleficence**: Involves avoiding the causations of harm; however, as a position or course of action may involve some degree of harm, the concept of non-maleficence implies that the harm resulting from any position or course of action should not be disproportionate to the benefits from any position or course of action.

Ethical concepts are general in nature and are separate to any specific ethical principles, guidelines, codes or legislation.

1.4.6 THE ETHICS SPIDER

TABLE 1.2 A summary of the five ethical concepts for psychology research

Concept	Description
Beneficence	Having a commitment to do good (and minimise risks and harms)
Integrity	Acting with honesty and transparency
Justice	Ensuring fair distribution of benefits, risks, costs and resources
Non-maleficence	Avoiding harm or ensuring potential harm is outweighed by benefits
Respect	Giving due regard to individual difference and ensuring the right to autonomy and choice

TABLE 1.3 A summary of the six ethical guidelines for psychology research

Guideline	Description
Confidentiality	Ensuring the privacy of participants' personal information
Debriefing	After the experiment, disclosing the aim, results and conclusions, answering questions and providing support
Informed consent procedures	Ensuring participants understand the nature, purpose and risks of the study before agreeing to participate
Use of deception in research	Concealing aspects of the study (only used when absolutely necessary, and must be accompanied by debriefing)
Voluntary participation	Ensuring there is no coercion or pressure to participate
Withdrawal rights	Allowing participants to discontinue involvement in an experiment, without penalty

1.4.7 ETHICS

- **Respect**: Involves consideration of the extent to which living things have an intrinsic value and/or instrumental value; giving due regard to the welfare, liberty and autonomy, beliefs, perceptions, customs and cultural heritage of both the individual and the collective; consideration of the capacity of living things to make their own decisions; and when living things have diminished capacity to make their own decisions, ensuring that they are empowered where possible and protected as necessary.

Source: VCE Psychology Study Design (2023–2027), p. 20

Read the exam question carefully to see whether it asks for 'ethical concepts' or 'ethical guidelines'. Don't get these mixed up. Where the question asks for something vaguer like 'ethical considerations', you should specify in your response which of your responses are concepts and which are guidelines.

The ethical guidelines, as listed in the VCE Psychology Study Design, are:

- **Confidentiality**: The privacy, protection and security of a participant's personal information in terms of personal details and the anonymity of individual results, including the removal of identifying elements.
- **Debriefing**: Ensures that, at the end of the experiment, the participant leaves understanding the experimental aim, results and conclusions. Any participant questions are addressed and support is also provided to ensure there is no lasting harm from their involvement in the study. Debriefing is essential for all studies that involve deception.
- **Informed consent** procedures: Ensure participants understand the nature and purpose of the experiment, including potential risks (both physical and psychological), before agreeing to participate in the study. Voluntary written consent should be obtained by the experimenter and, if participants are unable to give this consent, then a parent or legal guardian should provide this.
- Use of **deception** in research: Is only permissible when participants knowing the true purpose of the experiment may affect their behaviour while participating in the study and the subsequent validity of the experiment. The use of deception is discouraged in psychological research and used only when necessary.
- **Voluntary participation**: Ensures that there is no coercion of or pressure put on the participant to partake in an experiment and they freely choose to be involved.
- **Withdrawal rights**: Involves a participant being able to discontinue their involvement in an experiment at any time during or after the conclusion of an experiment, without penalty. This may include the removal of the participant's results from the study after the study has been completed.

Source: VCE Psychology Study Design (2023–2027), pp. 20–21

Safety

Health and safety are important considerations for practical exercises in all sciences. If you are undertaking your own practical research investigations, you must consider any relevant occupational health and safety guidelines. In Victoria, workplaces are covered by the Occupational Health and Safety (OHS) Act 2004. As some psychology experiments are conducted in laboratories, researchers must ensure safe laboratory practices when planning and conducting investigations by using **risk assessments**, supported by safety data sheets (SDS), and accounting for risks. SDSs are usually only relevant if you are using chemicals as part of your investigation (more relevant to the sciences other than psychology). If your research does not use chemicals, but requires participants to take some actions that may cause harm, you will need to complete a risk assessment form (Figure 1.23). Your school is likely to have one of these documents for you to complete when you conduct your experiment. If you are unsure of either the ethical or the health and safety aspects of your experiment, check with your teacher.

Knowledge

I/we have read and understood the potential hazards and standard handling procedures of all the equipment, chemicals and living organisms.
I/we have read and understood the (Material) Safety Data Sheets for all chemicals used and produced.
I/we have copies of the (Material) Safety Data Sheets of all the chemicals available in or near the laboratory.

Agreement by student(s)

I/we, Bill Wilkins, Mary Newt, Christina Lee, agree to conduct this experiment safely in accordance with school rules and teacher instructions.

Risk assessment

I/we have considered the risks of:

fire	breakage of equipment	electrical shock	radiation
explosion	cuts from equipment	escape of pathogens	waste disposal
chemicals in eyes	sharp objects	heavy lifting	inappropriate behaviour
inhalation of gas/dust	rotating equipment	slipping, tripping, falling	allergies
chemicals on skin	vibration and noise	falling objects	special needs
runaway reaction	pressure	heat and cold	other risks

Assessment by student(s)

I/we have assessed the risks associated with performing this experiment in the classroom on the basis of likelihood and consequences using the School's risk matrix, according to International Organization for Standardization Standard ISO 31000:2009 and the Risk Management Guidelines, HB 436:2013.

I/we consider the inherent level of risk (risk level without control measures) to be:

Low risk **Medium risk** High risk Extreme risk

Control measures:

Always point test tube away from any person.
Add hydrochloric acid slowly and carefully to avoid vigorous reaction and projection of material from test tube.
Dip matches and tapers in water to ensure extinguished before disposal.
Additional measures: safety glasses, gloves

With the specified control measures in place, I/we have found that all the risks are "low risk". Risks will therefore be managed by routine procedures in the classroom, in combination with the specified control measures.

Certification by teacher

I have assessed the risks associated with performing this experiment in the classroom on the basis of likelihood and consequences using the School's risk matrix, according to International Organization for Standardization Standard ISO 31000:2009 and the Risk Management Guidelines, HB 436:2013. I confirm that the risk level and control measures entered by student(s) above are correct and appropriate.

Name: .... **Signature:** .... **Date:** ....

Certification by Laboratory Technician

I have assessed the risks associated with preparing the equipment, chemicals and living organisms for this experiment and subsequently cleaning up after the experiment and disposing of wastes, on the basis of likelihood and consequences using the School's risk matrix, according to International Organization for Standardization Standard ISO 31000:2009 and the Risk Management Guidelines, HB 436:2013.

I consider the inherent level of risk (risk level without control measures) to be:

Figure 1.23 An example of a risk assessment form

KEY CONCEPTS 1.4

- A table records and organises data using horizontal rows and vertical columns.
- A bar chart uses bars to represent total numbers (or summary statistics) of distinct categories.
- A line graph displays a line that connects data points to show changes, trends and relationships in the data.
- A normal distribution of data is relatively symmetrical and takes the shape of a bell.
- A skewed distribution is a data distribution that is not symmetrical.
- Descriptive statistics are single numerical values that summarise a data set. They include measures of central tendency (mean, mode, median) and variability (standard deviation).
- Variability tells us how close or far apart scores are.
- A large standard deviation means there is high variability (scores are more spread out from each other) whereas a small standard deviation shows low variability (scores are closer together).

- The true value is the value or range of values that would be found if the quantity could be measured perfectly.
- Accuracy is the closeness of a measurement to the true value.
- Precision is how close a set of measurement values are to one another (not necessarily to the true value).
- Repeatability is the agreement of results carried out under the same conditions of measurement.
- Reproducibility is the agreement of results carried out under different conditions of measurement.
- Researchers should repeat experiments to ensure findings are robust.
- Findings are generalisable if they can be applied to other people and/or contexts.
- An investigation is internally valid if it investigates what it set out to investigate.
- Lack of internal validity implies that the results of the study deviate from the truth and therefore no conclusions can be drawn.
- An investigation is externally valid if the results of the research can be applied to similar individuals in a different setting.
- Lack of external validity implies that the results of the research may not apply to individuals who are different from the study population.
- Personal errors are mistakes, miscalculations and observer errors.
- Random errors affect the precision of a measurement.
- Systematic errors affect the accuracy of measurements.
- The scientific method attempts to reduce uncertainty.
- Outliers are data points that differ substantially from the rest of the collected data.
- Contradictory data are incorrect data.
- Incomplete data are missing data, such as questions without answers or variables without observations.
- Larger sample sizes improve data quality.
- The five ethical concepts in VCE Psychology are: beneficence, integrity, justice, non-maleficence and respect.
- The six ethical guidelines that underpin psychological research are: confidentiality, debriefing, informed consent procedures, use of deception in research, voluntary participation and withdrawal rights.
- When undertaking any research investigation, you must consider all relevant health and safety guidelines.
- Use risk assessments and safety data sheets to identify and account for risks in experiments conducted in a laboratory.

1 Chapter summary

KEY CONCEPTS 1.1

- A research question defines the question that a research investigation tries to answer.
- An aim is a broad statement about the goal of a research investigation.
- A hypothesis is a proposed answer to a research question (made before the investigation is conducted). It is usually a directional statement about the relationship between the variables and states the expected results of a research investigation.
- A research investigation is used to test a hypothesis.
- Data are any information collected in scientific investigations.
- Primary data are data we collect ourselves from a study.
- Secondary data are data collected by someone else that we use when conducting a literature review.
- Quantitative data are numerical (recorded in the form of numbers).
- Qualitative data are non-numerical (descriptive).
- Sampling is the selection of participants from a population to participate in a research investigation.
- Convenience sampling means recruiting participants that are readily available.
- WEIRD samples (Western, educated, industrialised, rich and democratic) under-represent a large proportion of the population.
- Random sampling gives every member of the population an equal chance of being recruited.
- Stratified sampling ensures that the sample contains the same proportions of participants from each stratum as the proportions in the population of interest.
- Using larger samples can improve representativeness.

KEY CONCEPTS 1.2

- Each type of research design has specific benefits and limitations.
- A controlled experiment involves experimental manipulation of a variable to determine the effect on an outcome(s) of interest.
- A correlational study is a non-experimental investigation of the relationship between variables.
- Correlations can be visually represented by plotting two variables on a graph.
- Correlation does not equal causation.
- A positive correlation is where high scores for one variable occur with high scores for another variable.
- A negative correlation is where high scores for one variable occur with low scores for another variable.
- A case study is an investigation of one particular example of an activity, behaviour, event or problem, to acquire knowledge about the process as a whole.
- Classification and identification are processes used to organise phenomena into categories and identify examples of that categorisation.
- Fieldwork involves observing and interacting with an environment in the real world.
- Modelling means creating a representation of an event, process or system of concepts.
- Product, process or system development is the design or evaluation of a process, system or artefact to meet a human need.
- A literature review is a report produced by reading scientific research on a particular area and summarising it.

KEY CONCEPTS 1.3

- Manipulation of a variable gives the controlled experiment the unique ability to find cause-and-effect relationships between variables.
- The independent variable (IV) is the variable that the researcher manipulates.
- The dependent variable (DV) is the variable that the researcher records to see whether it has been affected by a change in the IV.
- IVs and DVs are only relevant to controlled experiments.
- When plotted on a graph, the IV goes on the horizontal axis and the DV goes on the vertical axis.
- An extraneous variable (EV) may affect the DV and influence the results in an unwanted way.
- A confounding variable does affect the DV and influences the results in an unwanted way.
- A controlled variable is held constant in an investigation.
- A controlled variable is not the same as a control group.
- A between-subjects design is a controlled experiment where participants are allocated to different groups (for example, the experimental group or the control group).
- Between-subjects designs have no order effects.
- Random allocation of participants to groups helps to reduce confounding variable bias in between-subjects designs.
- A within-subjects design is a controlled experiment where each participant is exposed to both the experimental and the control conditions.
- In within-subjects designs, individual differences between people do not influence the results.
- Within-subjects designs can suffer from the order effect.
- Counterbalancing can overcome the order effect.
- A mixed design combines the between-subjects and within-subjects design, with multiple groups and data recorded at multiple times.

KEY CONCEPTS 1.4

- A table records and organises data using horizontal rows and vertical columns.
- A bar chart uses bars to represent total numbers (or summary statistics) of distinct categories.
- A line graph displays a line that connects data points to show changes, trends and relationships in the data.
- A normal distribution of data is relatively symmetrical and takes the shape of a bell.
- A skewed distribution is a data distribution that is not symmetrical.
- Descriptive statistics are single numerical values that summarise a data set. They include measures of central tendency (mean, mode, median) and variability (standard deviation).
- Variability tells us how close or far apart scores are.
- A large standard deviation means there is high variability (scores are more spread out from each other) whereas a small standard deviation shows low variability (scores are closer together).
- The true value is the value or range of values that would be found if the quantity could be measured perfectly.
- Accuracy is the closeness of a measurement to the true value.
- Precision is how close a set of measurement values are to one another (not necessarily to the true value).
- Repeatability is the agreement of results carried out under the same conditions of measurement.
- Reproducibility is the agreement of results carried out under different conditions of measurement.
- Researchers should repeat experiments to ensure findings are robust.
- Findings are generalisable if they can be applied to other people and/or contexts.
- An investigation is internally valid if it investigates what it set out to investigate.
- Lack of internal validity implies that the results of the study deviate from the truth and therefore no conclusions can be drawn.
- An investigation is externally valid if the results of the research can be applied to similar individuals in a different setting.
- Lack of external validity implies that the results of the research may not apply to individuals who are different from the study population.
- Personal errors are mistakes, miscalculations and observer errors.
- Random errors affect the precision of a measurement.
- Systematic errors affect the accuracy of measurements.
- The scientific method attempts to reduce uncertainty.

- Outliers are data points that differ substantially from the rest of the collected data.
- Contradictory data are incorrect data.
- Incomplete data are missing data, such as questions without answers or variables without observations.
- Larger sample sizes improve data quality.
- The five ethical concepts in VCE Psychology are: beneficence, integrity, justice, non-maleficence and respect.
- The six ethical guidelines that underpin psychological research are: confidentiality, debriefing, informed consent procedures, use of deception in research, voluntary participation and withdrawal rights.
- When undertaking any research investigation, you must consider all relevant health and safety guidelines.
- Use risk assessments and safety data sheets to identify and account for risks in experiments conducted in a laboratory.

Adobe Stock/by Eva

Unit 1

How are behaviour and mental processes shaped?

Area of study 1: What influences psychological development?

Area of study 2: How are mental processes and behaviour influenced by the brain?

Area of study 3: How does contemporary psychology conduct and validate psychological research?

Psychological development 2

Key knowledge

- the interactive influences of hereditary and environmental factors on a person's psychological development
- the biopsychosocial approach as a model for considering psychological development and mental wellbeing
- the process of psychological development (emotional, cognitive and social development) over the course of the life span
- the role of sensitive and critical periods in a person's psychological development

Key science skills

Develop aims and questions, formulate hypotheses and make predictions

- identify independent, dependent and controlled variables in controlled experiments

Plan and conduct investigations

- design and conduct investigations; select and use methods appropriate to the investigation, including consideration of sampling technique and size, equipment and procedures, taking into account potential sources of error and uncertainty; determine the type and amount of qualitative and/or quantitative data to be generated or collated

Generate, collate and record data

- record and summarise both qualitative and quantitative data, including use of a logbook as an authentication of generated or collated data
- organise and present data in useful and meaningful ways, including tables, bar charts and line graphs

Analyse and evaluate data and investigation methods

- process quantitative data using appropriate mathematical relationships and units, including calculations of percentages, percentage change and measures of central tendencies, and demonstrate an understanding of standard deviation as a measure of variability
- identify and analyse experimental data qualitatively, applying where appropriate concepts of: accuracy, precision, repeatability, reproducibility and validity of measurements; errors (random and systematic); and certainty in data, including effects of sample size on the quality of data obtained

Construct evidence-based arguments and draw conclusions

- evaluate data to determine the degree to which the evidence supports the aim of the investigation, and make recommendations, as appropriate, for modifying or extending the investigation

Analyse, evaluate and communicate scientific ideas

- discuss relevant psychological information, ideas, concepts, theories and models and the connections between them

Source: VCE Psychology Study Design (2023–2027), pp. 24, 12–13

2 Psychological development

Scientific investigation covers a vast array of topics; however, one topic that scientists have been fascinated with for a long time is our psychological development. There is a wealth of scientific research that tries to answer the question, how do we get to be the way we are? What biological, genetic, social, cultural and environmental factors will influence who we'll become?

p. 51

2.1 Interaction of hereditary and environmental factors

You have probably heard of the nature versus nurture debate: which do you think has a greater influence on our psychological development? Nature refers to hereditary factors that influence our development, such as genetic characteristics that you inherit from previous generations: are you born with characteristics that make you who you are? In contrast, nurture refers to the environmental factors that make us who we are: how do your experiences influence your psychological development? Or could it be a mixture of both?

iStock.com/FatCamera

p. 59

2.2 The biopsychosocial model

Do both nature and nurture contribute to our physical and mental wellbeing? The biopsychosocial model proposes that biological, psychological and social factors interact together, and this is what determines our mental wellbeing throughout our lifetime.

iStock.com/StefaNikolic

p. 63

2.3 Psychological development over the life span

Many development theorists agree that our early life experiences are extremely important in our psychological development, particularly an infant's relationship to their primary caregiver. There are lots of theories that feed into this, such as Piaget's theory, which looks at cognitive development and how our thinking processes develop.

Adobe Stock/famveldman

Dreamstime.com/Ferli Achirulli Kamaruddin

p. 85

2.4 Sensitive and critical periods in psychological development

Sometimes, timing really is everything. Throughout the early stages of our life span, it feels like we learn quickly. This is why we're encouraged to learn to read and write in the first years of school, when our brain is primed to learn new skills. These periods are known as 'sensitive' periods. However, there are some cognitive or motor skills that we need to learn during a 'critical' period, or these skills may never be acquired.

Psychological development continues all throughout the life span. There are many ways in which our cognitive, emotional and social capabilities may be affected, and you will continue to study how we become who we are as you explore the subject of psychology.

Slideshow
Chapter 2 slideshow

Flashcards
Chapter 2 flashcards

Test
Chapter 2 pre-test

Assessment
- Pre-test
- End-of-chapter exam

Revision
- Chapter map
- Key term flashcards
- Key concept summary
- Slideshow

Investigation
- Investigation: Principles of conservation
- Data calculator
- Logbook template: Controlled experiment

Worksheet
- Influence of hereditary and environmental factors on IQ scores

To access these resources, visit **cengage.com.au/nelsonmindtap**

Nelson MindTap

Know your key terms

Abstract thinking
Accommodation
Animism
Assimilation
Attachment
Biopsychosocial model
Centration
Classification
Cognition
Concrete thinking
Conservation
Critical period
Egocentrism
Generativity
Goal-directed behaviour
Hypothetico-deductive reasoning
Mental wellbeing
Nature
Nature versus nurture
Nurture
Object permanence
Psychological development
Psychosocial dilemma
Reversibility of thought
Schema
Sensitive period
Strange situation test
Symbolic thinking

Imagine you found out one day that there was another person exactly like you. Someone who shared your exact biological make up who had been living in another home, with another family in other state your entire life. Would you be similar to this person? Have the same interests, personality, intelligence and social skills? Well, this is precisely what happened to a set of twins known as the Jim twins.

Jim Springer and Jim Lewis are identical twins who were separated at 4 weeks of age and reunited at the age of 39 years, in 1979. When they were brought together, there appeared to be some uncanny similarities between the twins. Both worked as part-time deputy sheriffs, spent their holidays in Florida, owned dogs named 'Toy' and drove the same type of car. One twin named his son James Allan and the other named his son James Alan. Both men married and divorced women named Betty (Santrock, 1997). They shared a dislike of spelling, but liked maths. They both enjoyed carpentry and mechanical drawing, had almost identical smoking and drinking habits and chewed their fingernails down to the quick.

Jim and Jim are just one set of twins who took part in the 'Minnesota Study of Twins Reared Apart' in 1990. This study aimed to investigate the effect that environmental and genetic factors have on a person's psychological development. This chapter explores the many facets and complexities of an individual's psychological development.

Alamy Stock Photo/Yuri Arcurs

Figure 2.1 Psychological development is determined by the interaction of biological, genetic, social, cultural and environmental factors occurring throughout the life span

Psychological development is the process of growth and change in humans' cognitive, emotional, and social capabilities and functioning over the life span, from conception to old age. Psychologists have traditionally focused their attention on the psychological development occurring in the early stages of the life span because these changes generally provide a basis for future development and behaviour. However, development does not stop once an individual reaches a certain age; it continues throughout the life span. Psychological development is influenced by the interaction of biological, genetic, social, cultural and environmental factors (Figure 2.1).

2.1 Interaction of hereditary and environmental factors

The terms 'nature' and 'nurture' are used in psychology to describe two major forces in the development of humans. Psychologists have long debated whether nature (hereditary factors) or nurture (environmental factors) has a greater influence on development. Most psychologists today agree that there is an important interaction between both inherited factors and environmental factors in the development of an individual. It is not the case that one is more important than the other in contributing to an individual's psychological development.

Nature: hereditary factors

The term hereditary refers to the inborn, inherited genetic factors passed from biological parents to children that partly determine individual characteristics; this is the **nature** component within the **nature versus nurture** debate. An incredible number of individual features are set at conception when a sperm and an ovum (egg) unite and the genetic components of an individual are assembled. For example, a characteristic such as your eye colour is genetically determined by the genes inherited from your biological parents. Other physical traits passed down via our genes include blood type, body type, skin pigment, hair colour and susceptibility to certain diseases and disorders. Some of these genetic characteristics are almost completely determined by nature, such as eye colour; other characteristics, such as height, are also influenced by environmental factors.

How genes affect inherited characteristics

Genes are made up of sections of DNA that may code for specific traits or characteristics in an individual. Genes are in every cell in the body and occur in structures called chromosomes. Sometimes genes are referred to as being 'for' a specific characteristic or behaviour. This is not really how our genes work. Genes may contribute to different behaviours, but not as directly as this statement would suggest. For example, there are no genes 'for' aggression; rather, a gene may affect aggressive behaviour by promoting growth in brain systems that respond to external stimuli and that organise aggressive behaviour (Gray, 1994). A typical human cell has 46 chromosomes, or 23 pairs (Figure 2.2). In each pair, one chromosome is inherited from the biological mother and the other from the biological father. Males and females have the same chromosomes, except for the sex chromosomes: males have XY and females have XX chromosomes.

There are some well-studied genetic disorders that affects a person's psychological development, such as Down syndrome, which occurs because of atypical chromosome formation. People with Down syndrome, also called trisomy 21, have an extra chromosome 21, giving a total of 47 chromosomes instead of 46 (Figure 2.2). This extra chromosome leads to characteristic physical features, health development challenges and some level of intellectual disability.

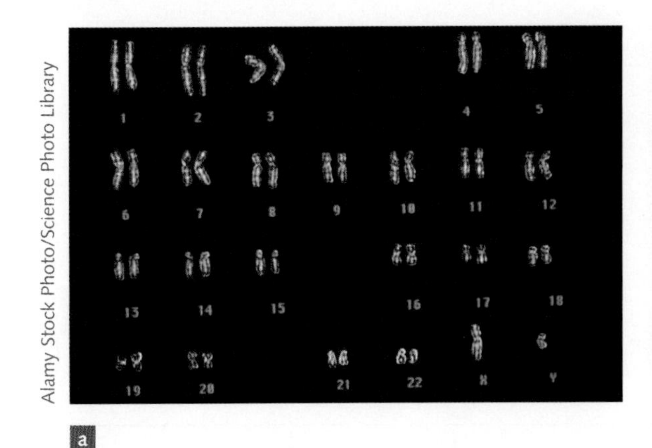

Alamy Stock Photo/Science Photo Library

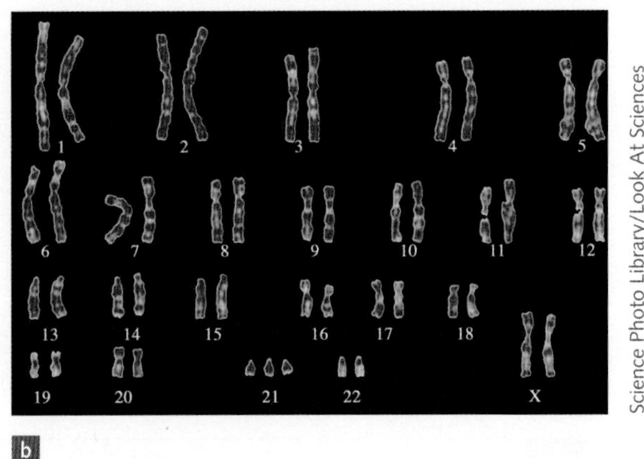

Science Photo Library/Look At Sciences

Figure 2.2 a The 22 numbered chromosome pairs in the human body and the sex chromosomes. b The chromosomes of an individual with Down syndrome

These characteristics vary in individuals with Down syndrome as each person has other genes that differ and different environmental factors that affect them (Down Syndrome Australia, 2022).

There are other diseases and disorders that some people are said to be genetically vulnerable towards contracting. Genetic vulnerability or susceptibility refers to an increased likelihood of developing a particular disorder or condition based on a person's inherited genetic features. Certain types of cancers, diabetes, heart disease and even some mental disorders such as schizophrenia have a genetic link. However, this does not mean that if both parents have one of these illnesses that they will certainly be developed. In most instances, environmental and lifestyle changes can play an important role in the heritability of these diseases, highlighting the interactive nature of hereditary and environmental factors on development.

2.1.1 ENVIRONMENT AND INFANT DEVELOPMENT

Nurture: environmental factors

Nurture refers to the effects that external biological and social environmental factors have on the development of our physical, psychological and behavioural characteristics. These include the experiences we have or the stimuli we are exposed to that influence or help to shape us even before our birth and over the life span. The care we are given as children in terms of food, education, love and support received from our environment are all examples of nurture. Other examples include the environment within the womb, exposure to environmental toxins and peer-group influences.

An individual's development can be affected by whether they are exposed to an 'enriched' environment or a 'deprived' environment. An enriched environment is one in which all a person's basic needs and more are met, such as nutrition, education and love. Their exposure to stimuli in the environment positively contributes to their growth and development. In a deprived environment basic needs are not met, and the individual is not exposed to the stimulation necessary for normal growth and development. For example, an individual who does not receive an adequately nutritious diet will not develop normally.

Research into the effects of nature and nurture on development suggests that the environment in which a child is raised can have a powerful impact on their physical, cognitive (intellectual) and emotional development. The lasting effect of early experiences has been demonstrated by studies of abused children who often exhibit lifelong emotional problems (Rutter, 1995). However, research also suggests that extra care given to an abused child can sometimes reverse the effects of a poor start in life (Bornstein, 1995). These findings suggest that although we may be born with certain capacities (genetics), environmental factors play a crucial role in determining how these capacities develop – or whether they develop at all.

Twin and adoption studies

The relative impact of nature and nurture is difficult to judge, given it is almost impossible to separate the effects of one from the other. Research using twin and adoption studies highlights the interactive nature of inherited and environmental factors on development. In a twin study, the development of identical twins is compared with the development of fraternal twins. Identical twins develop from a single fertilised egg that splits into two, with each developing into a genetically identical foetus. Fraternal twins develop at the same time from separate eggs and separate sperm. They are no more alike genetically than are brothers and sisters. In adoption studies, children from families who have one adopted child and one biological child are compared. These studies also include sibling pairs in which one child is adopted and the other reared by the biological parents. Comparisons of these children are then analysed.

Both twin and adoption studies provide powerful research tools for investigating how hereditary and environmental factors influence psychological development. These studies have been used to compare the relative strength of hereditary and environmental influences. For example, research has shown that the closer two people are on a family tree, the closer their intelligence quotients (IQs) are likely to be. The strongest correlation is found for identical twins who are reared together in the same environment. The correlation reduces as the genetic similarity lessens, supporting the proposal that there is a hereditary influence over intelligence. However, a shared environment (reared together) also increases the correlations in all cases (Figure 2.3) (Bouchard, 1983; Henderson, 1982).

Strong evidence for an environmental view of intelligence comes from adoption studies. With an adopted child, a parent can only contribute to the environment (Figure 2.4). If intelligence is highly genetic, the IQs of

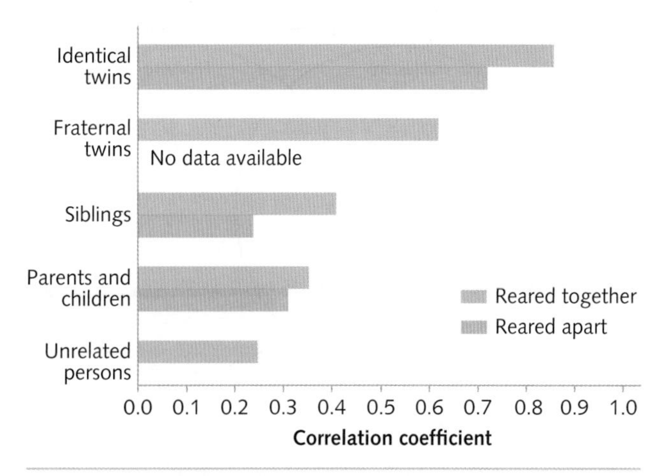

Figure 2.3 Hereditary influences: approximate correlations between IQ scores for people with varying degrees of genetic and environmental similarity (Bouchard 1983; Henderson 1982)

adopted children should be more like their biological parents than the IQs of their adoptive parents. However, studies show that all children reared by a mother resemble her in IQ to the same degree regardless of whether they share her genes (Weinberg, 1989; Kamin, 1981; Horn et al., 1979). This indicates that the environment and the mother's influence (her nurturing) has a greater influence on a child's cognitive abilities than genetics. Analysing research 2.1 (page 46) explores more twin and adoption studies in this area of research.

Worksheet Influence of hereditary and environmental factors on IQ scores

2.1.2 NATURE VS NURTURE

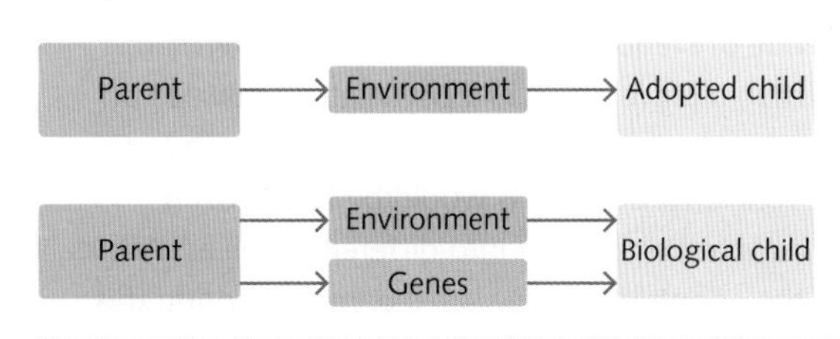

Figure 2.4 Comparison of an adopted child and a biological child reared in the same family

ACTIVITY 2.1 HEREDITARY INFLUENCES

Try it

1 Write down as many personal characteristics as you can in one minute. For example, this could include things like your hair, what hand you write with, and if you'd consider yourself an introvert or an extrovert.

Copy out the Venn diagram in Figure 2.5 and put your personal characteristics where you believe they belong. Compare with the person next to you. Are there any you have classified differently? Discuss your reasons for the placement of your characteristics.

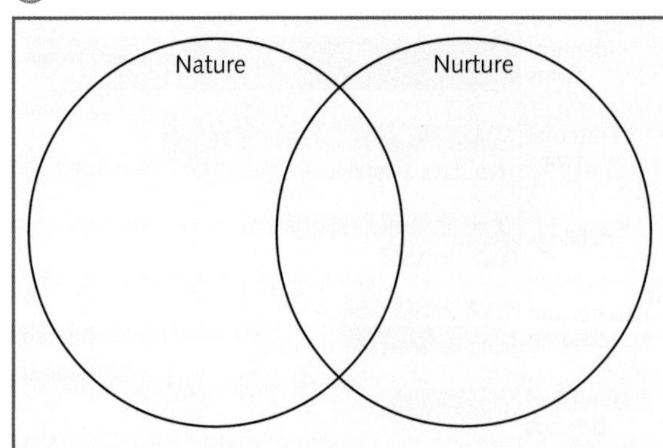

Figure 2.5

2 Think of three more personal characteristics that apply to you: one for nature, one for nurture and one that belongs in both. Add them to your Venn diagram.

Apply it

3 Consider identical twins Hal and Mal. If Hal develops schizophrenia, does that mean Mal will automatically develop schizophrenia? Explain with reference to heredity and environment.

Exam ready

4 Which of the following statements is true about psychological development?

A Only our genetics determine our psychological development.

B Only our environment and life experiences determine our psychological development.

C Our environment can influence the way in which our genes are expressed.

D There is no relationship between genetics and environment in our psychological development.

ANALYSING RESEARCH 2.1

Neubauer's twins study: identical strangers

Dr Peter B. Neubauer (1913–2008) was a child psychiatrist who was behind one of the most unethical twin studies on nature versus nurture in history. The study began in 1960, when one of Neubauer's colleagues, Viola Bernard, then a prominent child psychiatrist, advised an adoption agency, Louise Wise Services in New York City, that twins would be better off if adopted separately into different families. After the adoption agency started implementing this practice, Neubauer sought out the opportunity to study the twins who were raised in different homes. The adoptive parents did not know the child they were adopting was a twin. Over the next two decades Neubauer collected data by analysing school reports of the children, intermittently visited the children and parents to interview them and conducted psychological tests on the children. The parents were under the false belief that these visits were part of a routine evaluation of adoptive children. As these children grew to adults some became curious about their biological parents and through their investigations discovered they were twins. Another set of triplets used in the study discovered each other during a chance encounter (Hoffman & Oppenheim, 2019).

One set of identical twins from this study, who discovered each other in 2004 when they were 35 years old, were Paula Bernstein and Elyse Schein (Figure 2.6a). Upon their reunion, they discovered striking similarities in their lives, such as their mannerisms, interests and psychiatric history. They spent a great deal of time researching Neubauer's covert study to find out all of the details of his research. They even tracked him down and visited him, demanding answers to their questions about why they were separated and what the findings of his study were. They went on to write a memoir together, published in 2007, called *Identical Strangers*, which documents their lives and the information they collected (Boomsma, 2008).

Another better known group of participants from this study were the identical triplets Edward Galland, David Kellman, and Robert Shafran (Figure 2.6b). In 1980, these men were 19 years old when they discovered each other through chance encounters. Their story was featured in a 2018 documentary titled 'Three Identical Strangers', which documents their joyful reunion followed by the heartbreaking discovery they

Getty Images/New York Daily News Archive

AP via AAP/David Kellman

a b

Figure 2.6 Sets of twins and triplets who were unknowing participants in Neubeuer's experiment on nature versus nurture: a Paula Bernstein and Elyse Schein; b Edward Galland, David Kellman and Robert Shafran

were unwitting participants in Dr Neubauer's study and the painful aftermath of this encounter.

The stories told by the twins and triplets are incomprehensible accounts of loss and regret inflicted upon them by Dr Neubauer. The findings of Neubauer's research are not being released; they have been archived at Yale University and will remain restricted documents until 2065. Through long legal proceedings, the triplets and twins gained access to some of Neubauer's study materials. However, access to these materials still leaves many of their questions unanswered and does not make up for the trauma they experienced or the years these siblings lost.

Sources: Hoffman, L., & Oppenheim, L. (2019). Three Identical Strangers and The Twinning Reaction – Clarifying History and Lessons for Today From Peter Neubauer's Twins Study. *The Journal of the American Medical Association*, *322*(1), 10

Boomsma, D. (2008). Identical Strangers: A Memoir of Twins Separated and Reunited. *Twin Research and Human Genetics*, *11*(4), 478–479

2.1.3 TWIN AND ADOPTIONS STUDIES

Questions

1 What was the aim of Neubauer's study?
2 What are the variables of interest in this study?
3 Describe the type of data Neubauer collected for this study.
4 Why do you think Neubauer would be interested in studying identical and fraternal twins reared in different adoptive homes?

HOT Challenge

5 Imagine you are a participant on a present-day ethics committee and Dr Neubauer came to you with his research proposal. Compile a list of ethical guidelines and principles explaining why you would or would not approve his research on nature versus nurture using reared-apart adopted twins and triplets.

KEY CONCEPTS 2.1

- Most psychologists today agree that there is an interaction between inherited factors and environmental factors in the psychological development of an individual.
- Hereditary factors refers to inborn, inherited factors gained genetically from biological parents that partly determine individual characteristics.
- Environmental factors include external biological and social influences.
- Environmental influences include the care a child is given, the environment within the womb, exposure to environmental toxins and any peer-group influences.
- Twin and adoption studies allow researchers to study the interaction between genetic and environmental factors on development.
- In a twin study, the development of identical twins is compared with the development of fraternal twins.
- In adoption studies, families who have one adopted child and one biological child are compared.

Concept questions 2.1

Remembering

1 What is the nature versus nurture debate? r

2 What are hereditary influences on development? In your answer, provide at least three examples. r

3 What are environmental influences on development? In your answer, provide at least three examples. r

4 What have twin and adoption studies investigating the hereditary and environmental influences on intelligence (IQ) revealed? e

Understanding

5 Provide an example that demonstrates how genes affect our inherited characteristics. c

6 Describe how a deprived environment can be a negative influence on a child's psychological development. e

Applying

7 Provide an example of how inherited and environmental factors may influence change in human development in the following three areas c

a height of an individual

b musical ability

c incidence of mental illness.

HOT Challenge

8 The Jim twins case described at the opening of this chapter provides an example of hereditary and environmental influences on psychological development. Analyse the similarities of the two Jims from this text and the weblink given here. Which of the similarities between the two Jims do you believe can be attributed to hereditary influences? Which do you think are the result of other factors? Provide reasons for your answers. e

Weblink
Jim twins

First Nations peoples' perspectives

Aboriginal and Torres Strait Islander peoples should be aware that this section contains content that may be distressing.

Aboriginal and Torres Strait Islander peoples

Australia is home to two distinct broad groups of First Nations peoples. These are Aboriginal peoples and Torres Strait Islander peoples. Together, the phrase 'Aboriginal and Torres Strait Islander peoples' refers collectively to the group of Nations, cultures and language groups that live across mainland Australia and throughout the Torres Strait. These cultures pre-date British colonisation of Australia in 1788 by at least 60000 years. We speak of *peoples* (plural) to recognise that there are many different Nations, cultures and language groups, not just one single Aboriginal and Torres Strait Islander culture or identity.

Weblink
The AIATSIS map of Indigenous Australia

The adjective 'Aboriginal' refers broadly to the Nations and custodians of mainland Australia and most of the islands, including Tasmania, Fraser Island, Palm Island, Mornington Island, Groote Eylandt, Bathurst Island and Melville Islands.

The term 'Torres Strait Islander' is a broad term grouping the peoples of at least 274 small islands between the northern tip of Cape York in Queensland and the south-west coast of Papua New Guinea (Figure 2.7).

Throughout this text we use the term Aboriginal and Torres Strait Islander peoples or Australia's First Nations peoples when referring collectively to both groups. We use either Aboriginal peoples or Torres Strait Islander peoples when referring to a specific group.

Colonisation and its impacts

The colonisation of Australia massively disrupted the cultures of Aboriginal and Torres Strait Islander peoples. The continuing impacts of colonisation are the cause of the ongoing health and wellbeing challenges faced by many Aboriginal and Torres Strait Islander people today. Colonisation was based on a lie that Australia was a 'land belonging to no one', or '*terra nullius*'. Successive government policies saw Aboriginal and Torres Strait Islander peoples removed from their traditional lands. Colonists committed massacres and poisonings of communities (Ryan et al., 2022). They introduced diseases, including smallpox, which killed thousands of people. Colonial governments broke up communities and removed people to missions and reserves. Laws were introduced that banned the use of Language and cultural practices. At missions and reserves, Aboriginal and Torres Strait Islander peoples could no longer access traditional foods, and instead often ate a diet based on refined sugar and bread. Alcohol was introduced. Children were forcibly removed in their thousands from families to be 'protected' and 'assimilated' into a white Australia. These children and families are known as the Stolen Generations.

Colonisation continues to significantly impact the social and emotional wellbeing of many Aboriginal and Torres Strait Islander people today, negatively affecting their development and life expectancy. It is the responsibility of all Australians to know this history and the ongoing impacts. You can learn more by exploring the excellent resources in the weblinks provided.

Culture and kinship

In spite of the impacts of colonisation, Aboriginal and Torres Strait Islander peoples and communities remain strongly committed to maintaining and revitalising their cultures. First Nations communities who practice culture and instil a sense of cultural identity and pride in their children have much stronger health outcomes (Salmon et al., 2018; Shay & Sarra, 2021; Zubrick et al., 2014). There is much for all Australians to learn from the wisdom of Aboriginal and Torres Strait Islander cultures that have survived for at least 60 000 years.

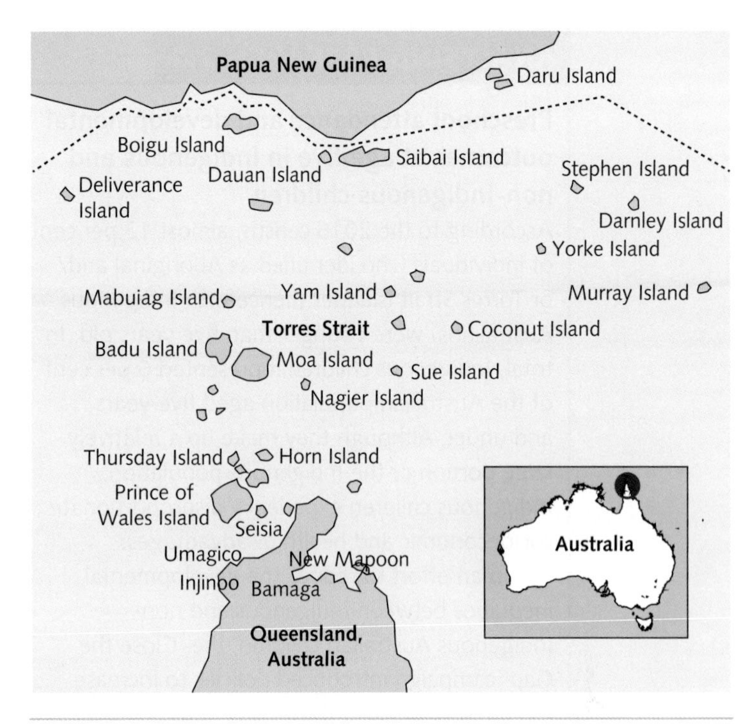

Figure 2.7 The Torres Strait Islands include many small islands between the tip of Cape York and Papua New Guinea

Development of people over the life span is one of the areas in which Aboriginal and Torres Strait Islander peoples' perspectives provide great insight. Within Aboriginal and Torres Strait Islander communities the development of an individual is embedded within a network of connections to family, broader kinship structures, communities, and to the lands and waters that support them. This is a holistic and ecological concept of development.

Weblinks
The continuing impact of colonisation

Our shared history

Truth-telling central to reconciliation process

One of the most distinctive aspects of Aboriginal and Torres Strait Islander cultures is their kinship structures. The kinship system defines relationships between people within a broad social network that extends far beyond Western concepts of the nuclear family and blood relatives. A person's location within the kinship network gives their identity a relative place, and defines their responsibilities to other people and for caring for Country. The kinship system is the main way that knowledge and teachings are passed on, with grandparents playing an important role in educating children.

You can learn more about Aboriginal and Torres Strait Islander kinship from the weblink.

Weblink
Family and kinship in Aboriginal culture

ANALYSING RESEARCH 2.2

Preschool attendance and developmental outcomes at age five in Indigenous and non-Indigenous children

According to the 2016 census, almost 12 per cent of individuals who identified as Aboriginal and/or Torres Strait Islander (henceforth Indigenous Australians) were younger than five years old. In total, Indigenous children represented 6 per cent of the Australian population aged five years and under. Although they make up a relatively large portion of the Indigenous population, Indigenous children experience disproportionate socioeconomic and health disadvantages.

In an effort to reduce the developmental inequities between Indigenous and non-Indigenous Australian children, the 'Close the Gap' campaign introduced policies to increase the number of Australian Indigenous children participating in preschool.

Researchers from the School of Population Health at the University of New South Wales conducted a large-scale study to understand the impact of attending preschool a year prior to starting school on the developmental outcomes of Indigenous and non-Indigenous children at age five. The researchers used data from the Australian Early Development Census (AEDC) for 7384 Indigenous and 95 104 non-Indigenous children who started school in 2009 or 2012 in New South Wales, Australia. They studied the relationship between exposure to preschool, home-based care or long-day care and measures of developmental vulnerability related to social skills, physical health, emotional maturity, and cognitive and language skills.

Findings showed that for both Indigenous and non-Indigenous children, developmental vulnerability was less common in children who attended preschool a year before school, compared to those who were solely in home-based care. However, the scale of benefits were smaller among Indigenous children compared to non-Indigenous children. Furthermore, it was shown that developmental vulnerability was more common among Indigenous children compared to non-Indigenous children regardless of the type of care received.

It was found that the gap in socioeconomic and health circumstances was the main reason for much of the developmental inequality between Indigenous and non-Indigenous children in all types of care.

Overall, the results in this study suggest that the Australian Government still needs to do more to improve the outcomes for Indigenous Australian children, including investment in culturally appropriate health and educational services.

Source: Falster et al. (2021). Preschool attendance and developmental outcomes at age five in Indigenous and non-Indigenous children: a population-based cohort study of 100357 Australian children. *Journal of epidemiology and community health, 75*(4), 371–379.

Fairfax Syndication/Louise Kennerley

Figure 2.8 Improving the quality and cultural safety of preschool education may help 'close the gap' in developmental outcomes between First nations and other Australian children

Questions

1 What was the aim of this study?
2 Identify the variables of interest in this study.
3 Summarise the results of this study.
4 What are the implications of this study?
5 What recommendations should be made to Australian Government policy makers to support the development of Aboriginal and Torres Strait Islander children?

2.1.4 INDIGENOUS GAMES USED FOR MOTOR DEVELOPMENT

2.2 The biopsychosocial model

The **biopsychosocial model** is an approach that proposes that health and illness outcomes are determined by the interaction and contribution of biological, psychological and social factors. Each factor on its own is not enough to lead to health or illness.

The biopsychosocial model was developed by George Engel in 1977. Engel challenged the previously dominant biomedical understanding that health and illness were solely the outcome of biological factors and there were no other contributing influences. In contrast, the biopsychosocial model recognises the contribution and complex interaction that biological (genes and hormones), psychological (cognitive and emotional systems) and social (social relationships and lifestyle) factors have on our physical and mental health.

The biopsychosocial model is also used to explain the factors contributing to mental wellbeing and psychological development. Psychological development includes the processes of growth and change in people's cognitive, emotional and social capabilities. The biopsychosocial model tries to explain this multifaceted interaction across the life span (Figure 2.9).

Mental wellbeing refers to a state in which an individual realises their own abilities, can cope with the normal stresses of life, can work productively and fruitfully, and is able to contribute to the community. It includes a state of mind characterised by emotional wellbeing, good behavioural adjustment, relative freedom from anxiety, and a capacity to establish constructive relationships and cope with the ordinary demands of daily life. Dealing with daily stressors and other developmental challenges throughout the life span can make it difficult to achieve a state of mental wellbeing, which is why it is ideal to have a model to help us best understand how to achieve this state through a holistic approach.

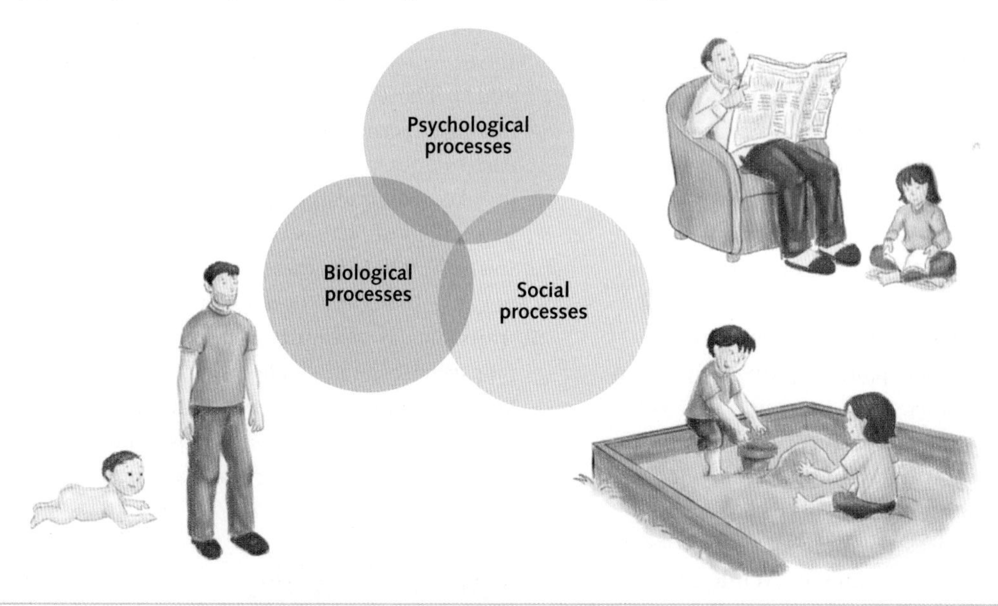

Figure 2.9 Changes in development are the result of biological, psychological and social processes. These processes are interrelated in the development of the individual throughout the life span

Biological factors

Biological factors contributing to development and wellbeing include all physiological factors. They include our genetics, hormones, immune function, stress response, brain function and other activities of our nervous system. For example, hypothyroidism is a disorder in which the body does not produce enough of a specific thyroid hormone to keep the body healthy. This hormone is critical for the development of the central nervous system, helps the body use energy, regulates temperature, and keeps the brain, heart, muscles and other organs functioning optimally. If a baby lacks thyroid hormone while in the womb, the growth of the cerebral cortex, proliferation of axons, branching of dendrites and myelination of neurons can all be impaired. If the deficiency is not treated in young children, they could suffer delay in reaching developmental milestones, such as sitting, standing and walking, as well as irreversible brain damage. In older individuals, deficiency of thyroid hormone can cause lethargy, decreased sweating, weight gain and low heart rate (Berne & Levy, 1996).

Biological factors that can stop us from achieving mental wellbeing can include our response to stress. When a person is stressed, the body's stress response is activated, which includes the release of a stress hormone called cortisol. Long-term effects of cortisol stimulation can disrupt a wide range of bodily process and can lead to digestive issues, anxiety and depression, headaches, muscle pain, cognitive issues and sleep disturbances, just to name a few. While we cannot control our genetic vulnerabilities and physiological systems, to avoid such effects on our body the biopsychosocial model recognises the interrelated factors that contribute to these illnesses. While there may be a genetic predisposition to an illness such as hypothyroidism, other factors such as diet may exacerbate or alleviate the onset of the illness. Similarly, how we manage our mood, equip ourselves with coping mechanisms and surround ourselves with a supportive network can all help us manage an overactivation of the stress response.

Psychological factors

The biopsychosocial model recognises psychological contributing factors to our development and mental wellbeing. These include cognitive processes such as our learning and memory systems, our self-concept, our emotional regulation, our perceptions, our thought processes (including our belief and attitudes), our personality traits and our coping skills.

In 2020–21 the Australian Bureau of Statistics (ABS) conducted its first National Study of Mental Health and Wellbeing to gain an insight into the mental health of Australians by measuring levels of psychological distress throughout the life span. Results revealed that 15 per cent of Australians experienced high or very high levels of psychological distress, with women more likely to report these symptoms compared to men (females 19 per cent; males 12 per cent). Stress was experienced at all ages, with the age group most often reporting high or very high levels of psychological distress being 18–35-year-olds (20 per cent), followed by the 35–64 age group (15 per cent), and lastly the 65–85-year-old group (9 per cent) (ABS, 2021). These data indicate that stress needs to be managed across the life span, particularly in the early stages.

According to the biopsychosocial model, psychological factors can heighten our response to stress. These factors include low self-esteem, negative thought processes and personality traits such as insecurity and self-doubt. The biopsychosocial model helps us address these factors and improve our mental wellbeing by using psychological approaches such as mindfulness practice, learning coping skills and developing positive thought processes.

Social factors

The biopsychosocial model recognises the influence of social factors on wellbeing and development. These factors include the interactions, relationships and social networks we have with others. They also include environmental factors such as our living conditions, access to health care and education, social disadvantage or deprivation, and our family support and circumstances.

Research focusing on the impact of social factors on health and wellbeing suggest that these factors influence biological and psychological development and can have long-term consequences for mental health. This is particularly true for environmental factors encountered in the early stages of the life span, such as parental care and children's living conditions. For example, results indicate that children living in poverty are more at risk of developing negative health and psychological outcomes throughout the life span. They are also more likely to face adversity in areas such as exposure to violence, unresponsive caregivers and increased family transitions (Black & Hoeft, 2015).

From the social perspective, one effective way to improve developmental outcomes and wellbeing for these children is to create strength and support in the family unit. One study found that the introduction of an 8-week family-based intervention training program improved brain function, cognition (specifically, selective attention) and behaviour in preschoolers from lower socioeconomic backgrounds. The training program involved equipping parents with skills to reduce stress in the family home, implement consistent fair discipline strategies, increase the responsiveness to their child's needs and improve the language used in their home (Neville et al., 2013).

The interrelated parts of the biopsychosocial model

The biopsychosocial model recognises that none of these three factors operate in isolation to cause a health or wellbeing outcome (Figure 2.10). Outcomes are the result of an interaction between biological, psychological and social factors.

For example, the biopsychosocial model recognises that individuals may have a genetic vulnerability (biological factor) that leads them to negative thinking (psychological factor), which increases their stress. Increased stress may lead to negative coping mechanisms such as increased alcohol intake that could result in liver problems (biological factor). Increased alcohol use and negative thinking could then lead to social withdrawal and loss of friendships or even employment (social factor).

2.2.1 BIOPSYCHOSOCIAL APPROACH

The biopsychosocial model should also be considered when trying to improve health and wellbeing outcomes. For example, if an individual suffering from acute stress develops an anxiety disorder, this may need to be managed with medication to stabilise neurotransmitters in the brain (biological factor). In addition to this, treatment should involve teaching the individual effective coping strategies to help them deal with the stress (psychological factor). The individual should also be encouraged to join social support groups to contribute to their recovery (social factor).

2.2.2 APPLICATION OF THE BIOPSYCHOSOCIAL MODEL

Biological factors: genes, hormones, immune function, stress response, brain function, nervous system function, genetic vulnerabilities

Psychological factors: cognitive processes: learning and memory systems, self-concept, emotional regulation, perceptions, thought processes including belief and attitudes, personality traits and coping skills

Social factors: relationships, social networks, environmental factors, living conditions, access to health care and education, social disadvantage or deprivation, family support and family circumstances

Figure 2.10 The biopsychosocial model demonstrates the complex interaction of biological, psychological and social factors that are all interrelated contributing factors to our psychological development and mental wellbeing.

ACTIVITY 2.2 USING THE BIOPSYCHOSOCIAL MODEL

Try it

1 Copy and complete Figure 2.11 by putting the terms in the correct area. The terms marked with an asterisk (*) belong in the intersection between two sets.

Peer group	Self-esteem
Coping skills	Culture
Physical health	Family relationships*
School	IQ*
Genetic vulnerability	Drug effects*

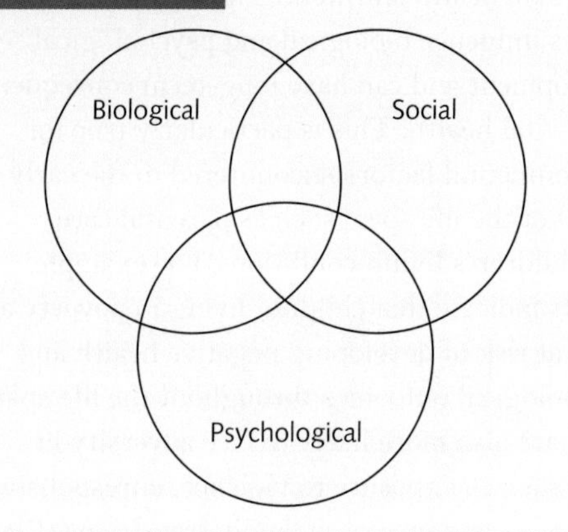

Figure 2.11

Apply it

2 Just like her mother, Xanthe suffers from anxiety. She has just started a new job and has to travel for 90 minutes to and from work each day. She is concerned that she will not cope with these challenges. Fortunately, she has a strong network of family and friends. Identify the biological, psychological and social factors that contribute to Xanthe's wellbeing.

Exam ready

3 According to the biopsychosocial model, which of the following correctly identifies a social factor influencing mental health?

A substance abuse
B self-esteem
C physical health
D loss of a significant relationship

KEY CONCEPTS 2.2

» The biopsychosocial model recognises the contributions of biological, psychological and social factors, and their complex interactions, to our physical and mental health.

» Biological factors contributing to development and wellbeing include our genetics, hormones, immune function, stress response, brain function and other activities of our nervous system.

» Psychological factors contributing to development and wellbeing include cognitive processes such as learning and memory systems, self-concept, emotional regulation, perceptions, thought processes including beliefs and attitudes, personality traits and coping skills.

» Social factors influencing our wellbeing and development include social interactions, relationships and networks.

Concept questions 2.2

Remembering

1 What is the biopsychosocial framework? r

2 Identify three biological contributing factors proposed by the framework. r

3 Identify three psychological contributing factors proposed by the framework. r

4 Identify three social contributing factors proposed by the framework. r

Understanding

5 Explain the relationship between the biopsychosocial framework and psychological development. e

6 Explain the relationship between the biopsychosocial framework and mental wellbeing. e

Applying

7 Consider the life circumstances from the following examples and identify each as biological, psychological or social factors. e

a Rob just broke up with his girlfriend. He has since become very depressed, stopped eating and has developed a stomach ulcer. He constantly ruminates about their relationship and what went wrong. As a result of this he has become quite distant from his group of friends.

b Lan is 16 years old and has just experienced the loss of her father. He had been battling cancer for the past 2 years, which has taken its toll on the whole family. Lan's grades have dropped since the diagnosis of her father's illness. Once a very driven student, she now struggles with her confidence and motivation. She also frequently experiences panic attacks, which cause severe heart palpitations. Lan's family do not have a lot of money so trying to access funds for all of her father's specialist appointments had been very stressful.

HOT Challenge

8 Cleo is 10 years old. She lives with her mother in a low socioeconomic area in Melbourne. Her mother was a victim of violence, some of which Cleo witnessed. Her mother tries her hardest to provide a safe and loving home for Cleo but is still constantly plagued by her own anxieties from the violent incident. Cleo's mother is often withdrawn and at times not responsive to all of her needs. According to the biopsychosocial model, analyse how Cleo's psychological development may be affected by her unstable upbringing. c

2.3 Psychological development over the life span

Over the course of the life span, psychological development can be influenced in various ways. Many developmental theorists focus on the impact early life experiences can have on the developing individual. Others focus on how experiences throughout all stages of the life span can influence development and wellbeing. In this next section we will explore three areas of psychological development – emotional, cognitive and social – and the major theories associated with each. These areas of psychological development can greatly influence our thoughts, feelings, behaviours and personality.

Emotional development

In developmental psychology, **attachment** refers to the close emotional bond or relationship between an infant and the mother or primary caregiver. Attachment theory suggests that:

» a number of important factors contribute to attachment

» attachment is universal to all humans; it appears in all races and cultures of people

» attachment has a biological basis, because its main function is to increase chances of survival by helping the child seek proximity to someone who will take care of their physical and emotional needs for security

- seeking attachment is an innate (inborn) behaviour; it is not learned through reasoning or teaching
- the attachments formed are hierarchical, occurring primarily between a mother and infant. Although multiple attachments often occur, for example with a father, siblings and friends, the mother–infant relationship is generally ranked highest
- there is a sensitive period (roughly the first year of life) during which attachment must occur for optimal development. Mothers may begin cultivating a parent–child bond within minutes after giving birth via skin-to-skin contact, when the baby is often placed onto the mother's bare chest (Figure 2.12)
- separation anxiety, which appears at approximately 8 months of age, is an indication that a parent–child bond has been formed. This is when infants may display characteristics of being visibly distressed when left alone or with a stranger.

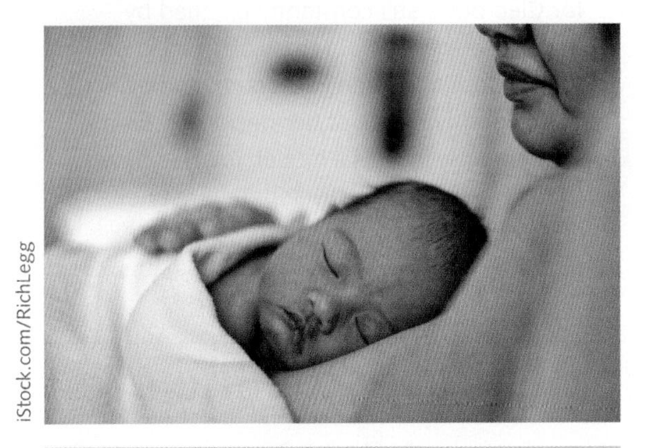
iStock.com/RichLegg

Figure 2.12 Mothers may begin to cultivate a lasting bond with their infant within minutes of giving birth though skin-to-skin contact.

Researchers such as Harry Harlow, Mary Ainsworth and John Bowlby have all researched emotional development and the importance of attachment. Their theories are outlined below.

Bowlby and attachment theory

In the 1940s, British psychiatrist John Bowlby (1907–1990; Figure 2.13) began work on attachment theory by observing children in institutions. Bowlby found that some institutions provided a higher level of care and nurturing than others. Some provided

Reproduced with permission from Sir Richard Bowlby

Figure 2.13 British psychologist John Bowlby, who was one of the first to propose attachment theory, believed infants were genetically pre-programmed to form an attachment to their primary caregiver (usually the mother) to ensure they are cared for.

for basic needs such as shelter and food, but lacked human contact with the children. Bowlby believed that children who suffered loss and failure in early relationships were left more likely to experience negative psychological consequences later in life, such as an inability to form close relationships. As a result, Bowlby proposed the key principles of his attachment theory. These were:

- infants are biologically pre-programmed to form an attachment with their primary caregiver (usually the mother)
- attachment is an evolutionary mechanism to ensure infants are cared for and all their needs are met
- there are sensitive periods during development for this attachment to occur
- to avoid long-term emotional difficulties for the infant, a mother and infant should not be separated for at least the first 2 years of the infant's life
- mothers are the best caregiver's to infants because they are biologically programmed to be the superior caregiver to an infant

» other caregivers play important roles in an infant's life (father, siblings, grandparents etc.) but they fall lower on a hierarchy compared to the mother.

In Bowlby's later studies (1969), he observed infants doing things to gain attention from their mothers. These were called signal behaviours, and they included crying and smiling as well as behaviours that physically maintained proximity, such as clinging to or following a caregiver (Vaughan & Hogg, 1998). As a result of these studies Bowlby proposed that there are four age-related phases in the development of attachment (Table 2.1).

Ainsworth and developments in attachment theory

Canadian psychologist Mary Ainsworth (1913–1999) expanded on Bowlby's claims by extensively observing infant attachments with their caregivers in an experimental setting she created (Figure 2.14). Ainsworth described a list of behaviours she observed as evidence of an infant's attachment to a specific caregiver (Ainsworth, 1972). These behaviours included:

» crying to attract the caregiver's attention

» smiling more at the caregiver than at others

Table 2.1 Bowlby's phases in the development of attachment

Phase	Developmental characteristics
Phase 1: Preattachment (birth through to 2 months)	The infant responds to people but does not discriminate between people.
Phase 2: Preliminary attachment (2–7 months)	The infant will begin to respond to familiar people with smiles and will prefer to be with the primary caregiver. An infant will express this by becoming upset or difficult to calm when not with the mother or familiar caregiver.
Phase 3: Clear-cut attachment (7 months – 2 years)	During this phase the infant is more mobile and can crawl or walk to be closer to their caregiver, using them as a secure base from which to explore their environment. If the caregiver is not there, the infant will most likely cry and become fretful.
Phase 4: Goal-directed partnership (2 years onwards)	The infant will begin to adjust to times when the caregiver is not present. They will become more willing to stay with other people or to be by themselves for short amounts of time.

Getty Images/JHU Sheridan Libraries/Gado

Figure 2.14 Mary Ainsworth (pictured centre) playing with a toddler during her research

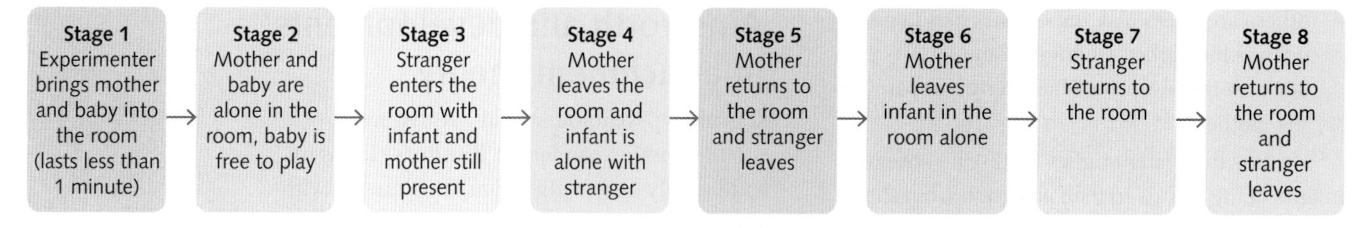

Figure 2.15 A summary of the eight events that form the Strange situation test

» vocalising more in the presence of a caregiver than when alone or with strangers
» crying when being put down
» following the caregiver
» fleeing to the caregiver when in danger (Peterson, 1989).

Ainsworth introduced the terms 'separation anxiety' and 'stranger anxiety'. Separation anxiety refers to the feelings of apprehension and worry experienced by an infant when they are separated from their caregiver; stranger anxiety is the same feelings experienced in the presence of a stranger.

2.3.1 ATTACHMENT

The Strange situation test

To test and measure the quality of an infant's attachment to the primary caregiver, Ainsworth created a laboratory experiment known as the **Strange situation test**. The test was typically performed with infants aged 9–18 months. The experiment took place in a room, where a fixed sequence of eight events (Figure 2.15) took place,

Figure 2.16 The mother returns after a short separation during the Strange situation test

and the infant's reactions were observed (through a two-way mirror) and recorded.

The observer would record three types of behaviours in the infant: any contact-seeking behaviour (where the child sought the mother or stranger), exploratory play and behaviour, and crying or distress. The quality of the infant's play and the level of distress when the mother was present was compared to when the mother was absent and the infant was left in the room alone or with a stranger.

Ainsworth's types of attachment

Ainsworth's observations of the Strange situation test (Figure 2.16) led her to develop three categories of infant attachment styles. She classified infants as having secure attachment, insecure resistant attachment or insecure avoidant attachment (Table 2.2).

The Strange situation test seems to be a good measure of attachment by infants to their mothers for societies such as Australia, New Zealand, Canada and the United States. The scenario is not unfamiliar to infants in these societies, because it is normal for infants to be left in the care of babysitters (family or childcare centre) or on the lounge room floor to play with toys. This is not common in other societies, where it is normal for infants to be carried most of the time and cared for primarily by their mothers. The results of placing infants from non-Western cultures in the Strange situation are completely different. For example, in a study using Japanese infants, more than 40 per cent of infants were classified as being type C type babies, or having an insecure resistant attachment to their mothers. These infants became very upset when their mothers left the room and refused to be comforted when she returned (Miyake et al., 1985). Even with these limitations, the Strange situation has provided a wealth of information about attachment of infants to their caregivers (Peterson, 1989).

Table 2.2 Ainsworth's types of attachment

Attachment type	Behaviours displayed by infants during Strange situation test	Characteristics of mothers with these infants	Percentage of infants
Insecure avoidant attachment (type A)	Not affected by mother's presence or absence. Rarely cried when mother left the room and showed little attention to her upon her return.	Neglectful, angry, hostile	20–22%
Secure attachment (type B)	Distressed when mother leaves the room but happy and comforted by her return. Open to exploration of the room when mother is present.	Loving, responsive, supportive	55–75%
Insecure resistant attachment (type C)	Very distressed when mother leaves the room and not comforted by her returning. Heightened anxiety before, during and after the test. Did not actively explore the room even when mother was present.	Non-affectionate, unresponsive	7–12%

KEY CONCEPTS 2.3a

- Attachment refers to the close emotional bond or relationship between an infant and the mother or primary caregiver.
- Attachment of an infant with their primary caregiver is crucial in their subsequent emotional development.
- John Bowlby was one of the first psychologists to develop attachment theory. He proposed that infants are biologically pre-programmed to form an attachment with their primary caregiver (usually the mother).
- Bowlby proposed that a mother and infant should not be separated for at least the first 2 years of the infant's life at the risk of causing long-term emotional difficulties.
- Mary Ainsworth developed the Strange situation test to observe the behaviours of infants with their caregivers and their reactions around strangers.
- Ainsworth classified infants into the following attachment categories: secure attachment, insecure resistant attachment and insecure avoidant attachment.

Concept questions 2.3a

Remembering

1 Define the term 'attachment'. In your answer, refer to why the formation of emotional development between caregiver and child is so important. **r**

Understanding

2 What are three examples of behaviours that show a child has formed an emotional attachment with their primary caregiver? **c**

3 Draw a flowchart summarising the four phases in the development of attachment according to Bowlby. **d**

4 Describe two key principles of Bowlby's attachment theory. **r**

5 Explain the procedure and purpose of the Strange situation test. **r**

6 Summarise the three attachment types proposed by Ainsworth. **r**

Applying

7 Read the following scenarios and identify the attachment type each infant is displaying with their primary caregiver (secure, insecure avoidant or insecure resistant). **e**

a Lesly is a 10-month-old infant. During the Strange situation test she becomes extremely distressed when her mother leaves and a stranger enters the room. She stops playing with the toys and looks around in a disorientated manner, crying hysterically. When her mother re-enters the room and tries to comfort her, Lesly struggles to be consoled by her. She continues crying despite the mother's best efforts to calm her down.

b Thomas is a 13-month-old infant. During the Strange situation test, he

»

is indifferent when his mother is in the room and does not react when she leaves. When a stranger enters the room, he does not become upset or cry. His observed temperament remains the same as it was before the stranger entered the room.

c Isabelle is a 15-month-old infant. During the Strange situation test, she happily plays in the room with her mother present. Once her mother leaves and a stranger enters, she becomes very upset and cries. Upon her mother returning, Isabelle is overjoyed and stops crying when her mother comforts her.

HOT Challenge

8 Imagine you are in charge of an adoption agency. Your goal is to ensure all adoptive parents have the best possible chance of forming a strong, loving emotional bond with their adoptive children. Based on the research on attachment theory, create an information leaflet for prospective adoptive parents. In your leaflet include the following: C
  a a summary of research in the area of emotional development
  b behaviours parents should exhibit to respond to the needs of their child
  c risk factors associated with insecure attachment.

Harlow's first attachment experiments

During the 1950s and 1960s, psychologist Harry Harlow conducted several experiments to investigate the factors influencing the development of attachments of infant rhesus monkeys to their mothers. In one of his best-known experiments, Harlow (1958) studied the role of feeding in infant–mother attachment. To do this he used eight infant rhesus monkeys.

The monkeys were taken from their mothers at birth and kept in separate cages containing two surrogate mothers made of wire mesh of the same size and shape as real monkey mothers. However, one of the surrogate mothers was covered in terry-towelling cloth (Figure 2.17) and the other was left uncovered. A feeding bottle was attached to each surrogate in the same area where a breast would be on a real mother. Four of the animals were placed in cages with the feeding bottle on the cloth surrogate and the other four in cages with the feeding bottle on the wire surrogate. The monkeys were not placed in cages together, as they may have comforted each other. If an infant's attachment to its mother was based primarily on feeding, the infant monkeys should have preferred and become attached to whichever surrogate mother had the bottle.

Harlow observed that, regardless of which surrogate provided the nourishment, the cloth surrogate was preferred to the wire surrogate. None of the monkeys would spend more than one or two hours in any 24-hour period on the wire surrogate. Although the infants in the two groups drank the same amount of milk and gained weight at the same rate, all eight monkeys spent far more time climbing and clinging to the cloth surrogate. By the age of about 3 weeks, all the monkeys were spending approximately 15 hours a day in contact with the cloth surrogate (Figure 2.18).

Alamy Stock Photo/Science History Images

Figure 2.17 A rhesus monkey with a cloth mother surrogate in one of Harlow's experiments

To examine what would happen if the monkeys were scared or stressed, Harlow introduced a 'stressor'. He placed a mechanical moving spider and a teddy bear that beat a drum in the cages with the monkey (Figure 2.19). The results were that the terrified monkey sought comfort from the cloth surrogate, regardless of whether or not it had the

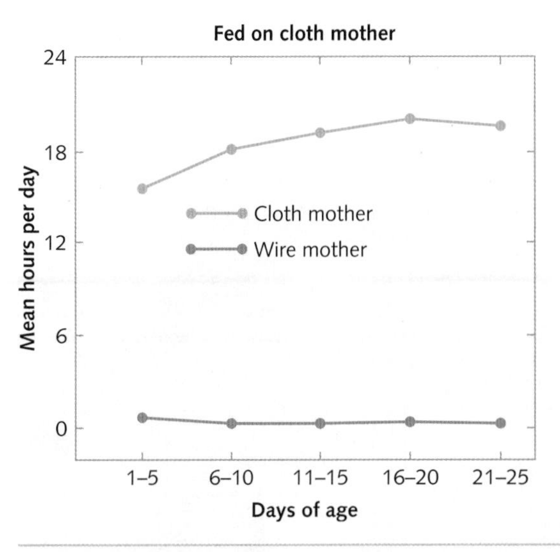

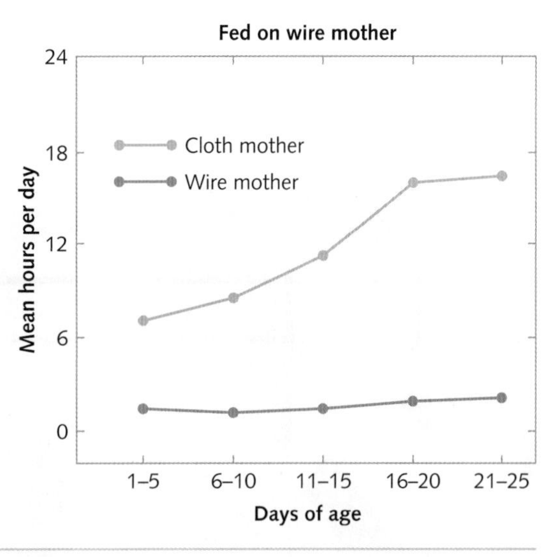

Figure 2.18 Time spent on cloth and wire mother surrogates

Alamy Stock Photo/Science History Images

Figure 2.19 A mechanical moving teddy bear was introduced as a fear stimulus

feed bottle, by rubbing its body against the cloth surrogate. If by chance the monkey fled in blind terror to the wire surrogate, it would soon leave it for the contact comfort of the cloth surrogate, even if the wire surrogate had the feed bottle.

Harlow concluded that 'contact comfort', which was provided by the softness of the cloth covering, was more important than feeding in the formation of an infant's attachment to its mother. He generalised his findings to suggest that contact comfort was also most likely to be a crucial factor in human infant–caregiver attachment.

Harlow's further attachment experiments

Harlow's subsequent studies focused on the effect social isolation can have on emotional development. Human social isolation can have devastating effects on an individual's emotional development. Children raised in neglectful, abusive families, and children raised in orphanages, institutions or in inadequate foster homes are all at risk of developing emotional disturbances, antisocial behaviours and maladaptive relationships, because of the deprivation experienced in these settings (Harlow et al., 1965).

In 1965 Harlow, Robert Dodsworth and Margaret Harlow (Harry's wife, herself a developmental psychologist) carried out a study investigating the effects of social deprivation on emotional development, using rhesus monkeys. This study involved removing infant monkeys hours after their birth and socially isolating them for either 3, 6 or 12 months (experimental group). During the period of isolation, the monkeys were kept in a stainless-steel chamber (Figure 2.20a) and were given no interaction with any human or animal. Following the period of isolation, their behaviours were compared with infant monkeys that had not been isolated (control group). Both groups were placed in a chamber called the 'social play room' (Figure 2.20b), with equipment to encourage play between the monkeys. They were placed in the play room for 30 minutes a day, 5 days a week, for 32 weeks and their interactions and behaviour was observed.

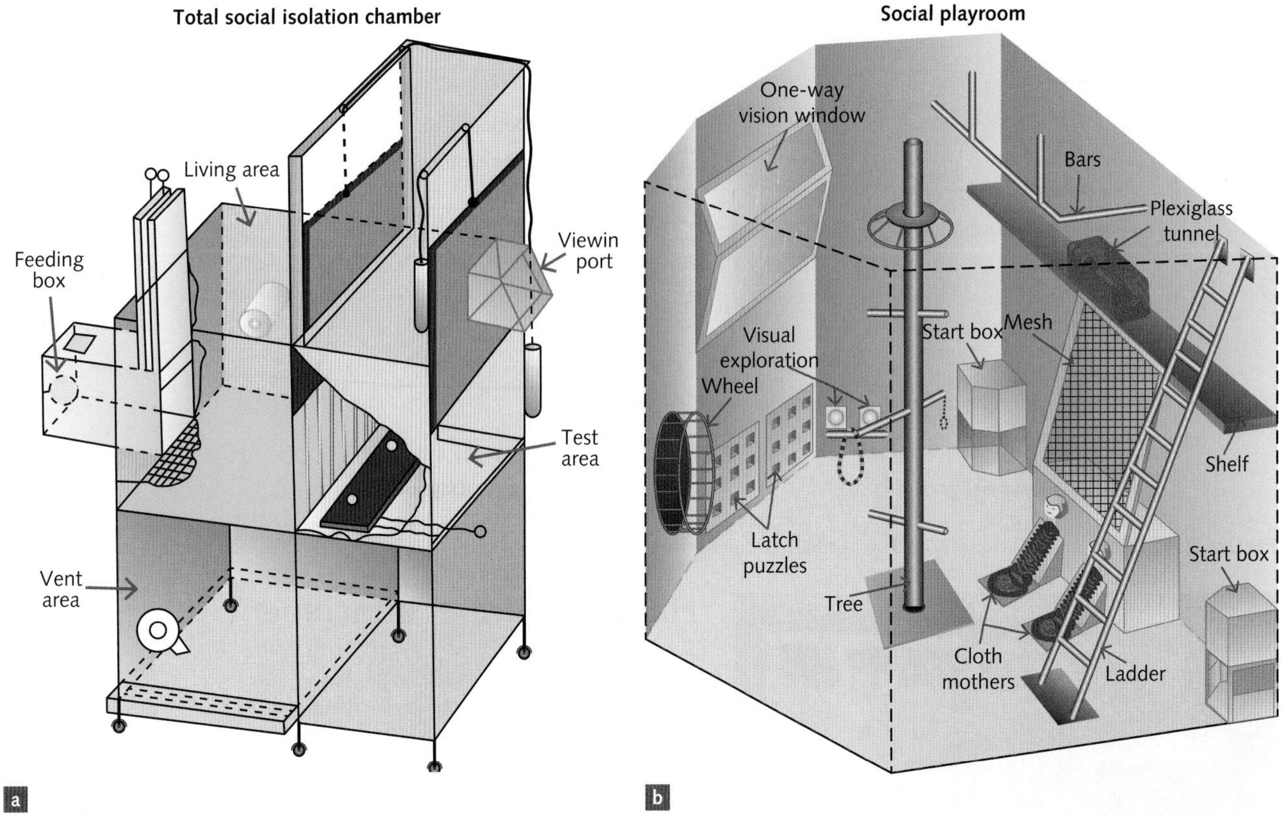

Total social isolation chamber

Living area
Feeding box
Viewin port
Test area
Vent area

a

Social playroom

One-way vision window
Bars
Plexiglass tunnel
Visual exploration
Start box
Mesh
Wheel
Latch puzzles
Shelf
Start box
Tree
Cloth mothers
Ladder

b

'FIG. 2. Total social isolation chamber' and 'FIG. 3. Playroom situation' from Harlow, Harry F., et al. (1965), JSTOR, http://www.jstor.org/stable/72996. Accessed 31 May 2022. Redrawn with permission from Harlow Primate Laboratory, University of Wisconsin.

Figure 2.20 a An isolation chamber where monkeys were kept for 3, 6 or 12 months; b a social play room, where the behaviour of isolated monkeys was analysed and compared with that of monkeys who had not been isolated

The results of this experiment revealed the debilitating effects social isolation can have on the social and emotional development of rhesus monkeys. When the experimental monkeys were first released from isolation they experienced emotional shock, displayed by self-clutching and rocking behaviours. They crouched in the corner of the play room and did not know how to interact with the control monkeys. They were socially withdrawn and sometimes fearful of the control monkeys. Other disturbing behaviours observed included self-mutilation and gazing expressionlessly around the play room. One monkey even died 5 days following its release because it refused to eat.

The key differences found between the monkeys isolated for 3, 6 or 12 months were the severity and longevity of these disturbing behaviours. The disturbing behaviours of the monkeys isolated for 3 months were reversible; they suffered no long-term effects and within a few weeks were interacting in a socially appropriate manner with the control monkeys. Monkeys that were isolated for 6 months eventually developed normal social behaviours, taking much longer to do so compared to those isolated for 3 months. The monkeys isolated for 12 months unfortunately never fully recovered from the effects of deprivation. While some improvements were made, they did not develop the capacity for 'normal' social interactions with the control monkeys (Harlow et al., 1965) (Figure 2.21).

Alamy Stock Photo/Science History Images

Figure 2.21 **Harry Harlow's work on attachment with infant rhesus monkeys was instrumental to our understanding of social and emotional development, but it did come at a cost to the welfare of the monkeys.**

KEY CONCEPTS 2.3b

- Harry Harlow's (1958) research with infant rhesus monkeys investigated whether food or contact comfort was more important for the attachment of infants to their caregivers.
- Harlow's findings revealed that contact comfort was more important than food in infant–caregiver attachment in monkeys. He generalised this finding to human infant–caregiver attachment.
- Harlow et al. (1965) investigated the effects of social deprivation on social and emotional development in infant rhesus monkeys.
- Their studies revealed that periods of social isolation in infancy can have debilitating effects on social and emotional development. The longer the period of isolation, the more severe the behaviours reported and the less likely to be reversed.

Concept questions 2.3b

Remembering

1 Answer the following questions on the research of Harlow's original experiments with rhesus monkeys (1958). r
  a What was the aim of this study?
  b Identify the independent and dependent variables.
  c Outline the procedure of this study.
  d Summarise the results.
  e What did Harlow conclude from this study?

2 Harlow introduced a stressor into the study. What effect did this have on the infant monkeys? r

Understanding

3 Evaluate Harlow et al.'s (1965) work on the social isolation of rhesus monkeys. r
  a What was the aim of this study?
  b Identify the independent and dependent variables.
  c Outline the procedure of this study.
  d Summarise the results.
  e What conclusions can be made from this study?

4 Harlow et al. (1965) used 'control' monkeys in these experiments. Which monkeys made up this group and why were they used? r

Applying

5 Consider how generalisable Harlow's studies are. e
  a Can the findings of Harlow's experiments be generalised to humans?
  b What is one limitation of generalising Harlow's experiments to humans?

HOT Challenge

6 Consider the issues surrounding Harlow's experiments with rhesus monkeys.
In your opinion, do the benefits of his research findings outweigh the harm caused to the infant monkeys? Justify your answer. i

Cognitive development: Piaget's theory

Cognition is a general term that refers to all mental (or thought) processes, including perceiving, conceptualising, remembering, reasoning, imagining, judging and problem-solving. Cognitive development refers to the growth and maturation of these thinking processes. This can include skills such as problem-solving, reasoning, evaluating and planning. Growth in these processes helps children make sense of the world around them and eventually develop more sophisticated thought processes as they move into adulthood.

Generally, children's thinking is less abstract than an adult's because they use fewer generalisations, categories and principles. This is usually because children have less experience and exposure to environmental stimuli. Children also tend to base their understanding on particular examples and direct experiences, such as objects they can see or touch. This is called **concrete thinking** and is used to describe thought processes that rely on experiences in the real world.

Once a child can learn to represent ideas mentally, they move away from concrete thinking to what Swiss psychologist Jean Piaget

(1896–1980; Figure 2.22) called **symbolic thinking**. This type of thinking involves the ability to mentally represent people, objects and events – even those that are not present.

Getty Images/-/Staff

Figure 2.22 Swiss psychologist Jean Piaget developed a theory that linked developmental stages with different types of thinking.

Beyond symbolic thinking, Piaget called the next level **abstract thinking**. This type of thinking is characteristic in older children and adults. It involves thinking in terms of general concepts rather than specific objects, experiences or events.

Cognitive development looks at how thinking processes change and mature over the life span. Piaget is famous for his theory on the ways in which children develop their cognitive abilities (Piaget, 1951, 1952) and studied children's reasoning abilities. Many of his ideas also came from observing his own children as they solved various thought problems.

The data Piaget collected led to the development of the theory that children experience a series of developmental stages from infancy through to adolescence. According to Piaget, at each of these stages children become able to reason in a qualitatively different and more advanced way as they shift from concrete, to symbolic, to abstract thinking abilities.

Piaget's key principles

Piaget proposed that the cognition (thinking) of children develops through exposure to their environment. He believed a child's intellect grows through a process of mental adaptation, by taking in new information, interpreting it and organising it in a way that makes sense to them.

Piaget used the terms 'assimilation' and 'accommodation' to describe these processes. **Assimilation** is when a person applies existing mental structures or processes to new experiences. For example, Sebastian's favourite toy is a plastic hammer. Sebastian holds the hammer properly and loves to hit blocks with it. For his third birthday, Sebastian gets an oversized toy spanner. If he uses the spanner for hitting, he has assimilated its use into his existing knowledge structure, by applying his understanding of what a hammer is used for.

Accommodation involves a person modifying their existing psychological concepts or processes to adapt their understanding and behaviour in response to new information or experiences. Using the earlier example, imagine Sebastian, who always played with a toy hammer, received the toy spanner and began to use it to pretend tightening and loosening nuts and bolts. His knowledge structure has changed to accommodate and produce a new response to new information. Accommodation is a more sophisticated cognitive process than assimilation, because it involves the child changing their mental idea or representation (Figure 2.23).

Shutterstock.com/granata68

Figure 2.23 A child may first try to hammer with a spanner (assimilation), but they will eventually learn that spanners are used to tighten and loosen nuts and bolts (accommodation).

Piaget used the term **schema** to describe the memory structure (or mental representation), developed from their experiences, that represents a person's general knowledge about different kinds of objects and events. As development progresses throughout childhood, a child's schema advances and matures.

Piaget's four stages of cognitive development

Piaget believed that all children pass through a series of distinct, age-related stages in cognitive development. As a child progresses through each stage, their thinking processes change and become more sophisticated. Piaget observed that as children advance through each stage, they achieve a set of accomplishments. He proposed a range of characteristics that accompany each stage, and these are discussed below.

The sensorimotor stage (0–2 years)

In the first 2 years of life, a child's intellectual development is mostly non-verbal (not speaking). Children are learning to coordinate purposeful movements with information from the senses by looking at, touching and mouthing objects in their immediate environment. For example, very young infants will instinctively grasp a rattle. They feel it, they see it, but initially they do not realise that if they shake their arm they can make the rattle produce a noise. Once they realise this, they may begin to purposefully shake everything they get hold of, trying to produce a sound. They start to distinguish between things that make noise and things that do not. This indicates that they are beginning to integrate sensory and motor information and to coordinate their motor responses. Children learn that there is a relationship between their actions and the external world, where they can manipulate objects and produce effects. Piaget called this **goal-directed behaviour**.

Another cognitive accomplishment of this stage is **object permanence**. This refers to an understanding that objects continue to exist even when they are out of sight. Before this stage infants will find a game of peekaboo appealing, because the child thinks that you are appearing as if out of nowhere (Figure 2.24).

Shutterstock.com/Creativa Images

Figure 2.24 The game peekaboo is so appealing to infants because they believe that objects that are out of sight no longer exist.

The preoperational stage (2–7 years)

During the preoperational stage, children begin to use language and engage in symbolic thinking. This cognitive skill enables children at this age to use symbols, gestures and images to mentally represent people, objects and events that may not be present. Often, this shows through pretend play. Piaget highlighted the importance of children learning through pretend play as a way to increase their mental representations of the world around them (McLeod, 2018).

Pretend play in children is also when another cognitive characteristic of this stage, called animism, is observed. **Animism** refers to the tendency of children to believe that any inanimate object possesses lifelike characteristics, such as feelings and emotions. This can be illustrated through pretend play where a young child may play 'doctor' and provide care to a teddy bear they think is in pain.

During this stage, the child is also quite egocentric. **Egocentrism** refers to the tendency of children to view situations and events only from their own perspective, with the belief that others will see things from the same point of view as themself (Figure 2.25). The concept helps us to understand why children appear to be selfish or uncooperative at times. Piaget suggested this is due to their limited understanding of mental states such as desires, beliefs, thoughts, intentions and feelings.

istockphoto.com/yellowsarah

Figure 2.25 Crossing a busy street can be dangerous for preoperational children. Because their thinking is still egocentric, younger children cannot understand why the driver of a car cannot see them if they can see the car.

Centration is another cognitive characteristic that occurs in this stage where children tend to narrowly focus on one aspect of a problem, object or situation at a time, while excluding all others. Centration is often referred to as the psychological example of tunnel vision and Piaget described it as the greatest limitation of young children's thinking. For example, a young child may become very upset if they see that their younger sister has received two slices of cake and they have received one. Whether the two slices their sister received are extremely thin and their one slice is much wider than their sister's two slices put together, the child can only focus on one feature (number of slices) and ignores all others (the size of each slice).

ANALYSING RESEARCH 2.3

The three mountain scene test

To test his theory that children below the age of 5 years are egocentric, Piaget conducted research in which several children under 5 years were presented with a papier-mâché model of a mountain range consisting of three mountains (Figure 2.26). The mountains had different characteristics; different colours, one had snow on the top, one a house and one a red cross on top. The child was allowed to walk around and explore the model. The child then sat on one side and watched an experimenter place a doll in a variety of locations around the mountain model. The child was then shown 10 pictures of different views of the model and asked to choose the one that represented how the doll would view it (Piaget & Inhelder, 1956).

Results indicated that 4-year-olds were completely unaware of viewpoints other than their own, always choosing a picture that matched their view of the model. Six-year-olds showed some awareness, but often chose the wrong picture. Only 7- and 8-year-olds consistently chose the picture that represented the doll's view.

According to Piaget, children below the age of 7 years are bound by the egocentric illusion. They fail to understand that what they see is relative to their own position, and instead take it to represent 'the world as it really is'. However, later research suggests that the children tested failed to take the doll's 'viewpoint' because the task was unfamiliar and too complex for them. When more familiar situations were used (such as Grover, a Sesame Street character, driving a car along a toy road), children could accurately report on this character's perspective (Flavell, 1977).

Questions

1 Write a hypothesis for Piaget's experiment.
2 What variable was Piaget interested in studying the effects of? What was his outcome variable?
3 Describe the procedure of the study and how it tested egocentrism in children.
4 Do the results of this experiment support Piaget's description of the cognitive abilities of children in the preoperational stage? Give reasons for your answer.

Figure 2.26 In the three mountain task, children are challenged to identify what can be seen from the doll's perspective.

The concrete operational stage (7–11 years)

The concrete operational stage follows the preoperational stage. An important development during this stage is the child learning the concept of **conservation**. This occurs when a child understands that physical properties such as mass, volume, number and length do not change (are conserved) even when their shape or appearance changes. To illustrate the conservation of volume: imagine a 5-year-old is shown two glasses full of liquid of the same volume, with one glass short and wide while the other is tall and narrow. The child will tell you that the taller glass contains more liquid (Figure 2.27). After the age of about 7 years, children can understand this situation, and they show a definite trend towards more logical, adult-like thought (Flavell, 1992).

Conservation of mass can be seen in children when they understand that rolling a ball of clay into a different shape, such as a 'snake', does not increase the amount of clay (Figure 2.28).

Reversibility of thought is another accomplishment in the concrete operational stage. This allows children to recognise that if $4 + 2 = 6$, then $2 + 4$ does, too. Younger children must memorise each relationship separately. Thus, a preoperational child may know that $4 + 10 = 14$, without being able to tell you what $10 + 4$ equals.

While he was testing children in this age group, Piaget was also interested in their ability to classify objects based on similar properties.

Figure 2.27 Children under the age of 7 years intuitively assume that a volume of liquid increases when it is poured from a short, wide container into a taller, thinner one.

Classification is the ability of children to be able to group together or categorise objects based on similar properties. Piaget believed that before the concrete operational stage, children have difficulty categorising and sub-categorising objects.

The formal operational stage (12 years and up)

Sometime after the age of 12 years, children begin to break away from concrete objects and specific examples, and develop thinking based more on abstract principles, such as democracy

Type of conservation	1 Initial presentation	2 Transformation	Question	3 Preoperational child's answer
Liquids	Two equal glasses of liquid	Pour one glass into a taller, narrower glass	Which glass contains more?	The taller one
Number	Two equal lines of checkers	Increase spacing of checkers in one line	Which line has more checkers?	The longer one
Mass	Two equal balls of playdough	Squeeze one ball into a long, thin shape	Which piece has more playdough?	The long one
Length	Two pencils of equal length	Move one pencil	Which pencil is longer?	The one that is further to the right

Figure 2.28 Different types of conservation

or honour. They also gradually become able to consider hypothetical possibilities. The two types of thinking that dominate this stage are abstract thinking and hypothetico-deductive reasoning.

2.3.4 PIAGET'S THEORY OF COGNITIVE DEVELOPMENT

Abstract thinking means thinking in terms of general concepts rather than specific objects, experiences or events. Individuals during this stage can think about things they are yet to experience or things they can't directly see or touch; for example, concepts such as love, freedom and morality. **Hypothetico-deductive reasoning** refers to being able to test a logical hypothesis by using abstract thought. For example, younger children typically might solve problems using trial and error. However, adolescents and adults using hypothetico-deductive reasoning will systematically plan how the problem should be solved by first thinking about possible strategies (Figure 2.29).

iStock.com/mediaphotos

Figure 2.29 Teenagers in the formal operational stage are able to think abstractly and hypothesise the probable cause of a range of problems.

Full adult intellectual ability is attained during the stage of formal operations. Older adolescents are capable of inductive reasoning (logical reasoning that begins with the specific and moves to the general) and deductive reasoning (logical reasoning that begins with the general and moves to the specific), and they can understand more advanced mathematics, physics, philosophy, psychology and other abstract systems.

Also, many adults can think formally about some topics, but their thinking becomes concrete when the topic is unfamiliar. This implies that formal thinking may be more a result of culture and learning than of maturation. In any case, after late adolescence, advances in thinking are based on gaining knowledge, experience and wisdom, rather than on any leaps in basic thinking capacity.

Piaget's four stages of cognitive development are summarised in Table 2.3.

Table 2.3 Piaget's stages of cognitive development

Stage	Age range	Cognitive accomplishment
Sensorimotor stage	Birth – 2 years	» Object permanence » Goal-directed behaviour
Preoperational stage	2–7 years	» Centration » Egocentrism » Symbolic thinking » Animism
Concrete operational stage	7–11 years	» Conservation » Classification » Reversibility of thought
Formal operational stage	12 years and beyond	» Abstract thinking » Hypothetico-deductive reasoning

Criticisms of Piaget's theory

Although Piaget was very influential in the field of cognitive psychology, he was also criticised for his methodology and research proposals. Criticisms include the following points.

- » Piaget may have underestimated children's abilities and that many accomplishments can be achieved by children before he specified.
- » Piaget gave too little credit to the effects of learning. For example, children of pottery-making parents can correctly answer questions about the conservation of clay at an earlier age than Piaget would have predicted (Bransford et al., 1986).
- » Piaget used his own children as research participants. He would record observations of their behaviour and cognitive accomplishments. There were also other issues surrounding the recruiting of his research participants (that is, his selection process).
- » Piaget did not measure cognitive development throughout adulthood, only during the early stages of the life span.
- » More recent research suggests that intellectual growth is not as related to age and stage as Piaget claimed.
- » According to other learning theorists, children continuously gain specific knowledge; they do not undergo stage-like leaps in general mental ability.

INVESTIGATION 2.1 PRINCIPLES OF CONSERVATION

Scientific investigation methodology

Controlled experiment

Data calculator

Aim

To investigate how children of different ages use principles of conservation

Introduction

This scientific investigation is related to the principles of conservation. You will need to generate appropriate qualitative and/or quantitative data, organise and interpret the data, and research a conclusion in response to your research question.

Conservation is the knowledge that physical properties such as mass, volume, number and length remain unchanged (are conserved) even when the shape or appearance of the material changes. Conservation is achieved during the concrete operational stage of Piaget's theory of cognitive development. Your investigation will test the performance of children of different ages on a number of conservation tasks.

Logbook template
Controlled experiment

Pre-activity preparation

You are required to apply the key science skills (listed at the start of this chapter) to:

- develop a research question
- state an aim
- formulate a hypothesis
- plan an appropriate methodology and method to answer the question, while complying with safety and ethical guidelines.

Procedure

You will need to recruit a number of Year 1 and Year 3 children to act as participants in this investigation. To do this, you will need to obtain informed consent from their parents or guardians.

This scientific investigation requires you to design a range of tasks to test the principle of conservation. You can investigate any or all of the following aspects of conservation: mass, length, number and volume. Here are some suggested approaches.

- You could investigate the conservation of mass using two equal ball-shaped amounts of plasticine. Once the participant has confirmed that each ball of plasticine is the same, one can be flattened into a pancake shape. The participant is then asked which has more plasticine and why. Record the participant's responses.
- You could investigate the conservation of length by first presenting the participant with two lengths of string stretched out on the table in front of them. Once the participant has confirmed that each string is the same length, bend one piece of string into a curvy shape. Ask the participant if the lengths of the string are the same and why. Record the participant's responses.
- You could investigate the conservation of number using 10 coins. Place the coins in two rows of five on the table in front of the participant. Ask the participant to confirm that there is the same number of coins in each row. Move one row of coins further part so the distance between the coins is greater for one row. Ask the participant if there is the same number of coins in each row and why. Record the participant's responses.
- You could investigate the conservation of volume by initially presenting the participant with two glasses with the same amount of water in each. Once the participant has confirmed that each has the same amount, pour the water from one glass into a taller, thinner glass or cylinder. Ask the participant if there is the same amount of water in each glass and why. Record the participant's responses.

You will be required to collect data, use evidence and transfer your findings to explain the relationship between age and performance on the conservation task(s) you study.

Results

Collate your results for all participants. Process any quantitative data that you have recorded, using appropriate mathematical calculations and units (for example, descriptive statistics such as a mean). Organise, present and interpret your data using an appropriate table and/or graph.

Discussion

1. State whether your hypothesis was supported or not.
2. Describe your observations about the relationship between a participant's age and performance on the conservation tasks.
3. Discuss the results with reference to different stages within Piaget's theory of cognitive development.
4. Draw a conclusion in response to your investigation question.
5. Identify whether it is possible to generalise any of your results to the wider population. Why or why not?
6. Discuss any potential extraneous variables and how they could have affected your data.
7. Suggest improvements you could make to the procedure to address the extraneous variables, if the investigation was to be repeated.

Adapted from: VCAA VCE Psychology Advice for Teachers https://www.vcaa.vic.edu.au/

ACTIVITY 2.3 TOY DESIGN

Try it

1 For each of Piaget's stages of cognitive development, design (or critique) one child's toy or game to suit a child's ability for that stage. Refer to the cognitive development achievement(s) for each stage. Present your answer in a table like the one below. The first one has been done for you.

Stage	Age	Toy	Cognitive achievement
Sensorimotor stage	0–2 years	Peekaboo Bear	Object permanence This toy will engage children under the age of 2 because when the bear hides behind the blanket, it will cease to exist for the infant, making it very exciting when the bear 'reappears' when the blanket is lowered.
Preoperational stage			
Concrete operational stage			
Formal operational stage			

Apply it

2 Declan is 9 and his sister Iris is 4.
A psychologist who is testing Piaget's theory of cognitive development gives the children the following test:
The psychologist presents the child with two jugs, both containing 100 mL of water. The psychologist pours one of the jug's contents into a short, wide glass. Next to this glass, the psychologist pours the contents of the second jug into a tall, thin glass. Each child is asked separately which glass contains more water.
Predict the responses given by Declan and Iris.

Exam ready

3 Baby Dylan is happily playing with a toy dog, but his mother notices some dirt on the toy and quickly hides it under a cushion. Dylan then plays with another toy without showing any concern.
It is very likely that Dylan has not yet achieved

A conservation of mass.
B egocentrism.
C animism.
D object permanence.

KEY CONCEPTS 2.3c

- Jean Piaget studied cognitive development, which explores how thinking processes change and mature over the life span.
- Piaget's key principles propose that children either assimilate new situations to existing ideas (schema), or accommodate new experiences by changing their ideas.
- Piaget developed four age-related stages he believes children progress through as their cognition develops, from concrete thinking to symbolic thinking, through to abstract thinking.
- By the end of the sensorimotor stage (birth – 2 years), children achieve two cognitive accomplishments: object permanence and goal-directed behaviour.
- During the preoperational stage of cognitive development (2–7 years) children begin to develop symbolic thinking.
- The preoperational child will tend to only focus on one aspect of a problem, object or situation at a time (centration), they tend to view situations and events only from their own perspective (egocentrism), and have the tendency to believe that any inanimate object possesses lifelike characteristics (animism).
- During the concrete operational stage (7–11 years) children will understand conservation, can reverse their thoughts back to a line of reason, and can classify and categorise objects based on common features.
- During the formal operational stage (12 years and beyond) children move to abstract thinking, and are able to formulate hypotheses when trying to solve problems.
- Piaget was criticised for some aspects of his theory. Many theorists now believe children can achieve cognitive accomplishments far earlier than Piaget predicted.

Concept questions 2.3c

Remembering

1 What is cognitive development? c

2 Explain what a cognitive schema is, using an example to support your response. r

3 Why might a child snatch a toy from another child at day care, even if it makes the other child upset? e

4 Explain the different types of conservation, using an example of each. r

Understanding

5 Compare and contrast assimilation and accommodation. e

6 Draw a table like the one shown and describe the four stages of cognitive development, including a description of each cognitive accomplishment and age ranges for each stage. r

Name of stage	Age range	Description of each cognitive accomplishment

7 Summarise the key criticisms of Piaget's theory of cognitive development. r

Applying

8 Using an example, explain how a child may use mental adaptation to assimilate and accommodate new information in their environment. c

9 A child is walking along, holding her teddy bear. She drops the bear on the ground. Immediately she picks it up and asks if it is ok. According to

Piaget's theory, why would the young child do this and how old might she be? **e**

10 What is the difference between symbolic thinking and abstract thinking? **r**

11 Using examples, identify why pretend play is so important for learning, for children during the preoperational stage of development. **c**

HOT Challenge

12 Imagine you are a developmental psychologist interested in testing the cognitive accomplishments described by Piaget. Design a test, not mentioned in this text, that can be used to measure whether a child has achieved a cognitive accomplishment in the preoperational stage of development. In your response you should include: **i**

- the age range of the children you will test
- a hypothesis, based on what you think the responses of the child would be
- a description of how you will carry out the test (procedure)
- a summary of the ethical principles you would have to implement to protect the rights of the participants.

Psychosocial development: Erikson's eight-stage theory

Erik Erikson (1902–1994; Figure 2.30) was a developmental psychologist and psychoanalyst who explored personality development. Erikson was a trained psychoanalyst and neo-Freudian. Neo Freudians are psychoanalysts who agree with many of the basic principles proposed by Sigmund Freud; however, they adapted his ideas to suit their beliefs, research and opinions. As such, Erikson adjusted a well-known theory of personality development devised by Freud and created an eight-stage theory of psychosocial development.

Erikson's psychosocial theory considers an individual's interaction with people around them (social) as a source of their personality development. A key difference between Erikson's theory and other developmental theories is that it explains the development of personality from birth through to death, rather than just focusing on the early stages of the life span.

Getty Images/Ted Streshinsky Photographic Archive

Figure 2.30 Erik Erikson was a developmental psychologist who developed theories about personality.

The journey of life: rocky road or garden path?

Erikson suggested that we face a specific psychosocial dilemma, or 'crisis', at each stage of life. A **psychosocial dilemma** is a conflict between personal impulses and the social world. Resolving each dilemma creates a new balance between a person and society. Erikson proposed that a string of 'successes' produces healthy development and a satisfying life. However, unfavourable outcomes throw us off balance, making it harder to deal with later crises. Life becomes a 'rocky road', and personal growth is stunted. It was using this premise that his eight-stage theory was created. During each stage of the life span, individuals are faced with a psychosocial crisis that will either be resolved or not.

Stage one (0–18 months): trust versus mistrust

Erikson believed that a basic attitude of trust or mistrust is formed during the first year of life. Trust is established when babies are given adequate warmth, touching, love and physical care. Mistrust is caused by inadequate or unpredictable care and by parents who are cold, indifferent or rejecting. Basic mistrust may later cause insecurity, suspiciousness or an inability to relate to others. Notice that trust comes from the same conditions that help babies become securely attached to their parents.

Trust and mistrust are not resolved once and for all in the first year of life; instead, the issue arises at each stage of development. Therefore, an infant who has an attitude of mistrust because they do not receive support and attachment in the first years of life may in later years trust a particular adult who instils a sense of trust, and can then overcome their earlier mistrust. Likewise, a child who starts out with an attitude of trust can also have their sense of mistrust activated by issues such as when their parents' divorce or become separated, causing conflict.

Stage two (18 months – 3 years): autonomy versus shame and doubt

In stage two, children express their growing self-control by climbing, touching, exploring and trying to do things for themselves. Parents can foster a sense of autonomy by encouraging children to try new skills. The child's first efforts can be crude, often resulting in spilling, falling, wetting and other 'accidents' (Figure 2.31). Thus, parents who ridicule or overprotect their children may cause them to doubt their abilities and feel shameful about their actions. It is important at this stage that parents or caregivers allow their children time to develop skills, realise that they will make mistakes and that they must acquire skills at their own pace. This includes things like speaking and using cutlery.

iStock.com/Susan Chiang

Figure 2.31 A child's first attempts at feeding themself may be messy, yet they will feel a sense of pride when they accomplish their goal.

Stage three (3–5 years): initiative versus guilt

In stage three, children move beyond simple self-control and begin to take initiative. Through play, children learn to make plans and carry out tasks. Parents can reinforce initiative by giving children freedom to play, ask questions, use their imagination and choose activities. Feelings of guilt about initiating activities are formed if parents criticise severely, prevent play or discourage a child's questions.

Stage four (6–12 years): industry versus inferiority

Many events of middle childhood are symbolised by the day when you first entered school. With dizzying speed, your world expanded beyond your family, and you faced a whole series of new challenges. Erikson describes the primary school years as the child's 'entrance into life'. In school, children begin to learn skills valued by society, and success or failure can affect a child's feelings of adequacy. Children learn a sense of industry if they win praise for productive activities, such as building, painting and studying. If a child's efforts are regarded as messy, childish or inadequate, feelings of inferiority result. For the first time, teachers, classmates and adults outside the home become as important as parents in shaping attitudes towards oneself.

Stage five (adolescence): identity versus role confusion

Adolescence is often a turbulent time. Caught between childhood and adulthood, the adolescent faces some unique problems. Erikson considered the need to answer the question 'Who am I?' the primary task during this stage of life. Mental and physical maturation brings new feelings, a new body and new attitudes. Adolescents must build a consistent identity out of their talents, values, life history, relationships and the demands of their culture (Douvan, 1997). People who fail to develop a sense of identity suffer from role confusion: an uncertainty about who they are and where they are going.

Stage six (young adulthood): intimacy versus isolation

In stage six, the individual feels a need for intimacy in their life. After establishing a stable identity, a person is prepared to share meaningful love or deep friendship with others. By intimacy, Erikson means an ability to care about others and to share experiences with them. In one 1970s study that aligned with Erikson's view, 75 per cent of university-age men and women ranked a good marriage and family life as important adult goals (Bachman & Johnson, 1979). And yet, marriage or sexual involvement is no guarantee of intimacy: many adult relationships remain superficial and unfulfilling. Failure to establish intimacy with others leads to a deep sense of isolation (feeling alone and uncared-for in life). This often sets the stage for later difficulties.

Stage seven (middle adulthood): generativity versus stagnation

In middle adulthood a person's focus shifts from narrow self-interest towards fulfilling family and wider social responsibilities to nurture the next generation. Erikson called this quality **generativity**. It is expressed by caring about oneself, one's children and future generations (Figure 2.32). Generativity may be achieved by guiding one's own children or by helping other children (as a teacher, religious leader or coach). Productive or creative work can also express generativity. In any case, a person's concerns and energies must broaden to include the welfare of others and society as a whole. Failure to do this is marked by a stagnant concern with one's own needs and comforts. Life loses meaning, and the person feels bitter, dreary and trapped (Peterson & Klohnen, 1995).

iStock.com/Susan Chiang

Figure 2.32 Example of generativity: a father working with his son and grandson to contribute to future generations

Stage eight (late adulthood): integrity versus despair

Old age is a time of reflection. According to Erikson, a person must be able to look back over life with acceptance and satisfaction. The person who has lived richly and responsibly develops a sense of integrity (self-respect). This allows the person to face ageing and death with dignity. If previous life events are viewed with regret, the elderly person experiences despair (heartache and remorse). In this case, life seems like a series of missed opportunities. If a person feels like a failure and knows that it is too late to reverse what has been done, ageing and the threat of death then become sources of fear and depression.

To squeeze all the stages of life into a few pages, this explanation ignores countless details. Although much is lost, the net effect is a clearer picture of the entire life span. Is Erikson's theory, then, an exact map of your future? Probably not. Still, the dilemmas described here reflect major psychological events in the lives of many people. Knowing about them may allow you to anticipate typical trouble spots in life. You may also be better prepared to understand the problems and feelings of friends and relatives at various stages in life.

Erikson's eight stages of psychosocial development are summarised in Table 2.4.

Weblink
Summary Erikson's theory

2.3.5 ERIKSON'S THEORY OF PSYCHOSOCIAL DEVELOPMENT

Table 2.4 Erikson's eight-stage theory of psychosocial development

Stage	Age range	Psychosocial crisis
1	Birth – 18 months	Trust versus mistrust
2	18 months – 3 years	Autonomy versus shame/ doubt
3	3–5 years	Initiative versus guilt
4	6–12 years	Industry versus inferiority
5	Adolescence	Identity versus role confusion
6	Young adulthood	Intimacy versus isolation
7	Middle adulthood	Generativity versus stagnation
8	Late adulthood	Integrity versus despair

KEY CONCEPTS 2.3d

- Erik Erikson proposed an eight-stage theory that explains psychosocial personality development throughout the life span.
- Each stage consists of a psychosocial dilemma. If the dilemma is resolved this will result in a healthy personality and fulfilling life. If the dilemma is not resolved, this can lead to an unsatisfying life and make future crises harder to deal with.
- In stage 1 babies are conflicted with trusting or mistrusting their caregivers and in stage 2 toddlers need to be given autonomy at the risk of developing feelings of shame and doubt.
- In stage 3 preschoolers need to be given the opportunity to take initiative rather than feeling guilt over their actions. Stage 4 children entering primary school should be challenged with a sense of industry and achievement to avoid feelings of inferiority and low self-worth.
- In stage 5 adolescents struggle with their identity and with role confusion, and try to figure out who they are. In stage 6 young adults seek intimacy through meaningful friendships and other relationships to avoid a sense of isolation.
- In stage 7 middle-aged adults focus their energy on supporting and caring for future generations. Stage 8 older adults begin a process of reflection and either feel integrity or despair for the life they have lived.

Concept questions 2.3d

Remembering

1 Define the term 'psychosocial dilemma'. r

2 According to Erikson, how does a person achieve a healthy personality and fulfilling life? r

Understanding

3 Copy the table below and use it to summarise Erikson's theory of the eight stages of psychosocial development. r

Stage	Age range	Psychosocial crisis	Description of psychosocial dilemma
1			
2			
3			
4			
5			
6			
7			
8			

Applying

4 Identify the psychosocial dilemma that has not been resolved in each of the scenarios below. Identify the stage this would be occurring. e

a An adolescent goes through a stage where they dye their hair bright red and get several body piercings.

b An older, childless man watches his friends around him helping and caring for their children and grandchildren while his life remains quiet and uneventful.

c A toddler tries to pour her own milk in her breakfast cereal each day. She often spills the milk all over the floor. Her mother becomes angry at her and doesn't let her do it anymore.

d A young woman has just started her first full-time job. She is really trying to fit in and make new friends with her colleagues; however, she is having a hard time doing so.

e A child at school is finding all of her school work very easy. She finishes all of her work well before the teacher expects. Her teacher doesn't give her extension work so she begins to play up in class.

HOT Challenge

5 Predict the personality development of the following individual. In your answer, ensure you refer to at least three stages of Erikson's theory of psychosocial development and discuss the outcomes of the crises at each stage. e

Gabriella is a 75-year-old woman. During the later years of her life she spent a lot of time reflecting on her life experiences and the decisions she made. Gabriella lived a difficult life. She lost her father at the age of 1, and for the next few years her mother was depressed, distant and not responsive to all of her needs. She spent a lot of time with her grandmother, as her mother often worked long hours. She feels like she never had a proper childhood, one where she was free to play as she wished. During her school years, she often felt like she didn't fit in to any friendship group. Her desire for meaningful relationships with peers continued into early adulthood and while she did make some friends she never felt true intimacy with those around her. She had a string of romantic relationships, none of which lasted more than a couple of years. After many failed attempts at a long-term relationship she instead poured all of her energy into her work, which brought her great satisfaction.

2.4 Sensitive and critical periods in psychological development

To psychologists, maturation is mainly about behaviours or activities that occur at particular times in an organism's development. These developmental activities depend on how ready, or mature, the central nervous system and the body are to demonstrate a specific behaviour. Having particular experiences at particular stages of the life span can be important for typical development to occur. Psychologists classify periods of development as either sensitive periods or critical periods. These periods of time in development typically occur in the early stages of the life span, when the brain is primed and ready for learning experiences. When an individual meets with the necessary learning experiences within these periods of time, development can proceed as normal. If these periods of time are missed, learning a particular cognitive or motor skill, or developing emotional competence, can be extremely difficult or even, in some cases, impossible.

Sensitive periods

A **sensitive period** is a stage during biological maturation when an organism is most able to gain a particular skill or characteristic. A sensitive period begins and ends gradually. It is a period of maximal brain growth, after which development within a particular area will take more effort, be slower, and be incomplete. It is a period of development when an organism is mature enough to demonstrate a behaviour, and it is the best time to be exposed to a particular environmental stimulus because the brain is developed to the degree that it is now able to process and respond to the stimuli.

A sensitive period is linked to the concept of 'experience-expectant learning', in that it is a stage in an organism's development where certain behaviours are expected to develop after exposure to certain stimuli. An organism may still acquire the learning at a later date, it just may take longer and require more effort to learn the behaviour than if the organism learned it during the sensitive period. Reading and writing are examples of this type of behaviour (Figure 2.33). Generally, we learn to read and write in the first years of school; however, some people do not. If these individuals took classes in reading and writing as adults, they could possibly learn these skills. However, the standard they reach and the effort required would both be affected.

In summary, during a sensitive period, the person has the cognitive capacity to do the task, and they must also have the opportunity to experience or be exposed to the task. Sensitive periods span a longer period than critical periods, but they are similar concepts.

iStock.com/track5

Figure 2.33 The sensitive period for reading and writing is during the child's school years. If this opportunity is missed, learning to read and write for the first time as an adult, though not impossible, is much more difficult.

Critical periods

Critical periods of development involve a stage, usually early in the life span, when an organism is most open to acquiring a specific cognitive or motor skill or socio-emotional competence, as a normal part of development, that cannot be acquired normally at a later stage of development. If the organism does not receive the necessary exposure to appropriate stimuli, further development of this ability will generally not occur.

A critical period involves a small window of opportunity to learn a behaviour. If this window closes, the learning becomes impossible or nearly impossible in the future. Critical periods start and end abruptly and usually have shorter time frames compared with sensitive periods.

2.4.1 COMPARING AND CONTRASTING CRITICAL AND SENSITIVE PERIODS

Research suggests that events that occur during a critical period can permanently alter the course of development (Bornstein, 1989). For example, research conducted by Hubel and Wiesel in the 1960s found that if a newborn kitten has its eye stitched shut early in its development (within 4–6 weeks), the kitten will become permanently blind in that eye. However, if the eye is sewn shut later in development (after 4 months), the kitten will not become blind in that eye. This study led the researchers to conclude that there must be a critical period for vision in kittens, and the neural pathway between the eye and the brain must have visual input at a particular time in development to develop properly (Hubel & Wiesel, 1962).

Often, certain events must occur during a critical period for a person to develop normally. For example, within the human nervous system, it appears that different areas of the brain used in learning develop fully at different times. The brain must go through a period of forming synapses and then pruning back unused connections to complete its development. Most of this synapse formation and pruning occurs in the first years of life, but the frontal lobe has one of the longest periods of development (see Chapter 5).

So why do sensitive and critical periods exist? Most sensitive and critical periods can be found in the first years of development of most species. Many scientists believe that the answer may be connected to the concept of brain plasticity. This refers to the brain's ability to modify its structure and function as a result of experience and in response to injury.

Research tells us that children have far more plasticity than adults. Their brains have more neural connections because their brains are growing. As they reach a certain age in development their neural connections begin to prune away. These connections will be lost forever if they are not used, which implies that sensitive periods are important in development. For example, if severe damage occurs in the language-dominant left hemisphere of a child's brain before the age of 2, the brain can shift language processing to the right side of the brain. If the damage occurs between the ages of 2 and 6, new language areas tend to develop in both hemispheres (Mueller et al., 1999). Such drastic shifts in language processing would be impossible for an adult brain, which has fewer neural connections compared to a young child's brain.

Sensitive periods and critical periods are summarised in Table 2.5.

Imprinting

One example of a critical period that occurs in the very early stages of life in animals is known as imprinting. Konrad Lorenz (1903–1989) was curious as to why baby geese follow their mothers. He observed the specialised learning ability of young animals to fix their attention on the first object they meet, and to follow that object thereafter. Chickens, ducks and geese all share the same characteristic of being able to walk once hatched. Because newly hatched chicks are able to wander around from birth, they can easily become separated from their mothers. Consequently, the species has evolved so that chicks form instant attachments to their mothers, and will recognise and stay near her.

2.4.2 ANALYSIS OF RESEARCH

Table 2.5 Sensitive periods versus critical periods

Period	Timing of experiences	Description
Sensitive period	Flexible and broad period of time	A stage during biological maturation when an organism is most able to acquire a particular skill or characteristic; a period of maximal plasticity, after which development within the domain will take more effort, and be slower and incomplete; the term tends to be used less restrictively than the term 'critical period', but they are similar concepts.
Critical period	Fixed and narrow period of time	A stage, usually early in the life span, when an organism is most open to acquiring a specific cognitive or motor skill, or socio-emotional competence, as a normal part of development, that cannot be acquired normally at a later stage of development.

Normally, the first large moving object a baby goose sees is its mother. Lorenz hatched geese in an incubator, so the first moving object they saw was Lorenz. From then on, these goslings followed Lorenz (Figure 2.34). They even reacted to his call as if he were their mother (Lorenz, 1937). If newly hatched ducklings do not imprint on their mother (or some other object) within 30 hours, they never will (Hess, 1959).

This example of attachment is different from human attachment: the period of time attachment needs to occur is extremely fixed and attachment of this kind is impossible after this period of time (30 hours). This is why human infant attachment is said to have a sensitive period rather than critical period for development.

Alamy Stock Photo/Science Photo Library

Figure 2.34 'Mother' Lorenz leads his charges. The goslings have imprinted on Lorenz because he was the first moving object they saw after they hatched.

Language acquisition

Language acquisition refers to the developmental processes through which children learn to understand and use their native language/s. First language acquisition usually begins in infancy and early childhood. As with other behaviours, language in humans is linked to both innate (maturational) and environmental (experiential) factors. Researchers generally agree that there is a window of opportunity for children to acquire their first (or second) language; however, the exact age when this time frame begins and ends is still up for debate. Some researchers in this area have explored the possible time frames of critical or sensitive periods for acquiring a first language and learning a second language (Figure 2.35). Others have compared the different areas of language development, such as phonetics or syntax, and proposed multiple 'critical' periods based on these different language skills (Abello-Contesse, 2009). This is why it is difficult to categorise language acquisition as either a sensitive period or critical period of development. The general consensus is that there certainly is a period of time that is optimal for learning language, but the complexities of learning language make it challenging to classify this key cognitive skill.

iStock.com/SerrNovik

Figure 2.35 Learning a second language is said to be much easier in childhood compared to adulthood, but up until what age? This is a question that has plagued linguistic experts for many years.

CASE STUDY 2.1

Feral children: Victor and Genie

The first research into language acquisition came from examples of 'feral' children and others who were deprived of language input during infancy and early childhood. A 'feral' child is a name commonly used to describe children found living in deprived conditions, and one that suggests that they are children who behave as if reared in the wild, by animals. One of the first cases of a child with an inability to develop language skills was reported in 1801 when a 'wild boy', known as Victor, was found in Aveyron, France. The child was found at the age of 12 or 13, having lived apparently by himself in the forests. Victor was unable to walk upright or speak, and had a high tolerance for cold because he had spent most of his life outside. After five or so years of trying to instruct Victor, the doctor taking care of him deemed him beyond education. Victor never learned language with any proficiency and was only able to engage in basic communication with his carers.

In 1970, another case of a feral child, a girl known as Genie (Figure 2.36), emerged. Her case was one of the worst examples of child abuse in history. Genie was a 13-year-old girl when she was found living in appalling conditions. It became apparent she had been subjected to extreme abuse, neglect and social isolation. She had been kept locked away and strapped to a toilet chair for most of her life because her father assumed she was mentally disabled at birth and had chosen to keep her isolated from others. She was severely malnourished, had little to no social contact with anyone and was often beaten for speaking or uttering noises.

When Genie was found, social workers believed she was only about 6 or 7 years old because of her appearance, but it was later discovered she was 13 years old. She could barely walk. Genie could make noises, but nothing approximating language. Researchers decided that it was an ideal opportunity to test the theory that a nurturing environment could somehow make up for a total lack of language before the age of 12. She spent the next several years in care and working with linguistic experts, in an attempt to develop her language skills. Although Genie was able to master some language skills, such as vocabulary, unfortunately she never became proficient at speaking and communicating and was unable to make up for the time spent deprived of language.

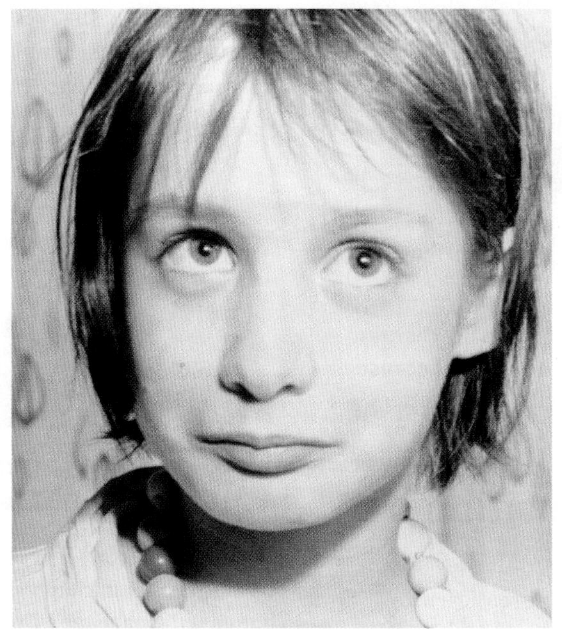

Figure 2.36 An image of Genie taken shortly after she was removed from isolation

Cases like Victor and Genie demonstrate the importance of the timing of experiences for language acquisition. If exposure to language has not taken place in the early years of life, then an individual may not be able to master this skill later in life.

Questions

1 Describe the case of Victor.
  a Why was he considered a 'feral' child?
  b At what age was he discovered?
  c What were his cognitive and behavioural capacities when he was discovered?
  d Was he ever able to acquire language?

2 Describe the case of Genie.
  a What were the living conditions during her period of abuse and isolation?
  b At what age was she discovered?
  c What were her cognitive and behavioural capacities when she was discovered?
  d Was she ever able to acquire language?

3 What differences might be observed for the language development of a feral child if they were discovered at 5 years of age, compared to at 15 years of age?

4 How do the cases of Victor and Genie support your understanding of the importance of the timing of experiences for language acquisition?

ACTIVITY 2.4 SENSITIVE PERIODS AND CRITICAL PERIODS

Try it

1 Environments can sometimes be described as 'deprived' or 'enriched'. For example, Genie (see Case study 2.1) grew up in an incredibly deprived environment. In contrast, an enriched environment such as a childcare centre or kindergarten is mentally stimulating.

Working with a partner, design your own playroom (complete with toys, obstacles, puzzles and so on) for a kindergarten. Share your design with the class and explain how your playroom would support the development of a young child.

Apply it

2 Each weekend, you and your family go for a walk by the river. This weekend, you notice a mother duck and its ducklings walking on the path. Your older brother asks, 'How do the ducklings know to follow their mother?'

Using relevant psychological terminology, how would you answer your brother's question?

Exam ready

3 With regard to the timing of experiences, critical periods differ from sensitive periods in that

A critical periods tend to begin and end gradually; sensitive periods have a narrow window where exposure must take place for the skill to be developed.

B critical periods are a narrow window where exposure must take place for the skill to be developed; sensitive periods tend to begin and end gradually.

C critical periods begin suddenly but remain open until the skill has been acquired; sensitive periods tend to begin and end gradually.

D critical periods are a narrow window where exposure must take place for the skill to be developed; sensitive periods begin gradually but end abruptly.

KEY CONCEPTS 2.4

- The timing of experiences in psychological development refers to the optimal period of time for learning a particular skill.
- A sensitive period of development is when an organism is most open to acquiring a particular skill or characteristic; the lack of appropriate experiences during a sensitive period will not make acquisition impossible, but will make it more difficult.
- Critical periods of development involve a stage early in the life span, when an organism is most open to acquiring a specific cognitive or motor skill, or socio-emotional competence that cannot be acquired normally at a later stage of development.
- Critical and sensitive periods are linked to the concept of brain plasticity and the reduction in plasticity of the human brain as we age, which makes it more difficult to develop new skills.
- The early stages of the life span are optimal times for acquiring language, although it is not possible to specify an exact timeframe because of the many complex characteristics of this cognitive skill.

Concept questions 2.4

Remembering

1 Distinguish between critical periods and sensitive periods of development. In your answer, include one similarity and at least two differences between them. r

Understanding

2 Why is the timing of experiences important in psychological development? r

3 Provide an example of a critical period of development. r

4 Provide an example of a sensitive period of development. r

Applying

5 The brain is thought to be able to change as a result of new experiences. Explain how this relates to critical and sensitive periods. i

6 Imagine a child was raised in social isolation on a deserted island. Their only human contact was their mother. Hypothesise any sensitive periods of development that may be missed. i

HOT Challenge

7 After consultation with Traditional Owners, a Victorian kindergarten (with children aged 3–5 years) is debating whether they should introduce a Victorian Aboriginal language program for their students. Some parents have suggested that the children wait until they are teenagers before they learn an additional language. Write an article for the kindergarten newsletter explaining the scientific evidence for critical and sensitive periods for language acquisition. e

2 Chapter summary

KEY CONCEPTS 2.1

- Most psychologists today agree that there is an interaction between inherited factors and environmental factors in the psychological development of an individual.
- Hereditary factors refers to inborn, inherited factors gained genetically from biological parents that partly determine individual characteristics.
- Environmental factors include external biological and social influences.
- Environmental influences include the care a child is given, the environment within the womb, exposure to environmental toxins and any peer-group influences.
- Twin and adoption studies allow researchers to study the interactive nature of the inherited and environmental factors on development.
- In a twin study, the development of identical twins is compared with the development of fraternal twins.
- In adoption studies, families who have one adopted child and one biological child are compared.

KEY CONCEPTS 2.2

- The biopsychosocial model recognises the contributions of biological, psychological and social factors, and their complex interactions, to our physical and mental health.
- Biological factors contributing to development and wellbeing include our genetics, hormones, immune function, stress response, brain function and other activities of our nervous system.
- Psychological factors contributing to development and wellbeing include cognitive processes such as learning and memory systems, self-concept, emotional regulation, perceptions, thought processes including beliefs and attitudes, personality traits and coping skills.
- Social factors influencing our wellbeing and development include social interactions, relationships and networks.

KEY CONCEPTS 2.3a

- Attachment refers to the close emotional bond or relationship between an infant and the mother or primary caregiver.
- Attachment of an infant with their primary caregiver is crucial in their subsequent emotional development.
- John Bowlby was one of the first psychologists to develop attachment theory. He proposed that infants are biologically pre-programmed to form an attachment with their primary caregiver (usually the mother).
- Bowlby proposed that a mother and infant should not be separated for at least the first 2 years of the infant's life at the risk of causing long-term emotional difficulties.
- Mary Ainsworth developed the Strange situation test to observe the behaviours of infants with their caregivers and their reactions around strangers.
- Ainsworth classified infants into the following attachment categories: secure attachment, insecure resistant attachment and insecure avoidant attachment.

KEY CONCEPTS 2.3b

- Harry Harlow's (1958) research with infant rhesus monkeys investigated whether food or contact comfort was more important for the attachment of infants to their caregivers.
- Harlow's findings revealed that contact comfort was more important than food in infant–caregiver attachment in monkeys. He generalised this finding to human infant–caregiver attachment.
- Harlow et al. (1965) investigated the effects of social deprivation on social and emotional development in infant rhesus monkeys.
- Their studies revealed that periods of social isolation in infancy can have debilitating effects on social and emotional development. The longer the period of isolation, the more severe the behaviours reported and the less likely to be reversed.

KEY CONCEPTS 2.3c

- Jean Piaget studied cognitive development, which explores how thinking processes change and mature over the life span.
- Piaget's key principles propose that children either assimilate new situations to existing ideas (schema), or accommodate new experiences by changing their ideas.
- Piaget developed four age-related stages he believes children progress through as their cognition develops, from concrete thinking to symbolic thinking, through to abstract thinking.
- By the end of the sensorimotor stage (birth – 2 years), children achieve two cognitive accomplishments: object permanence and goal-directed behaviour.
- During the preoperational stage of cognitive development (2–7 years) children begin to develop symbolic thinking.
- The preoperational child will tend to only focus on one aspect of a problem, object or situation at a time (centration), they tend to view situations and events only from their own perspective (egocentrism), and have the tendency to believe that any inanimate object possesses lifelike characteristics (animism).
- During the concrete operational stage (7–11 years) children will understand conservation, can reverse their thoughts back to a line of reason, and can classify and categorise objects based on common features.
- During the formal operational stage (12 years and beyond) children move to abstract thinking, and are able to formulate hypothesises when trying to solve problems.
- Piaget was criticised for some aspects of his theory. Many theorists now believe children can achieve cognitive accomplishments far earlier than Piaget predicted.

KEY CONCEPTS 2.3d

- Erik Erikson proposed an eight-stage theory that explains psychosocial personality development throughout the life span.
- Each stage consists of a psychosocial dilemma. If the dilemma is resolved this will result in a healthy personality and fulfilling life. If the dilemma is not resolved, this can lead to an unsatisfying life and make future crises harder to deal with.
- In stage 1 babies are conflicted with trusting or mistrusting their caregivers and in stage 2 toddlers need to be given autonomy at the risk of developing feelings of shame and doubt.
- In stage 3 preschoolers need to be given the opportunity to take initiative rather than feeling guilt over their actions. Stage 4 children entering primary school should be challenged with a sense of industry and achievement to avoid feelings of inferiority and low self-worth.
- In stage 5 adolescents struggle with their identity and with role confusion, and try to figure out who they are. In stage 6 young adults seek intimacy through meaningful friendships and other relationships to avoid a sense of isolation.
- In stage 7 middle-aged adults focus their energy on supporting and caring for future generations. Stage 8 older adults begin a process of reflection and either feel integrity or despair for the life they have lived.

KEY CONCEPTS 2.4

- The timing of experiences in psychological development refers to the optimal period of time for learning a particular skill.
- A sensitive period of development is when an organism is most open to acquiring a particular skill or characteristic; the lack of appropriate experiences during a sensitive period will not make acquisition impossible, but will make it more difficult.
- Critical periods of development involve a stage early in the life span, when an organism is most open to acquiring a specific cognitive or motor skill, or socio-emotional competence that cannot be acquired normally at a later stage of development.
- Critical and sensitive periods are linked to the concept of brain plasticity and the reduction in plasticity of the human brain as we age, which makes it more difficult to develop new skills.
- The early stages of the life span are optimal times for acquiring language, although it is not possible to specify an exact timeframe because of the many complex characteristics of this cognitive skill.

2 End-of-chapter exam

Section A: Multiple-choice

1 What do psychologists mean by the term 'nature'?
- A the environmental influences on development
- B the genetic influences on development
- C the cognitive influences on development
- D the social influences on development

2 Environmental influences on development can include
- A genes and hormones.
- B cognitive processes.
- C social interactions and life experiences.
- D nervous system functioning.

3 Identical twin and adoption studies are used to study the effects of hereditary and environmental factors on development because
- A similarities between identical twins adopted into different families can be attributed to heredity factors.
- B differences between identical twins adopted into different families can be attributed to environmental factors.
- C both A and B are correct
- D neither A nor B are correct

4 The rapid learning of permanent behaviour whereby an animal will form instant attachment to the first thing it sees at birth is called
- A maturation.
- B nervous system attachment.
- C imprinting.
- D plasticity.

5 Which of the following is an example of a sensitive period of development?
- A imprinting
- B infant–caregiver attachment
- C social development
- D none of the above

6 The formal operational stage of Piaget's theory involves the accomplishment of
- A object permanence.
- B conservation.
- C goal-directed behaviour.
- D hypothetical thinking.

7 According to Piaget, when children behave in an egocentric manner, they are in the __________ stage of development.
- A sensorimotor
- B preoperational
- C concrete operational
- D formal operational

8 The independent variable in Harlow's 1958 experiment on attachment was
- A whether the surrogate was made from wire or cloth and contained the feed bottle.
- B whether the monkey preferred the wire or cloth surrogate mother.
- C the amount of time the monkeys spent on the surrogates.
- D the attachment the monkeys formed for the surrogates.

9 Which of the following is not a psychological contributing factor according to the biopsychosocial framework?
- A self-concept
- B emotional regulation
- C perceptions
- D social networks

10 Stage seven of Erikson's theory of psychosocial development is called
- A integrity versus despair.
- B intimacy versus isolation.
- C initiative versus guilt.
- D generativity versus stagnation.

11 John Bowlby
- A believed that children have basic needs such as shelter and food, and as long as these are met, they will develop normally.
- B observed children in institutions and saw that they were well cared for.
- C believed that infants can thrive without their mothers as long as their physical needs are well taken care of.
- D observed that children who lacked relationships with others early in life were open to negative psychological consequences later in life.

12 Harlow's findings were that

A the monkeys spent more time on the wire surrogate and only spent time on the cloth surrogate to feed.

B contact comfort was not important; feeding capability was important.

C contact comfort was the most important factor in infant and mother attachment.

D the surrogates providing the food were the most important factor in the infant and mother attachment.

13 According to Erikson, if a person has conflicted experiences during stage five, they may suffer __________ later in life.

A signal behaviours

B spontaneous habituation

C role confusion

D mental breakdown

14 A sensitive period of development is a ________ timeframe where learning _______ occur in order for optimal development to proceed.

A fixed; should

B flexible; must

C flexible; should

D permanent; must

15 Piaget proposed children in the ____________ stage of development tend to narrowly focus on one aspect of a problem, object or situation at a time, and exclude all others. This is called ____________.

A preoperational; centration

B preoperational; conservation

C concrete operational; conservation

D concrete operational; centration

16 Ainsworth's category of 'insecure resistant' attached infants were characterised as showing

A great distress when their mothers left the room, and were not easily calmed when she returned.

B indifference to their mothers; they rarely cried when their mothers left the room and showed her little attention when she returned.

C great distress when their mothers left the room, and greeted her with pleasure when she returned.

D great distress when their mothers left the room, and were easily calmed when she returned.

17 Activation of the stress response involves the release of the hormone ____________. Stress can be managed by positive thinking strategies, which is a __________ contributing factor according to the biopsychosocial model.

A cortisol; social

B cortisol; psychological

C serotonin; social

D serotonin; psychological

18 Research into the hereditary influences on IQ have found that

A people who are unrelated usually have very similar IQ.

B people with higher IQ would usually have a greater chance of having family members with high IQs, too.

C there is no correlation between IQ and a person's genetics.

D children will not have an IQ similar to their parents.

19 A direct sign that an emotional bond has formed between mother and child appears at approximately 8 to 12 months of age when

A the mother and child are happy doing things together.

B the mother can leave the child without the child becoming angry.

C the child enjoys spending time with all caregivers equally.

D the child displays separation anxiety when left alone or with a stranger.

20 A young child looks at a dog and, using her existing schema, she points to it and says 'cat'. This is an example of

A assimilation.

B accommodation.

C mental adaptation.

D goal-directed behaviour.

Section B: Short answer

1 Explain the importance of both hereditary and environmental influences on development. [2 marks]

2 Identify and describe the three types of infant attachment that Ainsworth categorised in the Strange situation experiment. [3 marks]

3 Jean Piaget described three different types of thinking. For each of these (listed below), provide a description using an example to support your answer. Ensure you also refer to the stage of cognitive development where this type of thinking is characteristic.

a Concrete thinking [2 marks]

b Symbolic thinking [2 marks]

c Abstract thinking [2 marks]

[Total = 6 marks]

Defining and supporting psychological development

3

Key knowledge

- the usefulness, and limitations, of psychological criteria to categorise behaviour as typical or atypical, including cultural perspectives, social norms, statistical rarity, personal distress and maladaptive behaviour
- the concepts of normality and neurotypicality, including consideration of emotions, behaviours and cognitions that may be viewed as adaptive or maladaptive for an individual
- normal variations of brain development within society, as illustrated by neurodiversity
- the role of mental health workers, psychologists, psychiatrists and organisations in supporting psychological development and mental wellbeing as well as the diagnosis and management of atypical behaviour, including culturally responsive practices

Key science skills

Analyse, evaluate and communicate scientific ideas

- discuss relevant psychological information, ideas, concepts, theories and models and the connections between them
- critically evaluate and interpret a range of scientific and media texts (including journal articles, mass media communications and opinions in the public domain), processes, claims and conclusions related to psychology by considering the quality of available evidence
- analyse and evaluate psychological issues using relevant ethical principles and concepts, including the influence of social, economic, legal and political factors relevant to the selected issue
- use clear, coherent and concise expression to communicate to specific audiences and for specific purposes in appropriate scientific genres, including scientific reports and posters
- acknowledge sources of information and assistance, and use standard scientific referencing conventions

Source: VCE Psychology Study Design (2023–2027), pp. 24 & 13

3 Defining and supporting psychological development

We are all different. Not only in looks, height, hair colour and skin colour but also in our personalities and what we think. Our brains are all different too. The neurons in our brains are all connected differently; in fact, it would be odd if they were all the same. The important thing is that we accept and embrace the differences.

p. 101

3.1 Typical and atypical behaviour

We are all different and we all behave in different ways. When we talk about behaviour, we talk about typical and atypical behaviour. Typical behaviour is what most of us do, most of the time. Atypical behaviour is the rest. However, you need to be careful using these terms, because many things feed into behaviour. For example, think about cultural differences; what one culture sees as typical, another culture may not.

Alamy Stock Photo/dpa picture alliance

p. 107

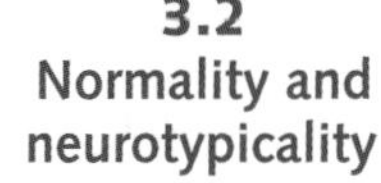

3.2 Normality and neurotypicality

Behaviour is what the neuronal connections within our brain are telling us to do. If our brain is neurotypical then we behave in a way that is typical of most people. If our brain is neurodivergent then we may display behaviours that may be considered atypical. Many very smart and famous people are neurodivergent.

Adobe Stock/baibaz

p. 114

3.3
Variations in brain development

Just like our bodies are different, our brains are different too. Our neurons are all connected differently and that makes us who we are. Depending on the connections formed, we might have a neurodivergent condition such as autism or dyslexia. Neurodiversity is a positive, not a negative.

Adobe Stock/rh2010

p. 118

3.4
Supporting psychological development and mental wellbeing

There is lots of support available to assist people who are having difficulty with their psychological development, atypical behaviour and mental wellbeing. Psychologists, psychiatrists and mental health workers can support people who are struggling with their mental health.

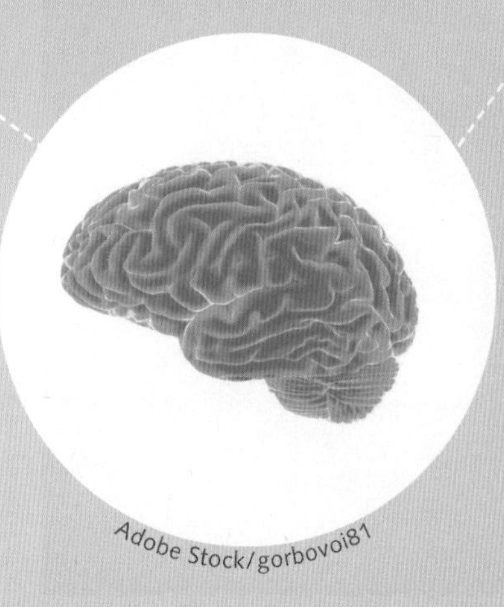

Adobe Stock/gorbovoi81

It takes a lot of people to make the world go round. It would be very dull if we were all the same. Everyone is different and that helps to fill all the niches that make our world a wonderful place.

Slideshow
Chapter 3 slideshow

Slashcards
Chapter 3 flashcards

Test
Chapter 3 pre-test

Assessment
- Pre-test
- End-of-chapter exam

Revision
- Chapter map
- Key term flashcards
- Key concept summary
- Slideshow

Investigation
- Investigation: Investigating the role of mental health professionals
- Data calculator
- Logbook template: Case study

Worksheet
- Comparing cognitive profiles of two individuals

To access these resources, visit **cengage.com.au/nelsonmindtap**

Nelson MindTap

Know your key terms

Adaptive behaviour
Atypical behaviour
Cognition
Cultural norms
Culture
Distress
Maladaptive behaviour
Neurodivergent
Neurotypical
Norm
Normality
Social nonconformity
Statistical rarity
Stress
Stressor
Typical behaviour

Alamy Stock Photo/Kathy deWitt

Figure 3.1 Oliver Sacks, neurosurgeon and author

Oliver Sacks (1933–2015) (Figure 3.1) was a British neurologist famous for publishing some of his case studies. One case study was of Dr P., who was not able to recognise objects or faces and saw faces when there was nobody there. Dr P. reported seeing a hat when he was looking at his wife. Other observations that Sacks recorded include Dr P. patting the heads of parking meters thinking

Shutterstock.com/Refluo

Figure 3.2 One person Sacks wrote about reported seeing the face of a black dragon when she looked at people or her computer screen.

»

they were children and talking to the knobs on the furniture, expecting the knobs to reply.

Another case presented to Sacks was a 52-year-old Dutch woman who had ongoing hallucinations. People's faces, walls and even her computer screen would turn into the face of a dragon. The dragon was black, with long pointy ears and a long nose (Figure 3.2). The eyes were described as large and were either yellow, green, blue or red. Tests, including an MRI scan, could not identify the cause of her hallucinations.

These two people are clearly different from most people in some ways. But how different do we have to be before we are recognised as needing medical help or psychological support? This chapter explores how we answer this question.

3.1 Typical and atypical behaviour

When something is considered normal, terms such as 'typical' or 'acceptable' are also commonly used to describe them. These terms can be ambiguous and can be based on personal opinion (that is, they are *subjective*), and require some level of value judgement. They can also have cultural and social variations, because what might be acceptable in one culture or society may not be acceptable or typical in another. However, most psychologists define **typical behaviour** as behaviour that represents the behaviour of most people in a population. **Atypical behaviour** refers to behaviours that are not displayed by the majority of people in a population. Atypical behaviours are behaviours that deviate from the 'norm' or what is deemed normal in a social context. The behaviour is judged by comparing it to others' behaviour within the same social context.

Typical behaviour can also specifically refer to an individual. Does a certain individual typically (normally) behave in a particular manner? We all have our own quirks and individual qualities, so judging whether a behaviour is typical or not (atypical) for that individual may alter the perception of the behaviour. If a person usually wears pyjamas to school, then wearing pyjamas would be normal behaviour for that person, but not typical behaviour for most people within society. Another example might be if your best friend, who would normally give you a hug or a handshake when you met them, started to avoid you and merely grunted 'hello' when they saw you. You might think this was atypical behaviour for them. However, this behaviour may be typical of your little brother, who is grumpy in the morning and hates you hugging him. Table 3.1 compares typical and atypical behaviour.

Generally, psychologists agree that **normality** refers to patterns of behaviour or personality traits that are typical, or that conform to some standard of acceptable ways of behaving. By this definition, wearing pyjamas to school even on a casual dress day may be considered abnormal, as it is not typical behaviour compared to most people. Likewise, behaviour that is considered unacceptable, such as smoking in a car with children, can also be considered abnormal.

Abnormal or atypical behaviours do not necessarily pose a problem to the individual or to society. There are many behaviours that could be defined as 'abnormal' or 'atypical' that are positively valued by society, such as being abnormally talented at playing a musical instrument or at playing golf. To determine whether a behaviour is harmful or has the potential to negatively impact on the individual or society, the terms 'adaptive' and 'maladaptive' are used.

3.1.1 WHAT IS TYPICAL BEHAVIOUR?

Table 3.1 Comparing typical and atypical behaviour

	Typical behaviour	Atypical behaviour
Definition	The behaviour of the majority of people in a population or behaviour that is consistent with the way an individual normally behaves	Behaviour that is not like the behaviour of the majority of people in a population or is inconsistent with the way an individual normally behaves
Example	Waving your arms in the air and singing loudly at a music festival	Waving your arms in the air and singing loudly in the supermarket

3.1.2 TYPICAL OR ATYPICAL?

What is considered typical or atypical behaviour can vary:

» between cultures and subcultures
» over time
» depending on the situation or context.

Psychological criteria that are used to categorise behaviour as typical or atypical include cultural background, social norms, statistical rarity, stress levels and maladaptive behaviour. These criteria are discussed below.

Cultural perspectives

Culture is one of the most influential contexts in which any behaviour is judged. **Culture** can be thought of as the distinctive beliefs, values, customs, knowledge, art and language that form the foundation of the everyday behaviours and practices of a society. Culture is passed from one generation to the next by spoken and written means. There is a fine line between what is classified as normal and abnormal behaviour in some cultural contexts because the expectations and standards that are considered normal can vary widely across cultures. For example, in some cultures it is considered normal to talk loudly and wave your arms in a spirited manner during a conversation. In some cultures, it is normal for women to remain housebound much of the time and not have paid work outside of the home. However, in modern-day Western cultures a woman with this behaviour might be diagnosed as suffering from agoraphobia or be considered significantly disempowered. In Australia it is illegal and considered abnormal to be married to more than one person at the same time, but in some Middle Eastern countries this is considered a tradition and seen as normal and acceptable behaviour.

Social norms: learning acceptable behaviour

A **norm** is a socially defined rule, standard or value that describes behaviours that are expected within groups. Behaviour can be considered typical according to whether the behaviour fits in with the norms of that society. For example, looking someone in the eye when having a discussion is considered normal in Australia. It indicates to the speaker that you are listening. However, for some Asian cultures eye contact is used to intimidate someone (Figure 3.3). To look at an elderly person is also a mark of disobedience in Chinese culture because it is considered to show a lack of respect. In one cross-cultural study on the perception of eye contact, Japanese participants made less eye contact than Finnish or American participants. In Japan, eye contact is a sign of disrespect and long eye contact is considered a challenge (Uono & Hietanen, 2015).

Shutterstock.com/Jennifer Lam

Figure 3.3 Eye contact during greetings is a social norm that varies in different cultures.

Socialisation is the process of learning to behave in a way that is acceptable to a society. It introduces us to **cultural norms** from early childhood. Socialisation is a major contributor to whether a person behaves in a typical or conforming way to social norms or whether they behave in an atypical or nonconforming way to these norms.

Impaired social learning can lead to social nonconformity and maladaptive, self-destructive behaviour (such as drug abuse) and emotional instability. **Social nonconformity** refers to disobeying the standards set by society for what is acceptable conduct. However, we must be careful to distinguish between unhealthy nonconformity and healthy creativity and uniqueness. People who dress or act differently from the majority are not necessarily nonconformist in an unhealthy way, nor is their behaviour necessarily unacceptable or maladaptive – unique individuals are generally emotionally stable and function normally in society (Figure 3.4).

Alamy Stock Photo/eddie linssen

Figure 3.4 Social nonconformity does not automatically indicate abnormality.

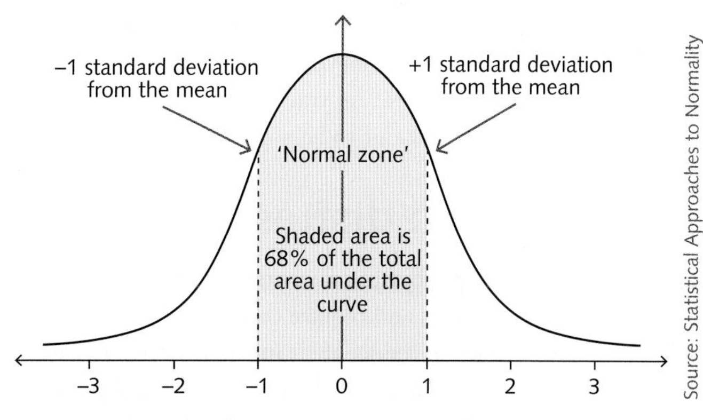

Source: Statistical Approaches to Normality by Graham Williamson, SLT Info, 1 March 2016, https://www.sltinfo.com/statistical-approaches-to-normality/

Figure 3.5 Statistical rarity may help to define what is normal.

Statistical rarity

If a person is atypical in some way and this leads to behaviour that is considered abnormal, they are considered to be a **statistical rarity**. For instance, if you have an extremely low or high IQ, you are deemed a statistical rarity.

Psychologists can classify a person's behaviour by comparing it with the behaviour of others. If the statistical majority display the behaviour, it is then considered normal. For example, when completing an IQ test, the average score is 100. If an individual obtains a score of 109 on an IQ test, this individual would be in the statistical majority and would therefore be considered as having a normal IQ.

Usually, the results of such tests will form a 'normal curve', which is shaped like a bell. A normal curve has the majority (68 per cent) of scores in the middle, tapering to very few extremely high and low scores (Figure 3.5). A person who deviates from the average might be considered abnormal because they are statistically rare.

One limitation of using statistics to determine normality is that statistics do not tell the difference between abnormal behaviour that is desirable and that which is not desirable. For example, an IQ test has an average score of 100. It is statistically abnormal for a person to score above 145 or below 55. In our society, it is often considered desirable to score at the high end of a normal IQ curve (Figure 3.6), and a score below 55 is considered undesirable; although both scores are actually judged to be abnormal, only one is considered undesirable.

Another limitation of using statistics to determine normality is that they may have trouble showing the difference between normality and abnormality. For example, we could measure the average (normal) amount of food eaten per day for people of a particular age and sex. Compared to this average, a person who eats four packets of biscuits, two loaves of bread and half a chicken in two hours clearly has abnormal eating, because this would be very much outside the norm. But as we look at behaviour that is closer to the norm, we face the statistical problem of where to draw the line that determines what is normal and abnormal. How far must people deviate to be considered abnormal? If it is deemed normal to eat two serves of fruit and five serves of vegetables a day, is a person abnormal if they eat three serves of fruit and six of vegetables?

Alamy Stock Photo/Everett Collection Inc

Figure 3.6 Some people estimate that Elon Musk, founder of Tesla and SpaceX, has an IQ around 155. Statistically, this is classified as abnormal.

3.1.3 USING STATISTICAL RARITY TO DETERMINE WHAT IS ATYPICAL

Or is the behaviour not deemed abnormal unless they eat five or six serves of fruit and eight or nine of vegetables? As you can see, statistical boundary lines tend to be somewhat arbitrary.

Personal distress

Stressors are objects or events that cause a feeling of stress and require coping mechanisms or strategies to reduce their impact. Stressors can be experienced internally, such as physical illness, or externally, such as losing employment. **Stress**, the physiological response to feeling threatened or challenged, is thought to be experienced when the demands on an individual exceed the necessary resources present to deal with a stressor, even if that stressor is not life-threatening or if it poses only a perceived (not an actual) threat. Examples of stress include sweating or blushing. **Distress** is a negative psychological response to a stressor. It is experienced when a person feels overwhelmed by the perceived demands of a situation, loss or threat. Distress can erode a person's sense of wellbeing and lead them to feel anxious for long periods of time. Distress can result from a person changing a healthy behaviour to an unhealthy behaviour. If this happens, the resultant behaviour may be considered abnormal because it is causing the individual distress.

The level of personal distress experienced can be linked to the number of stressors in a person's life (Figure 3.7). Delivering a talk at an assembly, asking a person out on a date and sitting an examination are all day-to-day stressors that are not life-threatening, but they can still cause a level of distress. If the person is feeling nervous or worried, but is still able to attempt these tasks, the level of distress and their behaviour would be considered normal. If, however, their level of distress caused them to be unable to deliver a talk at the assembly, this indicates that their level of distress and their behavioural responses are abnormal.

The many psychological effects associated with distress include increased levels of frustration, moodiness, anxiety, confusion and difficulty concentrating. A feeling of being unable to cope with normal activities (like getting to school on time or getting a proper night's sleep) or having a negative feeling about oneself are also signs of high levels of personal distress.

Maladaptive behaviours

An individual's ability to interact with others and adjust their behaviour to meet the demands of everyday living according to cultural and social expectations is a measure of their **adaptive behaviour**. They need to be able to complete everyday tasks such as attending to their personal hygiene, going to work and eating food. If they stay in bed all day even though they are not sick, do not regularly wash or eat, or are unable to hold a job, they would not be meeting their everyday needs and their behaviour would be considered maladaptive and abnormal.

Maladaptive behaviour is behaviour that interferes with an individual's ability to complete daily tasks and to function in their particular society. It can lead to emotional, social and health problems. Maladaptive behaviours can be learned early in life or they can develop after a major illness, traumatic event or life change.

Maladaptive behaviours cover a wide spectrum of behaviours. They can be minor behaviours that may only slightly impair an individual's overall functioning. For example, to alleviate the stress you feel because you are running out of time to complete an assignment you play video games rather than continue to work on the assignment. Maladaptive behaviours can also be more serious behaviours that interfere with an individual's practical ability to meet the demands of everyday living; for example, engaging in binge eating or excessive exercise as a way to alleviate stress. These maladaptive behaviours may temporarily relieve stress but they do not address the basic thoughts, fears and concerns associated with the stressor, and over time may result in additional stress and dysfunction.

Some behaviours that are commonly accepted by society, but not necessarily encouraged, are

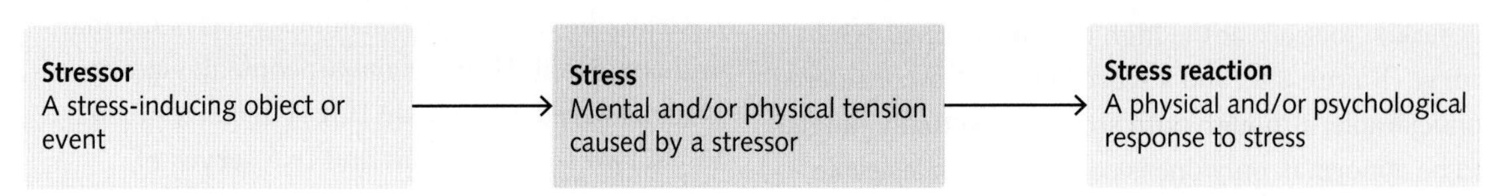

Figure 3.7 The relationship between a stressor, stress and a stress reaction

considered maladaptive once they begin to detract from the person's ability to undertake normal daily activities. For example, drinking alcohol in moderation is not considered maladaptive. If the person drinks alcohol excessively and they are unable to carry out their normal daily activities, it would be considered maladaptive. If someone refuses a social invitation in order to look after a sick relative, this would be considered to be normal, but if this is a strategy to avoid social interaction, it would be considered maladaptive.

The concept of whether a behaviour is maladaptive or not is a very important component in identifying problems such as drug abuse, mental illness and personality disorders. Once a behaviour persistently interferes with a person's ability to undertake normal daily activities, it is considered maladaptive. For example, if a person spends hours on social media every night and as a result they are unable to arrive at work on time, then the online activity is considered maladaptive.

3.1.4 ADAPTIVE OR MALADAPTIVE

Table 3.2 provides a summary of the psychological criteria that are used to categorise behaviour. As normality is very difficult to identify, using only one criterion could easily result in misunderstanding a person's behaviour.

Table 3.2 Summary of the criteria used to categorise behaviour

Criteria	Definition of criterion	Usefulness	Limitations
Cultural perspectives	Functioning or coping adequately in a particular society	Many behaviours can be explained when the cultural context is considered.	Different cultures often have different ideas about acceptable behaviour. Normality may not be consistent in every culture.
Social norms	Obeying popular or common standards in a particular social situation	There are some universal rules; e.g. all cultures consider it normal to be able to communicate with others and behave in a predictable fashion.	People defying social norms may be labelled as abnormal. The context (situation) of the behaviour may not be considered.
Statistical rarity	Fitting in with the most commonly occurring characteristics in a society	Commonly occurring characteristics or behaviours are usually seen by society as acceptable and positive.	Some extreme characteristics are desirable, but also classified as abnormal; is the average always the most desirable?
Personal distress	Experiencing a level of distress that doesn't prevent usual activities	It allows the person to evaluate how they are feeling without comparison to others.	It is subjectively measured; a complete lack of the behaviours that may cause distress results in boredom – a low level of distress is normal.
Maladaptive behaviour	Behaviours that stop a person from developing or functioning in their day-to-day life and limit their ability to adjust to changing circumstances	It can assist in the identification of problems such as substance abuse, mental illness and other psychopathologies.	It can be confused with bad or wrong behaviour.

ACTIVITY 3.1 CATEGORISING TYPICAL AND ATYPICAL BEHAVIOUR

Try it

1 Refer to Table 3.2. Copy and complete a table like the one below to provide an example of each of the psychological criteria that is used to categorise typical and atypical behaviour.

Psychological criteria	Example of this approach
Cultural perspectives	
Social norms	
Statistical rarity	
Personal distress	
Maladaptive behaviour	

Apply it

2 Anthony has been referred to a psychologist by his doctor. Below are the notes from the doctor who made the referral.

Anthony is a 25-year-old male of extremely high intelligence. Anthony works as an engineer but has struggled with work-related stress recently. Anthony has taken a number of days off and has reported that he struggles to get out of bed most days.

»

Recently, Anthony has reported that he tends to lash out at loved ones when they try to help him. Anthony has avoided seeing friends and has reported that he drinks alcohol almost every night.

Identify examples of statistical rarity, personal distress and maladaptive behaviours that are present in Anthony's report.

Exam ready

3 The condition obsessive-compulsive disorder (OCD) is characterised by intrusive thoughts (obsessions) and ritualistic behaviours (compulsions). This combination of thoughts and behaviours can interfere with a person's daily functioning, leading to

A cultural perspectives.

B statistical rarity.

C personal distress.

D social norms.

KEY CONCEPTS 3.1

- Typical or normal behaviour is behaviour seen in the majority of people in the population.
- Atypical or abnormal behaviour is different from the behaviour of the majority of the people.
- Psychological criteria can be used to determine whether a behaviour is normal or not.
- Criteria used to determine normality include the culture involved, social norms, statistical rarity, level of personal distress and maladaptive behaviour.
- Each criterion has limitations and should therefore not be used in isolation.

Concept questions 3.1

Remembering

1 What is normal behaviour? **r**

2 How are typical behaviour and atypical behaviour different? **r**

3 List three factors that can influence our perception of typical and atypical behaviour. **r**

Understanding

4 How does the concept of normality relate to typical behaviour? **e**

5 How is abnormal behaviour different from maladaptive behaviour? **e**

6 Provide one example of normal behaviour and one example of maladaptive behaviour. **c**

Applying

7 Give two examples of social norms that apply in Australia. For each example, explain how visitors to Australia could get confused by these norms. **c**

HOT Challenge

8 Consider the behaviours in the table below. Decide whether each behaviour is typical or atypical, considering the circumstances described. Copy and complete the table by placing a tick in the relevant box to show your choice. For each choice, provide a reason for your decision using the criteria in Table 3.2. **c**

Behaviour	Typical	Atypical	Reason for your decision
Being chased by bulls down Swanston Street in Melbourne			
Being chased by bulls down the main road of Pamplona, Spain, on 7–14 July			

Behaviour	Typical	Atypical	Reason for your decision
Not speaking to your work colleagues for a week even though you mix with them in the office every workday			
Not speaking to your work colleagues for a week because you are a Buddhist monk who has taken a vow of silence			
Being able to run 100 metres in under 10 seconds			
Being able to run 100 metres in over 5 minutes			
Singing in your bedroom for 3 hours every night to help relieve your stress			
Singing for 3 hours every night as a cabaret singer in a nightclub			

3.2 Normality and neurotypicality

What is normal bedtime for a teenager? Do I look normal when I go to out on the weekend and socialise with my friends? Normality (and, conversely, abnormality) is a generally accepted concept of the features of typical human behavioural, cognitive and socio-emotional development and functioning, influenced by cultural contexts. Popular conceptions of normality are derived from looking at what the majority of people do or typical behaviour. In psychology, normality has a more specific definition and relates to the capacity to cope with ordinary demands of life or to think and act in an effective way to achieve goals and meet needs (VandenBos, 2007). It is also a statistical term reflecting a normal distribution of values in a data set. As discussed in Section 3.1, normality is influenced by culture and social context. For example, in Western culture, it is considered normal that children learn to sleep in their own beds from a young age. In other cultures, such as Japanese and Korean, children sleep with their mothers at least to early childhood, if not through to their early teenage years (Berk, 2015) and this is considered normal.

Neurotypical is a specific term for normality that describes individuals who think, behave and feel in ways that are typical of most people. Neurotypical brain development refers to the expected development of brain structures and neurological connections over time. The term neurotypical is contrasted with the term **neurodivergent**, which refers to individuals who think, behave or feel differently from most other people. Neurodivergent individuals (Figure 3.8)

Alamy Stock Photo/Everett Collection Inc

Alamy Stock Photo/WENN Rights Ltd

Alamy Stock Photo/ZUMA Press, Inc.

Figure 3.8 Richard Branson (founder of Virgin), Daryl Hannah (actor and environmental activist) and Hannah Gadsby (comedian) are neurodivergent individuals.

include people with specific neurodevelopmental diagnoses such as autism, attention-deficit hyperactivity disorder (ADHD), dyslexia and Tourette syndrome. There is emerging evidence that people with neurodevelopmental diagnoses have differences in brain development, including in the brain's internal structures and neurological connections, but we are only just beginning to understand this through research.

'Neurodivergent' and 'neurotypical' are descriptive terms and are not formal diagnoses. We will investigate these terms with respect to emotions, behaviours and cognitions to understand potential differences. However, individuals who are considered neurotypical and those considered neurodivergent can have *adaptive* and *maladaptive* emotions, behaviours and cognitions.

Emotions

There is no such thing as a good or bad emotion, although some emotions feel more pleasant than others to experience. All emotions are designed to communicate a message. Most people are usually able to express their emotions using words and body language to communicate how they are feeling, as well as to recognise and understand the emotional states of others. This is considered emotional competence (Saarni, 2000).

Emotional competence

'Emotional competence' is a term used in psychology to refer to an individual's capability in identifying, labelling and managing their emotions as well as understanding the emotional states of others. Good emotional competence reflects typical or normal emotional expression and understanding. Being able to understand your own emotions and those of others is an important skill for social competence and getting along with others. It allows you to take cues from others' emotional reactions to interpret a situation. For example, if you see someone with a down-turned mouth and tears running out of their eyes, you understand that the person is sad.

Our emotional competence develops from early childhood, when there is attachment to a significant carer who provides an external reference for understanding how we feel and teaches us how to regulate emotions. If early attachment relationships are unhealthy, some individuals may not develop strong emotional competence. Emotional development extends through childhood to adolescence, by which time we are able to determine our own emotional state and develop awareness of more complex emotions, such as shame.

Emotional regulation

Recognition and expression of our own emotions is a precursor to understanding the feelings and emotions of others. Regulated and adaptive emotional expression allows us to communicate our internal states to meet our needs and engage in effective social interaction. Excessive and inappropriate expression of emotion originates from a diminished capacity to self-regulate, is maladaptive and interferes with an individual's capacity to function in their daily life.

For example, a person who can only communicate their anger through yelling and shouting will face ongoing issues in relationships, at school and in their employment. Some people with neurodivergent conditions such as ADHD may find it requires effort to express emotions in an adaptive way and they may need to be taught self-regulation strategies when they are young to develop this skill. Research suggests that people diagnosed with ADHD have difficulty inhibiting their behaviour. Behavioural inhibition is primarily driven by the front part of the brain, which controls our executive functions. **Executive functions** are those processes that allow us to make a plan and complete that plan. They direct, filter, control and monitor our thoughts and actions and are an essential component of our ability to self-regulate. It is now thought that ADHD is more a condition of disinhibition than attention (Barkley, 2006).

Interventions and therapies to help children with ADHD develop emotion regulation skills include working with parents and teachers to provide effective response strategies such as emotion coaching. Psychologists will also help the child to recognise and understand their emotions by identifying physical feelings and thoughts when they experience an emotion, and then develop strategies for how the child can respond more effectively.

Emotions are maladaptive when they persist beyond when they are needed in a particular context. For example, anxiety is adaptive when it signals to us that we need to perform, such as in an exam or a musical performance. It is accompanied by changes in our physiological and psychological arousal levels that help us to perform in an optimal way. These changes include increased heart rate and mental clarity. If our anxiety increases too much, we can head into 'fight-or-flight-or-freeze' mode, which can have a negative effect on our performance.

Emotion and neurodivergence

Some neurodivergent people have difficulty understanding and expressing emotions, as well as recognising and understanding emotional states in others. For example, people diagnosed with autism experience the full range of human emotions, but they may have difficulty understanding, identifying and expressing their emotions. This is referred to as 'alexithymia' or 'emotional blindness'. It often leads other people to think that people with autism are lacking in emotion, but this is not the case. Similarly, people with autism may find it difficult to understand emotions in others, resulting in them misreading social cues.

Neurodivergent people may struggle to connect an emotion with the words being used. They may need to rely on other cues including the tone of voice and facial expression (Figure 3.9) to decipher the emotion, but non-verbal cues can also be missed. Supporting people with autism to develop skills in emotion recognition is an effective component of therapy.

3.2.1 MEASURING EMOTIONS

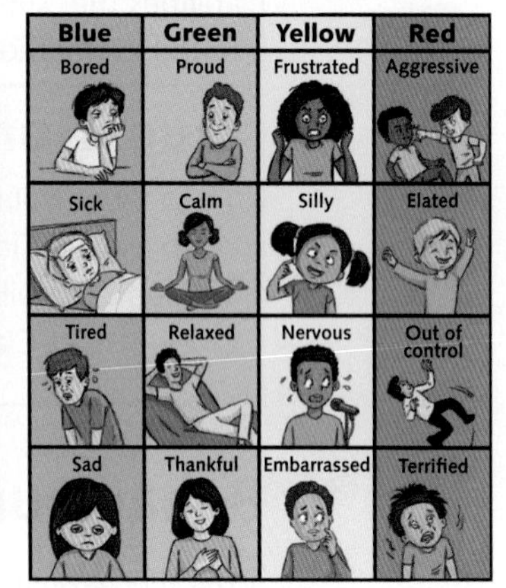

Adapted with permission of Kelsey Feyes, Neurodiverse classroom, https://www.teacherspayteachers.com/Store/The-Neurodiverse-Classroom

Figure 3.9 Emotions bingo is a resource used with primary school students to help teach them how to regulate their emotions through understanding similar emotions.

CASE STUDY 3.1

My brain is different, not broken

Katie Stavick

At the graduation of the class of 2022 at Mat-Su Central [a school in Alaska] it was student speaker Claire McDaniel who brought down the house and moved many with her speech. Speaking honestly about her struggles to find the best way to learn, McDaniel was frank in saying, 'homeschool saved my education.'

She told the audience of having attended a prestigious private school while living in Japan because public education simply wasn't an option. When she moved to the Mat-Su Borough, she said her parents knew that Mat-Su Central would be her best option to help with her neurodivergent brain.

'My brain is wired differently. I misinterpreted the most basic instructions, sure teachers tried to help … but only confused me further. Sometimes I needed extra help but the teachers couldn't because they were already overwhelmed balancing the needs of 30 other kids,' McDaniel told the crowd. 'Sometimes I didn't need extra help, all I needed was extra time … time that was hard because traditional schooling has such a tight schedule. All of the factors tied together mean I only got a "C" and I got further and further behind. I had these experiences at a prestigious 4 year private school.'

McDaniel spoke about how her homeschool journey was the best fit since she could study at her own pace and stay on subjects as long as she needed. She did use a tutor to give extra help and support as well.

It was with the help of her parents and tutor that McDaniel was able to find a way to learn and through homeschooling came to a self – realisation that was at the heart of her

9780170465052

speech – 'I don't have a disability. My brain is different, not broken,' a statement that was met with boisterous cheers and applause to the entire audience, many of whom can relate to McDaniel's struggles with learning.

'I was a student who, at best, was managing a "C" to graduating today with the highest honors.'

She praised homeschooling for giving her the time and opportunity to explore creative abilities that she may not have otherwise found in traditional schools. McDaniel also urged those not graduating to seek out non-traditional ways to learn and find new paths they otherwise may not know.

Adapted with permission from "My brain is different, not broken': One Mat-Su Central student speaks about her journey during graduation', by Katie Stavick, Frontiersman.com May 21, 2022, https://www.frontiersman.com/news/my-brain-is-different-not-broken-one-mat-su-central-student-speaks-about-her-journey/article_2a50d0f0-d89d-11ec-9dfb-df4875208ad3.html

Questions

1. Draw on your knowledge of neurodivergence to explain Claire McDaniel's comment that 'homeschooling saved my education'.
2. How does Claire describe her brain?
3. How could Claire's experiences assist other neurodivergent students to succeed in the school system?

Behavioural differences

3.2.2 PERSONAL STORIES

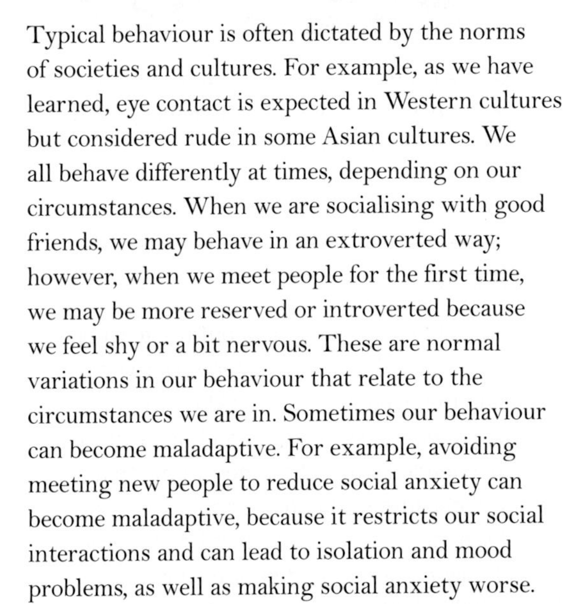

Typical behaviour is often dictated by the norms of societies and cultures. For example, as we have learned, eye contact is expected in Western cultures but considered rude in some Asian cultures. We all behave differently at times, depending on our circumstances. When we are socialising with good friends, we may behave in an extroverted way; however, when we meet people for the first time, we may be more reserved or introverted because we feel shy or a bit nervous. These are normal variations in our behaviour that relate to the circumstances we are in. Sometimes our behaviour can become maladaptive. For example, avoiding meeting new people to reduce social anxiety can become maladaptive, because it restricts our social interactions and can lead to isolation and mood problems, as well as making social anxiety worse.

Neurodivergent people may display behaviours that society considers atypical. This does not mean they are maladaptive. Research and new knowledge suggest that behaviours we see in neurodivergent people can be adaptive for them even if they seem different from normality.

A good example of this is 'stimming' in people with autism. Stimming, or self-stimulatory behaviour, is used by autistic people to help calm themselves or relieve intense feelings of anxiety. If the stimming activity is appropriate and not disruptive, such as bouncing a leg, then using it as a coping tool is adaptive. Some stimming behaviours can be more harmful, disruptive or cause social exclusion, such as head banging or scratching, and the person will require support to engage in less harmful and safer behaviours.

Atypical and maladaptive behaviours are often assessed as part of diagnosing neurodevelopmental conditions such as ADHD and autism. For example, when diagnosing ADHD, clinicians will consider some of the following behaviours as indicators of ADHD if they are causing problems or dysfunction in a person's life and would therefore be considered maladaptive:

- moving or fidgeting so much that it disrupts a person's ability to learn, interact with others or do a job
- calling out, interrupting or not waiting for their turn, which may limit social interactions and the development of friendships
- inappropriate climbing or running around in a situation such as in class.

Cognitions

Cognitions are thought processes used to interpret experience and respond adaptively, such as perceiving, conceptualising, remembering, reasoning, imagining, judging and problem-solving. It is difficult to define what typical or normal cognition is; however, psychologists are able to identify abnormal or maladaptive cognitions. Maladaptive thinking becomes a

Table 3.3 Thinking traps

Trap	Description	Example
Emotional reasoning	Assuming your emotions are the source of truth	I feel bad so I must have done something wrong.
Mind-reading	Assuming you know what other people might be thinking about a situation or you	They all think I'm stupid because I couldn't answer that question.
Labelling	Placing negative labels on yourself	I am stupid because I couldn't answer that question.
Filtering	Filtering out the positive in your thinking and only focusing on the negative	Today was a bad day; nothing good happened.
Over-generalising	Applying negative thoughts to all situations or to yourself as a person	I am always doing silly things.
Should statements	Putting undue pressure on how you should be behaving or thinking and not recognising that you might be going through a difficult time	I should be better than this.
Catastrophising	Assuming something bad will happen in the future	If I don't pass this exam, I won't get into university and my whole career path will be over.
Black and white thinking	Seeking perfect	Unless I do this perfectly, I will be a failure.

problem when negative and distorted thoughts dominate an individual's feelings and behaviour, and is associated with mental health issues such as anxiety and depression. Both neurotypical and neurodivergent people can experience maladaptive cognitions. For example, people may experience cognitive distortions that psychologists call 'thinking traps'. Thinking traps are maladaptive cognitions that are automatic and not obvious to the person experiencing them. Psychologists can help people to recognise their thinking traps and provide strategies to change them. Table 3.3 shows some common thinking traps and examples.

Neurodivergent individuals process information in a different way from neurotypical individuals. For example, people with autism can show:

» ability to be creative and think outside the box
» ability to respond to, and learn, music
» very high attention to detail
» independent thinking (they will not conform as a result of peer pressure)
» hyper- or hypo-sensitivity to extremes in temperature
» resistance to adapting and accepting change
» reduced tolerance for overstimulation from noise or crowds.

The neurological connections and the size of some brain structures in a neurodivergent person's brain can differ from those in the brain of a neurotypical person. This may explain why a neurodivergent person's cognitions and thoughts can show differences from those of neurotypical people. We know this through research that uses brain imaging and assesses cognitive functions, and through neurodivergent people talking about their cognitions. For example, when tested using standardised assessments, people with autism often demonstrate cognitive strength in attention and memory for detail (Baron-Cohen, 2006).

Worksheet
Comparing cognitive profiles of two individuals

Temple Grandin, a prominent researcher and advocate for the autistic community, shares her lived experience of autism. In Grandin's book *Thinking in Pictures* (Grandin, 2006), she says that her dominant processing mode to understand information and the non-autistic world is to create visual symbols and mind maps in her brain. However, Grandin also states that there is large variability in cognition between autistic people. As Dr Stephen Shore, a person with autism who is Professor of Special Education at Adelphi University, says: 'If you've met one person with autism, you've met one person with autism'. Dr Shore sees autism as a 'non-standard way of perceiving, processing and interacting with the world' (Flannery & Wisner-Carlson, 2020).

Many neurodivergent people have made outstanding contributions to our collective knowledge because of their different cognitive style or way of seeing the world, thinking and solving problems. These include evolutionary biologist Charles Darwin, physicist Albert Einstein, environmental campaigner Greta Thunberg and singer Billie Eilish.

ANALYSING RESEARCH 3.1

Neurodivergence is a positive, not a negative

Studies have found that people diagnosed with autism spectrum disorder (ASD) have strengths related to working with computer languages, mathematical systems and machines, and are better than control subjects at identifying tiny details in complex patterns (Baron-Cohen et al., 2009). Some technology companies have been aggressively recruiting people with ASD for occupations such as writing computer manuals, managing databases and searching for bugs in computer code (Wang, 2014).

People with dyslexia have been shown to possess global visual-spatial abilities, including the capacity to identify 'impossible objects' (von Károlyi et al., 2003), process low-definition or blurred visual scenes (Schneps et al., 2012) and perceive peripheral or diffused visual information more quickly and efficiently than participants without dyslexia (Geiger et al., 2008). Such visual-spatial gifts may be advantageous in jobs requiring three-dimensional thinking such as astrophysics, molecular biology, genetics, engineering and computer graphics (Paul, 2012 and Charlton, 2012).

In fact, neurodiversity may have been crucial to the survival of the human race. Such strengths suggest an evolutionary explanation for why these disorders are still in the gene pool. They may have conferred specific evolutionary advantages in the past as well as in the present (Brüne et al., 2012). The systemising abilities of individuals with autism spectrum disorder might have been highly adaptive for the survival of prehistoric humans. As one autism activist put it 'Some guy with high-functioning Asperger's developed the first stone spear; it wasn't developed by the social ones yakking around the campfire' (Solomon, 2008).

Similarly, the three-dimensional thinking seen in some people with dyslexia may have been highly adaptive in preliterate cultures for designing tools, plotting out hunting routes, and constructing shelters, and would not have been regarded as a barrier to learning (Ehardt, 2009). The key symptoms of ADHD, including hyperactivity, distractibility, and impulsivity, would have been adaptive traits in hunting and gathering societies in which people who were peripatetic in their search for food, quick in their response to environmental stimuli, and deft in moving toward or away from potential prey would have thrived (Jensen et al., 1997).

Source: 'The Myth of the Normal Brain: Embracing Neurodiversity' by Thomas Armstrong, PhD, AMA J Ethics. 2015;17(4):348–352. doi: 10.1001/journalofethics. 2015.17.4.msoc1-1504. Adapted with permission from Dr Thomas Armstrong

Questions

1 What methodology was used in this research?
2 Provide one example where neurodiverse skills could be an employment advantage.
3 Why has neurodiversity persisted in the gene pool?
4 Consider the statement 'Some guy with high-functioning Asperger's developed the first stone spear; it wasn't developed by the social ones yakking around the campfire'. Would the human race be as advanced as it is today without neurodiversity? Explain your answer.

ACTIVITY 3.2 NORMALITY AND NEUROTYPICALITY

Try it

1 'There is no such thing as a normal brain.' Brainstorm as many points as you can to support, and to challenge, this statement, and share them with the class.

Apply it

2 Below is an extract from a Year 11 student's report:

The student has shown very good application this semester. The student arrives on time,

»

with all required materials. The student's academic results are consistently excellent. The student is a valued member of the class. They contribute to discussions and support their peers when they find the work difficult. When stressed, this student takes breaks as needed and asks for support.

Based on the school report, would you classify the student as neurotypical or atypical? Ensure you refer to examples of emotional, behavioural and cognitive responses from the report in your answer.

Exam ready

3 Which one of the following statements would indicate neurotypical social functioning?

A The person speaks and moves in expected ways.

B The person is able to function well as a member of a group (for example, sports team, club or class group).

C The person feels uncomfortable in large crowds.

D The person learns new information at the same speed as people their age.

KEY CONCEPTS 3.2

- Neurotypicality describes individuals who think, behave and feel in ways that are typical of most people.
- Neurotypical brain development refers to the expected development of brain structures and neurological connections over time.
- Neurodivergent people may:
  - show differences in the way they process emotion
  - not hide their emotions and may not understand when neurotypical individuals do
  - avoid eye contact and physical contact
  - struggle with change and overstimulating environments.
- Cognitively, neurodivergent people can be very creative and have a high attention to detail.
- Neurodivergent people may learn, think and process information differently.

Concept questions 3.2

Remembering

1 Define neurotypicality. **r**

2 List three behaviours that may be observed in neurodivergent individuals. **r**

3 Describe two ways that neurodivergent individuals may excel cognitively. **c**

Understanding

4 Why are people diagnosed with autism sometimes considered to be lacking in emotion? **e**

5 Copy and complete a table similar to the one below to compare the differences between neurotypical and neurodivergent people in terms of emotion, behaviour and cognition. **e**

	Neurotypical	Neurodivergent
Emotion		
Behaviour		
Cognition		

Applying

6 Describe a task that a neurodivergent individual may do much better than a neurotypical individual. **e**

HOT Challenge

7 A new girl starts at your school and she is in your Year 11 Psychology class. You notice that she is very smart but a bit of a loner. She does not make eye contact with other members of the class and keeps her head down, looking at her books. She looks very uncomfortable when the class gets noisy at the end of the lesson. Describe two things that you could do to help this girl to settle into her new environment. **c**

3.3 Variations in brain development

Have you ever stood in an art gallery with a group of people and stared at an artwork on the wall? Are you and all the other group members seeing the same things in the artwork (Figure 3.10)? While you might be looking at the subject of the painting, some people may be focusing on the colours, medium, texture or tone. We readily assume that people see things, think and process information in the same way that we do. But our brains all develop differently and this means that another person will often not think, see and process material the way that you do.

Alamy Stock Photo/Shiiko Alexander

Figure 3.10 What do you see in this painting?

Normal variations of brain development

There is a typical path for brain development from birth through to adulthood. Several key phases of brain development have been discovered through research on animals and using advanced imaging techniques with humans. We have good knowledge about the processes of brain development across early childhood through to adolescence and beyond (Figure 3.11).

Furthermore, we know that brain development is influenced by our genes and our environment, including the introduction of cultural tools. For example, people born in this century have grown up with smart phones and other digital technologies, the cultural tools of their time. We are only beginning to understand the impact of this experience on brain development and the associated cognitive functions. For example, early studies have shown that teenagers who regularly multi-task with media (use multiple technologies simultaneously) show poorer performance on tests of academic achievement and short-term memory functions (Cain et al., 2016).

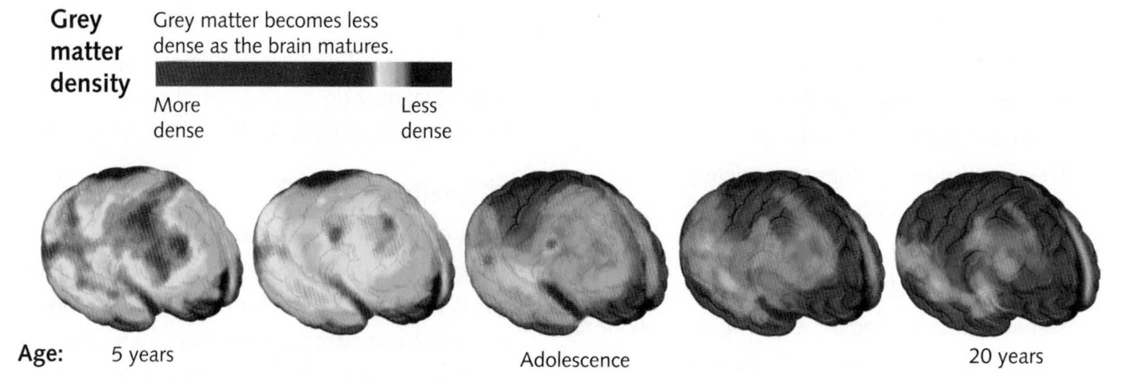

Source: Adapted from Gogtay et al. (2004, May 17). Dynamic mapping of human cortical development during childhood through early adulthood, Proceedings of the National Academy of Sciences, 101(21), 8174–8179 https://doi.org/10.1073/pnas.0402680101

FIGURE 3.11 It was once thought the brain was formed in early childhood. However, when scientists took MRI scans of children's brains as they grew from the age of 5 years to 20, they discovered large changes occurred during adolescence (Gogtay et al., 2004). Grey matter that is excess is pruned out as the brain becomes more efficient. The frontal regions that control decision-making are the last to mature.

Normal variations in brain development may include the rate at which the brain develops, with some individuals developing areas of the brain faster than others. Again, this rate of development is influenced by genes and early experiences. Similarly, areas of the brain may develop more significantly in some people than others, so that the size of brain structures and the brain itself will show diversity.

For example, research on expert musicians has shown that they have greater brain volume in areas that process sound (including musical pitch and speech sounds) compared to non-musicians (Criscuolo et al., 2022). Likewise, bilingual individuals show increased grey matter (the wrinkly outside of our brains) in the parts of the brain responsible for verbal fluency, sound processing and articulation. Therefore, within the neurotypical population, variations in brain development are demonstrated through different abilities and/or behaviours.

Neurodiversity

Neurodiversity is the natural neurological difference that occurs between people's nervous systems, particularly their brain, resulting in differences in learning, thinking and processing. The term 'neurodiversity' was created in the 1990s by Judy Singer, an Australian sociologist with autism. Singer wanted to challenge the assumption that all people see, feel, touch, hear and smell in basically the same way, and to promote the idea that we are all different: everyone is neurodiverse, to a degree. The term neurodiversity is applied across the population and reflects the broader concept of **diversity**, which can be defined as 'the degree of variability of a given variable in a given population' (Singer, n.d.). In this case, this refers to neurological diversity.

Singer considers everyone to be neurodiverse, hence the term is inclusive of individuals who have been diagnosed with autism, attention-deficit hyperactivity disorder (ADHD), Tourette syndrome, dyspraxia or dyslexia (Table 3.4), as well as neurotypicals.

Singer uses the term to indicate that diagnoses such as autism and ADHD are natural variations in brain development. This contrasts with how autism is described in medical classification systems such as the *Diagnostic and Statistical Manual of Mental Disorders, Fifth Edition, Text Revision* (also known as the *DSM-5-TR*™), which considers autism a disorder.

While there is some continued controversy about the word neurodiversity and its use, it is becoming more widely accepted and used in medical fields. Using the term helps to break down barriers and stigma associated with difference, particularly for people who are diagnosed with a neurodevelopmental condition. It also helps individuals with neurodivergence to understand their differences and see them as unique characteristics rather than impediments, and to identify their strengths (Figure 3.12). Viewing the self in this way protects and promotes greater self-acceptance and esteem.

3.3.1 WHO AM I?

Table 3.4 Examples of neurodivergence

Type of neurodivergence	Key features
Autism	» difficulty with social interactions » difficulty with changing focus from one activity to another » high attention to detail
Attention-deficit hyperactivity disorder (ADHD)	» inability to maintain focused attention » inability to sit still » difficulty maintaining self-control » often inventive and innovative
Tourette syndrome	» rapid, repetitive muscle movements, vocalisations and behavioural difficulties
Dyspraxia	» difficulty with muscle control and movement » very good at problem-solving
Dyslexia	» difficulty with reading, spelling and writing » good at seeing the big picture
Dyscalculia	» difficulty with maths and using numbers » intuitive thinkers

Shutterstock.com/Anton Shahrai

Figure 3.12 Neurodivergent people use a rainbow infinity sign as a symbol to represent that they are simply different, not broken.

Brain development and neurodiversity

3.3.2 EMBRACING NORMAL VARIATIONS

Neurodivergent individuals may show variations in their brain structures and functioning. Since the 1990s, Professor Simon Baron-Cohen has undertaken extensive research on people with autism. Based on his research, he suggests that people with autism demonstrate brain differences but not brain disorder.

Structurally, the autistic brain generally has areas that are larger than in neurotypicals, such as the amygdala. The amygdala (Latin for 'almond', reflecting the shape of this structure) is responsible for perception and regulation of emotion, especially fear and aggression, and for responding to environmental threats. In autistic people, the amygdala may be overactive, which provides some insight into why people with autism experience heightened anxiety more regularly and find recognising and processing emotion more difficult.

Baron-Cohen also suggests that the brains of people with autism can show differences in activity levels compared to the brains of neurotypical people while they are performing certain tasks under experimental conditions. For example, during tests of auditory perception, autistic brains show increased brain activity. This aligns with the tendency for some autistic people to experience hypersensitivity to sounds. These differences in structure and functioning indicate areas of efficiency as well as challenge for people with autism.

ANALYSING RESEARCH 3.2

Autistic and non-autistic people share more in common than we think

In three experiments, researchers analysed the link between autistic personality traits and thinking style. In the fourth experiment, researchers compared 200 autistic and over 200 non-autistic people. The results showed that autistic people thought as fast and as rationally as those who were non-autistic.

As stated by the study: 'Except for lower self-reported intuitive thinking, we found no unique contributions of autism to intuitive or deliberative thinking across all four studies. Overall, these studies indicate that intuitive and deliberative thinking is neither enhanced nor particularly impaired in relation to autism.'

The fact that society has been accustomed to the idea that autistic and non-autistic individuals process information very differently means that education, clinics, and people in general, might need to think about changing how they approach autism.

The researchers have emphasised ways in which organisations can accommodate autistic people and their families better. This includes mentioning how educational and commercial organisations have tried to accommodate neurodivergent people in the past, which, according to them, has not been an evidence-based example of support. The study suggests a change in social and sensory environments may be more impartial for autistic individuals to thrive.

Dr Punit Shah, Associate Professor of Psychology at the University of Bath, explained: 'There is a tradition of investigating mental difficulties in autism. While this can be important for developing clinical interventions, there is also a need to understand psychological similarities between different groups. The University of Bath is doing ground-breaking work on this, showing that there is often more that unites than divides us, and our new neurodiversity research is another step in this direction.'

'Study: Autistic and Non-Autistic People Share More In Common Than We Think' by Erin Bergman, BA, Autism Parenting Magazine, May 2, 2022, https://www.autismparentingmagazine.com/similaritiesbetween-

Questions

1 What is the aim of this study?
2 What population was used for this study?
3 What was the sample for this study?
4 What were the results of this study?
5 What are the implications of this study?

ACTIVITY 3.3 CELEBRATING NEURODIVERSITY

Try it

1 Neurodiversity Celebration Week is a worldwide initiative that challenges stereotypes and misconceptions about neurological differences. It aims to transform how neurodivergent individuals are perceived and supported by providing schools, universities and organisations with the opportunity to recognise the many talents and advantages of being neurodivergent, while creating more inclusive and equitable cultures that celebrate differences and empower every individual. Explore the weblink for ideas for your school.

Apply it

2 Research has shown that there are some parts of the brain that take longer to develop in children with ADHD (Hoogman et al., 2017). MRI scans of more than 3000 children and adults with and without ADHD indicated that there were a number of key areas of the brain that were smaller in children with ADHD compared to the children without ADHD. This does not mean that children with ADHD have lower intelligence, it just means that their brain can take longer to fully mature. The scans also showed that these differences were less marked when comparing adult subjects, suggesting that brain development for children with ADHD eventually catches up. What do these results indicate about neurodiversity?

Exam ready

3 Which of the following statements about neurodiversity is incorrect?

A Neurodiversity views brain differences as normal rather than deficits.

B Neurodivergent people experience the world in unique ways.

C Neurodiversity encourages people to reduce stigma around those who learn or think differently.

D Neurodiversity encourages people to focus on the challenges rather than the strengths that come from difference.

Weblink
Neurodiversity Celebration Week

KEY CONCEPTS 3.3

» People often tend to assume that neural development and processing occurs in the same way in all people.

» Neurodiversity refers to the neurological differences that occur naturally between people.

» Neurodiversity includes a series of diagnosable neurodevelopmental disorders, including ADHD, Tourette syndrome and autism.

Concept questions 3.3

Remembering

1 What is meant by neurodiversity? r

2 List three types of neurodivergent conditions. r

Understanding

3 What is the underlying explanation for neurodiversity? r

4 Discuss two ways that you could assist neurodivergent people in their day-to-day life. e

Applying

5 Study the artwork in Figure 3.13. Write 60 words to explain what you see in that artwork. Compare your explanation with others in your class. Did you all agree on what the artwork was about? c

Alamy Stock Photo/Panther Media GmbH

Figure 3.13

HOT Challenge

6 Prepare an argument that neurodiversity has been crucial to the development of civilisation and the survival of the human race. e

3.4 Supporting psychological development and mental wellbeing

Mental wellbeing is the psychological state of someone who is functioning at a satisfactory level of emotional and behavioural adjustment. This also means that they are functioning well in the usual domains of their everyday life: work, study and relationships.

Poor mental wellbeing occurs in response to stressors or environmental changes that affect a person's psychological state and result in impaired functioning in their day-to-day life. Everyone has the potential to experience poor mental wellbeing at times. Changes in our life situation, poor physical health or increased stress and pressure can mean that our mental wellbeing suffers, and we may need to seek support to cope and to restore our mental wellbeing. Similarly, children and adolescents may experience difficulties with their psychological development as they navigate life experiences that may impact their mental wellbeing.

People with neurodivergent conditions do not have a mental health condition but they are more at risk of suffering from poor mental wellbeing. This can be due to a lack of understanding and support from society and the pressure to avoid negativity or rejection by their peers (see Case study 3.2). There are a number of different healthcare professionals and organisations that are dedicated to the management and support of psychological development and mental wellbeing. These include mental health workers, doctors, psychologists and psychiatrists. These professionals also play a role in diagnosing and supporting people who experience maladaptive, atypical or neurodivergent behaviour that is having an impact on the quality of their life. Each of these professionals will work to understand the biological, psychological and social factors that are influencing the person. This includes consideration of the person's culture and the use of culturally responsive practices. Culturally responsive practices are often embedded in the code of ethics and practice standards for professional groups.

Culturally responsive approaches to health services

Cultural responsiveness is an attitude and approach towards working with people that demonstrates understanding that culture is central to people's experiences. Being culturally responsive requires that health workers continue to reflect on their approach, learning to recognise their own cultural biases and being open to understanding the cultural perspectives of others. It includes appreciating the diversity between various groups, families and communities. For example, working with Aboriginal and Torres Strait Islander peoples in ways that are culturally responsive requires learning about Aboriginal and Torres Strait Islander peoples' histories, cultures and perspectives and using this knowledge to inform practice. Culturally responsive practice is essential for providing services that are culturally safe – services that are free from racism and cultural bias (Indigenous Allied Health Australia, 2015).

Being aware of a person's culture is important because culture can affect how people experience mental health and wellbeing (Kleinman et al., 1978). For example, Mental Health First Aid (MHFA) USA identifies four potential influences of a person's culture on their mental wellbeing (MHFA-USA, 2019).

- *Cultural stigma.* Every culture has a different way of looking at mental health. For many, there is growing stigma attached to poor mental health, and mental health challenges are considered a weakness and something to hide. This can make it harder for people who are struggling to talk openly and ask for help.
- *Understanding symptoms.* Culture can influence how people describe and feel about their symptoms. It can affect whether someone chooses to recognise and talk about only physical symptoms, only emotional symptoms, or both.
- *Community support.* Cultural factors can determine how much support someone gets from their family and community when it

comes to mental health. Because of existing stigma, some people from cultural minorities are sometimes left to find mental health treatment and support alone.

» *Resources.* When looking for mental health treatment, you want to talk to someone who understands your specific experiences and concerns. It can sometimes be difficult or time-consuming for someone to find resources and treatment that take into account their specific cultural factors and needs.

In Australia, cultural responsiveness is particularly important when working with Aboriginal and Torres Strait Islander peoples or communities. And it is important that mental health professionals understand that using diagnostic tools designed for Western cultures may not be appropriate when working with Indigenous Australians (Westerman, 2021).

The mental health and wellbeing needs of Aboriginal and Torres Strait Islander peoples are best met using methods that are informed by First Nations Australians. Different communities may have different needs, depending on their distinct cultural and language groups, as well as whether they are located in cities, rurally, or in remote communities. For these reasons, a culturally responsive practitioner must be aware of the local context and needs, and be well connected with community leaders and Elders.

There is a growing number of First Nations psychologists working in Australia and they have been at the forefront of driving the training of all psychologists to be more culturally responsive. The Australian Indigenous Psychologists Association (AIPA) represents First Nations psychologists in Australia. You can find out more about AIPA at the weblink.

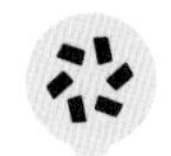

Weblink
Australian Indigenous Psychologists Association

The health and wellbeing of First Nations Australians is primarily supported by a specialised workforce of Aboriginal and/or Torres Strait Islander health workers and health practitioners. These health professionals are Aboriginal and/or Torres Strait Islander people who have completed specialised training to provide primary health care to First Nations communities, families and people of all ages. They enhance the amount and quality of clinical services provided to Aboriginal and Torres Strait Islander people and provide a crucial cultural link to help with communication between Aboriginal and Torres Strait Islander people and communities, and other health professionals. This reduces the anxiety that some people may feel when looking for health care and improves communication so that GPs and other health professionals can better understand their clients' needs.

You can learn more about the National Association of Aboriginal and/or Torres Strait Islander Health Workers and Health Practitioners and the service that their workers and practitioners provide via their website (see weblink).

Weblink
What A&TSI health workers and practitioners do

CASE STUDY 3.2

My autism journey

Louise Smith

My name is Louise and I am many things at once: I am a graduate student at the University of Oxford; I am a tutor, a rower, a feminist, a granddaughter, a daughter, a sister, a stepsister, a friend. I am also autistic.

I was diagnosed several years ago, aged 27. But, looking back, the signs were always there. I have always harboured intense 'special interests' that form something between a passion and an obsession. For instance, as a child, I was obsessed with collecting Barbie dolls, not to play with, but to create the 'perfect' Barbie doll home, complete with furniture made from cardboard cereal boxes and copious amounts of glue and glitter. Most neurotypical people have favourite interests, but theirs are more akin to hobbies, which they can put on hold if life is busy. For autistic individuals such as me, the opposite is true. We often need these special interests to stay sane in a world that can be so bafflingly complex – such interests can provide predictability, focus and great reward.

My interest in plastic people has since morphed into a deep fascination with understanding real people. Today I feel fortunate to study psychology as part of my PhD. Another of my special interests is literary fiction. Since I was small, I've read voraciously. What I found most enticing about literature was the possibility of learning social rules, expectations, how to cope with challenges and much more, all from the comfort of my armchair without the risk of saying the wrong

»

thing or making a mistake. Again, this is typical for many autistic people, particularly women but also many men, who learn about the social world explicitly through pursuits such as literature, but also soap operas, films and closely watching significant others. We then use what we have learnt in social situations, to 'camouflage' our lack of social instinct, and behave according to the social rules of the specific situation.

Unfortunately, immersing myself in literature did not equip me with all the understanding and skills I needed to cope with the complex social rules of teenage life. When I turned 13 and moved to senior school, that's when things went wrong for me. I didn't understand the social rules in the enormous concrete monolith that became my hell, and I began to be badly bullied. For instance, a girl once spat at me in the corridor, at which point I informed her that spitting on someone is considered an offence of common assault under the Criminal Justice Act. This prompted a lot of laughter from the girl and her friends, only escalating the situation. I thought it would deter them at the time, but looking back I didn't understand how to 'keep my head down' and stay out of harm's way.

The bullying left me highly anxious, constantly feeling as if the bullies were about to burst out of my wardrobe. I wouldn't go out in public if I could help it, and nightmares plagued my sleep.

The American author Paul Collins, whose son is autistic, wrote in *Not Even Wrong: Adventures in Autism* (2004) that: 'Autists are the ultimate square pegs, and the problem with pounding a square peg into a round hole is not that the hammering is hard work. It's that you're destroying the peg.' I can say from my own experience that the social pressure of growing up can be a toxic environment for us autists as we are forced to conform to the norms or stand out and risk bullying and trauma.

With hindsight, the next warning sign that I was autistic was my first experience of university, at a place I'd like to forget, to study English literature. I arrived with a car-full of books, and was shocked at the person who parked next to us unloading crates of alcohol. I struggled immensely with the social side of university including the loud bars and clubs, which assaulted my senses and left my ears ringing for days afterwards. I left after two terms.

'My autism journey: how I learned to stop trying to fit in' by Louise Smith, Aeon, 2 October 2019, https://aeon.co/ideas/my-autism-journey-how-i-learned-to-stop-trying-to-fit-in.

Questions

1. What were the signs that Louise was autistic?
2. What is one strategy autistic people use to try to fit into social situations?
3. How did Louise's treatment at school affect her mental wellbeing?
4. How do you think you would you treat a neurodivergent person like Louise if she became a student in your class?

General practitioner

A local doctor or general practitioner (GP) is often the first person to be involved when there is concern about atypical or neurodivergent development. A doctor will record any behavioural symptoms and family history, and investigate whether there are any environmental factors that may be encouraging or exacerbating the symptoms. A doctor will then decide whether to refer a patient to another mental health professional.

Mental health worker

Mental health worker is a broad description that covers a number of different types of workers who care for people with mental health issues. These can include mental health issues that are due to substance abuse, poverty and other social problems. Mental health workers assist psychiatrists, doctors and nurses by monitoring patients and providing therapeutic assistance. A mental health worker has a bachelor's degree in either nursing or social work as well as work experience in the field.

Mental health nurses are registered nurses with further training in psychological therapies. They can work in hospitals, clinics, nursing homes and mental health units. Their duties can include assessment of conditions, giving medications and visiting people in their homes.

Aboriginal and Torres Strait Islander mental health workers understand the mental health issues of First Nations Australians. They are able to provide culturally safe and accessible mental health services.

They have undertaken training in mental health and psychological therapies and can work in mental health clinics, hospitals and Aboriginal health services to support the mental wellbeing of Aboriginal and Torres Strait Islander peoples (Figure 3.14).

Newspix/Amos Aikman

Figure 3.14 Aboriginal health services support the mental wellbeing of Aboriginal and Torres Strait Islander people.

Psychologist

Psychologists are professionals who study mental states, and perceptual, cognitive, emotional and social processes and behaviour. In Australia, to practice as a psychologist you must be suitably qualified and registered with the Australian Health Practitioners Registration Authority (AHPRA). Registered psychologists are employed in hospitals, community health services, schools, courts, prisons or in private practice. A referral is not needed to see a psychologist (however, access to some services under Medicare do require a GP referral). Depending on the person's needs, a psychologist will use evidence-based therapies to support clients in their psychological development or to regain their mental health and wellbeing. Examples of therapies include client-centred approaches, cognitive behavioural therapy, acceptance and commitment therapy and interpersonal therapy. The psychologist will also monitor the person's progress and reduction in symptoms through standardised scales that measure the level of distress a person may be experiencing. As psychologists have not undertaken a medical degree, they cannot prescribe medication.

Psychologists are qualified to assess, diagnose, support and advocate for neurodivergent individuals. Psychologists consult the *Diagnostic and Statistical Manual of Mental Disorders, Fifth Edition, Text Revision* (also known as *DSM-5-TR*™) to classify and diagnose conditions such as autism, ADHD and dyslexia.

The process of diagnosis involves talking with the person to understand their experience, recording in detail their developmental history and undertaking cognitive assessments. If the person is young, the psychologist will also talk to their parents and teachers, consider academic and learning progress, and conduct observations of them at school. When diagnosing an individual with a neurodevelopmental condition, psychologists regularly consult and work with other health professionals, including psychiatrists, paediatricians, and speech and language therapists.

3.4.1 SUPPORTING MENTAL WELLBEING

Psychologists often use the *DSM-5-TR*™ classification system to diagnose specific mental conditions and to guide the selection of appropriate therapies. It provides a common language for therapists, researchers, social agencies and health workers. For example, when one health professional in Australia diagnoses someone as having autism, another health professional in the UK using the *DSM-5-TR*™ would also apply the same diagnostic criteria on similar symptoms. The *DSM-5-TR*™ is not the only system of classifying mental disorders, but it is the best known and most widely used system in mental-health settings.

Once a diagnosis has been made, psychologists can work with children, teenagers and adults to assist them to understand and accept their condition, identify and build on their strengths and to live successful and fulfilling lives.

Psychologists take a biopsychosocial approach in understanding their clients' needs and planning their therapy and support. This means they consider the impact of the person's biology (e.g. whether they were born premature), their psychology (e.g. their memory or learning capacity) and their social world (e.g. their family support) on their life and how these different areas can be made use of in therapy.

3.4.2 DIAGNOSIS OF ATYPICAL BEHAVIOUR

For example, when supporting a child with ADHD, a psychologist may work with the child's parents and teachers (social factor) to implement routines and structures that scaffold and develop the child's planning and organisational skills (psychological factor). These routines and structures might include visual schedules of the school day and reminder checklists for resources they need to take to school. Psychologists use a range of evidence-based therapies when working with neurodiverse clients. Increasingly they are taking a neurodiversity-affirmative approach, which comprises:

» identifying and addressing ableist attitudes. Ableism is discrimination against neurodivergent people in favour of neurotypical people
» focusing on individual strengths
» teaching the person how to advocate for themself

» considering how to modify the person's home, school and work environments to create better spaces to thrive (for example, by avoiding loud or bright patterns, having low light and low noise)
» helping them learn how to build healthy relationships
» helping them learn how to cope with stress.

3.4.3 LABELLING AND STIGMA

Psychologists in schools

Psychologists who work in schools support young people in their learning and academic achievement and promote and support good mental health and social–emotional wellbeing. Psychologists in schools may also provide counselling to students for mental health concerns. While all psychologists can work in schools, educational and developmental psychologists have extra training in their field. Their expertise includes assessment and intervention of learning difficulties, giftedness and cognitive functioning. They can also diagnose neurodevelopmental conditions such as autism.

School psychologists also support school staff and the broader school community, including parents, to promote good mental health in students, by educating them on typical and atypical psycho-social development and when to seek help for concerns. They also implement evidence-based programs for groups of students or the whole school population that support psychological development and wellbeing.

Psychiatrist

Psychiatrists complete 6 years of medical training to become a doctor. They then complete a postgraduate qualification that focuses on diagnosis and treatment of mental health issues. As doctors, psychiatrists can prescribe medication to treat or reduce the symptoms of a mental illness. They are also authorised to admit a person to hospital for treatment. A referral from a GP is needed to see a psychiatrist.

Psychiatrists are best placed to develop treatment and support interventions for people suffering from more severe mental health conditions that require medications to control the symptoms, such as schizophrenia. Figure 3.15 compares the roles of psychiatrists and psychologists.

Other support workers

Table 3.5 outlines the roles of some of the other mental health workers who work with people with a mental illness.

Organisations

There are several organisations in Australia that are dedicated to the diagnosis, support and management of atypical behaviour and neurodivergent individuals (Table 3.6).

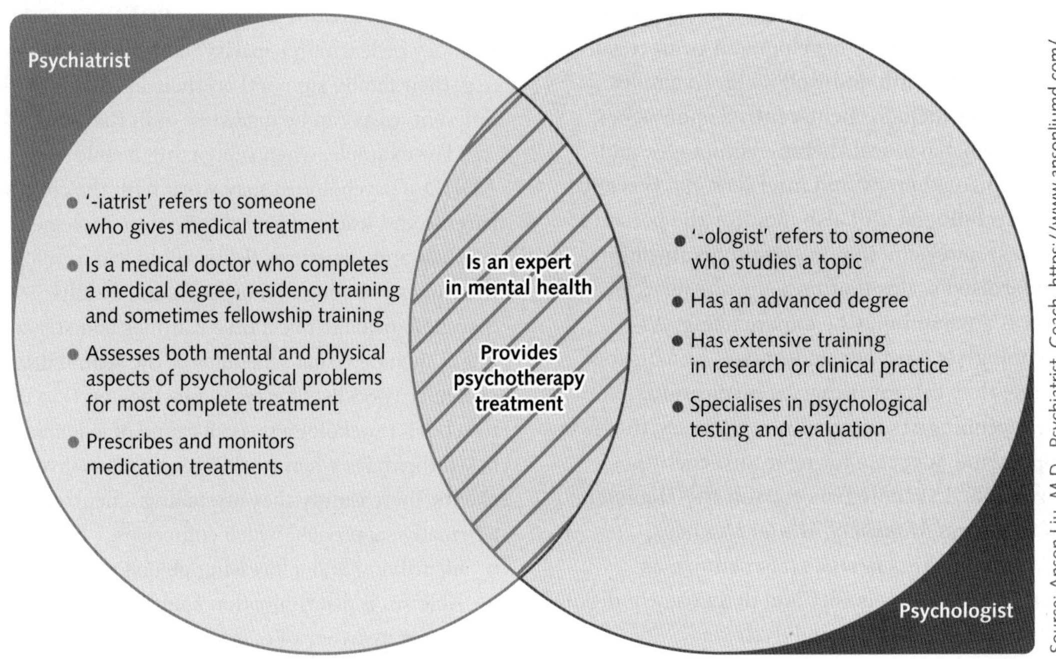

Source: Anson Liu, M.D., Psychiatrist, Coach, https://www.ansonliumd.com/psychiatry

Figure 3.15 A comparison of psychiatry and psychology

Table 3.5 The roles of some mental health workers

Type	Qualification	Role
Social worker	Bachelor of social work	Provide general support, including finding ways to cope, understanding the process of mental illness and how to cope at work
Counsellor	May have a Bachelor or Graduate Diploma of Counselling	Use cognitive behavioural therapies and other therapies to help a person work through a problem, understand themselves better, work through options and then support them to take action
Recovery and rehabilitation worker	May have a Certificate IV in mental health, or qualified as a social worker	Assist with a psychosocial issues caused by a mental illness, which may include supporting the person to rebuild social connectedness, rejoin employment, find a place to live and build their confidence

Table 3.6 Some of the many organisations that support neurodivergent people in Australia

Organisation	Description
headspace National Youth Mental Health Foundation	headspace is the National Youth Mental Health Foundation, providing early intervention mental health services to 12–25-year-olds. Each year, headspace helps thousands of young people access vital support through their headspace centres in communities across Australia, online and phone counselling services, vocational services, and presence in schools. headspace can help young people with mental health, physical health (including sexual health), alcohol and other drug services, and work and study support. Centre details, as well as factsheets and resources for young people and their families and friends, can be found on the headspace website.
REACHOUT	ReachOut is the leading online mental health service in Australia supporting young people during tough times. ReachOut helps young people feel better about today and the future, no matter what challenge they're facing. They provide a safe place where young people can openly express themselves, explore what's happening in their lives, connect with people who understand their situation, and find the resources to help them manage their challenges now and in the future
amaze	Amaze builds autism understanding in the community, influences policy change for autistic people and their families/supporters, and provides independent, credible information and resources to individuals, families, professionals, government and the wider community.
ADHD Australia	ADHD Australia's mission is to create systemic positive change for people living with ADHD through discovery, translation and advocacy. Their website provides information about ADHD, resources and support groups across Australia.
embrace multicultural mental health	Embrace Multicultural Mental Health (the Embrace Project) provides a national focus on mental health and suicide prevention for people from culturally and linguistically diverse (CALD) backgrounds. It provides a national platform for Australian mental health services and multicultural communities to access resources, services and information in a culturally accessible format.

Weblinks
headspace

ReachOut

Neurodiversity Celebration Week

Amaze

ADHD Australia

Embrace

3.4.4
MANAGING ATYPICAL BEHAVIOUR

ACTIVITY 3.4 WHO'S WHO IN MENTAL HEALTH CARE?

Try it

1 Copy and complete this table to describe the roles of the people who work in mental health.

	Make diagnosis	Provide psychological treatment	Prescribe medication treatment	Offer advice
Mental health worker				
Psychologist				
Psychiatrist				
Organisations (e.g. ReachOut)				

Apply it

2 PJ is having a bad week. PJ is experiencing feelings of anxiety and is having a lot of trouble sleeping. PJ is not sure who to see – a GP, a psychologist or a psychiatrist.Who should PJ see first?

Exam ready

3 One difference between a psychologist and a psychiatrist is

A a psychologist studies for longer than a psychiatrist.

B a psychologist can prescribe drugs for treatment but a psychiatrist cannot.

C a psychologist uses psychotherapies and a psychiatrist does not.

D psychologists are more likely to see people with a behavioural problem while a psychiatrist will see people who need a medical intervention.

CASE STUDY 3.3

Neurodiversity in the time of COVID

Erin Kavanagh-Hall

Jacob* is a 29-year-old nursery worker.

Research has shown that us ADHD folk can excel in times of crisis – like a pandemic. In my experience, that's absolutely true.

I was diagnosed with ADHD two years ago. I mainly have the inattentive sub-type (likely to struggle with limited attention span, distractibility and procrastination), but I also experience hyperactivity – if I'm not medicated, I go into Labrador puppy mode. It took me a long time to get a correct diagnosis. In the end, I paid for a private assessment, as the public health system had categorised me as bipolar and wouldn't entertain the idea of anything else. Before taking Ritalin, my life was messy and disorganised – now I like to look at it as organised chaos.

Lockdown was hard to start with. But then, things started settling down. It was interesting, I'm not sure if it was the adrenaline kicking in, but I almost feel like I stepped out of my body and could see the world burning – and it didn't faze me.

According to research, people with ADHD produce more theta brain waves which appear when you're nodding off to sleep – indicating deep relaxation. So, while other people's brains overload during a major event, ours feel relatively normal. Where there's a panic situation, we can become relaxed, laid back and under control. I find crises, like COVID-19, kickstart me into action – I've been able to lead others who've been falling apart, taking them under my wing and helping ground them. Instead of going crazy, I've felt more clear minded than ever – in fact, in the days leading up to lockdown, I was functioning so highly my medication was having almost no effect.

The first couple of days of lockdown, I was thinking of all these things I could do to occupy

»

and fill my time. But it became chaotic – I started a hundred projects and wore myself out. My mind is like spaghetti junction in rush hour, and it can be exhausting. So each day, I decide to do one thing – and anything after that is a bonus. I've learned not to expect too much of myself.

I have started cooking again rather than eating takeout every night, and have actually started to enjoy it. The sudden switch off has given me a chance to think about things I want to change about myself – in order to grow and move forward. I feel optimistic and excited about resuming a regular life.

*name has been changed

'A different headspace: Six people on being neurodivergent during lockdown' By Erin Kavanagh-Hall, The Spinoff, April 26, 2020, https://thespinoff.co.nz/society/26-04-2020/a-different-headspace-being-neurodivergent-during-covid-19

Questions

1 What diagnosis did Jacob receive two years ago?
2 How do the symptoms of Jacob's condition affect his life?
3 What medication does Jacob take to control his symptoms?
4 Which person would have prescribed Jacob this medication?
5 How has this medication affected his life?
6 What effect did the COVID-19 lockdown have on Jacob? Why?
7 What can you learn from Jacob's experiences?

INVESTIGATION 3.1 THE ROLE OF MENTAL HEALTH PROFESSIONALS

Scientific investigation methodology

Case study

Aim

To investigate the role of mental health professionals in supporting psychological development and mental wellbeing

Logbook template
Case study

Introduction

There is a range of different healthcare professionals and organisations who are dedicated to the diagnosis, management and support of normal brain variations including neurodivergence. These include mental health workers, doctors, psychologists, psychiatrists and organisations that support psychological development and mental wellbeing.

Data calculator

Your task

Your task is to complete a case study on the role of one mental health professional or organisation in supporting psychological development and mental wellbeing, as well as the diagnosis and management of atypical behaviour.

Use the questions below to help guide your case study. Make sure you record all your work in your logbook. Present your findings as a poster or multimedia presentation so they can be shared with other students.

Questions

1 Identify your selected mental health professional or organisation.
2 Outline the qualifications, training and expertise required to work in this professional field.
3 Identify the settings and locations they would commonly work in.
4 Provide a description of the services, strategies and treatments they use in their work to support psychological development and the diagnosis and management of atypical behaviour.
5 Does this mental health professional or organisation provide culturally responsive practices? If yes, describe what these are.
6 Describe the relevant ethical principles and concepts that the mental health professional or organisation needs to follow in their work.
7 Include labelled images of the work of the mental health professional to represent visually their role in supporting psychological development and mental wellbeing.
8 List your sources of information and assistance using standard scientific referencing conventions.

KEY CONCEPTS 3.4

- A doctor (GP) is often the first professional who is consulted when an issue with mental health arises. They can create a mental health management plan, which may include referrral to another professional.
- A mental health worker is a qualified nurse. Aboriginal and Torres Strait Islander mental health workers understand the mental health issues of Aboriginal and Torres Strait Islander peoples.
- A psychologist is a qualified mental health expert who can diagnose and treat clients with poor mental wellbeing.
- A psychiatrist is a medical doctor who has completed further training in mental health. Psychiatrists can prescribe medication.
- In Australia there are several organisations devoted to mental health and wellbeing, including headspace and ReachOut.

Concept questions 3.4

Remembering

1 What is meant by mental wellbeing? r

2 What resource do psychologists and psychiatrists use to help them diagnose atypical behaviour? r

Understanding

3 Explain why neurodivergent people are more at risk of developing poor mental wellbeing. e

4 List two ways in which psychologists differ from psychiatrists. r

Applying

5 You have a friend who is failing their tests and exams, is showing signs of erratic behaviour and is becoming withdrawn from their social group. You suspect that your friend might be abusing drugs. Who would you recommend your friend contact for help? c

HOT Challenge

6 A person presenting to a psychologist shows the following atypical behaviours: persistent difficulties in using verbal and non-verbal communication in social situations, and restricted repetitive patterns of behaviour. Using the *DSM-5-TR*™, what diagnosis would fit with this person's condition? Suggest some features that could be included in a management plan for this person. e

7 Go to the website of one of the organisations listed in Table 3.6 (page 117). Find out: e
- the organisation's purpose (what they do)
- the services they provide
- the community they service
- suggestions for how you can get involved either to access services or as a supporter (e.g. fundraising).

3 Chapter summary

CHAPTER 3 SUMMARY

KEY CONCEPTS 3.1

- Typical or normal behaviour is behaviour seen in the majority of people in the population.
- Atypical or abnormal behaviour is different from the behaviour of the majority of the people.
- Psychological criteria can be used to determine whether a behaviour is normal or not.
- Criteria used to determine normality include the culture involved, social norms, statistical rarity, level of personal distress and maladaptive behaviour.
- Each criterion has limitations and should therefore not be used in isolation.

KEY CONCEPTS 3.2

- Neurotypicality describes individuals who think, behave and feel in ways that are typical of most people.
- Neurotypical brain development refers to the expected development of brain structures and neurological connections over time.
- Neurodivergent people may:
  - show differences in the way they process emotion
  - not hide their emotions and may not understand when neurotypical individuals do
  - avoid eye contact and physical contact
  - struggle with change and overstimulating environments.
- Cognitively, neurodivergent people can be very creative and have a high attention to detail.
- Neurodivergent people may learn, think and process information differently.

KEY CONCEPTS 3.3

- People often tend to assume that neural development and processing occurs in the same way in all people.
- Neurodiversity refers to the neurological differences that occur naturally between people.
- Neurodiversity includes a series of diagnosable neurodevelopmental disorders, including ADHD, Tourette syndrome and autism.

KEY CONCEPTS 3.4

- A doctor (GP) is often the first professional who is consulted when an issue with mental health arises. They can create a mental health management plan, which may include referral.
- A mental health worker is a qualified nurse. Aboriginal and Torres Strait Islander mental health workers understand the mental health issues of Aboriginal and Torres Strait Islander peoples.
- A psychologist is a qualified mental health expert who can diagnose and treat clients with poor mental wellbeing.
- A psychiatrist is a medical doctor who has completed further training in mental health. Psychiatrists can prescribe medication.
- In Australia there are several organisations devoted to mental health and wellbeing, including headspace and ReachOut.

3 End-of-chapter exam

Section A: Multiple-choice

1 The difference between maladaptive behaviour and adaptive behaviour is that
- A maladaptive behaviour promotes good mental health.
- B adaptive behaviour promotes good mental health.
- C maladaptive behaviour is positively viewed and accepted by the community and adaptive behaviour is not.
- D adaptive behaviour cannot be learned.

2 An example of a stressor that would affect many people is
- A completing a yoga session each week.
- B listening to music that you like.
- C a walk along the beach.
- D delivering a speech in assembly.

3 In Tibet it is normal for a woman to be married to more than one man. This is called polyandry. In Australia, this would be considered abnormal according to which psychological criteria?
- A statistical rarity
- B personal distress
- C cultural perspectives
- D maladaptive behaviour

4 Health is characterised by
- A mental wellbeing.
- B a lack of bodily diseases.
- C social connectedness.
- D all of the above

5 Derek has his hair cut in a mohawk and dyed green. He wears torn clothes held together by safety pins and bright purple Doc Marten boots. Derek has a job at a cat protection society that he loves. Derek could be viewed as having
- A destructive social nonconformity.
- B unhealthy nonconformity.
- C healthy nonconformity.
- D maladaptive nonconformity.

6 Rani has an IQ of 70. The mean IQ is 100. Rani's IQ would be considered to be
- A a cultural norm.
- B a statistical rarity.
- C a normal curve.
- D maladaptive.

7 Which of the following professionals would be able to help a person access medications to manage their atypical behaviour?
- A a social worker
- B a psychologist
- C a psychiatrist
- D a counsellor

8 A cause of differences between neurotypical and neurodivergent individuals is that
- A they behave differently from each other.
- B they have different thought processes.
- C they are unable to tell someone's emotions just by looking at their face.
- D the neurons in their brain are connected in a different way.

9 The *Diagnostic and Statistical Manual of Mental Disorders, Fifth Edition* (*DSM-5-TR*™) is used
- A to decide whether a person needs to be admitted to hospital for treatment.
- B to treat people with mental disorders.
- C to diagnose, treat and manage people with atypical behaviour.
- D as a reference for doctors who are not trained in mental illness.

10 Neurotypical individuals display
- A normal psychological development but abnormal behaviour.
- B abnormal psychological development and abnormal behaviour.
- C normal psychological development and typical behaviour.
- D normal psychological development and atypical behaviour.

11 Neurodiverse individuals may also suffer from diagnosable conditions such as
- **A** autism, ADHD and dyslexia.
- **B** autism, ADHD and mental illness.
- **C** suicidal thoughts, poverty and homelessness.
- **D** cerebral palsy.

12 John is neurodivergent and often misses what his wife really means when she folds her arms in disapproval but says she doesn't mind. This is known as
- **A** introversion.
- **B** emotional blindness.
- **C** emotional detachment.
- **D** emotional autism.

13 Wearing a suit and tie to work is seen as typical behaviour but wearing a suit and tie to swim at the beach would be seen as
- **A** maladaptive.
- **B** atypical.
- **C** neurodiverse.
- **D** neurotypical.

14 Which one of the following is true about general practitioners (GPs)?
- **A** They are unable to refer patients to a psychiatrist.
- **B** They have completed a medical degree.
- **C** They have further qualifications in psychiatry.
- **D** They must write a referral if patients need to see a psychologist.

15 Sarah is preparing to give a talk to her class on neurodiversity. She does not look forward to doing this and is getting very anxious about it. Which one of the following is true about this situation?
- **A** Sarah is exhibiting stress about giving the talk to her class.
- **B** Sarah is exhibiting distress about giving the talk to her class.
- **C** The stressor is the number of people in her class.
- **D** Sarah's behaviour is maladaptive.

16 Mark has bought a new house and taken out a very large mortgage with the bank. He has just lost his job and has been asked by the bank manager to come in for an interview about his mortgage. The stressor in this situation is
- **A** the very large mortgage.
- **B** the loss of his job.
- **C** the interview with the bank manager.
- **D** all of the above

17 A person presented themselves at a psychologist's office. The person showed an inability to maintain focused attention, inability to sit still and difficulty maintaining self-control. The psychologist could use the *DSM-5-TR*™ to diagnose this person as suffering from
- **A** dyslexia.
- **B** Tourette syndrome.
- **C** ADHD.
- **D** dyspraxia.

18 Chrissie maintains her mental wellbeing by
- **A** having a strong social network.
- **B** staying physically active.
- **C** eating nutritious food.
- **D** all of the above

19 While shopping at the supermarket, Sammi took the last packet of Tim Tams off the shelf and put it into her trolley. Another customer confronted her (Figure 3.16), demanding she give him the packet of Tim Tams immediately. Sammi started crying and hitting the other customer. A possible explanation is that
- **A** she is a neurotypical individual and needed time to process the customer's expression.
- **B** she shows adaptive behaviour and needed time to process the customer's expression.
- **C** her behaviour is typical and she got to the Tim Tams first.
- **D** she is a neurodivergent individual and needed time to process the customer's expression.

Figure 3.16

20 Ariel is on the autism spectrum and has a job at the local supermarket, which she loves. The owner of the supermarket has made alterations at the supermarket to make it more accommodating for Ariel. These changes could include

A giving Ariel a time out room when she feels overwhelmed.

B putting flashing lights above where Ariel works.

C playing loud rock and roll music.

D all of the above

Section B: Short answer

1 Explain what is meant by social norms when considering typical and atypical behaviour. Provide an example to illustrate your explanation. [2 marks]

2 Distinguish between adaptive and maladaptive behaviours. [2 marks]

3 A person comes to school and shows the following behaviours:

» difficulty with social interactions

» inability to look people in the eye

» high attention to detail.

Outline the steps that you, as the school assistant-principal, would put in place to have this person assessed for possible diagnosis and their needs supported. [5 marks]

Unit 1, Area of Study 1 review

Section A: Multiple-choice

Question 1

Psychologists are interested in the relative contributions of nature and nurture to human development. In this context, nature refers to

A the type of parental care the baby receives during early stages of life; how natural it is.
B our genetic inheritance.
C environmental influences.
D the natural setting in which the individual develops.

Question 2

Louis and his twin brother were separated at birth. When they turned 21, Louis found his brother. Much to his surprise, both were construction workers, played AFL football and took an interest in action movies. Which statement is true of this scenario?

A Their behaviour is primarily due to nurture.
B Their behaviour is due to both nature and nurture equally.
C Their behaviour is due primarily to nature.
D Their behaviour is due neither to nature nor to nurture.

Question 3

Which of the following is true of the outcome of the nature-versus-nurture debate?

A Heredity is more important in shaping who we are.
B All psychological development is due to environmental factors only.
C Our psychological development is due to both nature and nurture as well as their interaction with one another.
D The debate is still ongoing, and no conclusion has been reached.

Question 4

Which of the following is true of epigenetics?

A It is the interaction of our genetics and how it impacts on our way of thinking.
B It is the interaction of the environment in our gene expression.
C It is how our genes may predispose us to certain psychological disorders.
D It is how the biopsychosocial framework interacts to impact on our gene expression.

The following information relates to questions **5–7**.

Jessica has been feeling overly anxious lately and cannot find a direct cause. It is starting to impact on her everyday life, and she is avoiding going to school because she is so overwhelmed all the time, leading to her failing classes. Her mum suffers from anxiety, and she and Jessica are starting to think that Jessica may also suffer from this.

Question 5

Which of the following is a biological contributing factor to Jessica's anxiety from the scenario above?

A lack of sleep
B genetic predisposition to anxiety
C hormonal imbalances because she is going through puberty
D poor coping mechanisms

Question 6

Which one of the following relates best to Jessica's behaviour?

A It is adaptive because she is acting in a way that is impacting on her ability to function.
B It is atypical because she is acting normal for herself.
C It is typical because she is acting in a way that is different from her normal behaviour.
D It is maladaptive because it is impacting on her ability to function in everyday life.

Question 7

Who could Jessica see to help her to cope with her anxiety?

A a psychiatrist so they can prescribe her antidepressants if she suffers from anxiety
B a general practitioner so they can diagnose her mental disorder and offer psychoeducation
C a psychologist so they can help her with her anxiety regardless of whether it is a diagnosable disorder or not
D a psychiatrist first then a psychologist if medication is needed

Question 8

Which of the following is not an example of a social factor impacting on mental wellbeing?

A coping strategies
B support from family and friends
C traumatic events
D community support groups

Question 9

Which of the following is an example of a psychological factor contributing to an individual's mental health?

A taking medication

B the functioning of the nervous system

C a person's response to a stressor

D how strong an individual's friendships are

Question 10

The peer group has a significant effect on the individual's development in which lifespan stage?

A childhood

B early adulthood

C adolescence

D middle age

Question 11

Higher-order cognitive development begins in which stage of the lifespan?

A childhood

B adolescence

C early adulthood

D middle age

Question 12

Which one of the following is true of cognitive development over the lifespan?

A It transitions from concrete thinking to logical thinking.

B It transitions from logical thinking to concrete thinking.

C It doesn't begin until age 12.

D It occurs in the exact same way for everyone.

Question 13

Language acquisition is an example of

A a critical period.

B a sensitive period.

C nature.

D heredity.

Question 14

Which one the following is true of sensitive periods?

A It is a time when an individual is more primed to learn everything.

B It is a time when individuals can more easily acquire skills, but they can still develop them outside of this time.

C They start and end suddenly.

D It is when an individual must learn a particular skill, or else it will never develop.

Question 15

Which one of the following does not apply to critical periods?

A Once the period passes, it is nearly impossible to acquire the skill.

B They start and end suddenly.

C Imprinting is an example of a critical period.

D Brain plasticity is at its lowest during critical periods.

Question 16

Genie was neglected for years, tied to a chair in her room. Which of these describes her language acquisition?

A She was able to completely master all language.

B She couldn't speak at all.

C She was able to gain some speech but not filler vocabulary.

D She could only say pronouns and couldn't use adjectives or verbs.

Question 17

What is the *DSM-5-TR*™?

A a diagnostic guide to determine whether development is normal or not

B a diagnostic guide to normality; if people fall outside of this then they are considered mentally ill

C a diagnostic guide to mental disorders

D an outdated measure that is no longer used by psychologists

Question 18

How is typical behaviour determined through the statistical rarity approach?

A Behaviour is typical if it is considered socially acceptable.

B Behaviour is typical if it is not causing any form of personal distress.

C Behaviour is typical if the individual can function in everyday life.

D Behaviour is typical if it is congruent with the majority of society, statistics wise.

Question 19

What is a limitation of using criteria to determine atypical behaviour?

A It is too subjective.

B It looks at individuals objectively and rejects individual differences.

C It fails to describe symptoms of a disorder.

D It doesn't provide advice in how to cope with the repercussions of behaviour.

Question 20

Which of the following is an example of an adaptive behaviour?

A doing so well in your job that you get a promotion

B getting to work at 8.30 a.m. as you usually do

C being unable to get out of bed so you are late to work all the time

D behaving as you usually do

Question 21

Keke is highly focused on doing well in her Psychology assessment, so she is spending all her time studying for it. On top of this she is juggling part-time work and all her other homework. She manages to get this all done and gets an A on her test. Which statement is true of Keke's behaviour?

A It is adaptive because she is coping well in her daily life.

B It is maladaptive because she is unable to focus on her other subjects.

C It is atypical because students usually cannot juggle all of that.

D Her behaviour is normal because she is doing all of her duties.

Question 22

Which statement does not describe being neurotypical?

A do not have any developmental disorders

B standard brain functioning

C includes people who suffer from ADHD

D generally, function well in social contexts

Question 23

Which of the following is not a characteristic of being neurodivergent in early childhood?

A poor eye contact

B no words spoken by 9 months

C no two-word phrases by two

D no smiling or social responsiveness

Question 24

Which of the following is not a developmental disorder leading to neurodivergence?

A ADHD

B autism

C anxiety

D dyslexia

Use the following information to answer questions **25–27**.

Luca is 2 years old and struggles to communicate. He can say a few words such as 'mum' and 'dad', but can't place two words together to form sentences. He also struggles socially, doesn't like interacting with strangers and avoids eye contact.

Question 25

Which is most likely the case for Luca?

A He is neurotypical; this is normal development for his age group.

B He is neurodiverse because he has accelerated development for his age group.

C He is neurodiverse because he is slower than typical to develop speech and is showing social struggles.

D It is too early to tell if he is neurotypical or neurodivergent.

Question 26

If Luca does not develop speech by this age, will he ever be able to speak?

A Yes, because it is only a sensitive period and learning to speak later may be difficult, but can still be done

B Yes, because it is a critical period

C No, because it is a critical period and if not learned by then, it never will be

D No, because it is a sensitive period so will be too difficult to learn later

Question 27

How could a psychologist help Luca?

A teach Luca skills in coping in social situations

B give him medication to help treat his social anxiety

C force Luca to make new friends

D discuss with his parents why they aren't doing enough to support Luca's development

Question 28

Which statement is false of the role of psychologists in the management of atypical behaviour?

A help diagnose mental disorder

B provide advice on how to best cope with the behaviour

C give support to the individual

D prescribe medication to help treat the behaviour

Use the following information to answer questions **29–30**.

Jennie has recently been experiencing intense hallucinations and delusions. She hears voices and sees things that are not there and can no longer tell what is real from what is not.

Question 29

Who is the best person for Jennie to see about her problems?

A mental health worker
B psychiatrist
C psychologist
D general practitioner

Question 30

What is a possible diagnosis Jennie may be given?

A anxiety
B depression
C schizophrenia
D narcissistic personality disorder

Section B: Short answer

Question 1

Aya is 15 years old and is struggling to keep up with her schoolwork. She has been feeling depressed lately and has stopped showing up to some of her classes, Instead, she is staying in bed all day.

a Is Aya's behaviour adaptive or maladaptive? Why? [2 marks]

b According to the biopsychosocial model, state two social aspects that may help with Aya's feelings of depression. [2 marks]

c Is Aya's behaviour considered socially normal? Justify your response. [2 marks]

[Total = 6 marks]

Question 2

a What is the nature-versus-nurture debate? [2 marks]

b What conclusion was drawn from this debate? [2 marks]

[Total = 4 marks]

Question 3

Describe how social development differs in children compared to adolescents. [2 marks]

Question 4

Name and describe one strength and one limitation of using psychological criteria to categorise mental health. [4 marks]

Question 5

Differentiate between neurotypical and neurodivergent behaviour. [2 marks]

The role of the brain in behaviour and mental processes

4

Key knowledge

» different approaches over time in understanding the role of the brain in behaviour and mental processes
» the roles of the hindbrain, midbrain and forebrain, including the cerebral cortex, in behaviour and mental processes

Key science skills

Analyse, evaluate and communicate scientific ideas

» use appropriate psychological terminology, representations and conventions, including standard abbreviations, graphing conventions and units of measurement
» discuss relevant psychological information, ideas, concepts, theories and models and the connections between them
» analyse and explain how models and theories are used to organise and understand observed phenomena and concepts related to psychology, identifying limitations of selected models/theories
» use clear, coherent and concise expression to communicate to specific audiences and for specific purposes in appropriate scientific genres, including scientific reports and posters
» acknowledge sources of information and assistance, and use standard scientific referencing conventions

Source: VCE Psychology Study Design (2023–2027), pp. 25 & 13

4 The role of the brain in behaviour and mental processes

Your brain is responsible for every thought, action, memory, feeling and experience you have. It's a complex organ that oversees complex behaviours. This complexity may help to explain why our understanding of the brain in behaviour and mental processes has changed and developed so much over time.

p. 139

4.1 How the brain controls behaviour and mental processes

How does the brain, a physical organ that transmits electrical and chemical signals all over the body, cause us to think and act the way we do? Philosophers and scientists have been trying to answer this question since ancient times, coming up with different theories. Only in the early 1800s did they start to apply the scientific method to test these theories.

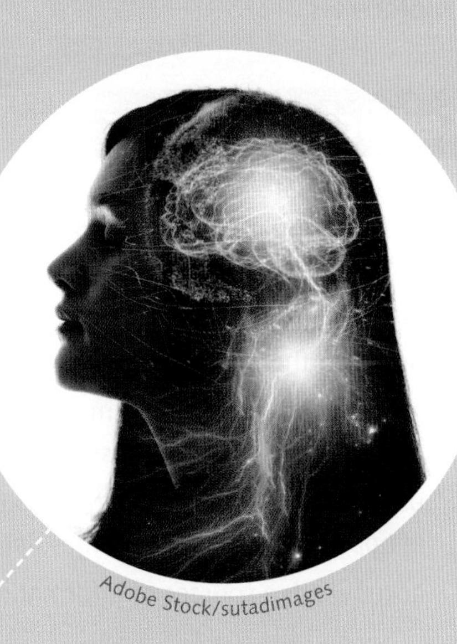

Adobe Stock/sutadimages

p. 149

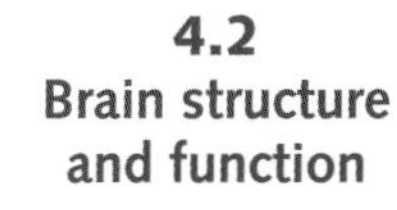

4.2 Brain structure and function

As scientists began to experiment on and investigate the brain, they soon learned that it could be subdivided into three main regions: the hindbrain, the midbrain and the forebrain, each responsible for different functions. Damage to the brain has provided opportunities to understand which mental processes and behaviours the damaged region is responsible for.

Shutterstock.com/Yurchanka Siarhei

iStock.com/CasarsaGuru

p. 155

4.3
The cerebral hemispheres

The cerebrum is a structure in the forebrain divided into two hemispheres. The left hemisphere is responsible for verbal and analytical functions, and it oddly enough controls the right side of the body. This is the hemisphere that helps us to communicate using spoken or written language. The right hemisphere is responsible for non-verbal functions, and it controls the left side of the body. It allows us to process the information around us.

p. 167

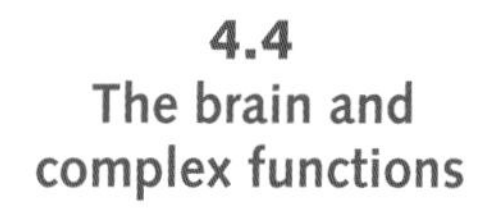

4.4
The brain and complex functions

Each task that we carry out daily may involve a range of mental processes and behaviours. Even when you get up in the morning, you need to lift yourself out of bed, choose what to wear and make sure not to bump into anything as you move around the room. These complex behaviours require different regions of the brain to work together.

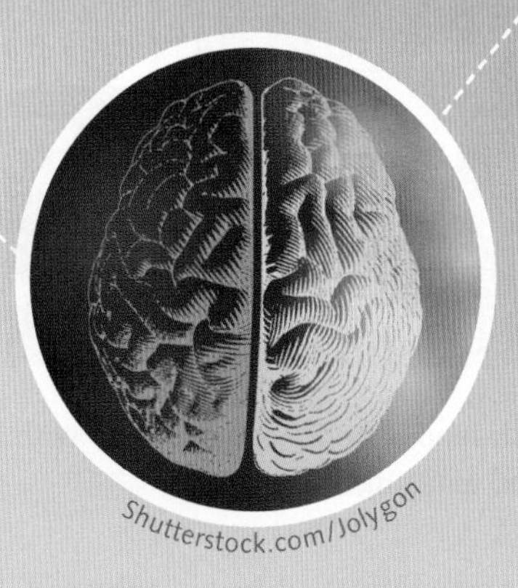
Shutterstock.com/Jolygon

It's amazing how our brains adapt and change as they respond to the world around us. Both early experiments and damage to the brain have helped us to understand how different parts of the brain are responsible for different behaviours and mental processes. These regions work together to allow us to carry out more complex tasks.

Slideshow
Chapter 4 slideshow

Flashcards
Chapter 4 flashcards

Test
Chapter 4 pre-test

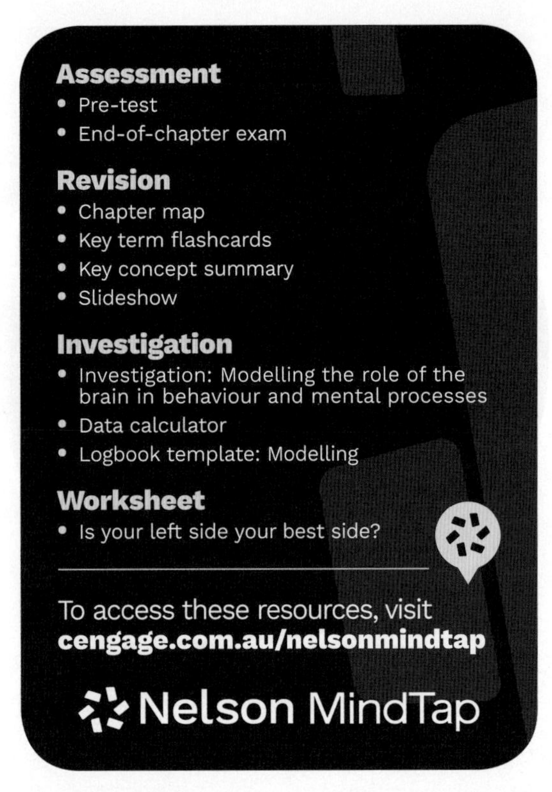

Know your key terms

Amygdala
Basal ganglia
Broca's area
Cerebellum
Cerebral hemispheres
Cerebrum
Corpus callosum
Decision-making
Emotional regulation
Emotions
Endocrine system
Executive functions
Forebrain
Frontal lobe
Geschwind's territory
Hemispheric specialisation
Hindbrain
Hippocampus
Hypothalamus
Limbic system
Lobotomy
Medulla
Midbrain
Neuroimaging
Occipital lobe
Parietal lobe
Philosophy
Pons
Primary auditory cortex
Primary motor cortex
Primary somatosensory cortex
Primary visual cortex
Reticular formation
Split-brain operation
Temporal lobe
Thalamus
Wernicke's area

Laurence Kim Peek (Figure 4.1) was born in the USA in 1952. Born with an abnormally large head, his cerebellum was malformed and he was missing the corpus callosum, the bundle of nerve tissue that connects the brain's hemispheres. These abnormalities impaired Peek's physical coordination, made ordinary reasoning difficult and limited his social abilities. Peek showed developmental difficulties from a very early age. He wasn't able to walk until the age of 4, and even then did so with a strange, sidelong gait. He couldn't button his own shirt or brush his teeth without help and motor skills were difficult. When enrolled in school, he was expelled after one day for disrupting class.

Shutterstock.com/ITV

Figure 4.1 Laurence Kim Peek was a man with an unusual and amazing brain.

But Peek also showed flashes of unusual abilities. He received part-time tutoring from the age of 7 and completed a high school curriculum by 14. According to his father, Peek was able to memorise things from the age of 16–20 months. He read books, memorised them, and then placed them upside down on the shelf to show that he had finished reading them, a practice he maintained all his life. He could speed through a book in an hour. He could remember almost everything he had read in subjects ranging from history and literature, geography and numbers to sports, music and dates. Peek read by scanning the left page with his left eye, while reading the right page with his right eye. He could accurately recall the contents of at least 12 000 books. 'He was the Mount Everest of memory,' said Dr Darold A. Treffert, an expert on savants (people with an intellectual disability who demonstrate exceptional, usually isolated, cognitive abilities) who knew Peek for 20 years.

According to his father, Peek had memorised so many Shakespearean plays and musical compositions and was such a stickler for accuracy that they had to stop attending performances. 'He would stand up and correct the actors or the musicians. He'd stand up and say: "Wait a minute! The trombone is two notes off," his father said.

The brain is responsible for every thought, action, memory, feeling and experience we have of the world. Much of what we know about the human brain comes from studies of individuals like Kim Peek who were born with abnormal brains. As a result of their brain anatomy, their thoughts, behaviours and experience of the world is very different from the norm. In this chapter we will explore how our understanding of the brain has changed over time. We will look at basic brain anatomy and the roles the various structures and areas of the brain play in our behaviour and mental processes.

Adapted from: "Kim Peek: savant who was the inspiration for the film Rain Man". The Times. December 23, 2009. Retrieved December 23, 2009; Darold A. Treffert; Daniel D. Christensen (December 23, 2009). "Inside the Mind of a Savant"; Scientific American. https://www.nytimes.com/2009/12/27/us/27peek.html

4.1 How the brain controls behaviour and mental processes

The brain is composed largely of billions of nerve cells that receive and transmit electrical and chemical signals all over the body. It is the largest and most important part of our central nervous system (CNS). It is linked to our senses, and thus our body parts, through the CNS's connection to the peripheral nervous system (PNS).

To fully understand how the brain controls behaviour, we must understand the relationship between the brain and the mind. Theories relating to the nature of this relationship began centuries ago and emerged from the world of **philosophy** (Figure 4.2). These theories provided insight into the brain, mind and behaviour, but they lacked evidence. However, during the last few centuries the application of scientific method to the study of brain structure and function has provided psychologists and neuroscientists with evidence that the brain is indeed the source of all behaviour.

PHILOSOPHY		SCIENCE
• Makes predictions about human behaviour based on logic and observation	versus	• Makes predictions about human behaviour based on empirical evidence
• Does not have empirical evidence to support its claims		• Is reliable
• Is not reliable		

Figure 4.2 Philosophy versus science

Changes in theories of brain structure and function

Our understanding of brain structure and function has changed over time. Some of the more notable stages in our understanding of brain structure and function are summarised in Table 4.1. These include the brain-versus-heart debate, the mind–body problem and phrenology.

Table 4.1 Early brain theories

Brain-versus-heart debate	Ancient philosophers argued over whether the brain or the heart controlled our thinking, intelligence, emotions and movement.
Ancient times	**The Egyptians** When the ancient Egyptians mummified their dead, they removed the brain through the nostrils and discarded it, believing it was of little importance (Figure 4.3). However, they preserved the heart because they believed it was the most important body organ and that the mind, or soul, was located in the heart. 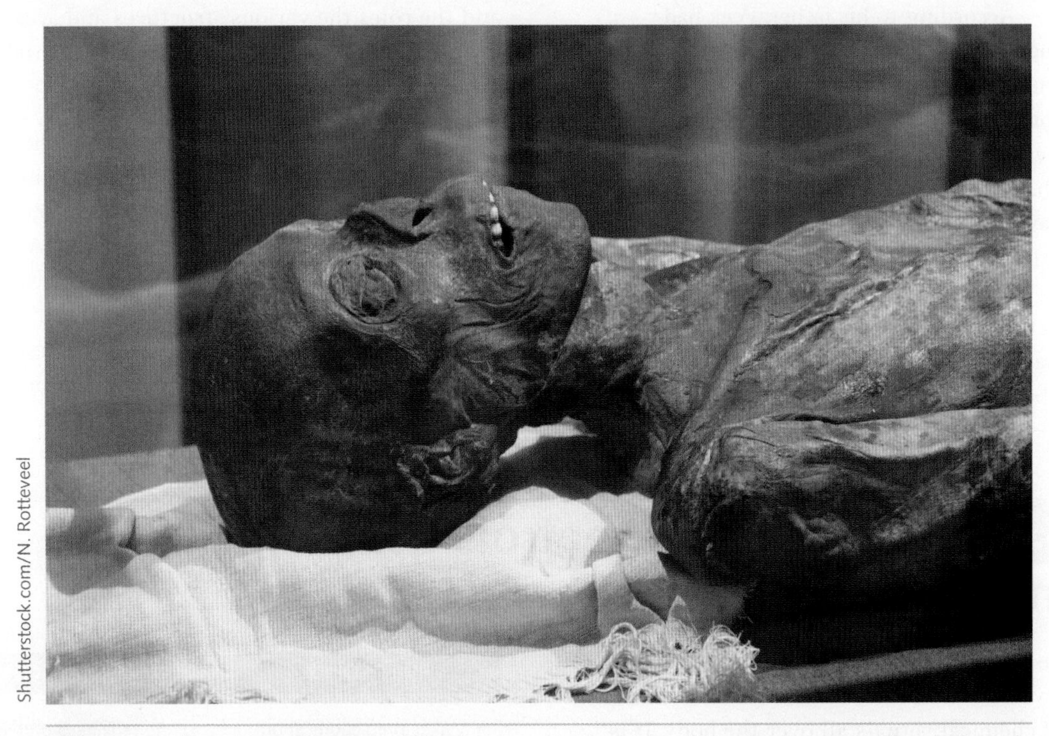 Shutterstock.com/N. Rotteveel **Figure 4.3 The ancient Egyptians did not mummify the brain of the dead person. They believed the brain was an unimportant organ.** **Greek philosopher Aristotle (384–322 BCE)** Studies of animal anatomy led Aristotle to conclude that: » the heart, not the brain, was the most important bodily organ » the heart is a three-chambered, hot, dry organ that pumped warmth and vitality to the body; all other organs exist to cool the heart » the heart is more responsive than the brain during strong emotional experiences, and this indicates that the heart is the site of mental processes » the body and mind are inseparable and interact through the heart » the mind is located in the heart, therefore it is a function of the body; so, when the body (and heart) died, so did the mind.

The mind–body problem **Ancient times**	Ancient philosophers argued over the relationship between the biological (body and brain) and the psychological (mind).The focus was: are the mind and the body separate or the same? **Greek philosopher Plato (429–347 BCE)** According to Plato (Figure 4.4): » humans have a body and a soul (a mind) and they are two separate and distinct entities » the brain is the physical location of the soul (mind) » the body is a physical entity that is observable, measurable and has a finite life span » the soul is a mental entity (controls intelligence and wisdom) that has no physical representation, so it cannot be observed or measured » when the body dies, the mind or soul is released because it cannot decompose; therefore, the body can die but the mind (or soul) cannot.
2nd century CE	**Greek physician and philosopher Aelius Galen** Galen moved our understanding of the brain's role in behaviour from philosophy to experimentation. According to Galen: » the mind and body were the same » mental activity occurred in the mind, not the heart » the brain, not the heart, controls all muscle movement by means of its connection to the spine and peripheral nerves – a theory that is still accepted today. Galen conducted vivisections (experimental surgery) on live animals to observe how different organs or body parts functioned and then generalised his findings to humans. For example, he found that when he tied off the vocal cords of a live, squealing pig, the pig stopped squealing. When he released the vocal cords, the squealing continued. When he severed the laryngeal nerve that connected the pig's brain to the muscles surrounding the vocal cords, the pig became silent.
17th century	**French philosopher René Descartes (1596–1650)** According to Descartes: » the body and mind are separate » the mind is located in the pineal gland (a small gland located in the centre of the brain) and it directs all bodily activities (Descartes had no evidence for this) » the mind is a non-physical substance and the brain is a physical substance » the mind controls the body, and therefore behaviour » to understand the mind, we must study behaviour. Descartes' theory was known as dualism.

Alamy Stock Photo/World History Archive

Figure 4.4 Plato argued that the mind and brain were separate entities.

Phrenology
19th century and early 20th century

Phrenology

Phrenology was one of the earliest techniques that tried to *predict* human behaviour. It involved feeling the bumps and depressions on a person's skull and using this information to assess their personality traits, talents and intelligence (Figure 4.5).

Phrenology was popularised in the 19th century by Franz Gall (1758–1828), a Viennese physician who believed that:

- different areas of the brain are responsible for specific psychological traits that control complex mental faculties (or personality characteristics) such as cautiousness, combativeness and agreeableness, and simpler functions such as memory, calculation ability and colour perception
- the mind is composed of 40 different abilities and their development affects the size and shape of the skull
- the more you used a specific ability, the more the area of the brain that controlled the ability would grow in size and the skull would mould around it in a bump; the lesser-used faculties would shrink, causing troughs in the skull
- if you examined the shape of the skull, you could analyse the bumps and depressions and determine a person's personality and intelligence
- memory is located in the brain, just behind the eyes; Gall based this conclusion on the observation that his friends who had bulging eyes appeared to have superior memories.

Phrenology's claim that mental functions are localised in areas of the brain is still accepted today.

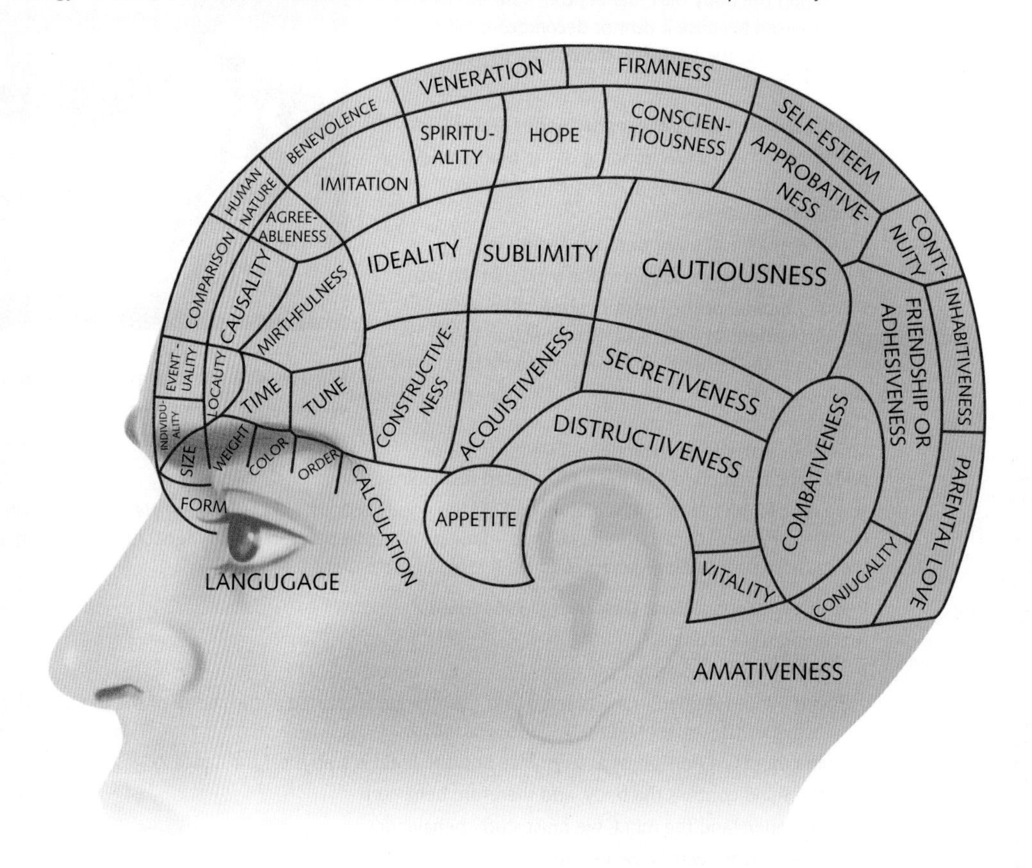

Figure 4.5 Phrenology suggested that bumps on a person's head indicated their personality type and intelligence level.

As an adult, Gall examined his friends' skulls and noted that those who had bumps on similar locations had similar personality traits. From these observations and further examination of the skulls of criminals and mental patients, Gall created a map of the skull that linked a specific personality characteristic and intellectual ability with a specific bump. He assumed, incorrectly, that the bumps and depressions of the skull reflected the amount of brain tissue lying under them.

Phrenology was very popular throughout Europe, including Germany, in the early 1900s. The Nazis used phrenology to promote racism and eugenics (the study of methods of improving the quality of the human race, especially by selective breeding) and to support their claim of German superiority and Jewish inferiority.

Though phrenology lacked scientific evidence, it did spark interest in which parts of the brain controlled specific behaviours. Gall's assumption that different parts of the brain control different aspects of behaviour has proved true, but the links he made between specific brain areas and specific behaviours has not.

Weblink
Phrenology: the weirdest pseudoscience of them all?

KEY CONCEPTS 4.1a

- Debate about the relationship between the mind, the brain and behaviour dates back to ancient times; some ideas were based on logic and reasoning but lacked scientific evidence.
- The brain-versus-heart debate focused on whether the heart or the brain controlled behaviour.
- The mind–body problem focused on whether the mind and body were separate entities or the same entity.
- In the 2nd century CE, Galen used vivisection and experiments to demonstrate that some behaviours were controlled by the brain.
- In the 17th century, Descartes argued that the separate entities of the mind and body interacted with each other through the brain's pineal gland.
- In the 19th century, Gall popularised the pseudoscience of phrenology, where a person's intelligence and personality could be determined by analysing the bumps and indentations of their skull.

Concept questions 4.1a

Remembering

1 What was the brain-versus-heart debate? r

2 What was the mind–body problem? r

3 Identify one of Galen's theories that is still accepted today. r

Understanding

4 Gall and other phrenologists held many incorrect beliefs about the brain and mental abilities. Identify one belief held by phrenologists that has proven to be correct. r

Applying

5 Identify two reasons why phrenology became known as a **pseudoscience** (a fake or false science). e

HOT Challenge

6 Revisit the information on Aelius Galen's vivisection experiment in Table 4.1. i

a What was the aim of this experiment?

b Write a hypothesis for this experiment.

Early brain experiments

Only in the last two centuries have physicians and scientists begun to apply scientific method to systematically research the brain and its function and provide evidence of how the brain works.

Experimental ablation

Around 1825, the leading brain expert of the time, Pierre Flourens (1794–1867), began using new experimental methods to determine the relationship between brain areas and behaviour and to investigate whether phrenology's claims were true. To do this, Flourens surgically removed or electrically stimulated specific parts of animal brains so he could study the behavioural effects. The method of removal of part of the brain was termed 'experimental ablation'. Flourens also electrically stimulated the brains of some of his human patients during surgery.

The results of these experiments demonstrated that specific brain regions were responsible for specific functions. For instance, when he removed an animal's cerebellum (a structure at the bottom of the brain that sits on top of the brain stem), the animal lost all balance and motor coordination. When Flourens removed one of its cerebral hemispheres, the animal's perceptual ability and behaviour were severely affected. Based on these results, Flourens concluded that:

- phrenology had no scientific or biological basis
- the cerebral hemispheres were responsible for all behaviour
- the cerebellum regulates and integrates movement
- the brain stem controls vital functions such as circulation, respiration and general bodily stability.

Flourens was unable to find specific locations for complex behaviours such as memory and cognition. This led to the further conclusion that while some behaviours can be ascribed to specific brain regions, other behaviours are represented throughout the brain (Sabbatini, 1997).

4.1.1 HISTORY AND CONCEPTS OF BRAIN RESEARCH

Lobotomy

In the 1890s, the Swiss physician Dr Gottleib Burckhardt argued that severing portions of the brain could change the behaviour of individuals with severe mental illness. Burckhardt tested his theory on several schizophrenic patients and found that many of them became much calmer after this psychosurgery.

Years later, in 1931, the physician-scientist Antonio Moniz began experimenting with psychosurgery on some of his uncontrollably emotional and violent patients. His experimental procedure involved:

» placing his patients into a coma and hammering an ice-pick-like instrument through each eye socket into their brain
» wiggling the ice-pick to sever the nerves connecting the frontal lobes and the emotion-controlling centres of the inner brain (Figure 4.6).

In the 1930s, the American neurologist Walter Freeman performed a modified version of Moniz's procedure, which was named the **lobotomy**. He also used an ice-pick-like tool to disturb the section of the frontal lobe that sits above the eye sockets. Although lobotomy succeeded in calming some disturbed patients, these patients generally became permanently lethargic, acted immaturely and were often impulsive (Caruso & Sheehan, 2017). Lobotomy was often performed without the patient's consent. Because of its devastating effects on a person's personality and ability to function independently, it is now viewed as an inhumane and unethical treatment for psychological disorders. It is now only rarely performed.

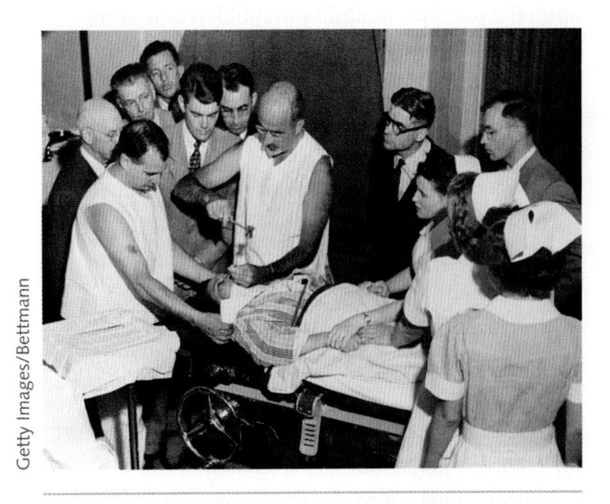

Getty Images/Bettmann

Figure 4.6 Dr Walter Freeman performing a lobotomy in 1949 by inserting an ice-pick-like instrument under the upper eyelid of the patient and cutting nerve connections in the front part of the brain

Brain ablation and brain lesioning

Modern brain **ablation** (also known as brain lesioning) is a surgical procedure used to treat various neurological or psychological disorders. For example, it is used to remove or destroy brain tumours, or change dysfunctional brain circuits (such as those associated with schizophrenia and epilepsy). Ablation uses procedures such as lasering, surgical removal or vaporisation. The standard ablation procedure involves:

» drilling holes in the skull and inserting an electrode or small tube (a canula) into a targeted area of the brain
» electricity being conducted through the electrode to destroy abnormal brain tissue or make irreversible brain lesions (tissue or organ change)
» observing the person's behaviour so conclusions can be drawn about specific functions of different parts of the brain (Franzini et al., 2019).

Split-brain studies

In the 1950s, neuropsychologists Roger Sperry (1914–1994) and Michael Gazzaniga (1939–) conducted research involving **split-brain operations** (called a corpus callosotomy) which involved cutting an area of the corpus callosum. The **corpus callosum** is a thick band of about 200 million nerve fibres (white matter) that connect the brain's left and right hemispheres. It is the structure responsible for communication between the two hemispheres because it transfers information registered in one hemisphere to the other hemisphere for processing.

Following split-brain surgery, the areas of the brain are still connected by many other structures that lie below the cerebral cortex, but the flow of information between the left and right hemispheres is interrupted. So, information in one hemisphere is unable to be transferred to the opposite hemisphere for processing. As a result, the brain cannot integrate information registered separately in each hemisphere. For example, as the 'speech centre' for most people is in the left hemisphere, information that enters the brain's right hemisphere can only be verbalised if that information is transferred to the speech centre in the left hemisphere. For this transfer to happen, the corpus callosum must be intact.

Split-brain operations were done usually to control the spread of seizures from hemisphere to hemisphere that occurs with severe epilepsy.

Effects of a split-brain operation

Following a split-brain operation, patients would be tested to determine the effect on their behaviour of severing their corpus callosum and interrupting communication between their brain hemispheres. A standard test would follow these steps.

- The patient would sit at a table with a screen with a black dot in the centre in front of them. They could fit their hands under the screen to reach objects on the other side, but they could not see the objects or their own hands.
- They were asked to fix their eyes on the black dot in the centre of the screen.
- A projector flashed images or words onto the screen to only one hemisphere. (It is important to note that each hemisphere controls the opposite side of the body.) Information presented to only the left visual field (the left side of the patient's vision) was sent to the right hemisphere; information presented to only the right visual field (the right side of the patient's vision) was sent to the left hemisphere.
- To ensure patients could actually see the image or word presented on the left, several items that represented the image or word were hidden and accessible through a hole, and the patient would be able to identify the word or image through feeling the hidden object.
- Patients were asked, 'What did you see?' If flashed on the right, the information travelled to the left hemisphere and patients would accurately recall seeing the image or words; if flashed on the left, the information travelled to their right hemisphere. These patients would not be able to accurately recall what they had seen and in some cases denied seeing anything.

Conclusions from split-brain studies

Sperry and Gazzaniga concluded that:

- the corpus callosum is essential in allowing information to travel between the hemispheres
- the two hemispheres have different abilities and functions. The left hemisphere is dominant in speech and language. The right hemisphere is dominant in terms of visual-motor tasks.

Although we might imagine that someone with a 'split brain' might perform actions that continually conflict with each other, most patients act completely normal. This is because both halves of the brain have the same experience at the same time. Information is usually available to both the left and right sides of the body and travels at the same rate to each hemisphere. For instance, when you watch television, images enter both eyes at the same time and travel to each hemisphere at the same rate, arriving at the same time.

Also, if a conflict arises, one hemisphere usually overrides the other. For example, if you were watching a scary movie involving a mystery to be solved, a conflict may arise over whether to respond to the frightening emotion and stop watching; or continue watching in order to solve the mystery. One hemisphere will dominate, and a decision will be made.

Advances in treatments and medication now enable many people to control their condition without having to undergo surgery. However, split-brain surgery is still used in extreme cases for uncontrollable forms of epilepsy when frequent seizures affect both sides of the brain. Split-brain research has provided valuable insight into the fields of psychology and neuroscience. For example, it has added to our knowledge of the corpus callosum's role in transmitting information and how the hemispheres are specialised for behaviour.

WB 4.1.2 SPLIT-BRAIN OPERATIONS

ANALYSING RESEARCH 4.1

Split brain does not lead to split consciousness

In 2017 a research study found strong evidence that split brain does not cause two independent conscious perceivers in one brain.

Split brain is a lay term to describe the result of a corpus callosotomy, a surgical procedure first performed in the 1940s to alleviate severe epilepsy among patients. During this procedure, the corpus callosum, a bundle of neural fibres connecting the left and right cerebral hemispheres, is severed to prevent the spread of epileptic activity between the two brain halves. While mostly successful in relieving

epilepsy, the procedure also virtually eliminates all communication between the cerebral hemispheres. The result is a 'split brain'.

This condition was made famous by the work of Nobel laureate Roger Sperry and Michael Gazzaniga. Sperry and Gazzaniga discovered that split-brain patients can only respond to stimuli in the right visual field with their right hand and vice versa. This was taken as evidence that severing the corpus callosum causes each hemisphere to gain its own consciousness.

For their study, the research team, led by UvA psychologist Yair Pinto, conducted a series of tests on two patients who had undergone a full split-brain operation. In one test, the patients were placed in front of a screen and shown various objects displayed in several locations. The patients were then asked to confirm whether an object appeared and to indicate its location. In another test, they had to correctly name the object they had seen.

'Our main aim was to determine whether the patients performed better when responding to the left visual field with their left hand instead of their right hand and vice versa,' says Pinto. 'This question was based on the textbook notion of two independent conscious agents: one experiencing the left visual field and controlling the left hand, and one experiencing the right visual field and controlling the right hand.'

To the researchers' surprise, the patients were able to respond to stimuli throughout the entire visual field with all the response types: left hand, right hand and verbally. According to Pinto, 'The patients could accurately indicate whether an object was present in the left visual field and pinpoint its location, even when they responded with the right hand or verbally. I was so surprised that I decided repeat the experiments several more times with all types of control.'

Pinto's results present clear evidence for unity of consciousness in split-brain patients. 'The established view of split-brain patients implies that physical connections transmitting massive amounts of information are indispensable for unified consciousness; i.e. one conscious agent in one brain. Our findings reveal that although the two hemispheres are completely insulated from each other, the brain as a whole is still able to produce only one conscious agent. This directly contradicts the established view of unified consciousness.'

In the coming period, Pinto plans to conduct research on more split-brain patients to see whether his findings can be replicated. 'These patients, who are rapidly decreasing in numbers, are our only way to find out what happens when large subsystems in the brain no longer communicate with each other. This phenomenon raises important questions that cannot be investigated in healthy adults because we have no technique to isolate large subsystems in healthy brains.'

Adapted from Universiteit van Amsterdam (UVA). "Split brain does not lead to split consciousness." ScienceDaily. ScienceDaily, 25 January 2017. <www.sciencedaily.com/releases/2017/01/170125093823.htm>.

Questions

1 What was the aim of Pinto's study?
2 Who were the participants in this study?
3 What was the result of Pinto's study?
4 What conclusions did Pinto draw?

Neuroimaging techniques

To examine a living, functioning brain early neuroscientists had to wait to operate on people who had sustained brain injuries. Or they could study the brain of a recently deceased person, by waiting for people to die in order to conduct an autopsy. However, neither of these two options allowed the brain to be studied in action.

Fortunately, technological advances and the introduction of **neuroimaging** techniques now allow scientists to study the structure *and* function of the human brain in ways that were not previously possible. Today, researchers can electrically, chemically or magnetically stimulate various parts of living brains (both intact and damaged) and see the effects on behaviour. These techniques are discussed in more detail in Chapter 5.

ACTIVITY 4.1 APPROACHES TO UNDERSTANDING THE BRAIN

Try it

1 Select one of the following approaches to understanding brain structure and function and research the questions below online. Different approaches:

» Lobotomy

» Brain ablation

a What is the name of the treatment?

b When was this treatment first used?

c Outline the procedure of the treatment.

d What was the basis for this treatment? What did physicians believe would be the benefit of the treatment?

e What were the side-effects?

f Is this treatment method still used today? If so, for what purpose?

Apply it

2 A recent scan has revealed that Michael has a tumour growing in his brain. His specialist has stated that they will need to operate to prevent the tumour spreading. Suggest whether a lobotomy or brain ablation would be the preferred option to treat Michael's tumour.

Exam ready

3 A key finding of experimental ablation is that

A researchers discovered that phrenology was largely accurate.

B specific brain areas are responsible for particular functions.

C lobotomies are a targeted way to improve the mental health and wellbeing of individuals.

D ablation only works in studies using monkeys, making it difficult to generalise to humans.

ANALYSING RESEARCH 4.2

Stories may change the brain

Researchers at Emory University have used a neuroimaging technique called functional magnetic resonance imaging (fMRI) to demonstrate that reading a novel may cause brain changes that last for a few days after reading the story. Previous research using fMRI data had identified brain networks associated with reading stories while participants read them; that is, while they were in the fMRI scanner. The Emory study, however, focused on the lasting neural effects of reading a novel.

Twelve female and nine male subjects between the ages of 19 and 27 participated in the study. Each morning for 19 consecutive days, the participants underwent an fMRI scan of their brains in a resting state. For the first 5 days and the last 5 days of the study, participants did not perform any task other than to have their brains scanned. On the middle 9 days, their task was to read a section of a novel (1/9 each day) during the evening. Before their scan the next morning, participants also completed a quiz and self-report to confirm that they had done the reading and report on how arousing the reading was.

The chosen novel was *Pompeii: A Novel*, a 2003 thriller by Robert Harris, based on the eruption of the volcano Mount Vesuvius, which buried the cities of Pompeii and Herculaneum in ancient Italy. The researchers chose this novel due to its dramatic plot: its main protagonist is outside the city of Pompeii when he notices steam around the mountain. Sensing something is amiss, he tries to return to Pompeii to save the woman he loves. Meanwhile, nobody in Pompeii realises the volcano is about to erupt.

The researchers found that, following the reading assignments, even though participants were not actually reading while they were in the scanner, the fMRI showed heightened connectivity in the left temporal cortex, an area of the brain associated with language. The primary sensory motor region of their brain also showed increased connectivity. These increases in connectivity persisted for several days after completion of the reading.

'The neural changes that we found associated with physical sensation and movement systems suggest that reading a novel can transport you into the body of the protagonist,' says

»

neuroscientist Gregory Berns, the lead author of the study. 'We already knew that good stories can put you in someone else's shoes in a figurative sense. Now we're seeing that something may also be happening biologically.'

Berns says that the neural changes were not just immediate reactions, because they persisted the morning after the readings and for the five days after the participants completed the novel.

'It remains an open question how long these neural changes might last,' Berns says. 'But the fact that we're detecting them over a few days for a randomly assigned novel suggests that your favourite novels could certainly have a bigger and longer-lasting effect on the biology of your brain.'

Sources: Berns, G. S., Blaine, K., Prietula, M. J., & Pye, B. E. (2013). Short- and Long-Term Effects of a Novel on Connectivity in the Brain. Brain Connectivity, 3(6): 590–600, doi:10.1089/brain.2013.0166; Clark, C. (2013). A novel look at how stories may change the brain. eScienceCommons. Emory University. Retrieved from http://esciencecommons.blogspot.com.au/2013/12/a-novel-look-at-how-stories-may-change.html

Questions

1 What was the aim of this research?
2 Create a hypothesis for this research.
3 What was the purpose of the first series of resting-state fMRI scans?
4 What were the results of the study?
5 What conclusion can be drawn from the results?

KEY CONCEPTS 4.1b

» During the 19th and 20th centuries, brain research began using a more scientific approach.
» Flourens used an early form of ablation and electrical stimulation to demonstrate that phrenology was baseless. He provided some evidence that the cerebral hemispheres control behaviour.
» During the 19th and 20th centuries lobotomy became a popular surgical technique for calming psychologically dysfunctional people. Results of lobotomies suggested that the brain's frontal lobe controls cognition, personality and emotion.
» Lobotomy was often carried out without the patient's consent and resulted in devastating effects on personality and the ability to function independently. Therefore, it is now considered unethical and barbaric.
» The modern surgical technique of brain ablation and lesioning is a more ethical way of treating a range of neurological and psychological disorders.
» Since the 1950s, split-brain operations, which sever the corpus callosum, have been used to treat extreme cases of epilepsy, although they are now used less frequently because of improvements in other treatments.

Concept questions 4.1b

Remembering

1 What is the corpus callosum? r
2 What is the purpose of brain lesioning? r

Understanding

3 Identify one similarity and one difference between a lobotomy and an ablation. e

Applying

4 Provide examples to show that Flourens' work followed a scientific approach when compared to previous work on the brain. c

HOT Challenge

5 Burkhardt, Moniz and Freeman all performed surgical procedures on mentally disturbed patients, which involved destroying a portion of their frontal lobe and permanently altering their behaviour. Identify and explain three current ethical principles and/or guidelines with their procedures. i

4.2 Brain structure and function

The brain is the body's master information-processing and decision-making organ. It receives, processes and interprets information received from the body's sensory systems, which it integrates and forms a response to. It then sends motor messages out to all parts of the body so that coordinated and appropriate responses can be made.

Structurally, the brain can be subdivided into three main regions:
» the hindbrain
» the midbrain
» the forebrain.

Each region (Figure 4.7) has its own purpose that is vital for everyday functioning and information processing.

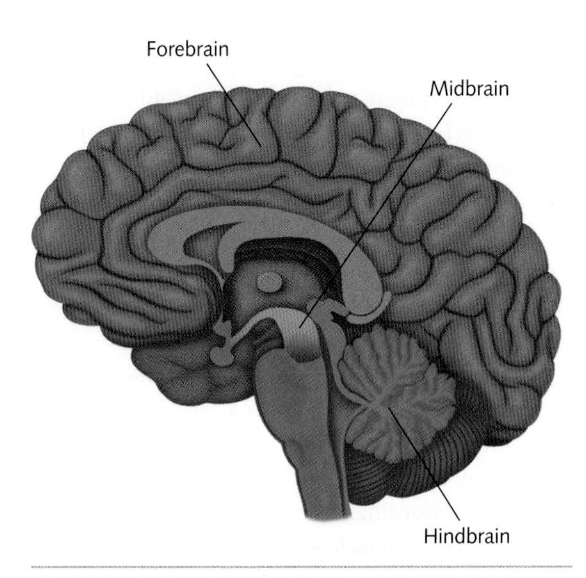

Figure 4.7 The three main regions of the brain

The hindbrain

The **hindbrain** is often referred to as the 'brain stem'. It is located at the base of the brain near the back of the skull. The hindbrain is responsible for lower-brain functions that occur without any conscious effort. These functions include:
» control of basic autonomic survival functions such as heart rate, breathing, sleep and arousal; these functions are automatic and do not require conscious effort
» coordination of voluntary muscle movements, balance, posture and reflexive actions such as coughing, swallowing and vomiting.

The hindbrain is made up of the following structures:
» the pons
» the medulla
» the cerebellum.

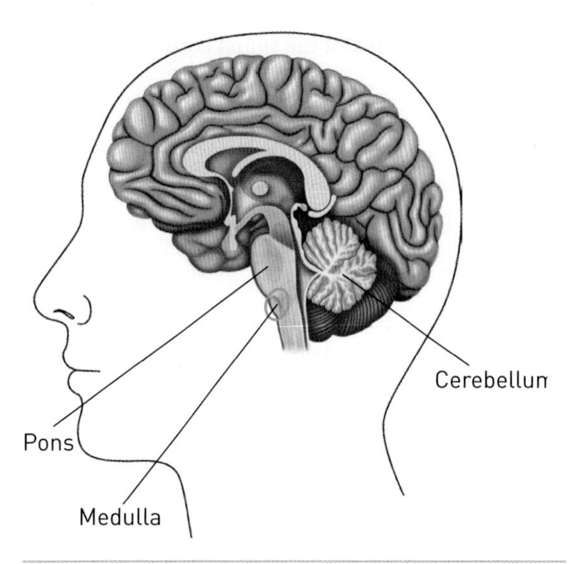

Figure 4.8 The location of the pons, medulla and cerebellum

The pons

The **pons** is a group of nerves that connects the cerebral cortex (the outer layer of the cerebral hemispheres) with the medulla. Four of the 12 cranial nerves, which enable a range of activities such as tear production, chewing and blinking, begin in the pons.

Location

The pons is located beneath the midbrain and above the medulla (Figure 4.8).

Functions

The pons:
» helps to transfer neural messages between various parts of the brain and the spinal cord
» is involved in arousal, sleep, daydreaming, waking, breathing and coordination of some muscle movements and motor tone.

Effects of damage to the pons

Injury to the pons may result in sleep disturbances, sensory problems, arousal dysfunction, coma, difficulty swallowing, walking and speaking, and paralysis.

The medulla

The **medulla** (also known as the medulla oblongata) is the lowest part of the brain. It connects the brain and the spinal cord.

Location

The medulla is located at the base of the brain stem, in front of the cerebellum (Figure 4.8). It sits below the pons and above the spinal cord.

Functions

The medulla:

» relays information between the spinal cord and the brain

» regulates vital involuntary bodily functions (such as swallowing, digestion, breathing, heart rate, blood pressure, vomiting, salivating, coughing, gagging and sneezing) by communicating with the autonomic nervous system (ANS), the part of the nervous system responsible for energising and arousing the body, and directing various body systems to change their level of activity according to the needs of the body. For example, when you are faced with a potentially threatening situation, the medulla responds to the perception of the threat by directing the ANS to send signals to your heart to increase its beating in order to flood the body with oxygen. This helps you deal effectively with the potential threat by either fleeing or fighting it.

Effects of damage to the medulla

Damage to the medulla:

» causes interrupted transmission of neural information between the spinal cord and the brain. This in turn causes some form of physiological dysfunction, such as problems with breathing and balance, tongue dysfunction and difficulty swallowing, vomiting, loss of muscle control, uncontrollable hiccups, loss of feeling in the limbs, body or face, and loss of the gag, sneeze, or cough reflex

» can be fatal, because the medulla controls the vital organs and if this function is affected, a person may need life support machines to regulate such things as breathing and heart rate.

A variety of drugs can affect the functioning of the medulla. Sometimes, drug overdoses cause the medulla to stop performing its survival functions.

The cerebellum

The **cerebellum** is often referred to as the 'little brain'. This is mainly due to its wrinkly appearance, similar to the outer layer of the entire brain (cerebral cortex). The cerebellum has several functions relating to coordination and movement.

Location

The cerebellum is located at the rear of the brain stem (Figure 4.8), beneath the occipital and temporal lobes.

Functions

The cerebellum:

» helps coordinate voluntary movement and balance by relaying motor information to and from the cerebral cortex

» receives information from the spinal cord, sensory systems and other parts of the brain and combines this information so we can adjust our posture, muscle tone and muscular coordination to fit our circumstance; for example, to produce the posture and balance needed for various yoga positions, such as the tree pose (vrikshasana; Figure 4.9)

» helps to coordinate the timing and force of the different muscle groups that act together during a voluntary movement, so that we have smooth limb and body movements.

» is believed to play a role in motor learning, where motor skills are improved through practice. For example, a football player who learns to kick a football and regularly practices will remember the specific movements they need to make in order to master this skill. They may need to exert a certain amount of muscular strength in their leg when dropping the ball on their foot to kick the ball, as well as directing their foot on the correct angle in order for them to kick the ball through the goal posts. This process involves voluntary movement and balance, and in order to do this process well, motor learning has often occurred.

Effects of damage to the cerebellum

The cerebellum can be damaged by viruses that affect the nervous system, or by head trauma or stroke. Damage to the cerebellum:

» results in reduced motor control, and difficulty coordinating balance and maintaining equilibrium
» has symptoms that include frequent stumbling; impaired coordination of the arms, legs and back; unsteady walking; and dizziness.

The cerebellum is also affected when you drink alcohol, which explains why drunk people often have difficulty balancing and coordinating their movements.

Figure 4.9 The cerebellum functions to maintain posture and balance and to coordinate movement; for example, when practising the tree pose in yoga.

The midbrain

The **midbrain** is a small area of the brain that connects the hindbrain and the forebrain. It is located below the cerebral cortex and at the top of the hindbrain. The midbrain plays a crucial role in processing information related to hearing, vision, movement, pain, sleep and arousal. Its systems help to keep us alert, awake and attentive. The reticular formation is one important structure located in the midbrain.

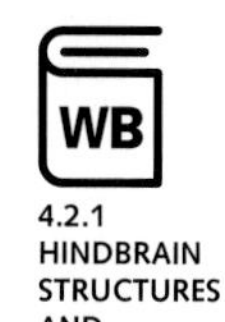

4.2.1 HINDBRAIN STRUCTURES AND FUNCTIONS

The reticular formation

The **reticular formation** is a complex network of neurons responsible for coordinating the function of many vital brain systems necessary for survival.

Location

The reticular formation extends throughout the length of the brainstem, from the spinal cord to the midbrain (Figure 4.10).

Function

The reticular formation:

» stimulates the brain by bombarding it with important sensory information, which keeps the cerebral cortex active and alert (we experience this as a state of conscious awareness). This helps us selectively focus our attention because unimportant information is ignored and doesn't receive any further processing. For example, if you were asked to look around the room and identify all the blue objects, your cortex would be able to attend to these objects and identify them. If you were then asked to close your eyes and recall all the brown objects in the room, you may have difficulty doing this because your reticular formation was directing your cortex to attend to blue objects only when your eyes were open. Brown objects have been filtered out of your attention system as this information was considered unimportant. This filtering function stops the brain from being overloaded
» has side branches of the nerve fibres from sensory neurons leading into it, and these branches filter incoming sensory information, sorting it into two categories: important and unimportant. They also stimulate the reticular formation to send its own nerve impulses up towards the cortex, arousing the cortex to a state of alertness and activity
» controls our physiological arousal and awareness and muscle tone by regulating the function of the autonomic nervous system (ANS); therefore, it helps to regulate the sleep–wake cycle.

Effects of damage to the reticular formation

Damage to the reticular formation can:
- » disrupt the sleep–wake cycle
- » cause loss of control of attention, because it is no longer able to filter out unimportant stimuli
- » cause problems with pain management and balance
- » result in the person going into an irreversible coma.

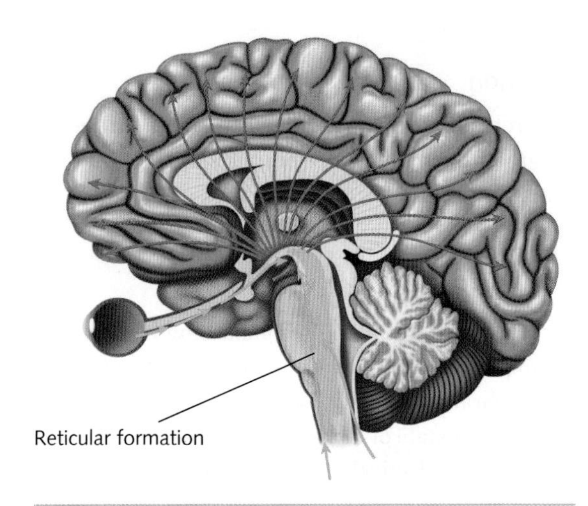

Figure 4.10 The location of the reticular formation

The forebrain

The **forebrain** is located above the midbrain towards the top of the brain. It is the largest, most complex and highly developed region of the brain. It contains a variety of structures that are responsible for our most complex processes. These include emotions, motivations, sensations, perceptions, learning, memory and reasoning. The forebrain contains three distinct areas:
- » the hypothalamus
- » the thalamus
- » the cerebrum.

The hypothalamus

The **hypothalamus** is a peanut-sized structure that weighs approximately 4 grams.

Location

The hypothalamus is located just below the thalamus (Figure 4.11).

Function

The hypothalamus:
- » maintains homeostasis in the body, the state of internal physiological stability
- » regulates the release of hormones that help us achieve a particular physiological state by connecting the nervous system to the **endocrine system** (a network of glands that produce and secrete hormones). For example, the pituitary gland releases cortisol to energise the body so it can cope when a person is confronted with a stressful situation such as studying for an upcoming exam. Once the exam is over, the hypothalamus directs the pituitary gland to stop releasing cortisol. When cortisol release stops, the body can return to homeostasis
- » by the release of hormones, influences behaviours associated with basic biological needs, such as hunger and thirst. For example, leptin and ghrelin are hormones made by fat cells that influence appetite. When leptin is released, your appetite increases. When ghrelin is released, your appetite decreases
- » controls the brain's internal 'body clock', which helps us regulate our circadian rhythms and coordinate our sleep–wake cycle
- » regulates our appetite, thirst and body temperature.

Effects of damage to the hypothalamus

Damage to the hypothalamus may cause disruptions in body temperature regulation, growth, eating habits and weight control, emotions, sexual behaviour and motivation, and sleep cycles.

The thalamus

Location

The **thalamus** consists of two small egg-shaped structures (thalami) joined together, which are positioned in the centre of the brain, on top of the brain stem (Figure 4.11).

Function

The thalamus has several functions.
- » It acts as a relay system for sensory messages on their way to the cerebral cortex (outer layer of the brain). Information from all senses travels up the spinal cord and into the thalamus (except smell [olfaction], which has its own route to the brain). The thalamus analyses the information and directs it to appropriate sensory areas of the cerebral cortex. For example, when you experience a thunderstorm,

information about the sound of the thunder, the pounding of the rain, the darkening sky, the flash of lightning and the moisture in the air are all sent to the thalamus. These pieces of information are then relayed to the areas of the cortex responsible for registering all those different sensations, then sent back to the thalamus. This flow of information is so rapid that the numerous sensory inputs from the storm are noted, received, registered, processed and integrated all at once, to provide a perception of the intense experience that is a thunderstorm.

» It conducts motor signals and relays information from the brain stem to the cortex.
» It coordinates shifts in consciousness such as waking up and falling asleep.
» It is capable of regulating itself, and consequently allows us to automatically focus attention on those stimuli or inputs that are more important, while filtering out unnecessary information. The thalamus does this by working with the reticular formation. This attention function is like an on-off switch, and is related to the level of activity in the rest of the nervous system. For example, when you are asleep, the thalamus may 'switch off' this attention function so that your cortex is not bombarded with sensory information, allowing you to have a restful night's sleep.

Effects of damage to the thalamus

Damage to the thalamus can lead to:

» deafness, blindness or light sensitivity, or loss of any other sense, except smell
» sensory issues, such as tingling, numbness, hypersensitivity and pain
» problems with movement and tremors
» attention problems
» sleep–wake problems, such as insomnia
» in the case of severe damage, a coma.

The cerebrum

The **cerebrum** is the largest and most highly developed part of the brain. It is located at the top of the forebrain (Figure 4.12) and lies over and around most brain structures. It is responsible for most of our conscious actions.

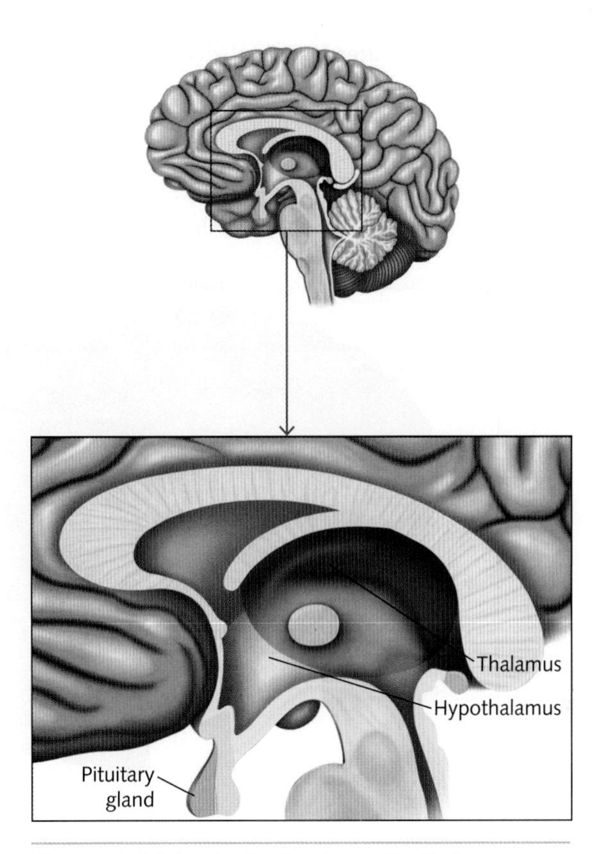

Figure 4.11 Forebrain structures: the thalamus and the hypothalamus

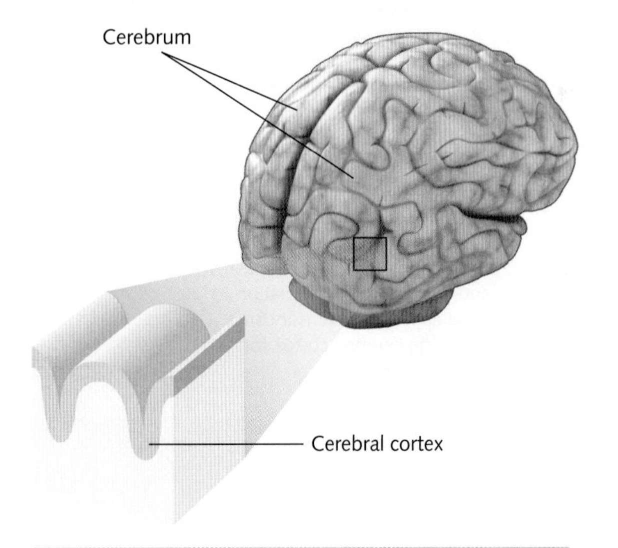

Figure 4.12 The location of the cerebrum. The outer layer of the cerebral hemispheres is the cerebral cortex, and it has separate functional regions that control specific higher cognitive processes (such as consciousness, thought, memory, language, learning, perception and planning), as well as sensory motor processing.

4.2.2 FOREBRAIN STRUCTURES AND FUNCTIONS

4.2.3 THE BRAIN COMMUNICATES WITH THE ENDOCRINE SYSTEM

4.2.4 STRUCTURES CONNECTING THE HINDBRAIN AND FOREBRAIN VIA THE MIDBRAIN

The cerebrum's outer layer is the cerebral cortex, the brain area responsible for higher cognitive functions, voluntary movement, emotions and personality.

The cerebrum can be divided into two parts: the left and right hemispheres. Both hemispheres are mostly symmetrical in their anatomical structure; however, they differ in their functions (see Section 4.3).

Figure 4.13 is a summary of the structures and functions of the human brain.

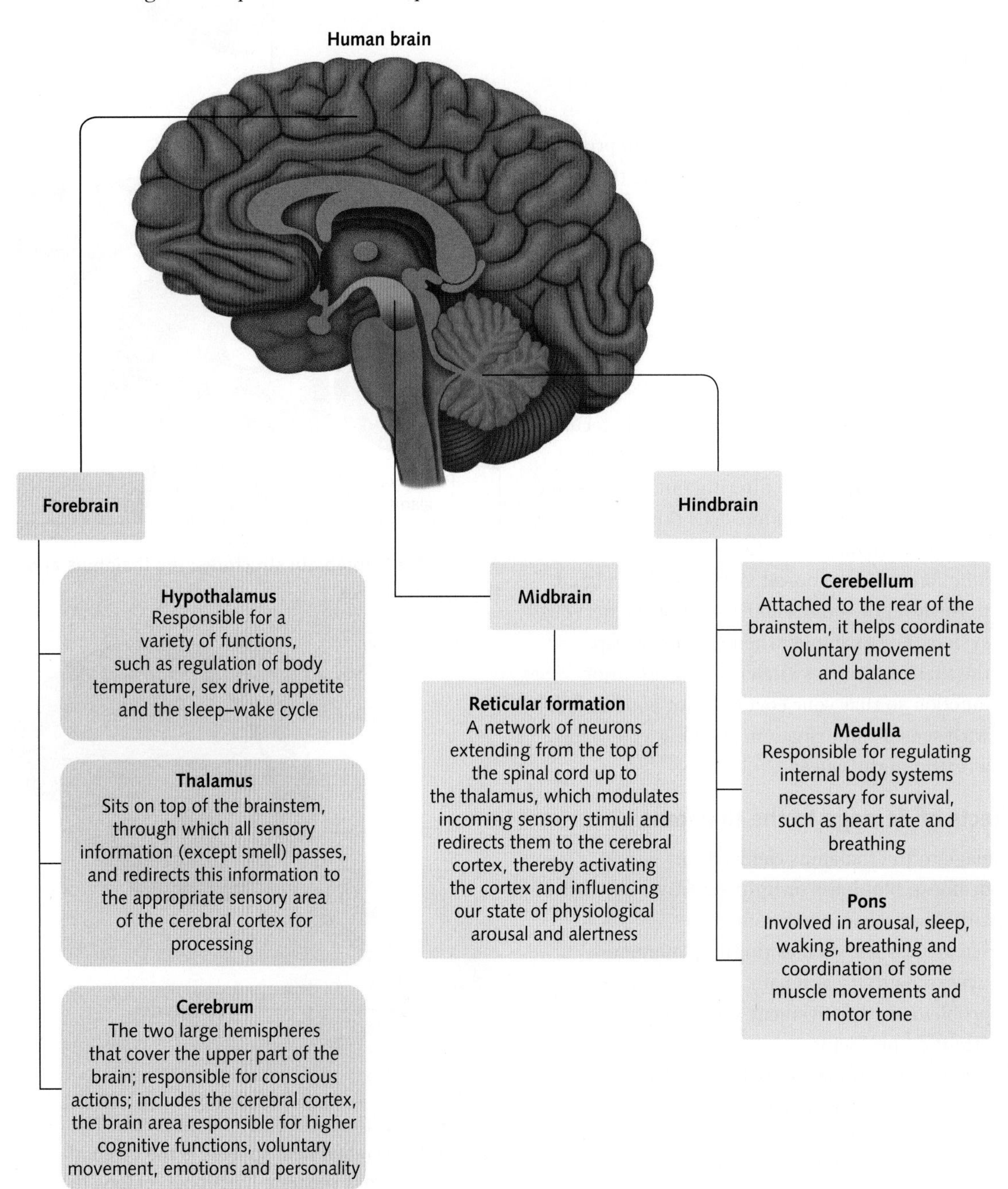

WB 4.2.5 FOREBRAIN, MIDBRAIN AND HINDBRAIN STRUCTURES

Figure 4.13 A summary of the structures and functions of the human brain

KEY CONCEPTS 4.2

» The brain has three main regions: the hindbrain, the midbrain and the forebrain.
» The hindbrain consists of three structures: the pons, medulla and cerebellum, which are responsible for autonomic survival functions and coordination of voluntary muscle movement, balance and posture, and reflexive actions.
» The midbrain consists of several structures involved with processing sensory information, movement, sleep, arousal and attention.
» The reticular formation is a major midbrain structure. It helps us to selectively focus our attention and alertness, and it controls physiological arousal.
» The forebrain is the largest and most highly developed part of the brain. It is divided into two hemispheres.
» The forebrain has three distinct structures: the hypothalamus, thalamus and cerebrum. These parts interact to control our most complex processes, including emotions, motivations, sensations, perceptions, learning, memory and reasoning.

4.2.6 FUNCTIONS OF THE HINDBRAIN, MIDBRAIN AND FOREBRAIN: SUMMARY

Concept questions 4.2

Remembering

1 What is the hindbrain responsible for? r
2 Identify two functions of the reticular formation. r
3 What brain structure is responsible for balance, posture and coordination of voluntary movement? r
4 What is the area located at the lower back part of the brain called? r

Understanding

5 Explain how the thalamus functions to help us have a restful night's sleep. e

Applying

6 The finger-to-nose-test is used to assess balance and coordination. The person being tested has to close their eyes and extend their arms outward from the sides of their body, and then touch the tip of their nose with alternating movements by each arm.
If the person has a damaged cerebellum, what would be the result of this test? Explain the reasons why. c

HOT Challenge

7 Salvatore, a retired boxer, was experiencing a number of physical and psychological problems so he went to see his doctor. After Salvatore described his symptoms, his doctor suspected that Salvatore's thalamus might be damaged, so he ordered a PET scan.
Describe two physical symptoms and two psychological symptoms that Salvatore may have described that would lead his doctor to this conclusion. c

4.3 The cerebral hemispheres

As mentioned in the previous section, the cerebrum is located at the top of the forebrain and is divided into two hemispheres. The two **cerebral hemispheres** have particular functions and specialties (Figure 4.14), and this is called **hemispheric specialisation**.

» The left hemisphere is usually dominant in verbal and analytical functions. It also receives sensory information from the right side of the body and controls voluntary movement on the right side of the body.
» The right hemisphere specialises in non-verbal, artistic and spatial abilities. It receives sensory information from the left side of the body and controls voluntary movement on the left side of the body.

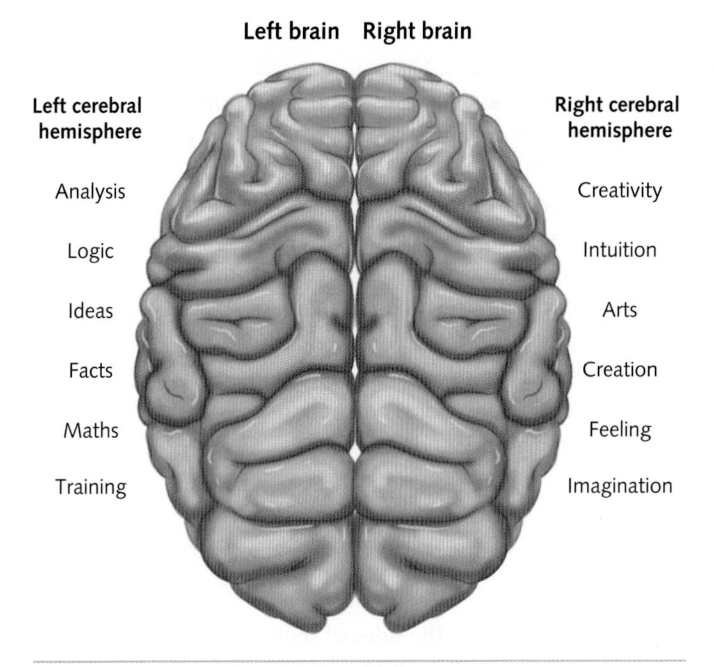

Figure 4.14 **The left and the right cerebral hemispheres, showing which abilities each hemisphere predominantly controls**

The left hemisphere: verbal and analytical functions

The left hemisphere controls the functions of the right side of the body and is responsible for the verbal and analytical functions.

Language function

The left hemisphere controls language function in these ways.

» Approximately 95 per cent of people have their language centres (Broca's area and Wernicke's area) in the left hemisphere.
» The ability to communicate our awareness using spoken or written language depends on left hemisphere function. Most adults use the left hemisphere for speaking, writing, reading and understanding language.

Analytical function

The left hemisphere controls analytical function in these ways.

» It breaks information into parts and processes it sequentially (Springer & Deutsch, 1998).
» It focuses on small details, so its focus is local (Heinze et al., 1998; Hellige, 1993; Hübner, 1998).
» It is involved in a person having superior ability at mathematics, judging time and rhythm, and coordinating the order of complex movements, such as those needed for speech.

The right hemisphere: non-verbal functions

The right hemisphere controls the functions of the left side of the body and is responsible for the non-verbal functions.

Information processing

The right hemisphere controls information processing in these ways.

» It processes information simultaneously (at the same time) and holistically (all at once) (Springer & Deutsch, 1998).
» It is involved in a person having superior ability at assembling pieces of information into a coherent image, because it identifies overall patterns and general connections and it has a global focus (Heinze et al., 1998; Hellige, 1993; Hübner, 1998).

Non-verbal communication

The right hemisphere is involved in non-verbal communication in these ways.

» For 95 per cent of adults, the right hemisphere can only produce the simplest language and numbers. To answer questions using functions of the right hemisphere means using non-verbal responses, such as pointing at objects or drawing them.
» It contributes to the understanding of language by allowing us to see the overall context in which something is said (Beeman & Chiarello, 1998). This helps us understand nuances of language such as jokes, irony and sarcasm.
» It is more dominant in detecting and expressing emotion non-verbally (Borod et al., 1998; Christianson et al., 1995).

Spatial skills

The right hemisphere is dominant in spatial skills and visual skills such as recognising patterns, faces and melodies; putting together a puzzle; reading a map or interpreting a painting.

Summary of left and right hemisphere function

Left and right hemisphere functions are summarised in Table 4.2.

Some tasks require the use of one hemisphere more than the other. However, in most real-world activities the two hemispheres work together holistically to complete a task. Each hemisphere shares its information with the other hemisphere via the corpus callosum.

4.3.1 HEMISPHERIC SPECIALISATION

Table 4.2 Specialised functions of the left and right hemispheres

Functions	Left hemisphere (verbal and analytical)	Right hemisphere (non-verbal)
Sensory functions	» Receives sensations from the right side of the body	» Receives sensations from the left side of the body
Motor functions	» Controls movement in the right side of the body » Helps in production of speech (Broca's area)	» Controls movement in the left side of the body
Cognitive functions	» Reading » Writing » Analytical thinking » Sequential processing » Logical reasoning » Mathematical ability	» Musical ability » Spatial ability – design, movement, dance » Emotional expression » Direction of emotion
Perceptual functions	» Understanding of language (Wernicke's area)	» Facial recognition » Pattern recognition

CASE STUDY 4.1

Karley Miller

'I am strong, I can get through this': The 17-year-old girl who had half her brain removed – so she could finally live a normal life

Belinda Cleary

Karley Miller, 17, of Cowra in central west NSW underwent a hemispherectomy to remove the right side of her brain which had been severely damaged when she suffered a stroke in utero at just 18 weeks.

The stroke caused Karley to be born with cerebral palsy, suffer from epilepsy and live with a vision impairment as it had killed half of her still-developing brain.

'I was told the risks of the operation were infection, death, loss of speech and loss of mobility,' she said.

The brave decision came after she suffered through a massive nine-and-a-half hour seizure.

'I just had to do it. I couldn't live having seizures anymore.

'I thought, I am strong I can get through this.'

Her mother, Nikki Miller, said Karley would remain conscious throughout her fits, which made them even more unbearable.

'I couldn't go anywhere without mum being a few steps behind, I couldn't even have a shower with the door locked in case I had a seizure and no-one could get to me.'

Mrs Miller admitted the operation was a scary prospect.

But 'it was all worth it' because four months after the operation her daughter has been given the 'all clear'.

She has now been taken off one of her anti convulsion medications, and is being weaned off the other.

"I am strong, I can get through this': The 17-year-old girl who had half her BRAIN removed - so she could finally live a normal life" by Belinda Cleary, MailOnline, 07/02/2016

»

»

Living with half a brain: seeing life anew through the camera lens

Melanie Pearce with Luke Wong

After having radical, but life-changing surgery to remove half her brain, a young woman from regional New South Wales has found new purpose as a photographer and wants to help change the lives of other young people with disabilities.

Ms Miller has Sturge-Weber syndrome, which is a rare medical condition that resulted in her having seizures, a facial port wine birthmark and vision impairment through glaucoma.

She also has cerebral palsy.

Before surgery she was having almost constant seizures and had a brain herniation protruding from her forehead.

'The right side of her brain was basically killing her. She had to have this surgery or she wouldn't have made it here today,' Ms Miller's mother, Nikki, said.

Surgeon, Mark Dexter, told the family the right side of Karley's brain may as well be removed in radical surgery called a hemispherectomy.

The surgery, in September 2015, at Sydney's Westmead Hospital involved cutting Ms Miller's head from ear to ear, taking away some of her skull and removing the right side of her brain.

The seizures stopped the day after the surgery and she has not had one since.

Her headaches have also decreased to only occasional ones and she is able to function and look forward to the future.

Several months of recovery and rehabilitation followed the surgery and then Ms Miller and her family began to look at what she would do with her newfound lease on life.

A friend suggested photography and so for her 18th birthday early this year she was given a camera, which is now almost never out of her hands.

Unbeknown to Ms Miller, her mother contacted the Sebastian Foundation, created by pop star Guy Sebastian and his wife Jules, and the teenager was given the opportunity to go on work experience of a lifetime.

With her mother she travelled to the United States and received tutoring from award-winning photographers including Jerry Ghionis and Colin Smith.

Ms Miller is now working with the heART Project, which is a collaboration between the Sebastian Foundation and Australian photographer Karen Alsop, to take portraits of children with disabilities.

'It doesn't matter if things get tough, there are things that you can push forward,' she said.

'No matter what life throws at you, you can do it.'

Questions

1. What was the cause of Karley Miller's epilepsy?
2. Karley's stroke severely damaged her right hemisphere. List five functions of the right hemisphere.
3. List the ways Karley's stroke affected her.
4. Karley Miller's story is an example of a case study. What is a case study?
5. Identify one advantage and one limitation of a case study as a method of data collection.
6. When Alexander was 5 years old he began having epileptic seizures. These seizures increased in severity and frequency over the years. PET and MRI scans showed that the seizures were originating in his left hemisphere. Identify one similarity and one difference between a PET scan and an MRI scan of Alexander's brain.

The corpus callosum

The left and right hemispheres are separated by a deep groove – a longitudinal fissure – but connected by a thick band of nerve fibres known as the **corpus callosum** (Figure 4.15).

The key features of the corpus callosum are summarised in Table 4.3.

Table 4.3 The corpus callosum

Definition	The largest tract of nerve fibres (white matter) providing the main connection between the left and right cerebral hemispheres
Location	Deep in the centre of the brain between the two hemispheres
Dimensions	Approximately 10 centimetres long and 7.5 centimetres thick
Functions	» Physically connects the two hemispheres » Transmits information registered in one hemisphere to the opposite hemisphere for processing » Exchanges and integrates information continuously from both hemispheres

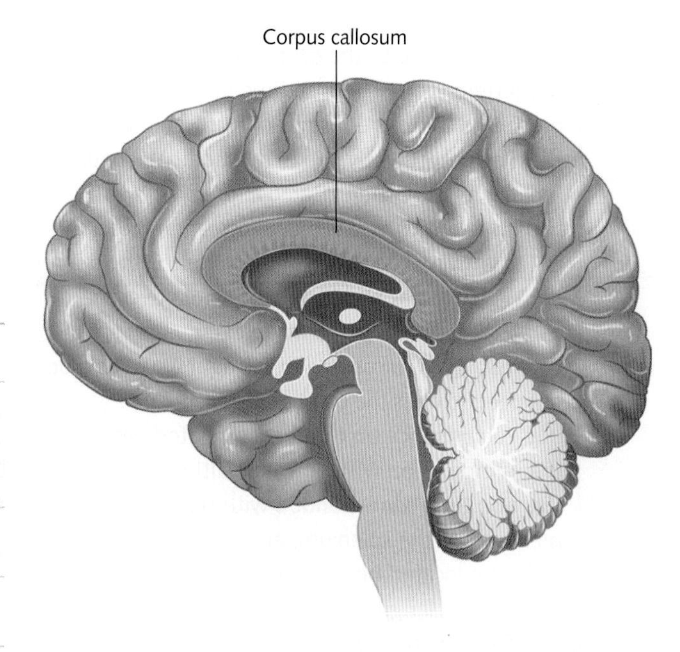

Figure 4.15 The corpus callosum is a thick bank of nerve fibres in the centre of the brain that connects the brain's left and right hemispheres. By transmitting information registered in one hemisphere to the other hemisphere for processing, it allows the hemispheres to communicate.

ANALYSING RESEARCH 4.3

Concussion alters how information is transmitted within the brain

Damage from concussion alters the way information is transmitted between the two halves of the brain, according to a new study. Previously, research has shown that the corpus callosum (a bundle of nerve fibers that carries signals between the left and right hemispheres of the brain) is vulnerable to damage from mild traumatic brain injury. This is commonly known as concussion.

Researchers at New York University (NYU) School of Medicine compared the condition of the corpus callosum in 36 patients with recent concussion to that of 27 healthy controls. To study the participants' brains, the researchers used a neuroimaging technique called MRI that uses measures of water diffusion to provide a microscopic view of the brain's signal-carrying white matter. 'Looking at how water molecules are diffusing in the nerve fibers in the corpus callosum and within the microenvironment around the nerve fibers allows us to better understand the white matter (myelin covered axons) microstructural injury that occurs,' said study co-author Melanie Wegener.

The MRI findings were combined with results from the study's second innovative advance, called an Interhemispheric Speed of Processing Task. This test evaluates how well the two hemispheres in the brain communicate with each other. Participants sat in a chair and focused their gaze on the letter X that was displayed on a screen directly in front of them. Three-letter words were flashed to the right or the left of the X and participants were asked to say those words as quickly as possible.

When the researchers evaluated this reaction time in both the patients with concussion and the healthy controls, they noticed an interesting phenomenon. 'There is a definite and reproducible delay in reaction time to the words presented to the left of the X compared with words presented to the right visual field,' Dr. Wegener said. 'This shows it takes time for information to cross the corpus callosum from one hemisphere to the other. This is measured by the difference in response time between words presented to different sides of our visual field.'

This delay is likely due to the fact that language function is most often located in the brain's left hemisphere. This means that information presented to the left visual field is first transmitted to the brain's right visual cortex and then has to cross over the corpus callosum to get to the left language center. In contrast, words presented to the right visual field do not need to cross the corpus callosum.

Performance on the test correlated with brain findings on MRI. In the healthy controls, reaction time corresponded with several diffusion measures in the splenium, an area of the corpus callosum located between the right visual cortex and the left language center. No such correlation was found in the concussion patients, suggesting microstructural changes relating to injury.

'We saw a correlation between white matter microstructure injury and the clinical status of the patient,' Dr Wegener said. 'This information could ultimately help with treatment in patients who have mild traumatic brain injury.'

For instance, Dr Wegener said, patients could undergo MRI immediately after a concussion to see if they experienced any clinically important white matter injury and thus may benefit from early intervention.

'Another thing we can do is use MRI to look at patients' brains during treatment and monitor the microstructure to see if there is a treatment-related response,' she said.

Adapted from Radiological Society of North America. "Concussion alters how information is transmitted within the brain." ScienceDaily. ScienceDaily, 3 December 2019. <www.sciencedaily.com/releases/2019/12/191203082910.htm>.

Questions

1 What was the aim of this study?
2 Identify the study design used in this study.
3 Identify the population of interest for this study.
4 Identify the participants in this study.
5 Analyse the design of the study and identify variables that were manipulated experimentally with an IV and DV, and variables that were used to correlate with these.
6 Identify the results of this study.
7 What did the researchers conclude?

The cerebral cortex

The cerebral cortex:

- is the approximately 3 millimetres thick outer layer of the cerebrum
- is the largest brain structure and the most highly developed part of the brain
- contains 70 per cent of the nervous system's neurons
- consists of approximately half the weight of the total brain mass
- has a wrinkled appearance, consisting of bulges (gyri) and deep furrows (sulci). These folds increase the cortex's surface area and, therefore, the number of neurons and neural connections possible. If you spread the thin layer of cortex out flat, it would cover an area of more than half a square metre. The fact that humans are more intelligent than other animals is related to the large surface area of the human cortex
- is the ultimate control and information-processing centre in the brain.

Location

The cerebral cortex is on the surface of the brain.

Functions

The cerebral cortex functions in the following ways.

- It initiates, plans and controls voluntary body movements.
- Its specialised areas receive, process and integrate sensory information.
- It is responsible for higher order mental abilities. For example, when you are choosing a new blanket for your bed, your brain must first coordinate the voluntary movement of your hands to touch the blanket. Your brain will then receive and process the sensory information received by your hand about the texture of the blanket. It will then apply cognitive thought processes to visualise the blanket on your bed at home. Now your brain will decide whether the colour will match your bedroom décor, as well as process the sensory information about the texture of the blanket to determine whether or not it will be warm enough.

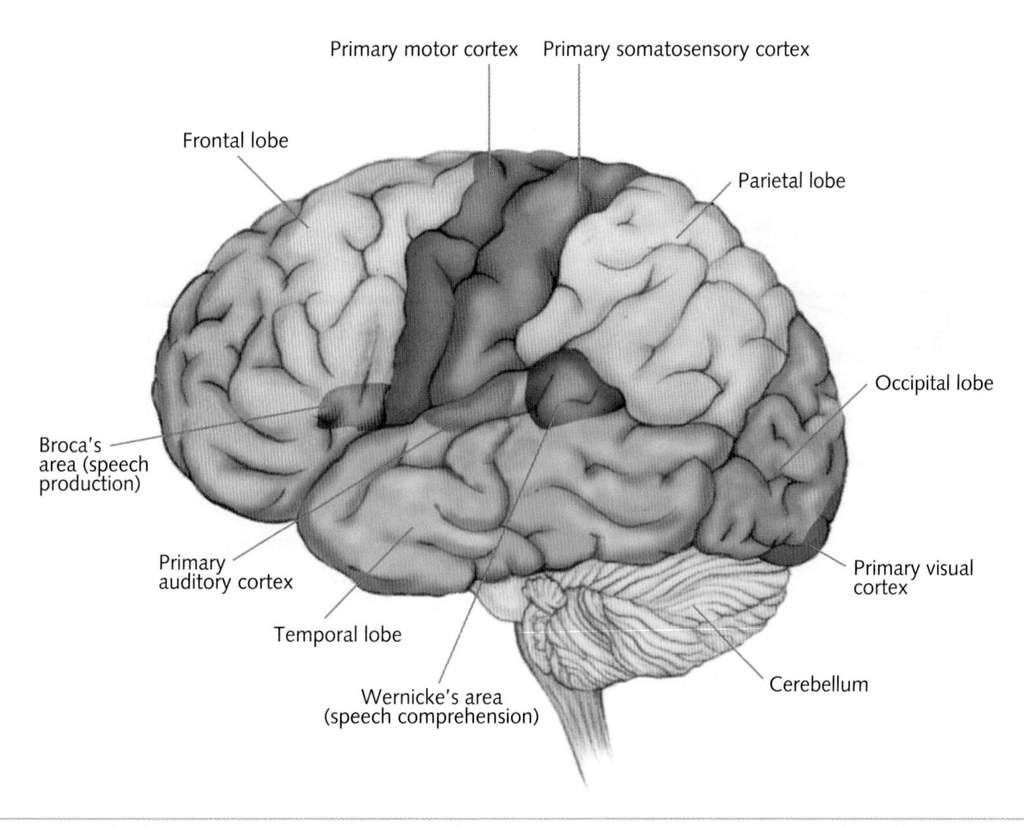

Figure 4.16 The brain, showing all four lobes, the sensory and motor cortex areas, and the language centres

Effects of damage to the cerebral cortex

If the cerebral cortex is damaged, the loss of function depends on where the damage occurs and how severe it is. Damage could cause problems with cognition, sensation, movement and/or behaviour.

The lobes of the cerebral cortex

The various functions of the cerebral cortex are localised in different areas of the brain. These areas are often divided into four distinct lobes in each hemisphere (Figure 4.16):

» the frontal lobe
» the parietal lobe
» the temporal lobe
» the occipital lobe.

The four cortical lobes are bordered by major fissures (grooves) and defined by their functions. The lobes are named after the particular bones of the skull under which they lie.

The frontal lobes

The **frontal lobes** are the largest of the four lobes. Cortical areas known as association areas cover the front of each lobe and are found throughout the cerebral cortex. They do not have a specific sensory or motor role. Instead, their role is to:

» integrate information received from other brain areas or structures
» give meaning to it
» determine an appropriate response.

Association areas are responsible for our most complex mental behaviours and cognitive functions. These include higher level executive functions.

Location

The frontal lobes are located on the upper front half of each hemisphere.

Functions

The frontal lobes are responsible for a variety of functions, including motor function, higher cognitive skills and executive functioning, and many aspects of personality and emotional responses. These functions are most easily understood if we divide the frontal lobe into three distinct areas, each having a specific brain function. These are (Figure 4.17):

» the prefrontal cortex
» the primary motor cortex
» Broca's area.

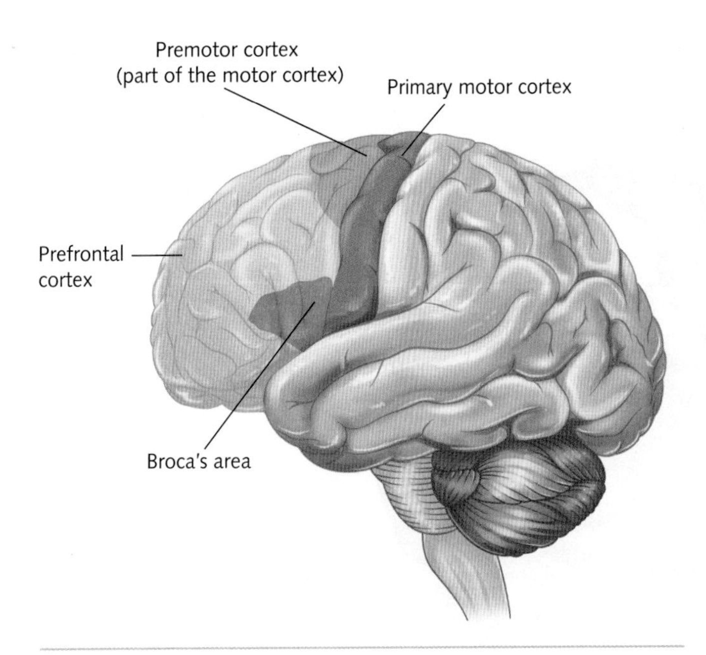

Figure 4.17 The structure of the frontal lobes

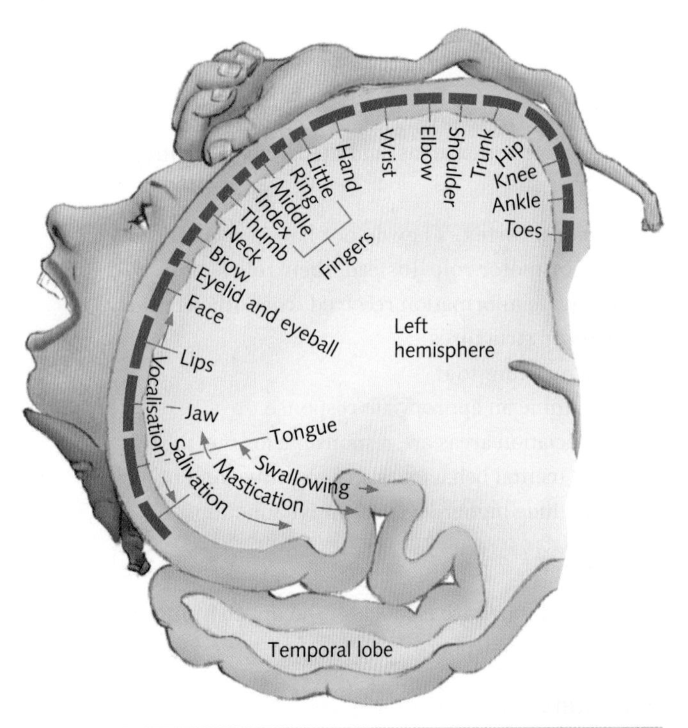

Figure 4.18 A homunculus is a diagram representing the portion of the brain responsible for controlling different bodily functions. In this homunculus, showing motor (movement) control, each of the body parts is drawn along the primary motor cortex so that their location and size corresponds to the number of muscles involved in moving that body part. The hands, for example, are represented by more area than the feet. If you've ever wondered why your hands are more dexterous than your feet, it's partly because a larger area of motor cortex is devoted to controlling the hands. Also, a large area of motor cortex is devoted to the lips, mouth and tongue. This is because dozens of muscles are involved in producing speech.

The prefrontal cortex

The prefrontal cortex is the forward part of the frontal lobe. It is responsible for most of our advanced cognitive and executive functions. It contributes to our personality, intelligence and social skills. The prefrontal cortex is the last brain area to mature, continuing to create and prune neural connections until a person reaches their mid-20s.

» **Executive functions** include higher level cognitive processes, such as decision-making, planning, analysing, motivation, attention, problem solving, organisation, initiating action, goal-directed behaviour, self-monitoring and controlling responses. These functions are associated with memory, abstract thinking, planning, and impulse control.
» The prefrontal cortex is responsible for our ability to predict possible consequences of our actions and to control socially unacceptable responses. In doing this it relies heavily on memory, previous experience and sensory information about the environment.
» **Emotional regulation** is the ability to apply procedures to control an emotion or set of emotions. The prefrontal cortex works with other complex brain structures, including the amygdala and the hippocampus, to generate and regulate emotions. Key features of the amygdala and the hippocampus and their role in emotional regulation are discussed later in this chapter (page 160).

The primary motor cortex

The **primary motor cortex** is an arch of tissue at the rear of the frontal lobe that extends over the top of the brain. It directs the body's skeletal muscles; therefore, it controls voluntary movement.

» The left primary motor cortex controls voluntary movement for the right side of the body.
» The right primary motor cortex controls voluntary movement for the left side of the body.
» Different parts of the body are represented at various locations on the motor cortex.
» The amount of motor cortex that controls each body part corresponds to how important the body part is and how often it is used, not to its size.
» Body parts are controlled by cortex areas that are in the opposite position to where they are located on the body. For example, the face and mouth are controlled by areas located towards the bottom of the motor cortex, while the feet and legs are towards the top (Figure 4.18).

Broca's area: speech production

In the 1800s, French physician Paul Broca (1824–1880) autopsied the brains of several patients who had at some time in their life suddenly lost their speech ability. In a classical study in 1861, Broca treated a 51-year-old man named Leborgne who was given the nickname 'Tan' because he had a severe speech disorder that made 'tan' the only syllable he could pronounce clearly. After Tan died, Broca autopsied his brain and found damage to a small area of his left frontal lobe. Broca concluded that this area of the brain controls the production of articulate speech, and it became known as **Broca's area**. Key features of Broca's area are summarised in Table 4.4.

How the various brain structures interact to regulate emotions, initiate voluntary movement and enable language is discussed later in this chapter.

Table 4.4 Broca's area

Location	In most people, Broca's area is located in the left frontal lobe, near the primary motor cortex (Figure 4.17)
Function	» Controls the production of articulate (clear and fluent) speech » Coordinates the activities of the muscles of respiration, the larynx, and the pharynx, as well as those of the cheeks, lips, jaws and tongue, so words are pronounced clearly and fluently
Key features	» Involved in formulating the structure of sentences and analysing the grammar of sentences
Connections	» Has neural connections to other brain areas involved in language, such as Wernicke's area
Effects of damage	» May cause **Broca's aphasia**, a speech disorder where the person is unable to produce clear and articulate speech (speech is deliberate and abbreviated and has very simple grammatical structure); for more detail see Chapter 5 page 201 » The person can still comprehend speech (verbal language) but, as well as having difficulty producing speech, may struggle with written language

Effects of damage to the frontal lobes

Damage to the frontal lobes can result in:

» difficulty with decision-making, planning, organisation and attentional control
» memory loss
» inability to understand social cues or regulate emotions
» loss of motor skills
» dramatic changes to an individual's personality and emotional control
» speech difficulties.

The parietal lobes

Touch, temperature, pressure, pain and other somatic (body) sensations relating to muscle and joint movement and the body's position in space are registered in the primary somatosensory cortex, found in the **parietal lobe**.

Location

The parietal lobes are situated at the top of the cerebral cortex, behind the frontal lobes.

Functions

The parietal lobes function to:

» process bodily (somatic) sensations (Figure 4.19)

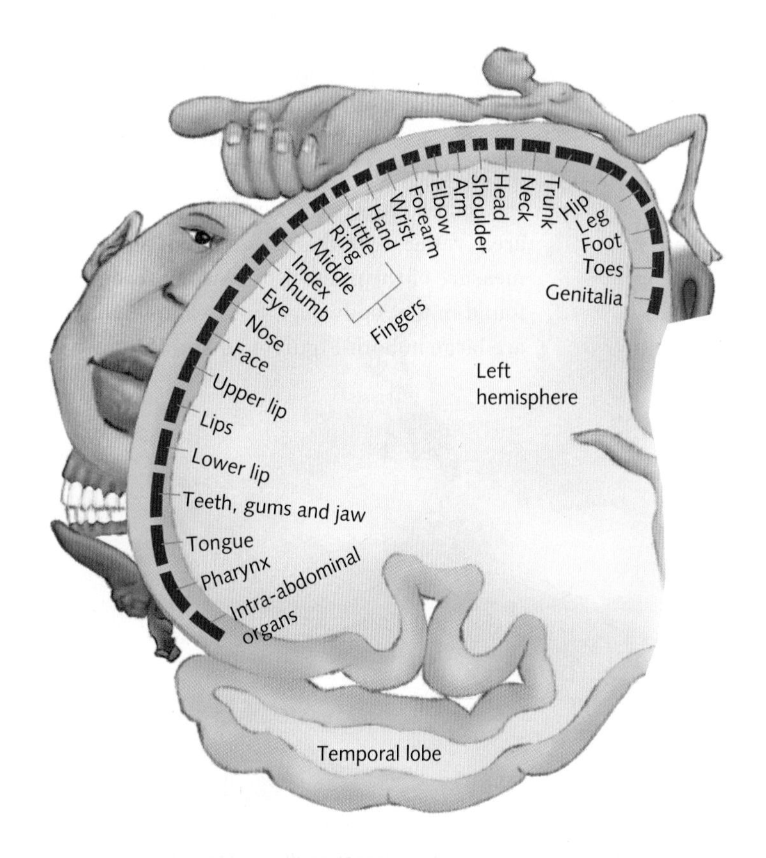

Figure 4.19 The somatosensory homunculus. Note the high proportion of primary somatosensory cortex devoted to the hand, fingers, mouth and tongue.

- receive and integrate information from other brain structures and areas (for example, our visual and auditory cortices), using its large area of association cortex
- via its integration function, enable us to determine where objects are located in space and where our body is positioned in space; therefore, it also helps coordinate our senses and our movement

Primary somatosensory cortex

The **primary somatosensory cortex** is a strip of neurons located at the front of the parietal lobe (next to the primary motor cortex). It has the following features and functions.

- It registers and processes sensations detected by the body's sensory receptors. As a result, we feel a sensation and form a perception of what it is and where it came from.
- Sensory information from the left side of the body is registered in the primary somatosensory cortex of the right parietal lobe. Sensory information from the right side of the body is registered in the somatosensory cortex of the left parietal lobe.
- As with the motor cortex, we find that the 'map' of bodily sensations on the somatosensory cortex is distorted. The homunculi in figures 4.19 and 4.20 show that the cortex reflects the sensitivity of body areas rather than their size. **Sensitivity** is a measure of the number of sensory receptors found in that body part. For example, the lips are large in both Figure 4.19 and Figure 4.20 because of their great sensitivity; the back and trunk, which are less sensitive, are much smaller. Each body part is represented on the somatosensory cortex according to its position on the body (adjacent body parts are adjacent on the somatosensory cortex). Parts of the body are represented on the somatosensory cortex in the opposite order to their position on the body. For example, sensations from the face, mouth and lips register towards the bottom of the somatosensory cortex and sensations from the feet and legs register towards the top.

TBC Science Photo Library/Peter Cardiner

Figure 4.20 These misshapen homunculi represent motor (left) and somatosensory (right) cortex organisation. The parts of the body that are enlarged are controlled by more of the motor cortex.

Effects of damage to the parietal lobes

Damage to the parietal lobes can result in:

- loss of sensation in a body area
- if on the right side of the parietal lobe, difficulty navigating spaces, even familiar ones. Movement may become clumsy and there may be difficulties with spatial skills. The person may confuse left and right, or may suffer from a neurological disorder known as **spatial neglect**, where they neglect one whole side of their world (usually the left), both inside their body and out (see Chapter 5, page 200).

The temporal lobes

The **temporal lobes** (Figure 4.21) are involved with hearing, language skills and social understanding, including perception of other people's eyes and faces.

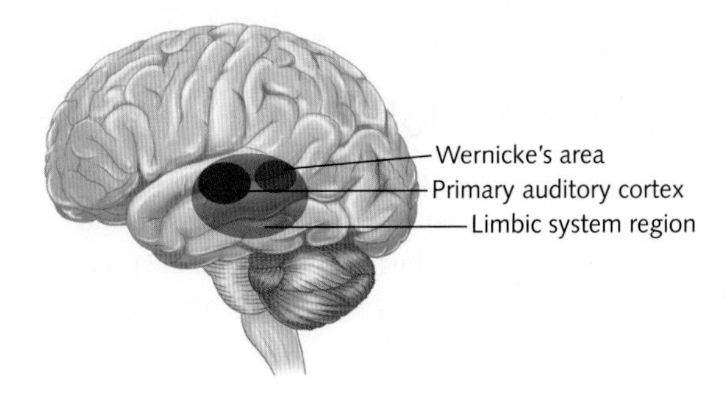

Figure 4.21 The structure of the temporal lobes

Location

The temporal lobes are located on each side of the brain.

Functions

The temporal lobes have functions that include processing sound and contributing to memory.

- The temporal lobes contain large amounts of primary auditory cortex. The **primary auditory cortex** registers and processes auditory (sound-based) information received by both ears and integrates it with information from other senses. If we were to stimulate the auditory area of a temporal lobe, our subject would 'hear' a series of sound sensations.
- The primary auditory cortex in the right temporal lobe is specialised to process non-verbal sounds (for example, the sound of a siren or a door slamming).
- The primary auditory cortex in the left temporal lobe is specialised to process verbal sounds that are associated with language.
- Association areas of the temporal lobe, through their link to the hippocampus, contribute to memory. This is particularly true for our ability to recognise faces and identify objects, and make appropriate emotional responses.

Wernicke's area: understanding language

In the 1800s, the German neurologist Carl Wernicke (1848–1904) discovered important brain structures in the left hemisphere responsible for understanding language. Wernicke examined patients who could hear words but could not recognise them or remember them. They spoke fluently but made little or no sense when they spoke, and they had difficulty comprehending speech. Upon their death, Wernicke autopsied their brains and found damage to the rear section of the left temporal lobe (the brain area just above the left ear). Wernicke believed this area was connected by nerve fibres to Broca's area, thus forming a complex system responsible for understanding and talking. He concluded that the rear section of the left temporal lobe was primarily responsible for language comprehension but was also involved in speech production. This area of the brain became known as **Wernicke's area** and the language impairment that resulted from damage to the area became known as Wernicke's aphasia (this is also discussed in Chapter 5, page 201). Key features of Wernicke's area are summarised in Table 4.5.

Effects of damage to the temporal lobes

The effects of damage to the temporal lobes may include the following.

- Damage to the left temporal lobe may result in loss of memory for verbal material (for example, facts or skills). Damage to the right temporal lobe may result in loss of memory for non-verbal material, for example, music and drawings.
- The person may forget events that were personally significant and may demonstrate inappropriate emotional responses to people or situations.
- Damage can result in an inability to recognise faces or objects even though they will still be able to describe them.
- Damage also disturbs auditory sensation and perception.

Table 4.5 Wernicke's area

Location	In most people, Wernicke's area is in the left temporal lobe, near the primary auditory cortex.
Function	» It identifies sounds as words so their meaning can be understood. » It accesses words stored in memory so it controls comprehension of speech and the formulation of meaningful sentences.
Connections	It has neural connections to other brain areas involved in language, such as: » Broca's area (responsible for speech production) » association areas in the left temporal lobe that receive the sounds It is associated with speech registered in the primary auditory cortex of both hemispheres.
Effects of damage	Wernicke's aphasia: a speech disorder where people have difficulty understanding written and spoken language although they can speak fluently. This is discussed in more detail in Chapter 5. The person: » can still speak fluently but their speech makes no sense » has difficulty communicating because they do not understand the meaning of words » has difficulty reading because, although they can see the words and pronounce them, they do not understand their meaning.

4.3.2 ROLES OF THE CEREBRAL CORTEX

The occipital lobes

The **occipital lobes** (Figure 4.22) register and process visual information. When the retina in an eye receives an image, it is divided into a right visual field and a left visual field. Information from the right visual field of each eye is transmitted to the primary visual cortex in the left occipital lobe. Information from the left visual field of each eye goes to the primary visual cortex in the right occipital lobe. This arrangement ensures that information from both eyes goes to both cerebral hemispheres for processing.

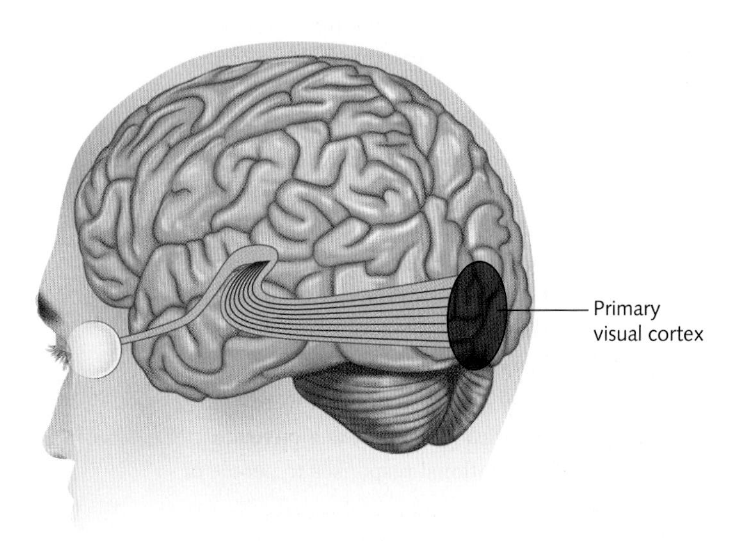

Figure 4.22 The structure of the occipital lobes

Weblink
Brain structure and function

Location

The occipital lobes are located at the back of the brain.

Functions

The occipital lobes are the major visual area of the cerebral cortex. They have the following functions and features.

» At the base of each occipital lobe is a large area of **primary visual cortex**, an area that registers and processes visual information transmitted from the retinas of both eyes via the optic nerve.
» The primary visual cortex contains a variety of neurons specialised to respond to specific features of visual information. For example, some detector cells only respond to colour, some to shape and some to motion.
» Specialised neurons in association areas select and integrate information from the primary visual cortex, and send visual information to other brain lobes. This helps the brain process a variety of different information at once and form an integrated, meaningful understanding of it. For example, an association area might combine visual information about the shape, distance and speed of a ball being thrown at you with sound information from the auditory cortex – 'Catch!' – and input from the frontal lobe to help you judge whether you can catch the ball.
» Visual information arrives at the primary visual cortex in many 'bits'. The visual cortex assembles the 'bits' of information into a whole image or pattern that can be given meaning.

Effects of damage to the occipital lobes

Damage to the occipital lobe may result in visual impairment even if the eyes and the optic nerve are uninjured. For example, individuals who experience a tumour in one of their occipital lobes may experience blind spots in their vision.

4.3.3 BRAIN POSTER

KEY CONCEPTS 4.3

» The cerebrum is divided into two hemispheres: the left hemisphere and the right hemisphere.
» The left hemisphere is dominant for verbal and analytical functions. It receives sensory information from and sends motor messages to the right side of the body.
» The right hemisphere is dominant for non-verbal functions and spatial skills. It receives sensory information from and sends motor messages to the left side of the body.
» The corpus callosum physically connects the hemispheres, allowing information registered in one hemisphere to be transferred to the opposite hemisphere for processing.
» The cerebral cortex is the outer layer of the hemispheres. It is divided into four lobes, each with specialised functions.

» The frontal lobe is the largest lobe. It is responsible for our most advanced cognitive and executive skills and many aspects of personality and emotional behaviour, and it controls voluntary movement.
» The parietal lobe registers and processes bodily sensations.
» The temporal lobe registers and processes auditory information and contributes to memory.
» The occipital lobe registers and processes visual information.

4.3.4 BRAIN LABELLING AND DISSECTION

Concept questions 4.3

Remembering

1 What functions is the left hemisphere responsible for? r
2 What is the role of the corpus callosum? r
3 What is the function of the primary somatosensory cortex? r
4 What is Wernicke's aphasia? r

Understanding

5 Provide an example to illustrate how different areas of the body are represented on the primary somatosensory cortex. e

Applying

6 Trevor fell off a ladder and hit his head. Before his accident Trevor was able to balance his accounts, make sensible decisions and plan ahead. After his accident, Trevor found even simple arithmetic difficult, his decision-making was poor, and he seemed unable to formulate any plans.
a What part of Trevor's brain was probably damaged in the fall? e
b Name two other changes that Trevor might have experienced as a result of his fall. c

HOT Challenge

7 After an area of Kenneth's corpus callosum was severed, his doctors conducted a number of tests to determine how his ability to process information was affected. In one test he sat at a table that had a computer screen on it. He looked at a dot in the middle of the screen. A picture of a dog was flashed to his right visual field and a picture of a shoe was flashed to his left visual field. A tray of unseen objects (a pencil, an apple, a key, a shoe, a toy car, a toothbrush and a paper clip) was then placed on the table and Kenneth was asked to identify what he had seen on the screen. How could he do this? e

4.4 The brain and complex functions

Many of our more complex functions, such as voluntary movement, processing language, regulating emotions and decision-making, require communication between and collaboration of multiple regions of the brain.

Voluntary movement

If you want to pick up the remote control and change the television channel, your brain needs to simultaneously process a variety of sensory and motor messages. It must register and process the visual information it received from the occipital lobe after looking at the remote so it can determine where the channel button is. Your parietal lobe must identify your body's position in space so you can accurately judge the distance between your body and the remote control. Your frontal lobe must send motor commands to your arms to enable them to accurately work the remote control and determine the strength needed to press the button.

The regions involved in executing voluntary movements include:

» the primary motor cortex
» association areas in the frontal and parietal lobes
» the cerebellum
» the thalamus
» the **basal ganglia**, clusters of nerve cells located in the forebrain, deep within the cerebral hemispheres (Figure 4.23).

Their contributions to voluntary movement are summarised in Table 4.6.

4.4.1 INITIATION OF VOLUNTARY MOVEMENT

Table 4.6 Brain regions involved in voluntary movement: primary motor cortex, association areas, cerebellum and basal ganglia

Brain region	Contribution to voluntary movement
Frontal lobe	» Association areas receive information about the body's position in space from other lobes and formulate a plan for motor movement. » Primary motor cortex transmits motor messages to the skeletal muscles that control voluntary movement.
Parietal lobe	» Provides information about the position and movement of body parts
Cerebellum	» Transmits motor information to and from the cerebral cortex to coordinate physical functions such as posture and balance
Thalamus	» Provides the relay system that enables motor messages to be transmitted between the cerebral cortex, cerebellum and basal ganglia so that voluntary movements can be carried out
Basal ganglia	» Transmit motor messages to and from the cerebral cortex » Work with the cerebellum to plan and manage smooth, coordinated movement (damage to the basal ganglia can cause movement problems like those in disorders such as cerebral palsy, Huntington's disease and Parkinson's disease)

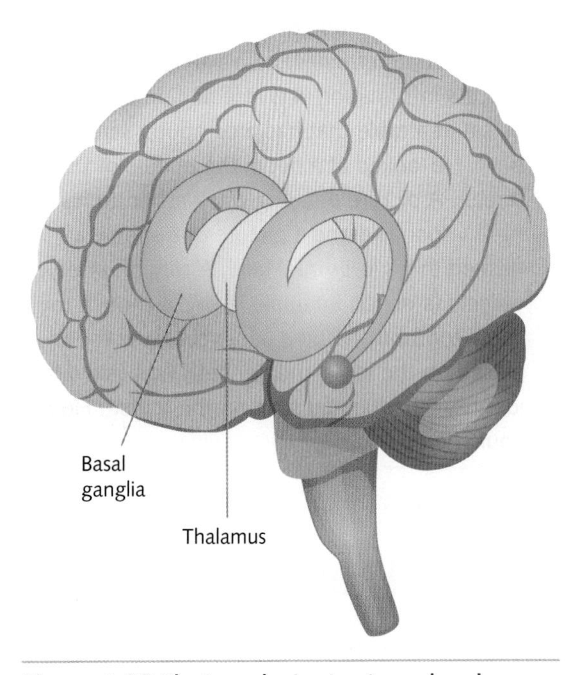

Figure 4.23 The inner brain structures: basal ganglia and thalamus

4.4.2 LANGUAGE PROCESSING

Language processing

The complex ability to produce speech and understand language (written and spoken) results from the interaction of specific brain areas dedicated to language processing. Traditionally, the brain areas primarily involved in language processing were thought to be the primary auditory cortex, Broca's area and Wernicke's area. However, since the 1960s a third area of importance for language has been added. This is known as **Geschwind's territory** (named after the American neurologist Norman Geschwind, who first suggested its existence). The key features of Geschwind's territory are summarised in Table 4.7.

Table 4.7 Geschwind's territory

Location	The lower part of the left parietal lobe, above Wernicke's area and adjacent to the occipital lobe (Figure 4.24)
Function	» Thought to have a multimodal function, which means it can process different kinds of stimuli simultaneously; for example, auditory, visual and sensory stimuli » With Wernicke's area, supports comprehension of spoken or written language by using its multimodal function to integrate the different components of a word; for example, sound, sight and meaning are integrated so we are able to understand the full meaning of a word (Carter, 2014)
Key features	» It is a large bundle of nerve fibres that connects Broca's area with Wernicke's area via a region in the parietal lobe of the cortex. » It is the last area in the brain to mature, the completion of its maturation coinciding with the development of reading and writing skills, so it is important for the acquisition of language in childhood.
Connections	» Connected to Broca's area and Wernicke's area by a bundle of nerve fibres known as the arcuate fasciculus
Effects of damage	» Reduced ability to produce and/or comprehend language

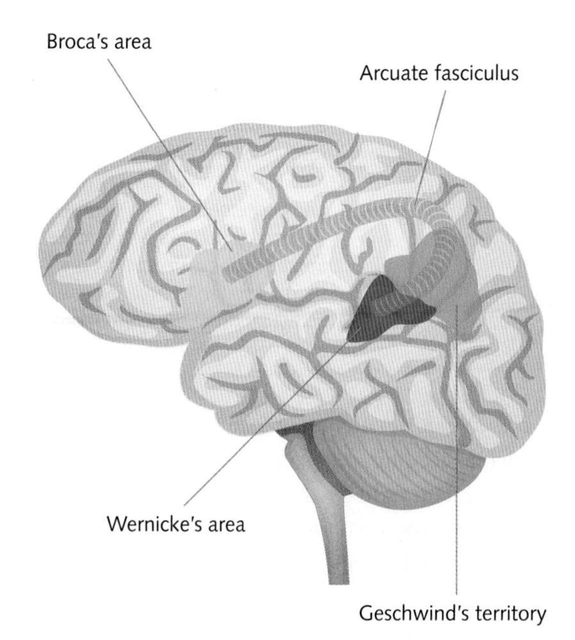

Figure 4.24 The left hemisphere of the brain, showing the location of Broca's area, Wernicke's area, Geschwind's territory and the arcuate fasciculus connecting these brain regions

In addition to the primary auditory cortex, Broca's area, Wernicke's area and Geschwind's territory, other brain regions contribute to language processing. For example, the **angular gyrus** transfers visual information to Wernicke's area so we make meaning out of visually perceived words. The **insular cortex** transfers visual information to Wernicke's area, in order to make meaning out of visually perceived words. The **basal ganglia** and cerebellum are involved in control of the muscles involved in speech production (Hall, 2010). These regions work together as a network to process words and word sequences to determine context and meaning. As a result, we are able to understand and produce language.

Regulation of emotions

Emotions are a set of complex and intense physiological, psychological and behavioural reactions to significant events that we experience as having a specific quality, such as happiness or fear. Emotions are often thought to be conscious experiences only; however, they do have a physiological basis. Generally, psychologists recognise three components to emotions:

» expressive behaviour
» physiological responses
» subjective feelings.

These components are summarised in Table 4.8.

4.4.3 BRAIN STRUCTURES INVOLVED IN THE REGULATION OF EMOTIONS

Table 4.8 Components of emotions

	Expressive behaviour (communicating the emotion)	Physiological responses (to the emotion)	Subjective feelings (about the emotion)
Definition	How an individual responds to the emotions they are experiencing	The unconscious biological experiences that occur in response to a specific event	How each individual experiences reactions to emotions
Key features	» Can be conscious or unconscious » Can be verbal or non-verbal	» Responses originate deep within the brain in the limbic system (the amygdala, hippocampus and hypothalamus). » The **amygdala** adds the emotional element to the event. » The **hippocampus** formulates and stores memories of the time and place of an experience. » The **hypothalamus** helps maintain homeostasis, and regulates the release of hormones from glands.	» The subjective evaluation of the physiological response leads to an emotional response in the brain's prefrontal cortex.

	Expressive behaviour (communicating the emotion)	Physiological responses (to the emotion)	Subjective feelings (about the emotion)
Examples	Slouching, smiling, frowning, waving arms, crying, laughing, throwing objects, hugging, yelling	Heart palpitations, sweaty palms, stomach lurching, irregular breathing, pupils dilating	Being asked to accept more responsibility at work may induce emotions such as joy and excitement in one person; another person may experience apprehension and anxiety; a third person may experience a combination of these emotions

WB 4.4.4 REGULATION OF EMOTIONS: MEDIA ANALYSIS

The limbic system

Emotions are created in an area of the brain known as the limbic system. The **limbic system** is an almond-shaped cluster of structures that lies on both sides of the thalamus. It is buried deep within the brain, underneath the cerebral cortex and above the brainstem. Structures in the limbic system help to regulate behavioural and emotional responses, especially those related to survival. These include feeding, reproduction, caring for our young, and fight-or-flight-or-freeze responses when we are faced with danger.

Limbic system structures are connected via a range of neural pathways. For example, neural pathways connect the limbic system to the prefrontal cortex. This enables a two-way communication between these brain regions. The result is that our thought processes (generated by the prefrontal cortex) can now regulate our physiological emotional reactions (generated by the limbic system) and vice versa. For instance, if a Melburnian who is afraid of flying is invited to a celebration in Sydney, they need to face their fear so they can board a plane. When the plane is getting ready for take-off, their amygdala will be activated, registering the situation as fearful. This increases their level of arousal and activates physiological responses such as increased heart rate and respiration rate. This arousal may be intensified if their prefrontal cortex assesses the situation as fearful – they may start thinking about the plane crashing and bursting into flames or the wheels falling off during take-off. However, if their prefrontal cortex evaluates the situation as positive and they start thinking about the plane landing safely in Sydney, the physiological responses associated with fear may begin to ease (Figure 4.25). In this instance their emotions have been regulated by the activity of the prefrontal cortex.

Worksheet Is your left side your best side?

The key features of the limbic system are:

- it contributes to unconscious physiological emotional responses to a particular event
- it is a major contributor to survival behaviours such as feeding, reproduction and caring for our young, and fight-or-flight-or-freeze responses
- its responses are designed to either push us away from danger (we experience the unpleasant emotion of fear) or push us towards reward (we experience the pleasant emotion of happiness)
- the intensity of the emotions experienced by each person also varies
- this subjective evaluation of the physiological response leads to an emotional response in the brain. This is when we consciously experience the emotion
- the prefrontal cortex is responsible for the subjective feeling in response to the body's physiological reaction. This is how our emotions are regulated.

The limbic system includes brain structures such as the thalamus and hypothalamus (discussed earlier in this chapter) and basal ganglia, a structure involved in reward processing, habit formation, movement and learning. However, the two major limbic system structures involved in emotional regulation are the **hippocampus** and the **amygdala** (Figure 4.26). The contributions of the amygdala and the hippocampus to emotional regulation are summarised in Table 4.9.

Decision-making

Decision-making involves a conscious choice between two or more options. It is a cognitive process that involves predicting the possible short-term and long-term outcomes of these options. Decision-making:

- is a complex executive function involving knowledge and past experience
- is influenced by a range of factors, such as current goals, anticipation of outcomes, emotional state, personality,

Table 4.9 The role of the amygdala and the hippocampus in emotional regulation

	The amygdala	The hippocampus
Location	Embedded deep within the temporal lobes, just in front of the hippocampus (there are two amygdalae – one in each cerebral hemisphere)	A horseshoe shaped structure in the inner region of the temporal lobe of each hemisphere (so there are two hippocampi, one in each temporal lobe)
Role	» Evaluates potential environmental threats and determines the emotional significance of a situation » Triggers the body's stress response so it is a major contributor to the emotion of fear (Carter et al., 2009). If it assesses a situation as dangerous, it sends signals to other brain structures to prepare the body to either face the situation (fight) or move away from it (flight); for example, it triggers increased heart rate, dilation of pupils, increased respiration. When this happens, we experience the emotional reaction of fear » Is involved in emotional learning and behaviour. Plays a key role in forming new memories, especially those related to fear, anxiety, pleasure and anger » Sends signals to other brain regions to experience appropriate emotional responses; for example, the feeling of pleasure is usually brought on by the release of dopamine, a mood-lifting hormone. Dopamine is released by the activation of the hypothalamus. When pleasure is experienced, the amygdala activity is reduced	» Receives input from and sends output to all association areas in the brain » Responsible for the formation of new memories (particularly fact- and event-related memories) and involved in relaying them for indefinite storage in long-term memory » Is involved in spatial navigation because it helps the brain map the layout of the environment » Helps form memories associated with sensory input; for example, when the smell of suntan lotion reminds you of a favourite beach holiday

Figure 4.25 A person's emotional reaction to fear is regulated by and based on the evaluation made by their prefrontal cortex. When the prefrontal cortex decides that a situation is not threatening, the amygdala will send signals to other brain regions to experience an appropriate, calm emotional response.

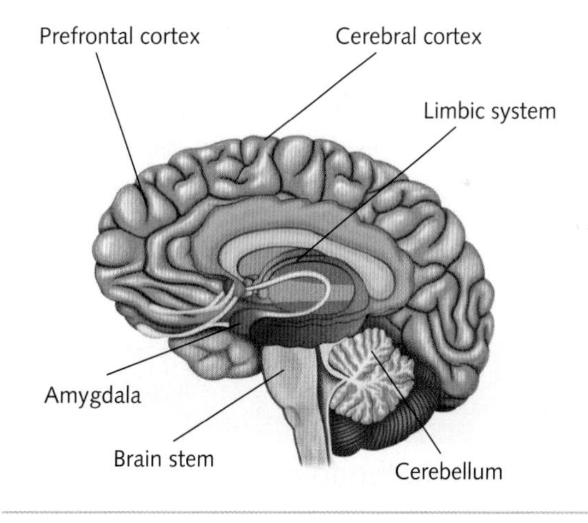

Figure 4.26 The limbic system and amygdala

values, age, commitment and culture (Rosenbloom et al., 2012)
» is goal (outcome) oriented (either solve a problem or reach a specific result)
» is the cognitive basis of behaviour.

Although decision-making is a cognitive process, it has a neural basis. The prefrontal cortex and the hippocampus have traditionally been considered the main neural structures involved in decision-making. However, new studies suggest that a complex neural network and a range of brain structures, including the amygdala, thalamus and cerebellum, are also involved (Broche-Pérez et al., 2016; Saberi Moghadam et al., 2019).

The prefrontal cortex

Our constantly changing environment requires us to make appropriate decisions about our response so we adapt our behaviour to suit these changes. These decisions involve the prefrontal cortex, specifically the **orbitofrontal cortex**, the area at the front of the prefrontal cortex that sits just

4.4.5 BRAIN DAMAGE CASE STUDIES

4.4.6 EVALUATION OF RESEARCH

above the eye sockets. As we saw earlier (page 162), the prefrontal cortex:

- directs the decision-making process, because it estimates whether our actions are likely to succeed in reaching a goal
- has dense neural connections to other brain structures and areas (Petrides et al., 2012; Fuster, 2015), which enable it to monitor and control their activity; for example, it can activate certain networks, inhibit other networks, and integrate interactions among networks.

People with damage to the prefrontal cortex experience no loss of sensation, perception or motor control, but they exhibit poor judgement, planning, and decision-making (Mesulam, 2000; Fuster, 2015).

The hippocampus

When making decisions, we draw on information from a variety of sources. One of these is our memory of events and prior experience. Research suggests that the hippocampus' involvement in the formation of new memories and their transfer to long-term memory for indefinite storage creates a record of the past. When we make a decision, we can draw on these memories and use them as a point of reference to guide our decision-making in the present (Palombo et al., 2015).

The roles of the amygdala, thalamus and cerebellum in decision-making are summarised in Table 4.10.

Table 4.10 The roles of the amygdala, thalamus and cerebellum in decision-making

Amygdala	» Assesses the environment for potential threats » Determines the emotional significance of a situation and triggers an autonomic response to emotional stimuli » During decision-making, when an initial choice is made, controls the emotional, bodily response to this choice based on its predicted outcome (for example, reward or punishment) (Gupta et al., 2011)
Thalamus	» Delivers sensory and motor information to the appropriate area of the cerebral cortex; this supports cognition and decision-making because it allows us to constantly update our mental representations of our environment (Saalmann & Kastner, 2011) » By delivering sensory and motor information to the cortex, directly influences our decision-making because it influences our level of awareness and attention
Cerebellum	» In addition to controlling movement, regulates a wide range of cognitive functions that includes decision-making (Konarski et al., 2005; Schmahmann and Sherman, 1998; Stoodley et al., 2012) » Because it is connected to the areas of the cortex involved in memory, language processing, visual and spatial analysis and emotion regulation, it helps the brain to create specific goal-directed behaviours in response to sensory inputs (Deverett et al., 2018; Rosenbloom et al., 2012)

INVESTIGATION 4.1 MODELLING THE ROLE OF THE BRAIN IN BEHAVIOUR AND MENTAL PROCESSES

Scientific investigation methodology

Modelling

Aim

To create a model that demonstrates how the brain enables humans to interact with the world around them

Logbook template
Case study

Data calculator

Introduction

In this task you will investigate and create a model (with labels and annotations) that demonstrates how the brain enables people to interact with the world around them. You could choose one of the topics listed below from the Unit 1 Area of Study 2 key knowledge for your model. Models could be made from plasticine, cardboard, paper mâché, clay, playdough, recyclables etc.

- The hindbrain (pons, medulla and cerebellum)
- The midbrain (reticular formation)
- The forebrain (hypothalamus and cerebrum)
- Hemispheric specialisation and the corpus callosum
- The frontal lobe

- The parietal lobe
- The temporal lobe
- The occipital lobe
- Regions of the brain involved in voluntary movement
- Regions of the brain involved in language processing
- Regions of the brain involved in the regulation of emotions

Key science skills

The key science skills you will need to demonstrate in this investigation are listed at the start of this chapter (page 135).

Criteria

Ensure your model includes the following:

- clear labels of brain structures
- descriptions of brain area functions
- descriptions of the impact of damage to the area or function investigated
- a list of all your sources of information, presented using standard scientific referencing conventions.

ACTIVITY 4.2 ENGAGING YOUR BRAIN – ONE REGION AT A TIME

Try it

1 **Exercise 1**

Stand with your feet together and close your eyes. You should be able to stay steady for 30 seconds.

Now stand on one leg and close your eyes. This should be more difficult.

Exercise 2

You ask your teacher for something but the teacher says no.

» What are three ways you could solve the problem?

» What are the consequences of each solution?

» Which solution has the best consequence?

Exercise 3

Go for a run across the school oval or up and down the stairs a few times. Soon you will start to sweat.

Exercise 4

Think about the last thing you wanted to buy. Did you all of a sudden start seeing the item everywhere you went?

Apply it

2 While there are many structures within the brain, these structures can be grouped into three main areas known as the midbrain, hindbrain and forebrain.

a The cerebellum, part of the hindbrain, coordinates muscle movements and regulates posture and balance. Which exercise in Try it engaged your hindbrain?

b The reticular activating system is a part of the midbrain and is responsible for filtering all of the incoming sensory data so that you can focus on the things that are important to you. Which exercise in Try it engaged your midbrain?

c The hypothalamus is a structure in the forebrain that is responsible for restoring homeostasis using its connections to the autonomic nervous system. If the hypothalamus registers that body temperature is too high, it will send a message to the sweat glands to produce perspiration to cool you down. Which exercise in Try it engaged your forebrain?

d Brain development continues into your mid-20s and the frontal lobe is the last part of the forebrain to mature. This part of the cerebral cortex is responsible for higher mental functions such as reasoning, planning and problem solving. Which exercise in Try it engaged your cerebral cortex?

Exam ready

3 The brain region that is mostly responsible for regulating hunger, thirst and body temperature is the

A midbrain.

B hindbrain.

C forebrain.

D cerebral cortex.

KEY CONCEPTS 4.4

- Complex functions require communication between and collaboration of multiple regions of the brain.
- Voluntary movement requires interaction between the frontal lobe, parietal lobe, cerebellum, thalamus and basal ganglia.
- The brain areas primarily involved in language processing are Broca's area, Wernicke's area and Geschwind's territory.
- Regulation of emotions requires interaction between the prefrontal cortex, amygdala and limbic system.
- Decision-making involves the interaction of the frontal lobe, hippocampus, amygdala, thalamus and cerebellum.

Concept questions 4.4

Remembering

1 Name the two language centres that connect to Geschwind's territory. r

2 What are emotions? r

3 What is the role of the structures that form the limbic system? r

Understanding

4 Explain how our prefrontal cortex helps us to survive. e

Applying

5 a Name three brain areas that would contribute to coordinating a voluntary movement such as throwing a ball. e

b Identify what each area contributes. r

HOT Challenge

6 When Amy was walking through the park she noticed a large dog ahead. She stopped walking, stood still and thought about what she should do. She decided to turn around and run away. Explain how her amygdala and cerebellum contributed to her running away. c

4 Chapter summary

KEY CONCEPTS 4.1a

- Debate about the relationship between the mind, the brain and behaviour dates back to ancient times; some ideas were based on logic and reasoning but lacked scientific evidence.
- The brain-versus-heart debate focused on whether the heart or the brain controlled behaviour.
- The mind–body problem focused on whether the mind and body were separate entities or the same entity.
- In the 2nd century CE, Galen used vivisection and experiments to demonstrate that some behaviours were controlled by the brain.
- In the 17th century, Descartes argued that the separate entities of the mind and body interacted with each other through the brain's pineal gland.
- In the 19th century, Gall popularised the pseudoscience of phrenology, where a person's intelligence and personality could be determined by analysing the bumps and indentations of their skull.

KEY CONCEPTS 4.1b

- During the 19th and 20th centuries, brain research began using a more scientific approach.
- Flourens used an early form of ablation and electrical stimulation to demonstrate that phrenology was baseless. He provided some evidence that the cerebral hemispheres control behaviour.
- During the 19th and 20th centuries lobotomy became a popular surgical technique for calming psychologically dysfunctional people. Results of lobotomies suggested that the brain's frontal lobe controls cognition, personality and emotion.
- Lobotomy was often carried out without the patient's consent and resulted in devastating effects on personality and the ability to function independently. Therefore, it is now considered unethical and barbaric.
- The modern surgical technique of brain ablation and lesioning is a more ethical way of treating a range of neurological and psychological disorders.
- Since the 1950s, split-brain operations, which sever the corpus callosum, have been used to treat extreme cases of epilepsy, although they are now used less frequently because of improvements in other treatments.

KEY CONCEPTS 4.2

- The brain has three main regions: the hindbrain, the midbrain and the forebrain.
- The hindbrain consists of three structures: the pons, medulla and cerebellum, which are responsible for autonomic survival functions and coordination of voluntary muscle movement, balance and posture, and reflexive actions.
- The midbrain consists of several structures involved with processing sensory information, movement, sleep, arousal and attention.
- The reticular formation is a major midbrain structure. It helps us to selectively focus our attention and alertness, and it controls physiological arousal.
- The forebrain is the largest and most highly developed part of the brain. It is divided into two hemispheres.
- The forebrain has three distinct structures: the hypothalamus, thalamus and cerebrum. These parts interact to control our most complex processes, including emotions, motivations, sensations, perceptions, learning, memory and reasoning.

KEY CONCEPTS 4.3

- The cerebrum is divided into two hemispheres: the left hemisphere and the right hemisphere.
- The left hemisphere is dominant for verbal and analytical functions. It receives sensory information from and sends motor messages to the right side of the body.
- The right hemisphere is dominant for non-verbal functions and spatial skills. It receives sensory information from and sends motor messages to the left side of the body.
- The corpus callosum physically connects the hemispheres, allowing information registered in one hemisphere to be transferred to the opposite hemisphere for processing.
- The cerebral cortex is the outer layer of the hemispheres. It is divided into four lobes, each with specialised functions.
- The frontal lobe is the largest lobe. It is responsible for our most advanced cognitive and executive skills and many aspects of personality and emotional behaviour, and it controls voluntary movement.
- The parietal lobe registers and processes bodily sensations.
- The temporal lobe registers and processes auditory information and contributes to memory.
- The occipital lobe registers and processes visual information.

KEY CONCEPTS 4.4

- Complex functions require communication between and collaboration of multiple regions of the brain.
- Voluntary movement requires interaction between the frontal lobe, parietal lobe, cerebellum, thalamus and basal ganglia.
- The brain areas primarily involved in language processing are Broca's area, Wernicke's area and Geschwind's territory.
- Regulation of emotions requires interaction between the prefrontal cortex, amygdala and limbic system.
- Decision-making involves the interaction of the frontal lobe, hippocampus, amygdala, thalamus and cerebellum.

4 End-of-chapter exam

Section A: Multiple-choice

1 According to the ancient Greek philosopher Aristotle, ____________ was responsible for governing behaviour.
A the mind
B the pineal gland
C the heart
D the soul

2 Which of the following terms describes psychosurgery in which a section of the frontal lobe is destroyed?
A a vivisection
B a lobotomy
C electrical
D phrenology

3 Brain lesioning results in
A brain changes that can be reversed.
B brain changes that cannot be reversed.
C increased brain tissue.
D no change to the brain's structure.

4 Which one of the following statements about phrenology is incorrect?
A Phrenology claims that the bumps on the skull reflect an individual's personality and intelligence.
B Phrenology is based on empirical evidence.
C Phrenology is regarded as a pseudoscience.
D Phrenology tries to predict human behaviour.

5 Phrenology
A suggests that the mind and body interact through the pineal gland.
B involves surgical removal of sections of the brain to alter emotional behaviour.
C suggests that the bumps along the skull indicate personality traits.
D is based on accepted scientific method.

6 During a split-brain operation, a portion of the ______________ is severed.
A corpus callosum
B frontal lobe
C cerebellum
D pons

7 What does neuroimaging mean?
A investigating the structure and function of the brain using imaging techniques
B investigating the structure of the brain using imaging techniques
C investigating the function of the neurons using imaging techniques
D investigating the structure and function of neurons using imaging techniques

8 What part of your hindbrain regulates your heartbeat?
A pons
B medulla
C hypothalamus
D thalamus

9 The cerebral cortex is the outer layer of the
A thalamus.
B cerebellum.
C cerebral hemispheres.
D medulla.

10 What connects the two hemispheres of the brain?
A pons
B cerebellum
C corpus callosum
D cerebral cortex

11 What are the three main regions of the brain?
A hindbrain, front brain and forebrain
B hindbrain, midbrain and forebrain
C postbrain, midbrain and forebrain
D hindbrain, midbrain and fullbrain

12 Which one of the following is part of the forebrain?
A thalamus
B pons
C medulla
D cerebellum

13 Jane stroked her cat with her right hand. Which part of her brain was responsible for detecting the feel of the cat's fur?
  A right frontal lobe
  B left frontal lobe
  C right parietal lobe
  D left parietal lobe

14 Damage to the left frontal lobe may result in
  A Wernicke's aphasia.
  B Broca's aphasia.
  C difficulty understanding speech.
  D difficulty remembering the meaning of words.

15 Specialties of the right hemisphere include
  A spatial thinking, appreciation of music and daydreaming.
  B creativity, speech production and appreciation of art.
  C creativity, speech comprehension and reading.
  D writing, spirituality and visualising places.

16 If Oscar was consistently shaving only the right side of his face and was unaware that the left side was unshaven, even though he was looking at himself in the mirror, which of the following lobes may be damaged?
  A right occipital lobe
  B left occipital lobe
  C right parietal lobe
  D left parietal lobe

17 Since Clarence's stroke he has had difficulty pronouncing his words clearly. This makes it difficult for others to understand him. What is the likely area and lobe that Clarence's stroke occurred in?
  A Wernicke's area; temporal lobe
  B Broca's area; parietal lobe
  C Wernicke's area; occipital lobe
  D Broca's area; frontal lobe

18 Which brain structure is responsible for producing the emotion of fear?
  A the thalamus
  B the amygdala
  C the hippocampus
  D the pons

19 Which brain structure plays a major role in memory formation?
  A the medulla
  B the hippocampus
  C the reticular formation
  D Broca's area

20 The cerebellum contributes to decision-making by
  A transforming sensory information into specific goal-directed behaviour.
  B relaying information between the pons and the medulla.
  C adding emotions to a decision.
  D transferring new memories to long-term memory.

Section B: Short answer

1 Name the three structures that comprise the forebrain. [3 marks]

2 Identify one psychological and one physiological change you might observe in someone who had frontal lobe damage. [2 marks]

3 Explain how a split-brain operation could stop epileptic seizures. [4 marks]

Brain plasticity and brain injury 5

Key knowledge

» the capacity of the brain to change in response to experience and brain trauma, including factors influencing neuroplasticity and ways to maintain and/or maximise brain functioning
» the impact of an acquired brain injury (ABI) on a person's biological, psychological and social functioning
» the contribution of contemporary research to the understanding of neurological disorders
» chronic traumatic encephalopathy (CTE) as an example of emerging research into progressive and fatal brain disease

Key science skills

Construct evidence-based arguments and draw conclusions
» distinguish between opinion, anecdote and evidence, and scientific and non-scientific ideas

Analyse, evaluate and communicate scientific ideas
» use clear, coherent and concise expression to communicate to specific audiences and for specific purposes in appropriate scientific genres, including scientific reports and posters
» acknowledge sources of information and assistance, and use standard scientific referencing conventions

Source: VCE Psychology Study Design (2023–2027), pp. 25 & 13

5 Brain plasticity and brain injury

Our brains are made up of billions of interconnected neurons. They are moulded as we grow and develop. They are influenced by nature (our genetics) and nurture (the world around us and the way we live and experience this world). Sometimes part of the brain is damaged and the body finds unique ways to overcome the injury; though sometimes, the brain is damaged too much, resulting in permanent damage.

p. 183

5.1 Neuroplasticity: rewiring the brain

Our brains have the unique ability to route and reroute neuronal pathways as a result of maturation and experience. This allows us to function to our best ability and to survive in this ever-changing world.

Adobe Stock/prockopenko

iStock.com/Pikovit44

p. 196

5.2 Acquired brain injury

Although the brain is encased in a tough skull it can still receive physical assaults that affect how it functions. This can impact the way we understand the world, the way we feel and how we talk and walk.

p. 203

5.3

Neurological disorders

The brain is a part of a much larger system – the nervous system. When disease strikes the nervous system, the effects can be widespread throughout our bodies and are just as damaging as a physical assault.

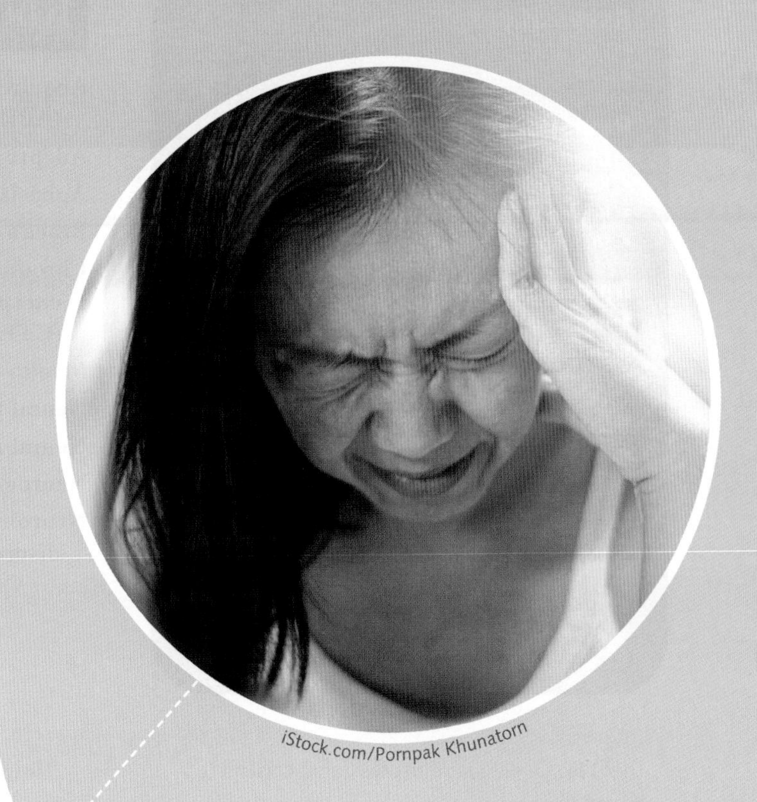

iStock.com/Pornpak Khunatorn

p. 210

5.4

Chronic traumatic encephalopathy

Repeated blows to the head over a long period of time, such as those received in sports such as football and boxing, can be just as damaging as diseases of the nervous system. We are only beginning to understand the impacts of serious repeated brain injury from playing these sports and to find ways to avoid or minimise them.

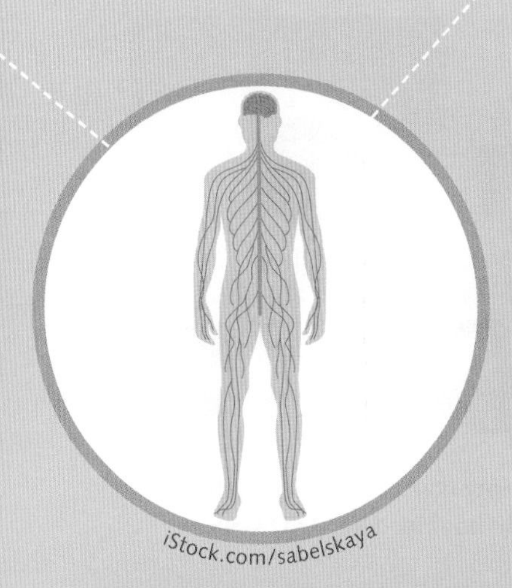

iStock.com/sabelskaya

Your brain is the most important organ in your body. It determines who you are, how you interact with other people and how you perceive the world. Any damage to the brain can affect these functions and in turn affect not only your life, but those around you.

Flashcards
Chapter 5 flashcards

Slideshow
Chapter 5 slideshow

Test
Chapter 5 pre-test

Assessment
- Pre-test
- End-of-chapter exam

Revision
- Chapter map
- Key term flashcards
- Key concept summary
- Slideshow

Investigation
- Investigation: Neurological disorders
- Data calculator
- Logbook template: Case study

Worksheet
- Brain development

To access these resources, visit **cengage.com.au/nelsonmindtap**

Nelson MindTap

Know your key terms

Acquired brain injury (ABI)
Adaptive plasticity
Aphasia
Broca's aphasia
Chronic traumatic encephalopathy (CTE)
Myelin
Neural maturation
Neural migration
Neurogenesis
Neurological disorders
Neuroplasticity
Non-traumatic brain injury (NTBI)
Rerouting
Spatial neglect
Sprouting
Synaptic pruning
Synaptogenesis
Traumatic brain injury (TBI)
Wernicke's aphasia

In 2021, Queensland professional triathlete Alexa Leary was 19 years old when she was thrown off her bike while training near Noosa. Travelling at more than 70 km/h, she clipped the wheel of the bike in front of her. The crash left her with traumatic brain injury (Figure 5.1). She couldn't remember her name, she didn't know where she was, and she couldn't move the right side of her body. She was not expected to survive. Yet, 111 days later she left the hospital, and she now hopes one day to return to training.

Alexa's remarkable recovery was due in part to brain plasticity. If something is plastic, it means that its shape can change. For example, playdough is plastic. It can be changed into whatever shape you want; one minute it can be a dinosaur and the next minute, a rocket ship. Research studies indicate that the human brain is plastic and in fact can change shape and grow new connections throughout your lifespan. In Alexa's case, new connections were made in her brain that enabled intact areas to take over the function of injured areas.

However, it is important to remember that the brain is not totally plastic. Certain key areas of the brain are largely responsible for specific actions, such as movement, language and speech. When there is an injury to these areas of the brain, some recovery may be possible, but other areas of the brain may not be able to fully take over the functions that were affected by the injury.

Russell Leary

Figure 5.1 Alexa Leary undergoes rehabilitation training at home with boyfriend Wil O'Dwyer.

5.1 Neuroplasticity: rewiring the brain

Neuroplasticity refers to the nervous system's ability to change its structure and function as a result of experience and in response to injury. 'Neuro' refers to the nerve cells (neurons) that are the building blocks of the brain and nervous system, and 'plasticity' refers to flexibility or ability to be shaped or changed. Neuroplasticity gives organisms an increased chance of survival.

To understand brain neuroplasticity, you need to understand what the brain consists of. At the basic level, the brain is made of 86 billion interconnected nerve cells, or neurons. Neurons communicate with each other, using electrical impulses and chemicals to transmit information. Each neuron is specialised to receive, transmit and process information. To carry out their complex task, neurons receive assistance from a range of supporting cells, known as glial cells.

Every neuron has a soma (cell body) that contains a nucleus, which controls the activities of the cell (Figure 5.2). Each soma has projecting branches. At one end, a single axon protrudes from the soma. At the other end, there may be up to 10 000 projections, known as dendrites.

Dendrites receive messages from other neurons and transmit the messages inwards towards the soma for processing. The more dendrites a neuron has, the more information it can receive. Axons are thin fibres that carry messages, in the form of electrical nerve impulses, away from the soma.

Some axons are coated with an insulating layer of myelin. The myelin sheath protects the axon and helps increase the speed of transmission of the electrical impulse down the axon.

When the electrical impulse (or action potential) reaches the end of the axon, the axon branches out into axon terminals that link with the dendrites of other neurons. This linking is important because it allows the impulse to pass from neuron to neuron. However, the axon terminals of one neuron do not physically touch or connect with the dendrites of another neuron. A microscopic gap, or synaptic gap, exists between the axon terminals of the sending neuron and the dendrites of a receiving neuron. This is how it works. When the action potential reaches the axon terminals, the axon terminal responds to the electrical impulse by producing chemicals known as neurotransmitters. The neurotransmitters are released and move across the synaptic gap to be received by the dendrites of the next neuron (Figure 5.2). The dendrite converts the chemical message into an electrical impulse, which then travels to the soma. This continues from neuron to neuron, enabling the message to be passed along.

Weblink
An introduction to the nervous system

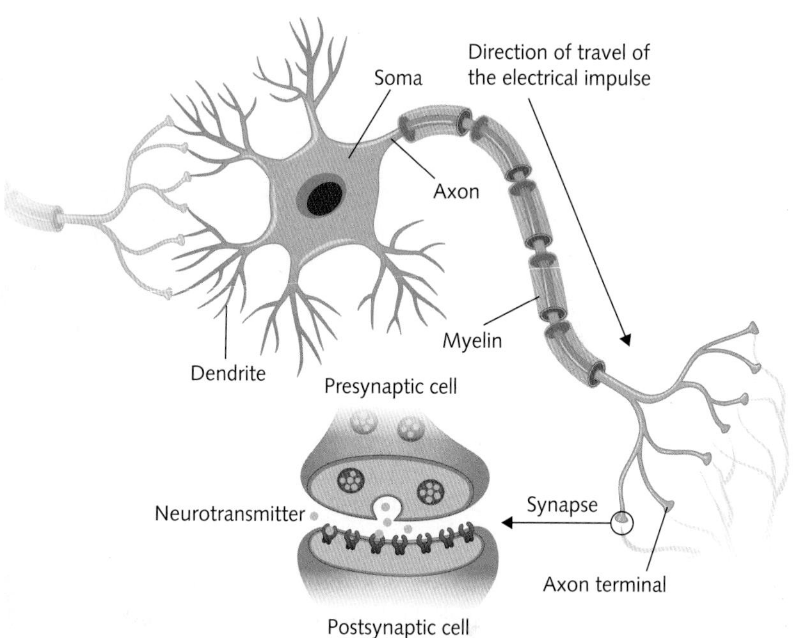

Figure 5.2 A neuron is made up of dendrites, a soma and a myelin-coated axon. An electrical stimulus travels along the neuron until it reaches the axon terminals. In response to the electrical stimulation, the axon terminals produce chemical neurotransmitters, which travel across the synapse. The neurotransmitters reach the dendrites of the next neuron where the message is converted into electrical stimulation and the process continues.

The developing brain

5.1.1 PARTS OF A NEURON

Brain plasticity occurs throughout our lifetime. The most rapid development and changes in brain structure and function occur during infancy and adolescence, diminishing with age. This is called developmental plasticity. As the immature brain grows, brain development progresses through a series of age-related stages, beginning with neurogenesis (Figure 5.3).

Neurogenesis

From the third week of gestation to (and during) adulthood, new brain cells are formed. The cells divide and multiply, in a process known as **neurogenesis**. Neurogenesis is followed by **neural migration**,

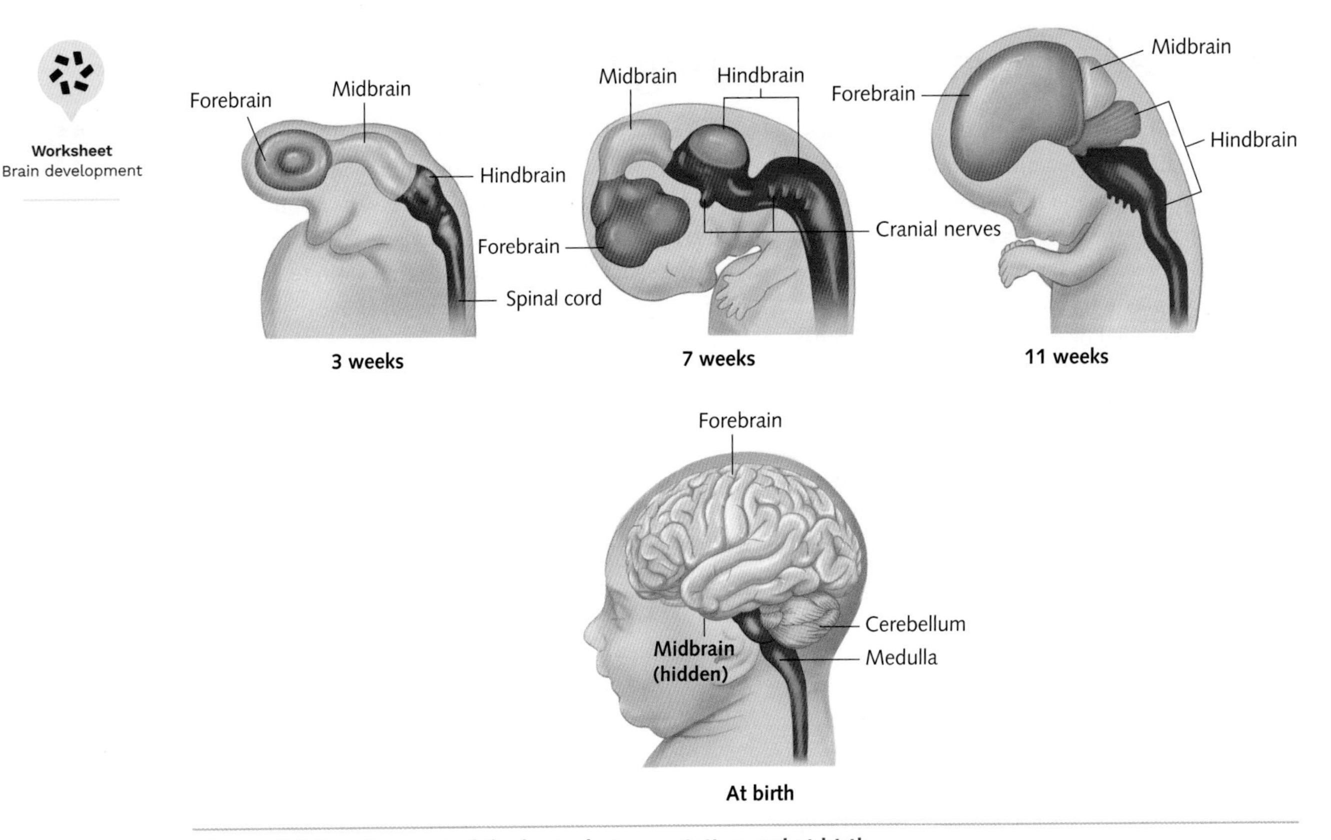

Figure 5.3 Development of the brain during gestation and at birth

Weblink
Axon guidance

where neurons travel to their final location within the nervous system. At the same time, dendrites grow and extend to axons of other neurons, which results in the formation of pathways between neurons (Figure 5.4). This is known as **neural maturation.**

Synaptogenesis

At birth, not all neurons have developed synaptic connections; rather, these connections form as our brains develop. This process is referred to as **synaptogenesis**. Synaptogenesis occurs throughout a healthy person's lifespan, but it occurs most rapidly during early brain development. This is when the brain overproduces both neurons and synaptic connections, with the peak of synapse formation beginning at about 2 months before birth and continuing until approximately 2 years after birth (Kolb & Gibb, 2011). This excess of neurons and synapses allows the infant to develop the range of sensory, motor (movement) and cognitive (thinking) skills necessary for development. During infancy, the brain is more 'plastic' or changeable; it can 'capture' experience more efficiently because there are more neurons available for the task than later in life. This is why learning new things such as language is easier in infancy compared to during

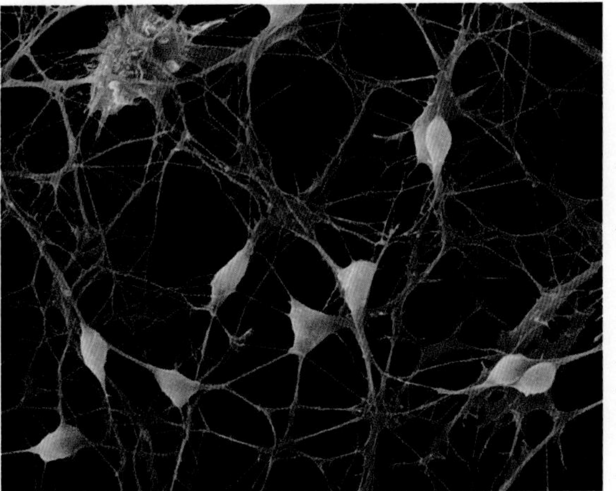

Alamy Stock Photo/Science Photo Library

Figure 5.4 Even when neurons grow in culture, they develop extensive networks and make connections with one another.

other life span stages. Under normal circumstances, assuming the infant is exposed to a variety of stimuli, the child should be able to achieve their optimal brain development during this time.

Throughout our life span, synaptogenesis establishes new neural pathways that allow different brain areas and structures to communicate. New synapses are constantly created and strengthened. This could be the result of learning, exposure to new experiences or even the formation of memories. The pathways most often stimulated are strengthened and are more likely to become a permanent part of the brain. Less active synapses are weakened and can ultimately be removed. The removal of synapses is referred to as synaptic pruning.

Synaptic pruning

In order to maximise its efficiency in transmitting information, the brain undergoes a process of **synaptic pruning** in which extra, weak or unused synaptic connections are removed (Figure 5.5). In the same way that pruning unhealthy branches from a tree results in stronger growth next season, synaptic pruning allows the remaining synaptic networks to strengthen and expand through repeated use. This ensures that neural transmission is as efficient as possible. In this way, the process of synaptic pruning and synaptogenesis follows the principle of 'use it or lose it'.

Synaptic pruning occurs throughout the lifespan. It is most prolific during infancy and childhood. Extensive pruning also occurs during early adolescence (Kolb & Wishaw, 2014; Spear, 2010). Synaptic pruning allows the brain to progressively develop more in-depth, efficient and complex functions because there are fewer connections and changes taking place. As a result, the brain becomes more flexible for future learning. The synapses to be pruned are determined by their strength. Synaptic strength relies on neurons being repeatedly activated. Clearly, the type of stimulation neurons receive and how regularly they are activated affects how our brains become 'hardwired'.

Synaptic pruning and synaptogenesis occur frequently in areas of the brain that are dedicated to sensory organs, such as the visual, auditory, gustatory (taste) and olfactory (smell) regions of the brain. These regions are constantly bombarded with sensory information and are continuously remodelled to adapt to environmental stimuli.

Myelination

Myelination of axons begins before birth with the most dramatic period occurring within the first 2 years of life. **Myelin** is a white, fatty substance that coats the axons of some neurons, giving them a whitish appearance (hence the name white matter). Myelin protects and insulates the axon (Figure 5.6). It also increases the speed at which the electrical

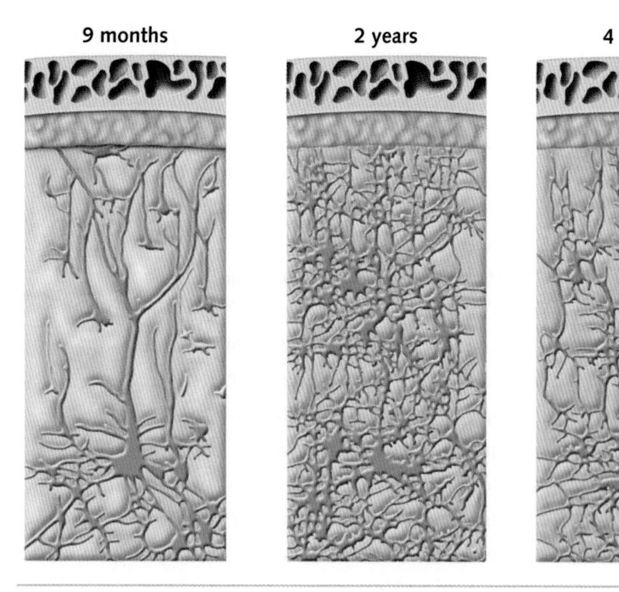

Alamy Stock Photo/BSIP SA

Figure 5.5 The increase in axons and dendrites is greatest during the first 2 years of life. During infancy, the brain develops an excess of neurons to maximise the likelihood of absorbing environmental stimuli. You can see that synaptic connections are increasing from 9 months to 2 years. There are more synaptic connections at 2 years than at 4 years, showing that some of the synaptic connections have been pruned.

impulse travels within the neuron. The greater the number of myelinated neurons, the more efficient and faster the neural pathways work in that part of the brain.

Weblink
Use it or lose it: the adolescent brain

5.1.2 BRAIN DEVELOPMENT

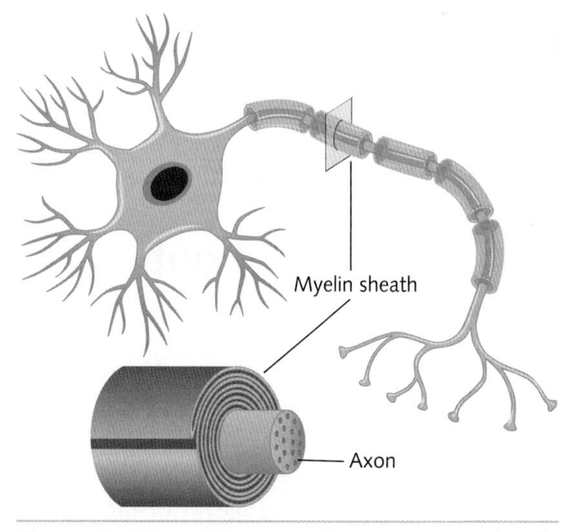

Figure 5.6 In some neurons the axon is coated with a protective layer of myelin.

Neuroimaging has shown that the more developed – or myelinated – the brain becomes, the more you are able to integrate information from multiple brain areas. Magnetic resonance imaging studies have indicated that developmental processes tend to occur in the brain in a back-to-front pattern, explaining why the prefrontal cortex develops last.

These studies have also shown that teens have less white matter (myelinated axons) in the frontal lobes compared to adults, and that myelin in the frontal lobes increases throughout adolescence. With more myelin comes growth of important neurocircuitry, allowing for better flow of information between brain regions.

Myelination greatly contributes to cognitive behaviours (such as planning, decision-making and organisation) and to emotional behaviour, personality and social behaviour. Research suggests that myelination begins before birth and can continue until around the age of 30, when the prefrontal cortex, the brain area responsible for 'executive' functions such as planning, decision-making and thinking logically, is finally myelinated (Sowell et al., 1999).

Table 5.1 summarises the stages of brain development.

Table 5.1 Stages of brain development

Stage	What happens?
Neurogenesis	Brain cells divide and multiply, forming new neurons.
Neural migration	Brain neurons travel to their final destination, which determines the neuron's function.
Neural maturation	Brain neurons extend their axons and grow dendrites to form synapses.
Synaptogenesis	Synapses are formed between neurons so information can be transmitted.
Synaptic pruning	Extra, weak or unused synaptic connections are removed.
Myelination	Myelin sheaths coat the neurons' axons.

Getting the best out of your brain

Like all parts of the body, our brains change during the ageing process (Table 5.2). For example, the brain's overall volume begins to shrink when people are in their 30s or 40s, with the rate of shrinkage increasing around age 60. Changes also occur at the neural level. The myelin sheaths that protect and insulate the axons begin to deteriorate. The number of connections, or synapses, between brain cells also drops. The normal ageing brain has lower blood flow, and this reduces the most in the frontal cortex. Adult neurogenesis, which supports important brain functions such as learning and memory, declines.

These structural and chemical changes lead to mild, subtle changes in cognition, including slower speed of thinking, decline in multi-tasking skills and decline in attentional control. They do not affect the person's ability to carry out everyday tasks such as shopping, cooking or driving, and they are considered a normal part of the ageing process. There is growing evidence that the brain is still plastic in healthy older individuals and that people can still meet new challenges and manage new tasks as they age.

There are strategies you can employ to delay disease and keep the brain healthy and functioning at peak performance well into old age. Ways to promote healthy brain functioning include maintaining mental stimulation, a healthy diet and physical activity, building social support and engaging in risk reduction.

Mental stimulation

Activities that are mentally engaging help to exercise the brain and reduce the risk of cognitive decline and dementia. Mental stimulation helps to increase **cognitive reserve** (the mind's resistance to damage of the brain) with novelty, variety and challenge. It involves using mental activity to build up and strengthen the reserve of connections in our brains. If our brains change with age, they can use the built-up reserve to better cope with the changes. In this way, mental stimulation helps to reduce the loss of synapses (Morrison & Baxter, 2012). Synaptic connections need to be used in order to stay strong. Mentally stimulating activities such as reading, playing or learning a musical instrument, playing chess or learning a new language challenge

Table 5.2 Brain changes during normal ageing

Structural changes	Cognitive changes
» Decrease in brain volume » Deterioration of myelin (slows processing and cognition) » Reduced synaptic connections » Thinning of the cerebral cortex » Decline in adult neurogenesis	» Slower processing of stimuli » Decline in multi-tasking ability » Decline in attention » Slower recall of words and names » Decline in ability to recall newly learned information

the brain to make new neural connections and build its reserves of brain cells (Figure 5.7). Learning new information promotes neuroplasticity and forms new synaptic connections.

Research suggests that people with greater cognitive reserves are better able to stave off symptoms of degenerative brain changes associated with dementia or other brain diseases, such as Parkinson's disease, multiple sclerosis or stroke. Mental stimulation is distinct from cognitive training, which involves specifically designed computerised training programs.

Shutterstock.com/Julio Rivalta

Figure 5.7 Engaging in mentally stimulating activities helps the brain to function normally because it increases the brain's ability to resist damage, by reducing the loss of synapses.

Diet

Just like any other organ in our bodies, our brains respond to a healthy diet (Figure 5.8). The brain needs glucose for energy, and the correct building blocks, supplied through diet, to repair damage and build more neurons, synapses and neurotransmitters. A healthy diet is well-balanced, low in cholesterol and saturated fats (such as fats from animals and some plants), and high in protein and unsaturated fats (such as fats from oily fish and nuts). It also involves low alcohol and coffee consumption.

Shutterstock.com/artem evdokimov

Figure 5.8 A healthy diet helps maintain brain function.

Physical activity

Research studies show that physical activity promotes neurogenesis (the formation of new neurons) (Akatemia Suomen, 2016). Exercise facilitates oxygen and growth factors in the brain by stimulating the release of chemicals that result in the production of new cells. This changes synaptic structure and increases synaptic plasticity. The functional changes that result have a broad effect on overall brain health (Cotman et al., 2007). Exercise helps to maintain cardiovascular health (blood pressure and cholesterol levels). It facilitates the release of neuromodulators such as serotonin, dopamine and noradrenaline, which helps to reduce depression. (A neuromodulator is a chemical substance released by a neuron that affects the transmission of signals between neurons.) Taking the stairs instead of a lift, getting off the train, bus or tram one stop before your stop and walking the rest of the way, and dancing have all been shown to improve brain health and function (Teixeira-Machado et al., 2019).

Social support

Building social support networks helps to reduce stress and challenge the brain. Research has illustrated that social support is positively associated with cortical thickness. It suggests that supportive relationships are critically linked to brain circuitry involved in emotional and social processing (Sherman et al., 2016).

Risk reduction

One of the most obvious risk reduction strategies is to take steps not to damage your brain. As discussed earlier, the cerebral cortex is prone to damage because it is near the outside of the brain and closest to the skull. If you wear a seatbelt when in a car and a helmet when riding a bike, motorbike or horse, or when playing contact sports, you reduce your risk of brain injury. There is also increasing evidence that stress hormones damage brain neurons. To limit your chance of damage from stress, use stress management strategies such as going for a walk or run, meditation or relaxation techniques.

KEY CONCEPTS 5.1a

- Neuroplasticity refers to the brain's ability to change its structure and function as a result of experience or in response to injury.
- The brain is made up of interconnected neurons.
- A neuron is made up of dendrites, a soma and an axon, which is often myelinated.
- Impulses move through a neuron via electricity and between neurons via chemicals called neurotransmitters.
- Developmental plasticity refers to changes in brain neurons and synaptic connections that occur as a result of experience during specific developmental stages.
- The stages of developmental plasticity are neurogenesis, neural migration, neural maturation, synaptogenesis, synaptic pruning and myelination.
- An adolescent's frontal lobes are not fully myelinated, so they have not yet developed the capacity to fully integrate information and think in the same complex, logical or abstract ways that an adult can.
- Neuroplasticity occurs throughout the lifespan.
- There are strategies to help people maintain their mental functioning across the lifespan.

Concept questions 5.1a

Remembering

1 What is neuroplasticity? r

2 What is the benefit of neuroplasticity? r

3 Without referring to your textbook, define developmental plasticity in 25 words or less. e

4 How is a message passed from one neuron to the next? r

5 Identify the six stages of brain development. r

6 How does synaptogenesis physically alter the brain? r

7 How does synaptic pruning physically alter the brain? r

Understanding

8 Use an example to explain why infants have an excess of brain neurons. Why is the excess of neurons later pruned? c

9 Explain how continuing myelination of axons enables an adolescent to perform more complex behaviours than a child. e

Applying

10 Victor arrived in Australia from Croatia when he was 4 years old. His family wanted to fit into the Australian lifestyle so they never spoke Croatian again. By the age of 6 years, Victor had completely forgotten how to speak Croatian. Name the process that caused this and explain how it led to Victor losing the ability to speak Croatian. e

HOT Challenge

11 In 2014, news sites reported the story of a US teen boy riding in the front seat of an SUV on the highway at night. He had a brilliant idea: if he set the driver's armpit hair on fire, it might impress the girls in the back. It did not turn out quite the way he planned. The SUV ran into a ditch, flipped over, and threw three of the five occupants – all teenagers, none of whom were wearing seat belts – out onto the highway. Miraculously, no one was killed. c

With reference to myelination, examine this event and explain the teen boy's decision.

The impact of experience

Research indicates that a person's environment and experience shapes their neural circuit development and impacts their brain development and plasticity throughout their life span. The types and variety of stimuli a person is exposed to influences the creation and pruning of brain synapses. The strength of these synapses contributes to how connected and efficient their neural networks will be. In turn, these networks influence how efficiently their brain transmits information and what behaviours they are capable of.

ANALYSING RESEARCH 5.1

The Bucharest project: the lasting impact of neglect

Kirsten Weir

In 1989, Romanian dictator Nicolae Ceauşescu was removed from power. The world discovered that 170 000 children were being raised in Romania's impoverished institutions. As the children's plight became public, Dr Nathan Fox, Director of the Child Development Laboratory at the University of Maryland, Dr Charles Nelson, Harvard Medical School and Children's Hospital Boston, and Charles Zeanah, MD, at Tulane University, realised they had a unique opportunity to study the effects of early institutionalisation on the developing brain.

The trio launched their project in 2000 and began by assessing 136 children who had been living in Bucharest's institutions from birth. They randomly assigned half of the children to live with Romanian foster families, whom the researchers recruited and assisted financially. The other half remained in care as usual. The children ranged in age from 6 months to nearly 3 years, with an average age of 22 months.

Over the subsequent months and years, the researchers returned to assess the children's development in both settings. They also evaluated a control group of local children who had never lived in an institution. They found many profound problems among the children who had been born into neglect. Institutionalised children had delays in cognitive function (thinking), motor development (movement) and language. They showed deficits in socio-emotional behaviours and experienced more psychiatric disorders. They also showed changes in the patterns of electrical activity in their brains, as measured by an electroencephalogram (EEG).

For children who were moved into foster care, the picture was brighter. These children showed improvements in language, IQ and social-emotional functioning. They were able to form secure attachment relationships with their caregivers and made dramatic gains in their ability to express emotions.

While foster care produced notable improvements, the children in foster homes still lagged behind the control group of children who had never been institutionalised. And some foster children fared much better than others. Those removed from the institutions before age 2 made the biggest gains. 'There's a bit of plasticity in the system,' Fox says. But to reverse the effects of neglect, he adds, 'the earlier, the better.' In fact, when kids were moved into foster care before their second birthdays, by age 8 their brains' electrical activity looked no different from that of the control group. The researchers also looked at structural differences, such as size of the children's brains. They found that institutionalised children had smaller brains, with a lower volume of both grey matter (which is made primarily of the soma of neurons) and white matter (made up of myelinated axons).

Source: Adapted from Weir, K. (2014). The lasting impact of neglect. *Monitor on Psychology, 45*(6). https://www.apa.org/monitor/2014/06/neglect

Weblink
Bucharest early intervention project

Questions

1 What was the aim of this study?
2 Identify the methodology used.
3 a Identify the population of research interest.
  b Identify the sample.
4 Why would these researchers have studied the population that they did when undertaking their research?
5 Identify the independent and dependent variables in this study.
6 What conclusion(s) can be drawn from this study?

HOT Challenge

7 Identify one ethical principle related to this study. Explain how this concept is involved in the study.

Response to learning

The brain alters the connections between synapses and reorganises neural pathways as a result of learning. We call this ability **adaptive plasticity** (Table 5.3). This response continues throughout the life span. During learning, neurons develop new connections and form new synapses, or existing synapses are strengthened. Information is then able to pass from one neuron to the next faster. If neurons are repeatedly stimulated, the transfer of information across the synapse is strengthened and the potential is enhanced for both transmitting and receiving neuron firing when stimulated.

Learning a new language, for example, causes extensive neuroplasticity in the brain as new connections and pathways are formed. In the early stages this mostly occurs in the frontal lobes (situated in the front of the brain). As the person becomes more fluent in this new language, the neural activity shifts to the parts of the brain that control more automatic motor processing. When this happens, they will be able to automatically answer a question or form a sentence without having to translate it first. The person has created new neural pathways (Grundy et al., 2017).

Weblink
Do blind people have better hearing?

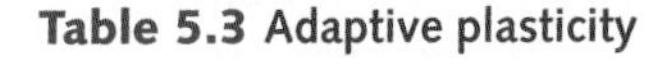

Table 5.3 Adaptive plasticity

Definition	The ability of brain neurons to alter the connections between synapses to best suit a person's environmental conditions when learning something new or when relearning something after brain injury
When it occurs	Occurs over the whole life span, but is greatest during infancy and early childhood
Impact on behaviour	Allows the brain to adapt to changing environmental circumstances or to loss of function following brain injury

5.1.3 ADAPTIVE PLASTICITY

Response to injury

Remarkably, adaptive plasticity allows the brain to also 'rewire' (or reorganise) itself after some types of damage to compensate for a lost function or to maximise existing functions. This rewiring can happen at any stage during the life span. Following brain injury, neurons in the undamaged brain regions can sometimes take over the sensory or motor functions that had been performed by the damaged areas. This happens after repeated stimulation. In effect, the brain works around the damaged neurons. It attempts to construct other neural pathways to compensate for functional loss, essentially remapping the brain. Depending on the type and extent of brain injury, plasticity may also allow the damaged brain neurons to repair or regenerate.

Adaptive plasticity mainly occurs as a result of two processes that take place at the synapse between neurons: sprouting and rerouting. **Sprouting** refers to the creation of new connections between neurons. Sprouting involves the generation of additional branches (either axon terminals or dendrites) so the neuron can make new connections with other active neurons. **Rerouting** involves an undamaged neuron that has lost connection with a damaged neuron connecting with another neuron. This effectively shifts a lost function from a damaged area to an undamaged area and creates an alternative pathway between active neurons. The loss of function has been compensated for and communication is reestablished (Figure 5.9).

Rerouting
New connections are made between active neurons to create alternative neural pathways

Undamaged neuron
New synapse
Damaged neuron
New primary circuit

Before
After

Neural pathway
Neural pathway
Terminated pathway

Sprouting
New axon and dendrite extensions allow existing neurons to form new connections

A B
A B

Neural connections by neuron A and neuron B prior to degeneration

Collateral sprouting of neuron B after damage to axon of neuron A

Figure 5.9 Structural changes in the brain can result from sprouting and rerouting of synapses between brain neurons to allow your brain to adapt to environmental changes and/or injury.

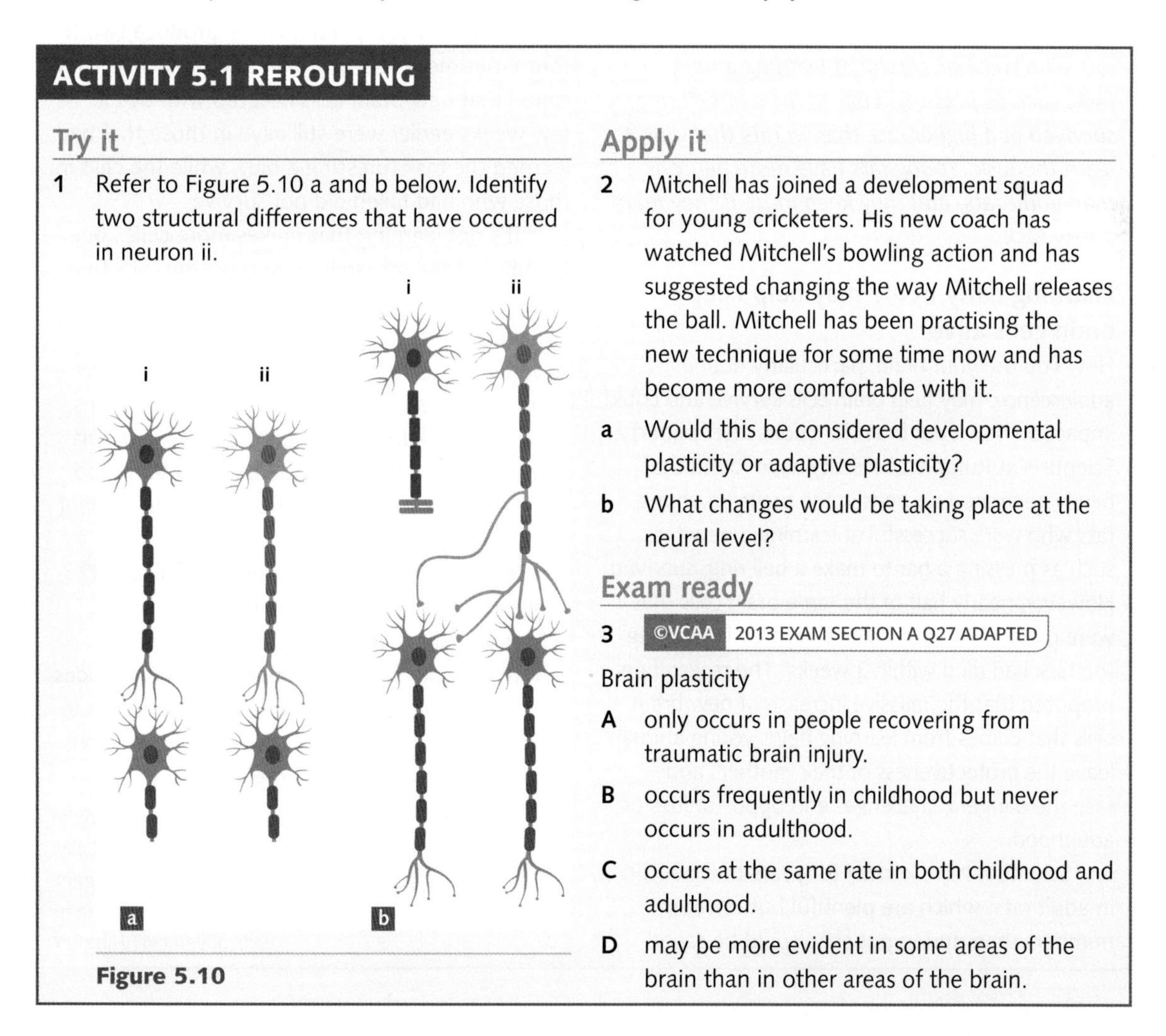

ACTIVITY 5.1 REROUTING

Try it

1 Refer to Figure 5.10 a and b below. Identify two structural differences that have occurred in neuron ii.

Figure 5.10

Apply it

2 Mitchell has joined a development squad for young cricketers. His new coach has watched Mitchell's bowling action and has suggested changing the way Mitchell releases the ball. Mitchell has been practising the new technique for some time now and has become more comfortable with it.

a Would this be considered developmental plasticity or adaptive plasticity?

b What changes would be taking place at the neural level?

Exam ready

3 ©VCAA 2013 EXAM SECTION A Q27 ADAPTED

Brain plasticity

A only occurs in people recovering from traumatic brain injury.

B occurs frequently in childhood but never occurs in adulthood.

C occurs at the same rate in both childhood and adulthood.

D may be more evident in some areas of the brain than in other areas of the brain.

It has been discovered that the brain can actually change its shape within a matter of weeks (Figure 5.11, page 196). This can be in response to certain mental and physical stimuli such as those that occur when we learn something new. Only recently have larger structural changes to the brain been observed.

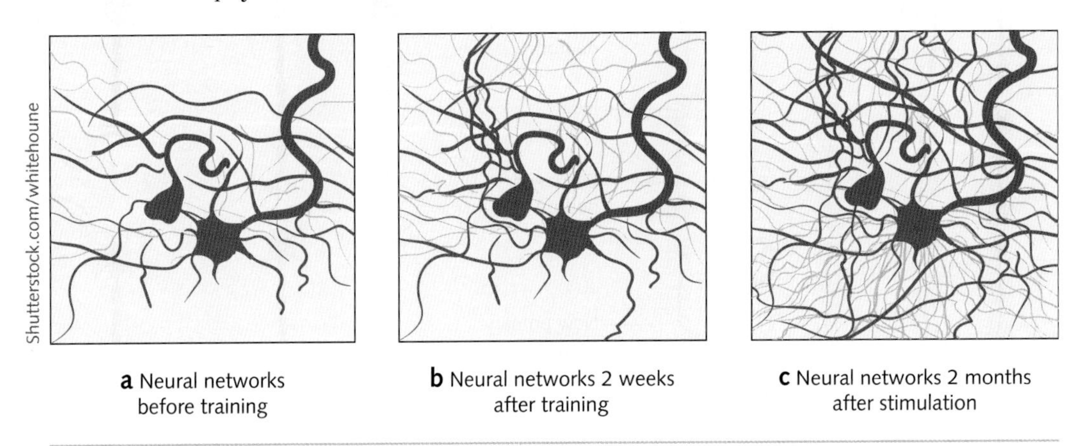

a Neural networks before training

b Neural networks 2 weeks after training

c Neural networks 2 months after stimulation

Figure 5.11 Adaptive potential plasticity allows the brain to recover and reduces the effects of structures resulting from disease or injury.

ANALYSING RESEARCH 5.2

In the following study, researchers found that newly generated neurons in the brains of young rats who were successful at learning a new task, such as pressing a bar to make a bell ring, survived at a higher rate than in rats that didn't learn the task. Young rats have more neurons than adult rats, and they keep more if they learn a new task.

Learning early in life may help keep brain cells alive

How you use your brain, particularly during adolescence, may help brain cells survive and could impact on how your brain functions after puberty. Scientists at Rutgers University have found that newly generated neurons in the brains of young rats who were successful at learning a new task, such as pressing a bar to make a bell ring, survived. However, nearly half of the same brain cells that were made in rats who didn't succeed in learning the task had died within 3 weeks. The researchers proposed that the massive increase of new brain cells that comes from learning helps young animals leave the protectiveness of their mothers and face the dangers, challenges and opportunities of adulthood.

Scientists have known for years that neurons in adult rats, which are plentiful but fewer in numbers than during puberty, could be saved with learning. They did not know if this would be the case for young rats, which produce two to four times more neurons than adult animals. They found that new brain cells injected with dye a few weeks earlier were still alive in those that had learned the task (pressing a bar), while the cells in those who had failed did not survive.

'It's not learning that makes more cells,' one member of the research team reported. 'It's that the process of learning keeps new cells alive that are already present at the time of the learning experience. Since the process of producing new brain cells on a cellular level is similar in humans, ensuring that adolescents learn at optimal levels is critical. You don't want the material to be too easy to learn or to be so difficult that the student gives up and doesn't learn.'

So, what does this mean for the adolescent? While scientists can't measure individual brain cells in humans, this study, on the cellular level, provides a look at what is happening. This provides a window into the amazing ability the brain has to reorganise itself and form new neural connections at such a transformational time in our lives.

Adapted with permission from 'Learning Early in Life May Help Keep Brain Cells Alive' by Robin Lally, Rutgers Today, May 26, 2014, https://www.rutgers.edu/news/learning-early-life-may-help-keep-brain-cells-alive#.U4TgJihg9rY

Questions

1 What was the aim of this study?
2 a Identify the population of research interest.
  b Identify the sample.
  c Why would these researchers have studied the population that they did when undertaking their research?
3 What was the key result from this study?

HOT Challenge

4 Analyse the results of the research study. In your response, include a discussion of how each condition affected the survival of brain cells, as well as a justification as to whether the results can be generalised to humans.

Research has shown that certain parts of the brain are function-specific and can grow and change shape when we learn. It is this type of adaptive plasticity that enables us to change our behaviour and adapt to changes in our environment.

The physical impacts of learning are clearly demonstrated by magnetic resonance imaging (MRI) scans of the hippocampus (a brain area involved in spatial navigation) shown in the results of a 2000 study of 16 male London taxi drivers. These scans use magnets and radio frequencies to create structural images of the brain. The taxi drivers had to memorise all London streets (a test known as the Knowledge) in order to obtain their taxi licence. After the training was completed, scans showed the end or posterior region of their hippocampus was significantly larger than in males of a similar age who were not taxi drivers (Figure 5.12b). In addition, they also showed that the more experienced the driver was, the larger their posterior hippocampus (Maguire et al., 2000). The study results suggest that the brains of the qualified taxi drivers changed to accommodate the knowledge of London's streets. It suggests that the human brain remains plastic throughout the lifespan, shaping and remodelling its structure to allow it to adapt when we learn new tasks.

Weblink
The hippocampi of taxi drivers – the people who have to remember 25000 streets

In his 2007 book, *The Brain That Changes Itself: Stories of Personal Triumph from the Frontiers of Brain Science*, Norman Doidge describes numerous examples of individuals who experienced changes in how they did things following brain injury. One story features a surgeon who suffered a stroke that paralysed his left arm. Part of his rehabilitation was not to use his good arm; to instead make himself use his left arm.

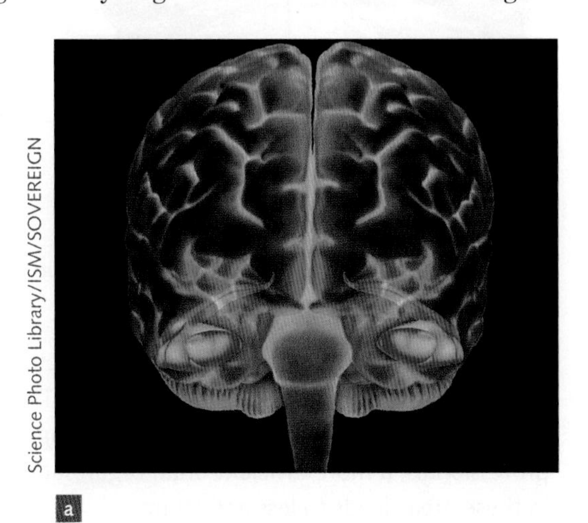

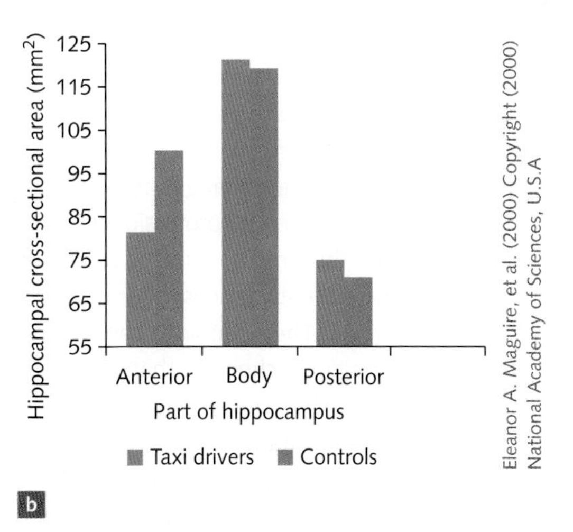

Figure 5.12 a A frontal image of the human brain showing the location of the two hippocampi (pink). **b** The mean volume measurements of anterior, body and posterior sections of the hippocampi of London taxi drivers compared to a control group of non-taxi drivers.

When the patient was asked to clean tables, at first the task was not possible because he was not allowed to use his right arm. Then, bit by bit, he began to remember how to move his left arm. Eventually, he learned to write and play tennis again (Doidge, 2007). In this case it would appear that the functions damaged by the stroke transferred themselves to regions of the brain that were unaffected by the stroke. His brain compensated for the lost ability caused by the brain damage by reorganising and forming new connections between intact neurons. In order to reconnect, the neurons had to be stimulated through repeated activity.

ANALYSING RESEARCH 5.3

How playing the drums changes the brain

People who play drums regularly for years differ from non-musical people in their brain structure and function. The results of a study using magnetic resonance imaging (MRI) data by German researchers from Ruhr University Bochum suggest that drummers have fewer, but thicker fibres in the main connecting tract between the two halves of their brain. In addition, their motor brain areas are organised more efficiently.

Neuroscientists have long understood that playing a musical instrument can change the brain via neuroplastic processes. The Bochum researchers were interested in drummers because their motor coordination far surpasses that of untrained people. 'Most people can only perform fine motor tasks with one hand and have problems playing different rhythms with both hands at the same time,' explains Dr Lara Schlaffke, head of the research team. 'Drummers can do things that are impossible for untrained people.'

The team intended to gain new insights into the organisation of complex motor processes in the brain by identifying brain changes associated with this training. They tested 20 professional drummers who had played their instruments for an average of 17 years and currently practised for more than 10 hours per week. They examined them using various MRI imaging techniques that provided insights into brain structure and function. They then compared the data with measurements of 24 non-musical control subjects. In the first step, both groups had to play drums to test their abilities and were then examined in the MRI scanner.

Drummers presented clear differences in the front part of the corpus callosum, a brain structure that connects the two hemispheres and whose front part is responsible for motor planning (Figure 5.13). Data indicated that drummers had fewer but thicker fibres in this important connecting tract between the brain hemispheres. This allowed them to exchange information between the hemispheres more quickly than the controls. The structure of the corpus callosum also predicted the performance in the drum test: the higher the measure of the thickness of the fibres in the corpus callosum, the better the drumming performance.

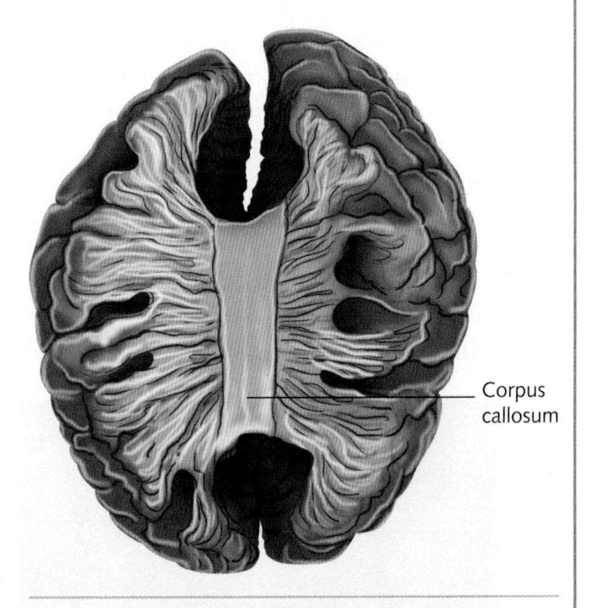

Figure 5.13 **Corpus callosum connects the two hemispheres of the brain**

Moreover, drummers' brains were less active in motor tasks than that of control subjects. This phenomenon is referred to as sparse sampling: in professionals, a more efficient brain organisation in these areas leads to less activation.

Source: Adapted with permission from press release, 'How playing the drums changes the brain', by Julia Weiler, Ruhr-Universität Bochum, 09 December 2019, https://news.rub.de/english/press-releases/2019-12-09-neuroscience-how-playing-drums-changes-brain

»

QUESTIONS

1 What was one possible aim of this study?
2 Construct a hypothesis for this research.
3 a Identify the population of research interest.
  b Identify the sample.
4 Identify the independent and dependent variables in this study.
5 State two findings from this study.

HOT Challenge

6 Write an evidence-based conclusion for this study.

KEY CONCEPTS 5.1b

» Adaptive plasticity occurs when the brain alters the connections between synapses and reorganises neural pathways as a result of learning, or following injury.
» After brain injury, neurons in an undamaged brain region can sometimes take over sensory and motor functions that had been performed by damaged areas. This happens after repeated stimulation.
» Sprouting occurs when new connections between neurons are created.
» Rerouting occurs when an undamaged neuron that has lost connection with a damaged neuron can make new connections with other active neurons.
» Certain parts of the brain that are function-specific can grow and change shape when we learn.

Concept questions 5.1b

Remembering

1 Without referring to your textbook, define adaptive plasticity in 25 words or less. r

Understanding

2 How are developmental plasticity and adaptive plasticity different? e
3 How are sprouting and rerouting different? e

Applying

4 Chantel received a laptop computer for her 10th birthday. The next day she began learning how to touch type. She has been practising every day for the last month and now she can type without having to look at the keyboard. Use your knowledge of brain plasticity to explain how learning to touch type has changed Chantel's brain structure and function. c
5 Explain how adaptive plasticity could help an adult who had sustained brain damage to recover some or all lost function. e
6 Can you teach an old dog new tricks? Provide reasons for your decision. e

HOT Challenge

7 Joe is a 15-year-old student. After school, he prefers to play his favourite computer game rather than complete the variety of homework tasks he is given. In terms of adaptive plasticity, explain why it is beneficial for Joe to complete the varied homework tasks he is given. Refer to synaptogenesis and synaptic pruning in your answer. c
8 Tim has been learning to speak English, Italian and French from an early age. English is spoken in his home, but Tim persists with learning Italian because he is able to speak to his nonna in Milan, Italy. He has not practised speaking French as frequently as Italian because he does not have a family member in France to talk with. Identify three ways in which Tim's brain has changed in structure and function as a result of his experiences with learning multiple languages. c

5.2 Acquired brain injury

An **acquired brain injury (ABI)** refers to brain damage that occurs after birth. This affects the functional ability of the brain's nerve cells. The result of this damage is some form of impairment or dysfunction, such as loss of speech. When a person has an ABI, a portion of their brain cells is destroyed or deteriorates.

Causes of acquired brain injury

Acquired brain injuries can occur suddenly or they can develop over time. They can be temporary or permanent and may result in partial or total cognitive, physical, emotional or motor impairment. Causes of ABI include disease, blows to the head, alcohol and drug use, or oxygen deprivation.

Acquired brain injuries may occur suddenly as a result of injury caused by an *external* force, such as a traumatic blow to the head that damages brain tissue or structures. This type of ABI is known as a **traumatic brain injury (TBI)**. They may result from events such as a fall, an assault, a car accident, shaken baby syndrome or during contact sports (Figure 5.14). A TBI is not the same as a head injury, since a person can experience an impact to the head and sustain damage to the face, scalp and skull without necessarily injuring their brain.

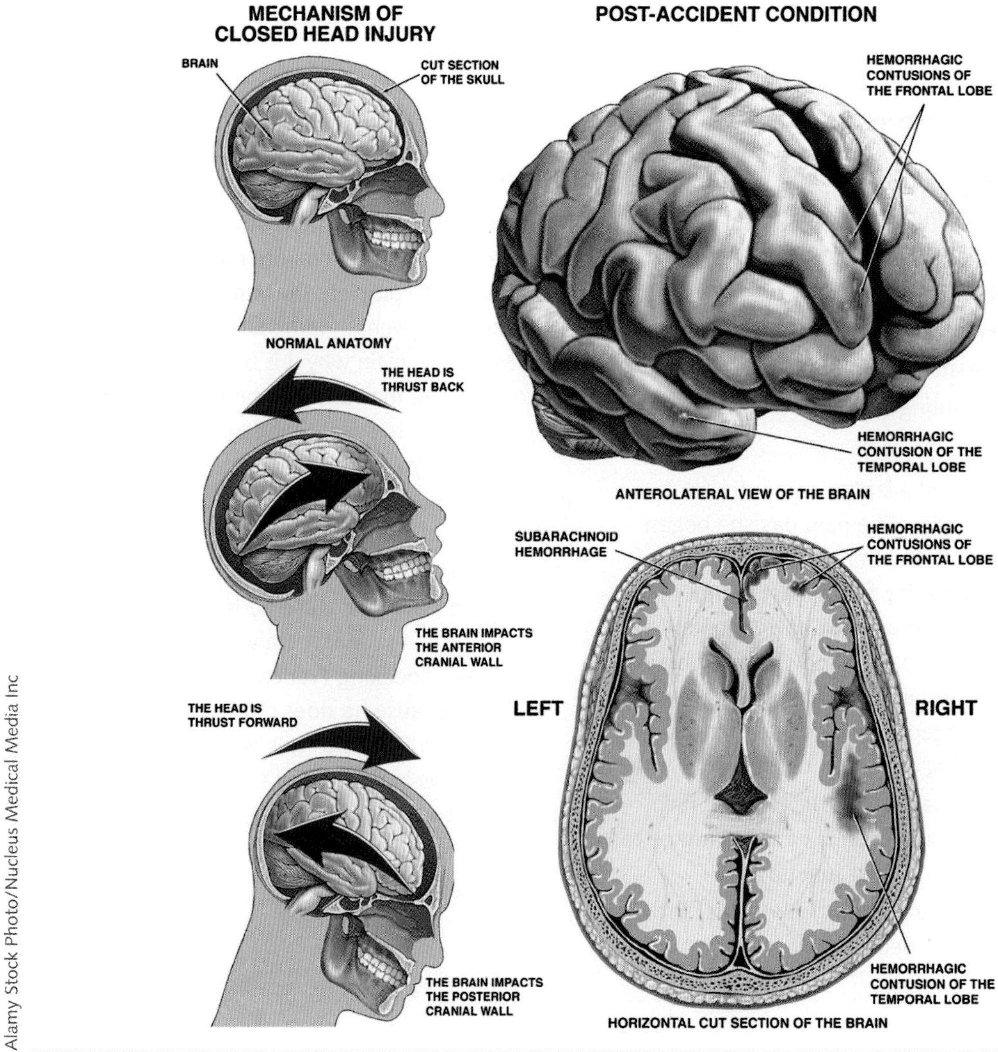

Figure 5.14 When a person receives a sudden, violent blow to the head, their head will jerk backward, forward, or in both directions. Their brain rapidly moves backward and then forward, hitting the sides of their skull. This can result in damage to brain tissue that increases the risk of traumatic brain injury (TBI). (Note: 'hemorrhagic' means bleeding, and a contusion is a bruise.)

Brain injury that occurs slowly over time as a result of *internal* factors is referred to as a **non-traumatic brain injury (NTBI)**. Causes of NTBI include:

» lack of oxygen to the brain (known as hypoxia or anoxia) that injures brain tissue
» exposure to toxins (such as excessive and prolonged alcohol or drug use)
» infectious disease that inflames the brain covering or the brain tissue itself
» tumours that damage surrounding brain tissue and structures
» a stroke as a result of a haemorrhage, or blockage to blood vessels that supply blood to the brain
» degenerative neurological diseases that cause abnormal changes to brain cells, such as Alzheimer's disease, multiple sclerosis or Parkinson's disease.

Figure 5.15 summarises the types of acquired brain injury.

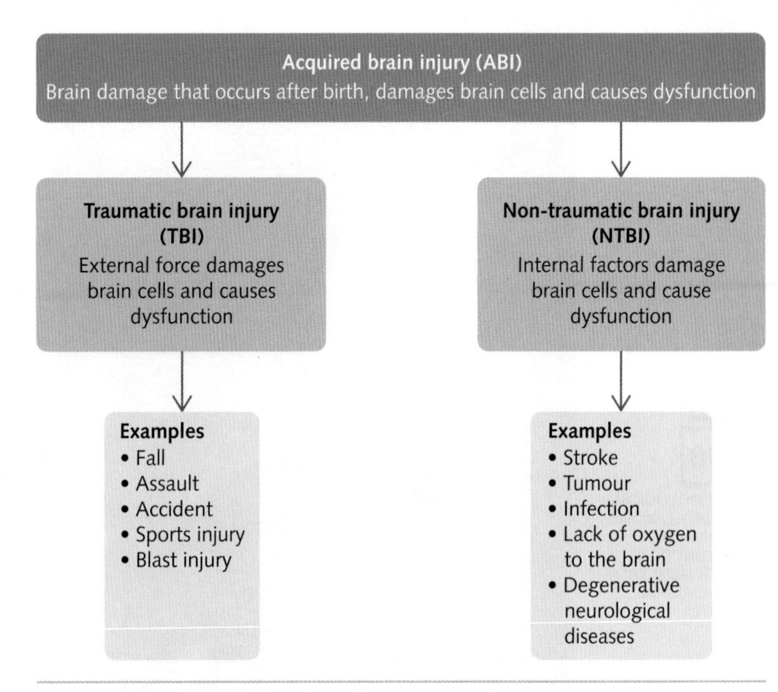

Figure 5.15 Types of acquired brain injury (ABI)

Effects of damage to the cerebral cortex

The upper part of the brain consists of two large cerebral hemispheres, each containing four lobes (see Chapter 4). These two hemispheres are united by a thick band of myelinated nerve fibres, the **corpus callosum**, which transfers information registered in one hemisphere to the other for processing. Thus, it allows the hemispheres to communicate. Together, the left and right hemispheres form the **cerebrum** (Latin for brain), which is the largest and most highly developed part of the human brain (Figure 5.16).

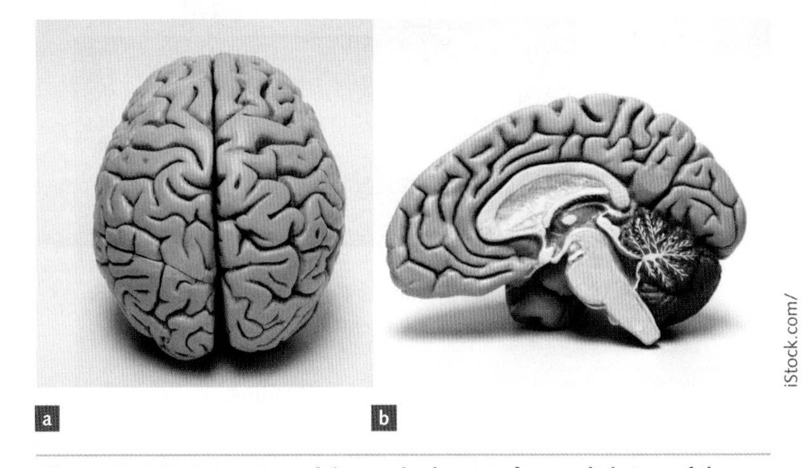

Figure 5.16 a A top view of the cerebral cortex; **b** a medial view of the brain shows the cerebrum and the corpus callosum (central white structure).

The outer layer of the cerebrum, the cerebral cortex, is approximately 3 millimetres thick and it is the largest brain structure. The cerebral cortex contains approximately 70 per cent of the neurons in the central nervous system, as well as the areas responsible for higher, more complex mental abilities such as intelligence, problem-solving, language, planning and organisation, judgement, movement, sensations, learning, memory, emotion and consciousness.

The cerebral cortex is prone to damage because it lies directly beneath the skull. When the cerebral cortex is damaged, the result can be some form of dysfunction. The degree of dysfunction and how long it will last varies according to which area of the brain was damaged and the level of severity of the damage.

The case of Phineas Gage

5.2.1 ACQUIRED BRAIN INJURY

On 13 September 1848 Phineas Gage was 25 years old. He was employed as the foreman of a crew blasting a cutting through a rocky outcrop in Cavendish, Vermont, USA. He was packing explosive powder into a hole using a tamping iron. The tamping iron struck the rock, causing a spark that detonated the explosive powder. The tamping iron, 110 centimetres long, 3 centimetres in diameter and 6 kilograms in weight, shot up and entered Phineas' left cheek, passing through his brain and skull and landing over 10 metres away (Figure 5.17). Remarkably Gage survived, and was up and walking within minutes.

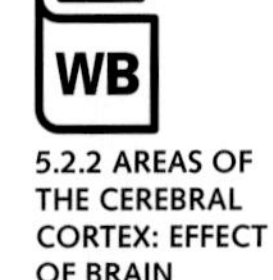

5.2.2 AREAS OF THE CEREBRAL CORTEX: EFFECT OF BRAIN DAMAGE

WB
5.2.3 BIOLOGICAL, PSYCHOLOGICAL AND SOCIAL DYSFUNCTIONS ASSOCIATED WITH THE CEREBRAL CORTEX

Alamy Stock Photo/GL Archive

Figure 5.17 Phineas Gage and the tamping iron that entered his left cheek and exited his skull.

Back at his lodgings the attending physician, John Harlow, cleaned the wound, removed small bone fragments, replaced larger bone fragments and closed the wound with adhesive strips. Although Gage eventually recovered physically, those close to him started to notice changes in his behaviour. Twenty years later his physician reported:

> *His contractors, who regarded him as the most efficient and capable foreman in their employ previous to his injury, considered the change in his mind so marked that they could not give him his place again. He is fitful, irreverent, indulging at times in the grossest profanity (which was not previously his custom), manifesting but little deference for his fellows, impatient of restraint of advice when it conflicts with his desires, at times pertinaciously obstinent [sic], yet capricious and vacillating, devising many plans of future operation, which are no sooner arranged than they are abandoned in turn for others appearing more feasible. In this regard, his mind was radically changed, so decidedly that his friends and acquaintances said he was 'no longer Gage'.*
>
> Harlow, J. M. (1848). Passage of an iron rod through the head. *Boston Med. Surg. J. 39*, 389–393. Viewed at https://neurophilosophy.files.wordpress.com/2006/12/harlowbmsj1860.pdf

The case of Phineas Gage is notable because it was the first time that people connected brain injury with personality. Gage's injuries affected his biological, psychological and social functioning.

» *Biological functioning* refers to the way the body and its various components operate. It refers to all the voluntary and involuntary physical behaviours that we engage in, such as combing our hair, walking and breathing.
» *Psychological functioning* involves our cognitive, perceptual and emotional abilities. These areas of behaviour underpin our ability to communicate our thoughts as well as understand others.
» *Social functioning* refers to our interactions with our environment and our ability to perform everyday social tasks appropriately and to maintain an adequate social life. Because we live in groups and interact with others on a daily basis, it is important that we are socially functional.

Brain injury causes the destruction or deterioration of brain neurons. The severity of brain injury varies according to the type and location. A mild brain injury may be temporary. A moderate brain injury may cause symptoms that last longer and are more pronounced. A severe brain injury may result in the person suffering life-changing and debilitating problems. People who are in a coma or in a minimally responsive state may remain dependent on the care of others for the rest of their lives.

The long-term effects of an ABI are difficult to predict and they will be different for each person. Initially, the physical disabilities following brain injury can be quite easy to see; however, disabilities that affect thinking, emotion and behaviour can be far harder to recognise (Table 5.4).

The biological, psychological and social impact and recovery from an ABI depends on a number of factors. These include the extent and location of the brain damage, the age and general health of the person, and the timing and the quality of treatment. Disorders such as Parkinson's disease and multiple sclerosis typically impact the body's ability to control movement. Other brain injuries may have a greater impact on cognition, personality and behaviour. An ABI may also lead to mental health problems. Mental health problems encompass a broad range of emotional and behavioural difficulties that alter mood, thinking, communication or behaviour (or a combination of these) to the degree that a person cannot effectively cope with the demands of day-to-day living. Compared with the general population, people with an ABI are more likely to suffer mental health problems, such as adjustment disorders, depression and anxiety, and drug and alcohol abuse.

There are some injuries to specific brain areas that cause distinct losses of function. These include spatial neglect and aphasias.

Table 5.4 Possible ABI symptoms

Biological	Psychological	Social
» Headaches » Speech deficits » Loss of motor skills » Changes in senses, e.g. loss of taste » Sleep disturbances » Loss of balance and coordination » Seizures » Stiffness and shakiness » Numbness or tingling » Ringing in ears	**Cognitive** » Difficulty processing information » Difficulty planning, organising and problem-solving » Problems with orienting to person, place or time » Loss of attention and concentration » Difficulty understanding language (written and spoken) and accessing memorised language **Emotional/Behavioural** » Increased aggression » Increased anxiety » Depression » Dulled emotions » Increased inhibition and impulsivity » Apathy » Mood swings	» Encroaching on personal space of others » Abnormal sexual behaviour » Inappropriate touching » Isolating from others » Change in relationships with others » Lack of self-awareness » Inability to identify social cues

ACTIVITY 5.2 PHINEAS GAGE

Try it

1 Reread the information about Phineas Gage. Watch the video in the weblink: Phineas Gage.

Apply it

2 As time progressed, Phineas Gage demonstrated a number of changes in behaviour. Identify the listed changes experienced by Phineas Gage following his accident as either biological (B), psychological (P) or social (S).

- Gage changed from friendly and quietly spoken to impatient and aggressive.
- Gage had trouble maintaining friendships.
- Gage changed from considerate to irresponsible and impulsive.
- Gage developed difficulties with problem-solving.
- Gage could no longer manage his job as a railway worker.
- Gage had impaired ability to sustain his attention.
- Gage had difficulty moving some facial muscles.

Exam ready

3 Select the correct option(s). Sal was in a serious accident and sustained a traumatic brain injury (TBI). A possible biological impact of Sal's brain injury could be:

A impaired memory.
B breakdown of family relationships.
C mood swings.
D impaired motor activity.

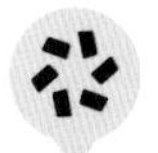

Weblink
Phineas Gage

Spatial neglect

Damage to the right parietal lobe can sometimes cause **spatial neglect**. This is a tendency to ignore the left or right side of visual space after a brain-damaging event, such as a stroke (Figure 5.18). Spatial neglect is more common following injury to the parietal lobe in the brain's right hemisphere, but it can also occur as a result of parietal lobe damage in the left hemisphere.

When blood flow to the right parietal lobe is blocked, neurons can be severely damaged. Often sufferers of spatial neglect will not eat food on the left side of a plate and some even refuse to acknowledge a paralysed left arm as their own (Springer & Deutsch, 1998). Spatial neglect is not the same as blindness, because the person's eyesight has not been damaged. Spatial neglect is primarily a disorder of attention whereby patients characteristically fail to respond to stimuli located on one side of their body. Although most people with spatial neglect may eventually recover, the evidence suggests that they continue to have significant cognitive impairments, particularly relating to attention (Li & Malhotra, 2015). Spatial neglect is a disabling condition for many who have it, making them unable to drive on the road because of the reduced awareness of their environment (Sarwar & Emmady, 2021).

Weblink
Aphasia: the disorder that makes you lose your words

5.2.4 SPATIAL NEGLECT

Aphasias

The left hemisphere contains the language centres, known as Broca's area and Wernicke's area. Broca's area, located in the left frontal lobe, is responsible for speech production. Wernicke's area, in the left temporal lobe, is responsible for understanding written and spoken language. Damage to either of these areas can result in **aphasia** (Table 5.5). Aphasia is an acquired language impairment that results from brain damage, usually in the left hemisphere. However, aphasia can be caused by damage to any one of the cortical areas involved in language. Some aphasia sufferers can speak fluently but they cannot read; others may understand what they read but they cannot speak; some can read but not write; some can read numbers but not letters; and others can sing but cannot speak.

In the late 1800s, French physician Paul Broca studied patients with damage to the lower left frontal lobe, close to the motor cortex. This area later became known as Broca's area. He discovered that while patients with damage to this area could understand what was said to them and knew what they wanted to say in response, they simply could not say it. They may have been able to move their tongue and lips, but they had no ability to say all or some of the words required. Broca called this condition **Broca's aphasia** (also known as expressive aphasia).

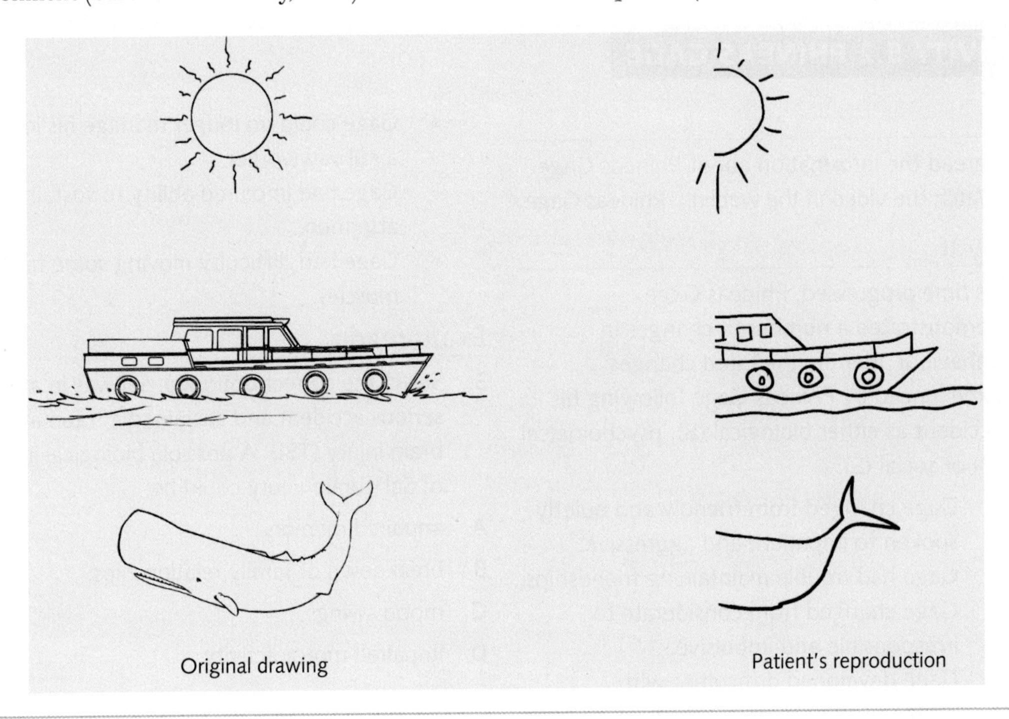

Figure 5.18 A person with spatial neglect tends to ignore one side of their visual space. The picture on the left shows what the person was asked to copy, and the picture on the right shows what they drew. You can see that the person completely ignored or neglected the left-hand side of the picture.

9780170465052

Table 5.5 Features of Broca's aphasia and Wernicke's aphasia

Broca's aphasia	Wernicke's aphasia
Inability to produce clear, fluent speech that others can understand	Inability to understand language (written and spoken)
Speech is slow and slurred, and words are not properly formed	Speech is clear and words are properly formed
Person can understand language	Words that are put together are not grammatically correct and sentences do not make sense
Person is usually aware of their inability to communicate effectively	Person is often unaware of their inability to communicate effectively

A typical patient with Broca's aphasia speaks slowly and laboriously and uses simple sentences. Usually only the concrete words are pronounced and the connecting words are omitted. They might say, for example, 'Boy went beach' instead of 'The boy went to the beach'. The 'sentence' makes sense in some ways, but it is not articulate (fluent) and is incomplete. Some patients with Broca's aphasia cannot speak at all. The ability to write is usually impaired as well.

Around the same time, German physician and psychiatrist Carl Wernicke studied patients with a different language deficit – the inability to understand speech or to produce coherent language. He identified a part of the brain in the left temporal lobe close to the primary auditory cortex as being the area responsible for language comprehension – Wernicke's area. The resultant language loss following damage to this area became known as **Wernicke's aphasia**.

While we know that Broca's area and Wernicke's area are crucial to speech production and understanding language, we also know that other areas of the brain are involved in language too. These include the angular gyrus in the parietal lobe, the insular cortex, the basal ganglia and the cerebellum (Figure 5.19). These areas form a network that allows us to process words and word sequences so we can understand their context and meaning. As a result, we are able to understand and produce language.

Weblinks
Wernicke's aphasia

Weblink
Broca's aphasia

5.2.5 FEATURES OF BROCA'S APHASIA AND WERNICKE'S APHASIA

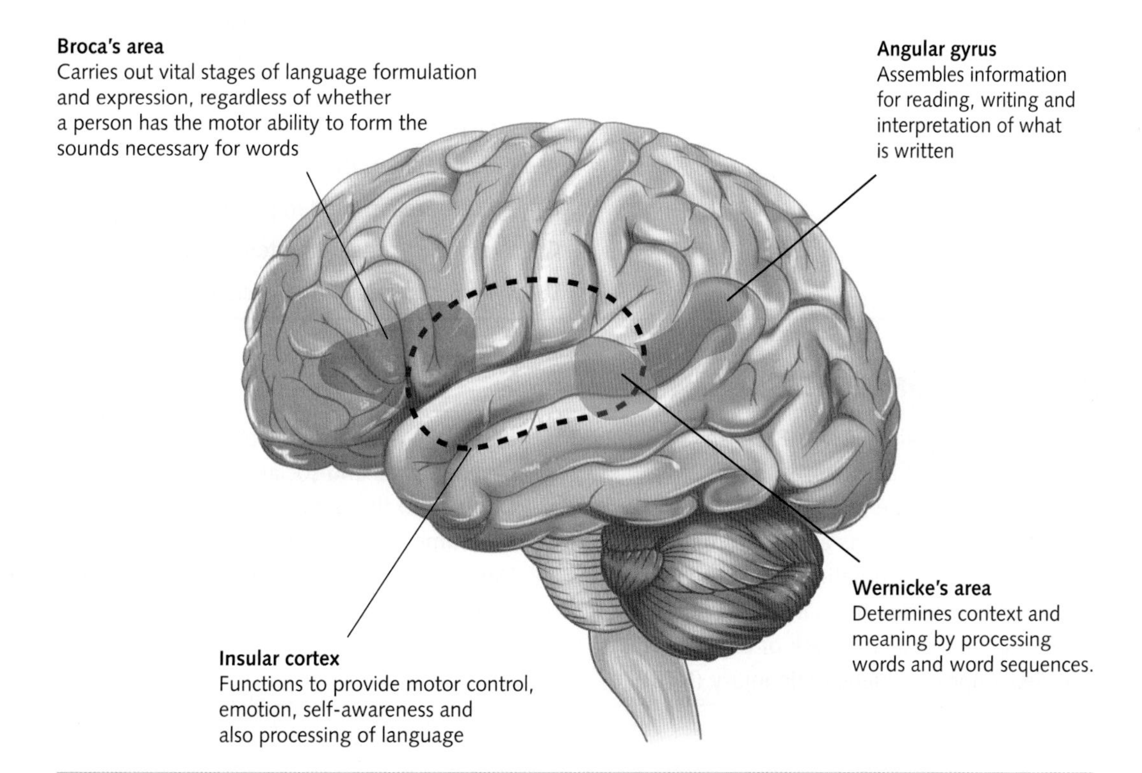

Figure 5.19 A number of brain areas work together as a network so we can understand and produce language.

5.2.6 BRAIN DAMAGE CASE STUDIES

ACTIVITY 5.3 TYPES OF APHASIA

Try it

1 Read the transcripts of Patients 1 and 2 below. In each case, the patient was shown a picture book of Cinderella and was asked to retell the story. Based on the transcripts, which types of aphasia do you believe Patient 1 and Patient 2 have?

Patient 1 – Speech sample	Patient 2 – Speech sample
'First I started with a little, small it was the lady's little which was thing that I wanted before I could remember, but I can't do it now. I look carefully about what he looked around but he couldn't really try it about there. At the same time, all these things, at least one, two, three people. Which were clever to the people. This, this and she supported to do that. I don't know, buy anyway, the say thinking.' Source: https://neupsykey.com/language-processing-disorders/	'Cinderella ... poor ... um adopted he ... scrubbed floor, um, tidy ... poor, um ... adopted ... Si-sisters and mother ... ball. Ball, prince um, shoe ... Scrubbed and uh washed and ... tidy, uh, sisters and mother, prince, no, prince, yes. Cinderella hooked prince. Um, um, shoes um, twelve o'clock ball ... finished.' Source: 'Linguistics 110 Linguistic Analysis: Sentences & Dialects - Lecture number one - what is language?', Bucknell University, https://www.departments.bucknell.edu/linguistics/lectures/aphasia.html

Apply it

2 Based on your answers to question, what indicators in each Cinderella story allowed you to reach your conclusion?

Exam ready

©VCAA 2002 EXAM 1 Q4 ADAPTED

3 Amber is able to speak clearly and she mostly uses long sentences. However, when she speaks, the words are mostly meaningless and she has difficulty comprehending what other people are saying. Amber's symptoms are consistent with brain damage to

A Broca's area in the frontal lobe.

B Broca's area in the temporal lobe.

C Wernicke's area in the temporal lobe.

D Wernicke's area in the occipital lobe.

KEY CONCEPTS 5.2

- An acquired brain injury (ABI) is any type of brain injury that occurs after birth.
- An ABI can be due to an injury from an external force, such as a blow to the head. This is called a traumatic brain injury (TBI).
- An ABI can be due to an injury from an internal factor, such as lack of oxygen. This is called a non-traumatic brain injury (NTBI).
- Brain injury can affect a person's biological, psychological and social functioning.
- Damage to the right parietal lobe can sometimes cause spatial neglect.
- Damage to the left hemisphere can produce impaired language ability known as aphasia.

Concept questions 5.2

Remembering

1 What is an ABI? r

2 List the two types of ABI. r

3 How is each type of ABI caused? r

4 What is spatial neglect? r

Understanding

5 Explain how Broca's aphasia and Wernicke's aphasia are similar but different. e

6 Explain why the cerebral cortex is the part of the brain that is most prone to injury. e

Applying

7 Based on the changes experienced by Phineas Gage, what does injury to the frontal lobe suggest about the role of this area of the brain? e

8 Create a table that shows whether the following symptoms of ABI are biological, psychological or social. e

Symptoms: inappropriate touching, changes in mood and personality, inability to follow group rules, involuntary muscle spasms, difficulty understanding written and spoken language, invasion of personal space of others, inability to smile, speech difficulties, difficulty recognising shapes.

HOT Challenge

9 Watch the weblink Australian BMX star Kai Sakakibara's brave battle following brain injury. Determine if, and how, Kai's ABI has affected his biological, psychological and/or social functioning. c

10 Explain the impact of Wernicke's aphasia on someone's life. e

Weblink
Australian BMX star Kai Sakakibara's brave battle following brain injury

5.3 Neurological disorders

Neurological disorders are diseases or events that affect the brain, the spinal cord and the nerves that connect them. All neurological disorders result from damage to the nervous system either due to genetic causes or after birth due to environmental causes. ABIs are an example of a neurological disorder.

Where the damage takes place and the severity of damage determines to what extent communication, vision, hearing, movement and cognition are impacted. These disorders have a range of biological, psychological and social/behavioural symptoms for the person with the disorder (Table 5.6).

Table 5.6 Examples of neurological disorders

Disorder	Symptoms	Cause(s)
Quadriplegia	Unable to move body trunk, legs and arms	Injury to the spinal cord at the neck, preventing messages getting past the injury site
Paraplegia	Unable to move the lower part of the body – legs, feet, stomach muscles	Injury to the spinal cord below the neck, preventing messages getting past the injury site
Acquired brain injury	See page 198	See page 196
Cerebral palsy	Lack of muscle control, affecting posture, reflex and balance; visual, learning, hearing, speech, epilepsy and intellectual impairments	Brain injury during pregnancy or shortly after birth
Epilepsy	Seizures of varying severity	Genetic causes; head trauma; infection; brain tumour
Motor neurone disease (MND)	Progressive weakening of muscles involved in moving, breathing, swallowing and speaking	Genetic and environmental factors cause degeneration of motor neurons (neurons that relay signals from brain to muscles)
Multiple sclerosis (MS)	Muscle tingling and spasms, paralysis of limbs, difficulty walking, loss of bladder control, vision problems	Malfunctioning immune system attacks the myelin coating around the axons of neurons

Disorder	Symptoms	Cause(s)
Huntington's disease	Involuntary jerking, muscle rigidity; abnormal eye movements; impaired gait; difficulty speaking or swallowing; difficulty prioritising tasks and learning new information; lack of impulse control; social withdrawal; feelings of sadness	Single gene defect inherited from mother or father
Lewy body disease	Visual hallucinations, movement disorders, poor regulation of bodily functions, cognitive and sleep difficulties	Protein deposits (Lewy bodies) develop in the nerve cells that make up the brain
Parkinson's disease	Difficulty in walking, talking, writing, dressing and other everyday tasks; tremors	Genetic and environmental causes. Due to the lack of the neurotransmitter, dopamine
Tourette's syndrome	Tics (sudden, intermittent movements or sounds) such as eye blinking, head jerking or grunting; other conditions such as attention deficit hyperactivity disorder (ADHD) and autism often associated	Genetic and environmental causes. Neurotransmitters such as dopamine and serotonin thought to be involved. Three to four times more likely in males than in females
Alzheimer's disease	Persistent memory lapses; difficulty concentrating, thinking, making judgements, decisions, planning; depression; apathy; social withdrawal; mood swings; delusions; wandering	Genetic, environmental and lifestyle factors. Brain proteins fail to function, which disrupts the work of neurons. Neurons lose their connections and die

5.3.1 PARKINSON'S DISEASE

CASE STUDY 5.1

Through my eyes: Epilepsy diagnosis in adulthood

Iva Moutelikova

There was absolutely no warning before the seizure came on. One moment, I was trying to open a door and feeling a bit weak, and the next thing I knew, I was being taken to the hospital in an ambulance. I lost almost an hour of my memory and have no recollection of what happened.

My colleagues told me that I fell into a perfect recovery position, and then, there was a lot of shaking. Once that stopped, I was sort of 'awake', walking around aimlessly and even giggling. It made me feel uncomfortable hearing this story …

In the hospital, they did a few tests, and I was told that it could have been an isolated seizure, which can happen to anyone once during their lifetime. To be honest with you, it didn't feel like a big deal …

One afternoon, about 10 months later, while I was walking outside, I suddenly started feeling really scared, like I was about to lose control over my body. I knew something really bad was about to happen, and I didn't know what to do.

I looked around, but there was nobody on the street … I felt a huge amount of panic and started walking really fast …

Now, knowing more about epilepsy, I know that this feeling of extreme dread was a symptom called aura that some people experience before a seizure. The smart thing would have been to sit down rather than keep walking, but I couldn't think clearly. I had never felt like this before.

I woke up about 50 minutes later – again, in an ambulance. Because I was walking so fast while the seizure started, I fell very hard, hitting my face on the hard concrete pavement …

The paramedics asked me whether I knew my name, what day it was, and the other usual questions … I was told that there were six people looking after me who called the ambulance and waited with me. Apparently, my seizure lasted about 25 minutes. I am very grateful to these kind strangers for getting me help …

I spent about 7 hours in the hospital that afternoon. I had an ECG [electrocardiogram: records electrical signals in the heart] and other tests … In the following weeks, I got referred for a CT scan and an MRI.

I vividly remember a conversation that I had with my epilepsy nurse ... We were discussing my condition, and she asked me what I knew about SUDEP … She explained that SUDEP means a

sudden unexpected death in epilepsy. She asked me whether I wanted to know more about it, insisting that every person with epilepsy has to know about it. That was one of the worst mornings I have ever had.

... [M]y neurologist finally gave me an epilepsy diagnosis, as suspected, and I was put on anti-seizure medication ... I decided to stop drinking alcohol completely, because ... I didn't want to mix it with my medication. Funnily enough, for some people I knew, my stopping drinking was the hardest thing for them to understand ... Whatever condition you have, there will always be people who won't take it seriously. They might dismiss it or act like you are being dramatic.

Within a few months, I started having horrible migraine. Adjusting to the side effects of my new medication, having regular migraine episodes, and dealing with the new diagnosis was a lot to handle, especially with no family nearby.

Around this time, I started having mental health problems. I developed anxiety, and every time I felt anxious, I kept thinking that I was about to have a seizure ... I decided to go to therapy ... [I] also changed my medication and decided to take a much lower dose, which agreed with me so much better.

It's been 6 years since my last seizure, and 6 years since I have been taking epilepsy medication ... Epilepsy hasn't stopped me from enjoying my life and doing the things I love.

Extracts republished from 'Through My Eyes: Epilepsy diagnosis in adulthood' by Iva Moutelikova (2021, November 30) by permission of *Medical News Today*, https://www.medicalnewstoday.com/articles/through-my-eyes-epilepsy

Questions

1 Summarise the case study in no more than 250 words. Make sure you identify the key points.
2 Identify the impacts on psychological, biological or social functioning for a person suffering from epilepsy.
3 Is epilepsy a neurological disorder? Give a reason(s) for your answer.
4 What was the purpose of the paramedics asking Iva Moutelikova questions such as 'What is your name?' and 'What is today's date?'

LOGBOOK TEMPLATE: CASE STUDY

INVESTIGATION 5.1 NEUROLOGICAL DISORDERS

Scientific investigation methodology

Case study

Investigate a well-known personality or sports person who has been diagnosed with one of the neurological disorders (listed in Table 5.6, pages 203–4). Construct a case study of this person by answering the following questions and, where possible, providing examples from your investigated personality or sports person. Make sure that all the information you present is scientific and is supported by evidence.

Logbook template: Case study

Data calculator

Questions

1 What is the neurological disorder?
2 How prevalent is it in society?
3 Is it inherited?
4 What causes it?
5 How is it diagnosed?
6 What are the symptoms?
7 Does it have different stages?
8 How does it impact on a person's psychological, biological and social functioning?
9 How is it treated?
10 What lifestyle changes can a person make to ease the symptoms?

Make sure you correctly reference all your sources using APA referencing format.

Contemporary research in neurological disorders

Modern research techniques and technology enable scientists and neurologists to study the brain in a non-invasive way (Figure 5.20).

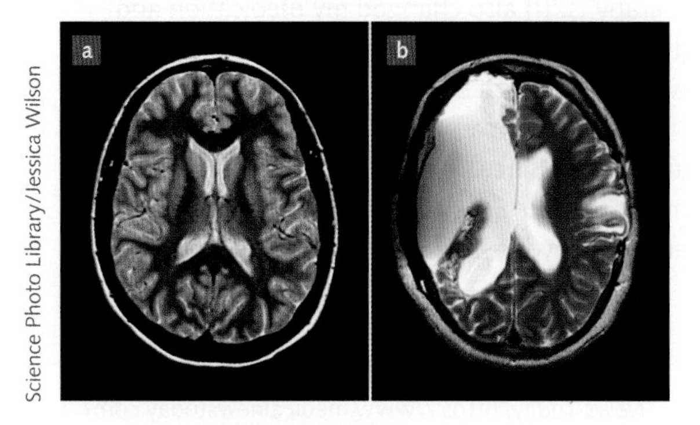

Science Photo Library/Jessica Wilson

Figure 5.20 MRIs are used to examine the brain. This pair of cross-sectional MRIs of the brain show a normal tissue and b tissue of a brain that has had almost complete removal of the right hemisphere to stop severe seizures.

In-vitro testing

A model is a representation of something where the real thing cannot be used because it is too big, too small or not ethical. Some models may be **in vitro**; that is, they are studies conducted in a test tube or controlled environment, rather than on a living organism. In-vitro testing allows scientists to isolate specific cells and focus their observations on specific mechanisms or processes. Work on rodent brains has taught us much of what we know about the synapse. This sort of work is difficult to do on living human tissue because of ethical issues. When using this method, scientists are not bound by the same ethical constraints they would be if live animal (human and non-human) participants were used. While in-vitro models have been useful in enhancing our understanding of neurological disorders, they also have their limitations. For example, cells grow differently in test tubes than in humans. They do not replicate the precise cellular conditions of a whole living body.

Animal models

As information is gathered and scientists want to learn more, they eventually need to test their ideas **in vivo**, or on a living animal (Figure 5.21). Animal studies are a form of in-vivo research and they are often used because they allow researchers to observe the overall effects of an experiment on a living subject. Animal models, while not a perfect substitute for human models, provide a closer approximation of human responses because of the many biological similarities shared by humans and animals. Many animals – for example, mice and monkeys – have the same organs and systems as humans, and they perform in much the same way.

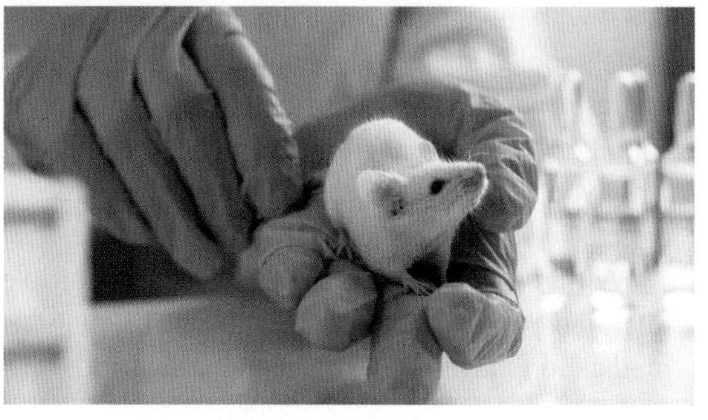

iStock.com/Evgenyi_Eg

Figure 5.21 Mice are the most frequently used model in scientific research.

Use of an animal model in research enables researchers to carry out experiments that would be ethically prohibited with humans. When animal models are used, researchers must apply the ethical concepts of non-maleficence (any action should not be disproportionate to the benefits) and respect (all living things deserve respect and, where possible, are protected). Using animal models enables researchers to test the effects of a range of possible treatments for neurological disorders before those treatments are given to humans.

Historically, animal models have been key in the discovery of drugs for a range of mental disorders, including anxiety, depression and schizophrenia, and neurological disorders such as Parkinson's disease, Huntington's disease and multiple sclerosis.

Neuroimaging: a window to the brain

Neuroimaging provides a non-invasive way to study the brain and is used to detect structural damage and assess functional changes in the brain due to disease or injury (Figure 5.22). Both functional and structural imaging have contributed dramatically to our understanding of the causes of various neurological disorders and their diagnosis and management.

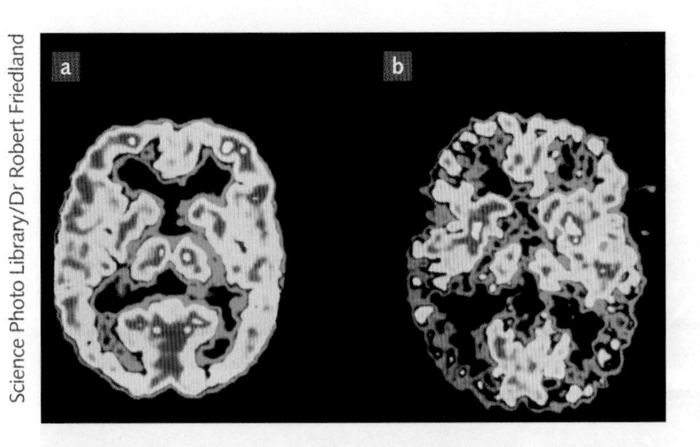

Figure 5.22 Positron emission tomography (PET) scans showing **a** a healthy brain and **b** a brain affected by Alzheimer's disease, which indicates lower blood flow and function

The most widely used neuroimaging technologies for investigating brain structure and function and neurological disorders are the following.

» *Electroencephalograms* (EEG) measure the electrical activity of the brain. They are used to detect abnormal electrical activity within the brain, such as that produced during an epileptic seizure (Figure 5.23).
» *Computerised tomography* (CT) scans are X-rays of the brain that produce an image of brain structure; they are less expensive than MRIs and PET scans and more commonly used (Figure 5.24a).
» *Magnetic resonance imaging* (MRI) scans use strong magnetic fields and radio waves to produce a detailed image of brain structure (Figure 5.24b).
» *Functional MRI* (fMRI) scans provide active images of the brain's metabolic functioning, such as blood flow (Figure 5.24c).
» *Positron emission tomography* (PET) scans require the patient to take in (drink or inject) a radioactive dye prior to the scan; this is used to show the metabolic functioning of tissues and organs. Tissues with high metabolic rates will show up as bright spots on a computer screen. The PET scan in Figure 5.24d used radioactive glucose injected into the bloodstream to track blood flow to the brain to produce a colour-coded image of brain activity.

More sophisticated neuroimaging techniques provide extremely sensitive measures of changes in brain structure and function. These include diffusion tensor imaging (DTI), magnetic resonance spectroscopy (MRS), the 'trimodal' model and magnetoencephalography (MEG).

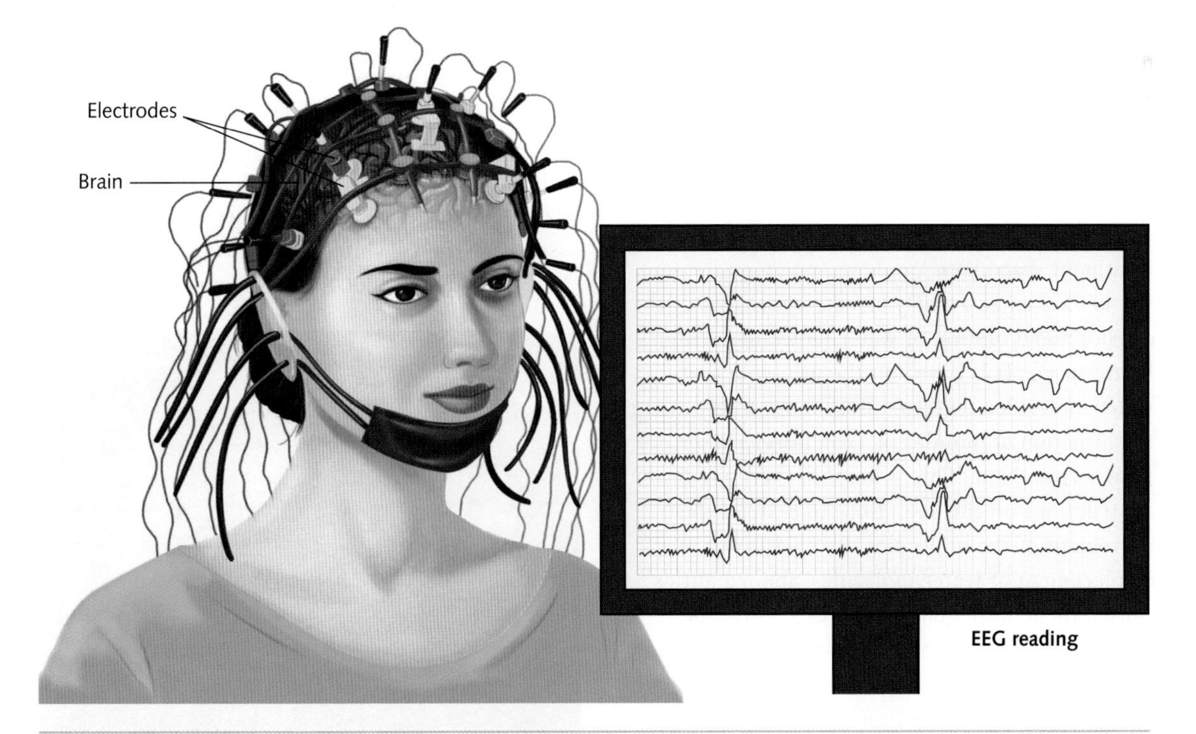

Figure 5.23 An electroencephalogram detects electrical activity in the brain.

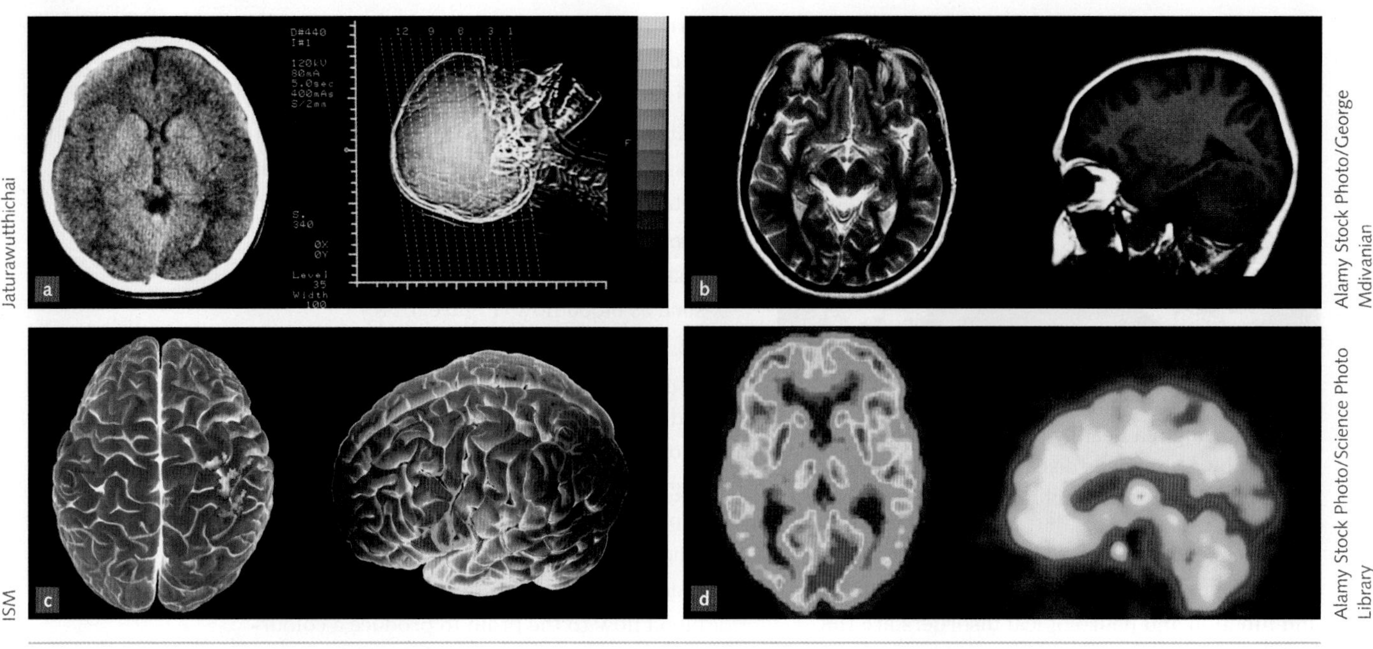

Figure 5.24 Top (axial) and side (sagittal) views of the human brain using **a** CT scan, **b** MRI scan, **c** fMRI scan and **d** PET scan.

Diffusion tensor imaging

Diffusion tensor imaging (DTI) is a relatively new and advanced form of MRI. It is used to investigate the structure of white matter and can detect microstructural damage, such as damage to axons, which is not visible on standard MRI and CT images. It also provides information about white matter tracts and the connectivity of different brain regions. Because DTI is non-invasive, it can be used in vivo (in a living organism) to evaluate changes in brain tissue caused by a neurodegenerative disease, the progression of the disease, and possible treatment responses.

DTI is an extremely sensitive measure of the direction and motion of water molecules within living tissue. It is based on the principle that water molecules tend to diffuse more freely along the direction of bundles of axon fibres rather than across them. This is because there are fewer obstructions on the axon fibre to restrict the movement of the molecules. By measuring the displacement of water molecules, abnormalities caused by neurodegenerative diseases can be identified. These abnormalities include loss of fibres, demyelination, damage within fibres and damage to support tissue around them (Beaulieu, 2002).

The information obtained from a DTI image can be condensed into either one number or four numbers. Condensation into four numbers is often used to display the image with an R (red), a G (green), and a B (blue) colour relating to the orientation of axons, and a brightness value (Figure 5.25). This information can be used to determine the brain's cellular organisation and how intact the structure of a brain network is.

DTI has successfully been used to study patients with neurodegenerative disorders including multiple sclerosis, epilepsy and Alzheimer's disease. It has also been widely used to assess the brain's white matter structure after TBI (MacDonald et al., 2007; Kinnunen et al., 2011).

Figure 5.25 DTI scan of white matter fibre tracts in the adult human brain

Magnetic resonance spectroscopy

Magnetic resonance spectroscopy (MRS) is a relatively new, non-invasive tool used to measure the chemical composition of tissues in vivo. It can identify changes in brain metabolism (how the brain uses energy to support and sustain cellular processes). When neurons increase their firing rate, they use more energy. This energy is supplied by chemical reactions in and around the neurons. The increased metabolic activity increases the demand for the basic fuels of metabolism (glucose and oxygen). This increased use of glucose and oxygen is associated with increased blood flow to the region, so blood flow is an indicator of neural activity (Figure 5.26). Signals indicating increased blood flow in the brain can be used to construct detailed images of the brain that reveal brain activity (Ramsey, 2012).

Many traumatic brain injuries and neurodegenerative disorders result in cognitive dysfunction for the person affected. However, this change in brain function is often manifested slowly. When neurodegeneration is suspected, MRS can be used to identify changes in brain metabolism. Treatments that may have a more favourable outcome for the person can then be offered (Eisele et al., 2020).

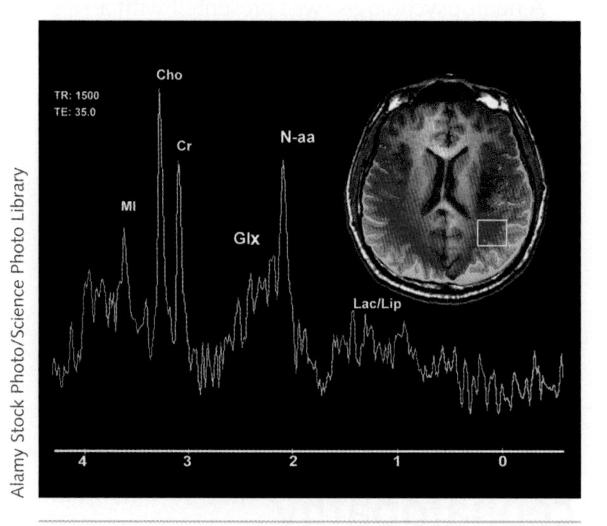

Alamy Stock Photo/Science Photo Library

Figure 5.26 Magnetic resonance spectroscopy (MRS) can detect in vivo the biochemical changes that occur in the brains of people with specific neurodegenerative diseases, such as brain cancer (shown here).

The trimodal method

The newly developed trimodal method combines three brain imaging techniques to provide more detailed images of the brain (Figure 5.27). This technique can show the exact timing and location of brain responses to a stimulus. The method combines functional MRI (fMRI), EEG and a new technique called EROS (event-related optical signal). The fMRI shows where things are happening in the brain, the EEG show when things happen in the brain and EROS accurately assesses the time of brain responses. Together they produce remarkably clear and detailed images of what is happening in the brain, where it is happening and when it is happening.

UNIVERSITY OF ILLINOIS NEWS BUREAU/L. BRIAN STAUFFER

Figure 5.27 A new approach to brain imaging combines functional MRI, electroencephalography and a technique known as EROS, which shines near-infrared light on the scalp to capture brain activity.

Magnetoencephalography

A magnetoencephalography (MEG) scanner can look inside the brain and detect the extremely weak magnetic signals that are generated by electrical currents flowing through active neurons. The scanner maps the brain activity and can be used to precisely identify the location that is active during an epileptic seizure (Figure 5.28).

Science Photo Library/James King-Holmes

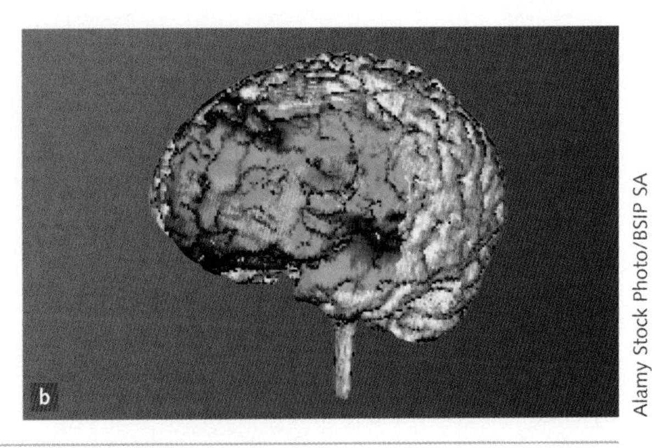

Alamy Stock Photo/BSIP SA

Figure 5.28 a Scanning the brain using a MEG machine; b images from MEG show the sites of brain activity during epileptic seizure (red areas) in a patient.

KEY CONCEPTS 5.3

- All neurological disorders are the result of damage to the nervous system.
- Neurological disorders affect the brain, the spinal cord and the nerves that connect them.
- Neurological disorders can affect biological, psychological and social functioning.
- Research may be in vitro or in vivo.
- Researchers often use animal models to investigate the causes, progress and treatment of neurological disorders in humans.
- Neuroimaging technologies such as CT scans, MRI scans and PET scans provide images of the physiological impacts of neurological disorders.
- New neuroimaging technologies such as DTI, MRS, the trimodal model and MEG can provide even more detailed images of the workings of the brain.

Concept questions 5.3

Remembering

1 What is a neurological disorder? r

2 List three examples of a neurological disorder. r

3 List three advantages of using neuroimaging tests to investigate a neurological disorder. r

4 Name three types of neuroimaging technologies used to investigate neurological disorders. r

Understanding

5 Explain the difference between in-vivo research and in-vitro research. e

6 Explain at least two advantages and two limitations of using animal models. e

Applying

7 A neuropsychologist was presented with a patient who had a suspected ABI. Explain how an MRI scan would provide information useful for diagnosis. e

HOT Challenge

8 Create a table to compare four different brain-imaging technologies. Decide on at least three different criteria to use for comparison. e

5.4 Chronic traumatic encephalopathy

Chronic traumatic encephalopathy (CTE) is a progressive (gradually increasing), degenerative (worsening) and fatal brain disease associated with repeated blows to the head over a long period. This disease is characterised by cognitive, motor and affective (emotional) dysfunction.

The term 'encephalopathy' is a broad term for any brain disease that alters brain function or structure. It originates from the Ancient Greek *en* (meaning in), *kephale* (meaning head) and *patheia* (meaning suffering). If a person suffers a chronic (long-lasting) and traumatic encephalopathy they experience a permanent, severe and lasting injury to their brain that can lead to serious mental, emotional and behavioural problems, which may be life-threatening.

Over a century ago, CTE was called 'punch-drunk syndrome' because it was originally found in the autopsied brains of professional and amateur boxers who had received repeated blows to the head and multiple concussions (Buckland et al., 2019). Recently, however, CTE has been found in athletes who play contact sports, military veterans who have been in active combat and received multiple head traumas, accident victims who have suffered head injuries and others with a history of repetitive head impacts. A research study conducted in 2017 at Boston University (Mez et al., 2017) examined 202 brains from deceased football players and found that 177 of them (87 per cent) had signs of CTE from repeated blows to the head (Figure 5.29). In Australia, cases are now being identified in players from professional sports such as rugby and Australian rules football.

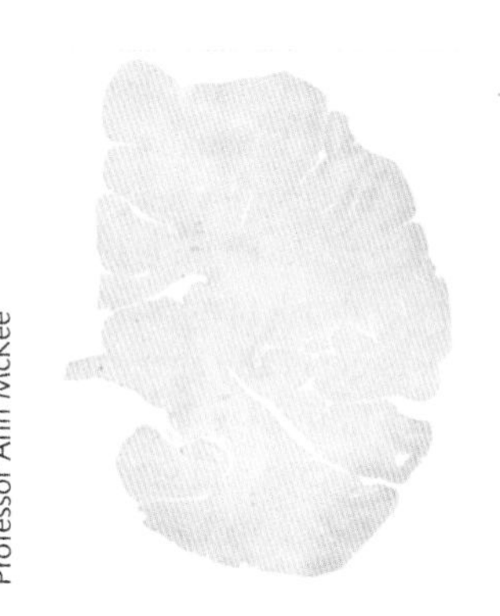

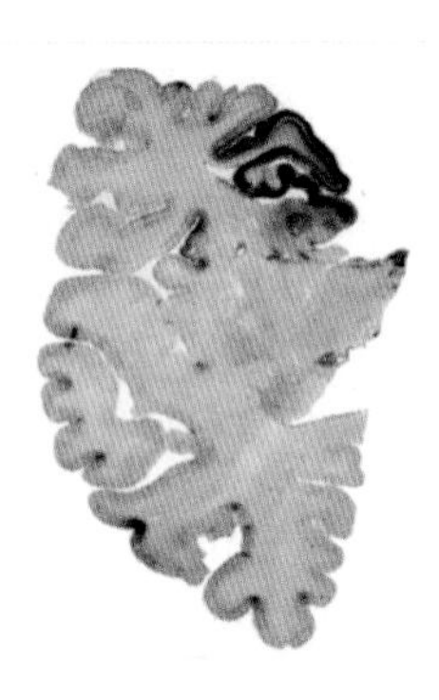

Figure 5.29 The normal brain on the left shows no signs of degeneration compared to the brain of a person in stage 4 of CTE. The image of the affected brain shows how CTE damages brain neurons and tissue, and leads to severe atrophy of the brain.

What causes CTE?

It is currently understood that the only risk factor for CTE is repeated TBI (Buckland et al., 2019). If you experience a forceful blow or repetitive blows to the head, your head suddenly moves rapidly back and forth in a whiplash-like fashion (Figure 5.14, page 196). During this movement, your brain bounces off the back of your skull and twists around, stretching and damaging the delicate cells and structures inside your brain. This movement can trigger ongoing deterioration of brain tissue, which results in CTE. CTE is thought to cause areas of the brain to waste away (atrophy), which results in the person losing brain mass. Injuries to the sections of nerve cells that conduct electrical impulses affect communication between cells.

Concussion

Concussion is a temporary injury caused by any violent jolt to the head (or upper body) that delivers an impulsive shock to the brain. This action can come from different directions and causes the brain to rock back and forth in the skull or twist on its axis (Figure 5.14, page 196). Concussion results in the disturbance of brain function but it is generally not considered to be a structural injury. It will not show up conclusively under X-ray, CT scan or MRI, or in tests of blood and saliva; it can only be diagnosed by observing overt symptoms such as dizziness and confusion, nausea and unsteadiness. Symptoms can appear immediately or hours later. Symptoms can also linger for months – or disappear within minutes. The effects of concussion are usually temporary, but can include headaches, nausea, fatigue, confusion, blurred vision, dizziness, and problems with concentration, memory, balance and coordination. Some concussions result in loss of consciousness, but most do not.

A subconcussive head impact is a bump, blow or jolt to the head that does not cause symptoms (unlike concussion, which always causes symptoms). A collision while playing sport is one way a person can get a subconcussive head impact. However, occasional hits to the head, such as the bumps and tumbles that children experience when learning to walk, do not cause CTE (Montenigro et al., 2015). While repetitive concussive and subconcussive injuries are believed to accelerate the progression towards CTE (Bailes et al., 2013), research continues to explore the possibility that biological, environmental or lifestyle factors could also contribute to the brain changes found after death in people who were diagnosed with CTE (Iverson et al., 2018).

Effect of tau protein clumps and tangles

Another aspect of CTE is that some brain areas, following repetitive blows to the head, gradually accumulate an abnormal form of a protein called tau (Figure 5.30). Tau protein is a substance that, in normal amounts, helps provide structural support to brain neurons.

The internal skeleton (cytoskeleton) of a nerve cell has a tube-like shape through which nutrients and other essential substances travel to reach different parts of the neuron. Normally, tau is inside the nerve cell, where it contributes to the cytoskeleton and also helps to hold the cell shape. However, under abnormal circumstances, such as after trauma, when the nerve cells are damaged, the tau separates from the cytoskeleton and starts to clump up or form tangles in brain cells. Eventually these tau clumps and tangles cause the cells to become defective and lose their ability to function. If enough tau builds up over time the affected parts of the brain shrink, adversely affecting the person's cognition and behaviour, and usually resulting in dementia. Tau clumps and tangles can develop years before a person starts to develop symptoms of brain damage, such as memory loss (Gorgoraptis et al., 2019).

5.4.1 UNDERSTANDING BRAIN INJURY

Alzheimer's disease also involves a build-up of abnormal amounts of tau protein in brain cells, which eventually leads to dementia, and in the past it was not uncommon for a CTE sufferer to be misdiagnosed with Alzheimer's disease.

However, recent research has found that in CTE, the pattern of abnormal build-up of tau protein around the blood vessels is different. Using new imaging techniques, the differences between CTE and Alzheimer's disease can be seen. Research continues into the brain changes that occur in CTE and how it is related to dementia.

Weblink
How Alzheimer's Changes the Brain

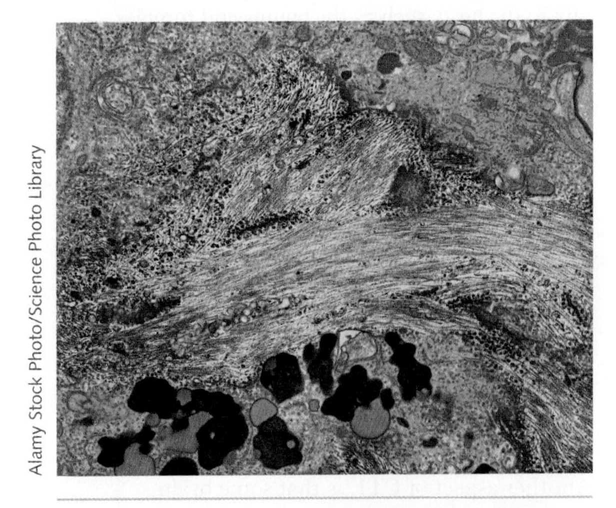

Alamy Stock Photo/Science Photo Library

Figure 5.30 An abnormal amount of tau protein (brown) in brain cells eventually leads to dementia.

Symptoms and stages of CTE

The symptoms of CTE may not be apparent until many years after a person experiences a brain injury and they may appear in different ways in different people. A person can experience CTE at any age. The condition often goes undiagnosed because the first symptoms can occur years or decades after brain injury, making CTE extremely difficult to diagnose. Symptoms are thought to be progressive, starting with confusion and mood disturbance, and progressing over time to dementia, speech difficulties and motor decline (Jordan, 2013). Young age onset is primarily characterised by mood and behavioural disturbances. Older age onset is primarily characterised by cognitive decline (Stern et al., 2013). How quickly the symptoms will begin to present is variable, ranging from a few months to several decades following brain trauma (DeKosky et al., 2013).

According to a 2012 Boston University study (McKee et al., 2013), CTE typically follows a pattern of phases that reflect the development of the disease over a period of years or decades. CTE usually begins very mildly and significant changes may not be noticed at first. Although research on CTE diagnosis, cause(s), symptoms and risk factors is still in the early stages, neurologists generally agree that there are four stages of severity in the development of CTE, with stage 1 being the mildest and stage 4 being the most severe (McKee et al., 2015).

The general pattern for progression of symptoms is as follows.

- **Stage 1:** headaches, loss of attention and concentration, mild short-term memory deficits
- **Stage 2:** anxiety, depression, suicidality, irritable mood and increased aggression, lack of impulse control
- **Stage 3:** cognitive impairment and problems with executive functions, specifically planning, organisation, multi-tasking and judgement, more severe memory loss and apathy
- **Stage 4:** a form of dementia (e.g. profound language deficits, psychotic symptoms, motor deficits, memory and cognitive impairments severe enough to impact daily living)

CASE STUDY 5.2

CTE in contact sports

Late Collingwood great Murray Weideman diagnosed with CTE

Collingwood premiership captain Murray Weideman has become the fourth known VFL/AFL player to be posthumously diagnosed with a debilitating neurological disease linked to head trauma.

Weideman, who led the Magpies to a famous upset grand-final win over Melbourne in 1958, joins Danny Frawley, Graham 'Polly' Farmer' and Shane Tuck in having chronic traumatic encephalopathy detected during the last two years.

Weideman's family on Saturday revealed the findings of the Australian Sports Brain Bank's report.

The 180-game Magpies star died in February, a day after his 85th birthday. After noticing serious changes to Weideman's personality in recent years, his family spoke with him about donating his brain.

'I said 'Dad, we have got to do this, we have got to help',' his son Mark Weideman told News Corp.

'The more science can build up and get evidence, the better things will become in the future.'

Farmer, a legendary ruckman from Western Australia who became a star with Geelong, was the first footballer diagnosed with CTE back in February 2020.

Former Richmond midfielder Tuck was assessed as having the 'worst seen case' of CTE when results were revealed by the brain bank in January.

Source: Extract from Australian Associated Press (2021, October 9). Late Collingwood great Murray Weideman Diagnosed with CTE. *The Guardian*. https://www.theguardian.com/sport/2021/oct/09/late-collingwood-great-murray-weideman-diagnosed-with-cte>

Weideman CTE diagnosis should prompt sport to ask hard questions: concussion expert

Peter Ryan

The discovery of chronic traumatic encephalopathy (CTE) in former Collingwood champion Murray Weideman, who died in February aged 85, should encourage administrators of contact sports to consider modifying rules for junior sport according to neuroscientist and concussion expert, Dr Alan Pearce.

The Age confirmed that the Australian Sports Brain Bank diagnosed Stage 2 CTE and intermediate Alzheimer's when Associate Professor Michael Buckland examined Weideman's brain posthumously.

Pearce, who works for the Victorian branch of the Australian Sports Brain Bank, said the findings indicated that sports had a responsibility to look after those suffering the impact of head knocks. He said sporting bodies needed to also tackle the issue at junior level.

'One thing we do know about CTE is that it is a disease of exposure so we may have to ask the hard questions on modifying sport for juniors before taking it up to full contact into mid-to-late teenagers even,' Pearce said.

The AFL has appointed Associate Professor Catherine Willmott as head of concussion innovation and research and Rachel Elliott as head of concussion and healthcare governance, while making rule adjustments at the elite level. They also introduced a 12-day concussion protocol which forces players to sit on the sidelines for at least 12 days if they suffer a concussion.

Pearce is an advocate for players being forced to take a 30-day break if concussed rather than 12, saying research based on symptoms was an inadequate assessment with physiological and biological-based research demonstrating a 30-day break was necessary.

Source: Extract from Ryan P. (2021, October 9), Weideman CTE diagnosis should prompt sport to ask hard questions: concussion expert. *The Age*. https://www.theage.com.au/sport/afl/murray-weideman-cte-diagnosis-should-prompt-sport-to-ask-hard-questions-concussion-expert-20211009-p58ymm.html

»

Questions

1 What was the cause of Murray Weideman's CTE?
2 List Weideman's symptoms.
3 How was CTE diagnosed?
4 A longitudinal study could be used to investigate the link between concussion sustained while playing football and CTE. Answer the following questions about such a study being used for this purpose.
   a What is a longitudinal study?
   b Identify one advantage and one limitation of a longitudinal study.
   c Write a hypothesis for a longitudinal study on this topic. Identify the independent variable, the dependent variable and the population of interest.
5 Fifteen-year-old Aziz has just started playing rugby league football but he is worried that he will get concussed or possibly experience brain trauma as a result of playing a heavy contact sport. Identify three things Aziz could do to minimise his risk of brain injury.

Diagnosis and treatment

5.4.2 UNDERSTANDING THE ANATOMY OF CTE

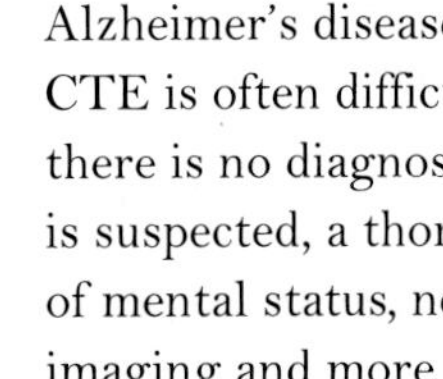

Some symptoms of CTE are similar to those of other degenerative brain conditions, such as Alzheimer's disease and Parkinson's disease, so CTE is often difficult to diagnose. Currently, there is no diagnostic test for CTE. When CTE is suspected, a thorough medical history, testing of mental status, neurological examination, brain imaging and more may be used to rule out other causes. However, despite technical advances in medical imaging, CTE cannot be accurately diagnosed in a living person. The only way to definitively diagnose many types of fatal brain diseases, such as CTE, is by conducting an autopsy of the brain of a dead person suspected of having CTE. This is because diagnosis requires brain tissue samples that reveal evidence of degeneration and deposits of tau and other brain proteins that indicate the changes associated with CTE (Concussion Foundation, 2020).

5.4.3 ANALYSING DATA

Because CTE can only be diagnosed after the person has died, treatment can be challenging. As with many other types of brain disease, treatment for people who have symptoms of CTE is based around supportive treatments. These treatments include behavioural therapy to deal with mood swings, pain management therapy to relieve discomfort, regular exercise and good nutrition, and memory exercises to strengthen the ability to recall daily events. People who learn ways to deal with the symptoms of CTE often have an improved quality of life, including less pain, improved memory and fewer mood swings. Although presently there is no cure or treatment for CTE, certain medicines may be used to temporarily treat the cognitive (memory and thinking) and behavioural symptoms.

ACTIVITY 5.4 CHRONIC TRAUMATIC ENCEPHALOPATHY

Try it

1 Table 5.7 shows cross-sections of the brains of people diagnosed with CTE. For each box, identify whether the person was in stage 1, stage 2, stage 3 or stage 4 of the disease.

Table 5.7 Brain dissections of people with CTE

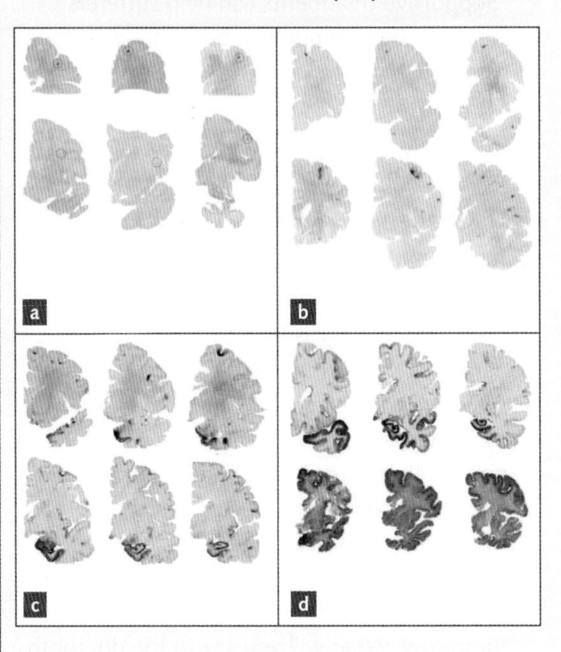

Apply it

2 In 2015, the *Journal of the American Medical Association* published a study of high school soccer players that showed that heading the soccer ball accounted for 30 per cent of concussions in male players and 25 per cent of concussions in female players (Comstock et al., 2015).

Based on the findings of this study, and similar research conducted in Europe, the US Soccer Federation implemented a number of changes at the junior level. As of 2016, the following rules were put in place:

» Under 11s: players aren't allowed to head the ball during practice or in the game

» Under 12s and 13s: players aren't allowed to head the ball more than 20 times in a week.

Similar rules have now been put in place in England, Scotland and Northern Ireland.

Explain how these rule changes could lessen the prevalence of CTE in young people who play soccer.

Exam ready

3 Which statement about CTE is not correct?

A In older people, CTE is difficult to distinguish from conditions such as Parkinson's disease or Alzheimer's disease.

B There are four stages of severity in the development of CTE.

C CTE can be accurately diagnosed while the person is alive.

D CTE is caused by repetitive blows to the head.

WB
5.4.4 MEDIA ANALYSIS

KEY CONCEPTS 5.4

- Chronic traumatic encephalopathy (CTE) is a progressive, degenerative and fatal brain disease associated with repeated blows to the head over a long period of time.
- A forceful blow to the head causes your brain to move back and forth inside your skull. This damages brain tissue and disrupts communication between brain cells.
- Damaged nerve cells accumulate abnormal clumps of tau protein, which makes them dysfunctional.
- Behavioural and cognitive symptoms of CTE may not appear for years after brain injury.
- CTE symptoms are similar to other degenerative brain diseases, making CTE difficult to diagnose.
- Currently, a definitive diagnosis of CTE can only be made by an autopsy of the brain.
- Supportive treatments can help sufferers manage their symptoms, but there is no cure for CTE.

Concept questions 5.4

Remembering

1 Define chronic traumatic encephalopathy (CTE). r

2 What causes CTE? r

3 Identify five symptoms of CTE. r

Understanding

4 Explain how a series of blows to the head can lead to CTE. e

Applying

5 Create a list of three ways a person playing contact sports could minimise their chance of acquiring CTE. c

HOT Challenge

6 Explain the relationship between repeated blows to the head and the build-up of tau protein in brain cells. e

7 Sam took up boxing when he was 11. Sam is now 31 and has been boxing professionally for 10 years. Recently Sam's family have noticed that his behaviour and memory have changed significantly. Doctors have suggested that Sam is in the early stages of Alzheimer's disease but you suspect Sam is suffering from CTE. You are meeting with Sam's doctor tomorrow. What will you say to the doctor to explain your suspicions? c

5 Chapter summary

KEY CONCEPTS 5.1a

- Neuroplasticity refers to the brain's ability to change its structure and function as a result of internal or external stimuli.
- The brain is made up of interconnected neurons.
- A neuron is made up of dendrites, a soma and an axon, which is often myelinated.
- Impulses move through a neuron via electricity and between neurons via chemicals called neurotransmitters.
- Developmental plasticity refers to changes in brain neurons and synaptic connections that occur as a result of experience during specific developmental stages.
- The stages of developmental plasticity are neurogenesis, neural migration, neural maturation, synaptogenesis, synaptic pruning and myelination.
- An adolescent's frontal lobes are not fully myelinated, so they have not yet developed the capacity to fully integrate information and think in the same complex, logical or abstract ways that an adult can.
- Neuroplasticity occurs throughout the lifespan.
- There are strategies to help people maintain their mental functioning across the lifespan.

KEY CONCEPTS 5.1b

- Adaptive plasticity is when the brain alters the connections between synapses and reorganises neural pathways as a result of learning, or following injury.
- After brain injury, neurons in an undamaged brain region can sometimes take over sensory and motor functions that had been performed by damaged areas. This happens after repeated stimulation.
- Sprouting occurs when new connections between neurons are created.
- Rerouting occurs when an undamaged neuron that has lost connection with a damaged neuron can make new connections with other active neurons.
- Certain parts of the brain that are function-specific can grow and change shape when we learn.

KEY CONCEPTS 5.2

- An acquired brain injury (ABI) is any type of brain injury that occurs after birth.
- An ABI can be due to an injury from an external force, such as a blow to the head. This is called a traumatic brain injury (TBI).
- An ABI can be due to an injury from an internal factor, such as lack of oxygen. This is called a non-traumatic brain injury (NTBI).
- Brain injury can affect a person's biological, psychological and social functioning.
- Damage to the right parietal lobe can sometimes cause spatial neglect.
- Damage to the left hemisphere can produce impaired language ability known as aphasia.

KEY CONCEPTS 5.3

- All neurological disorders are the result of damage to the nervous system.
- Neurological disorders affect the brain, the spinal cord and the nerves that connect them.
- Neurological disorders can affect biological, psychological and social functioning.
- Research may be in vitro or in vivo.
- Researchers often use animal models to investigate the causes, progress and treatment of neurological disorders in humans.
- Neuroimaging technologies such as CT scans, MRI scans and PET scans provide images of the physiological impacts of neurological disorders.
- New neuroimaging technologies such as DTI, MRS, the trimodal model and MEG can provide even more detailed images of the workings of the brain.

KEY CONCEPTS 5.4

- Chronic traumatic encephalopathy (CTE) is a progressive, degenerative and fatal brain disease associated with repeated blows to the head over a long period of time.
- A forceful blow to the head causes your brain to move back and forth inside your skull. This damages brain tissue and disrupts communication between brain cells.
- Damaged nerve cells accumulate abnormal clumps of tau protein, which makes them dysfunctional.
- Behavioural and cognitive symptoms of CTE may not appear for years after brain injury.
- CTE symptoms are similar to other degenerative brain diseases, making CTE difficult to diagnose.
- Currently, a definitive diagnosis of CTE can only be made by an autopsy of the brain.
- Supportive treatments can help sufferers manage their symptoms, but there is no cure for CTE.

5 End-of-chapter exam

Section A: Multiple-choice

1 The human brain starts to develop
- A at conception.
- B at birth.
- C during the third week of gestation.
- D during the third trimester of gestation.

2 Plasticity of the brain refers to the brain's
- A ability to remain stable over time.
- B ability to set like 'plastic' when learning occurs.
- C ability to adapt and change when required.
- D inability to adapt and change when required.

3 The process by which the brain disposes of excess neurons is called
- A homeostasis.
- B myelination.
- C synaptogenesis.
- D synaptic pruning.

4 The correct sequence in the stages of developmental plasticity is
- A neurogenesis, neural migration, neural maturation, synaptogenesis, synaptic pruning, myelination.
- B neurogenesis, neural maturation, neural migration, synaptogenesis, synaptic pruning, myelination.
- C neural migration, neurogenesis, neural maturation, synaptogenesis, synaptic pruning, myelination.
- D myelination, neurogenesis, neural migration, neural maturation, synaptogenesis, synaptic pruning.

5 An example of an in-vivo experiment would be
- A growing microorganisms in a test tube.
- B experimenting on a living organism.
- C autopsying a dead body.
- D attempting to fertilise eggs in a culture dish.

6 Synapse formation can occur
- A any time in a healthy person's life.
- B only during infancy.
- C only during adolescence.
- D only after brain damage.

7 Diagnosis of chronic traumatic encephalography (CTE) is challenging because
- A symptoms may not show for many years after the person acquired the brain damage.
- B definitive clinical diagnosis can only be made post-mortem.
- C the symptoms of CTE are very similar to Alzheimer's disease.
- D All of the above

8 Following brain damage ____________________ allows the brain to reroute neural circuits so a different brain area can take over a function that was lost when another brain area was damaged.
- A synaptic pruning
- B neurogenesis
- C sprouting
- D adaptive plasticity

9 Diffusion tensor imaging (DTI) allows neuroscientists to examine
- A the chemical composition of brain tissue.
- B the electrical activity of the brain.
- C the blood flow in the brain.
- D the circuits of myelinated neurons in the brain.

10 Brain injury can affect a person cognitively, physically and emotionally. Cognitive consequences do not include
- A slower processing of information.
- B organisational problems and impaired judgement.
- C inability to multi-task.
- D depression.

11 Which of the following statements is true about acquired brain injury (ABI)?
- A ABIs can only be caused by external forces.
- B ABIs can result in a range of psychological, physical and social symptoms.
- C People over 65 are more likely to acquire a brain injury.
- D ABIs always result in permanent loss of function.

12 A person with spatial neglect will
- A have damage to their left or right temporal lobe.
- B have damage to their corpus callosum.
- C only recognise one side of their visual space.
- D not pay attention to stimuli in the environment, because they have frontal lobe damage.

13 Complete the following text.
Broca's aphasia involves difficulty in ____________ caused by damage to the ____________ lobe while Wernicke's aphasia involves difficulty in ____________ caused by damage to the ____________ lobe.
- A producing articulate speech; left frontal; understanding speech; left temporal
- B producing articulate speech; left temporal; understanding speech; left frontal
- C understanding speech; left temporal; producing articulate speech; left frontal
- D understanding speech; left frontal; producing articulate speech; left temporal

14 Executive functions such as problem-solving, planning, decision-making and organisation are associated with
- A the amygdala.
- B the primary somatosensory cortex.
- C prefrontal cortex.
- D the primary auditory cortex.

15 Which of the following disorders is primarily a language-based disorder?
- A aphasia
- B spatial neglect
- C chronic traumatic encephalography (CTE)
- D epilepsy

16 What is the most common cause of chronic traumatic encephalopathy (CTE)?
- A car accidents
- B sports-related injuries
- C falling off a ladder
- D acts of violence

17 Which of the following is not a characteristic of Broca's aphasia?
- A a limited ability to write
- B slow and non-fluent speech
- C ability to comprehend simple language
- D an inability to perceive objects on one side of your visual space

18 What type of brain scan uses X-rays?
- A fMRI
- B MRI
- C CT scan
- D PET

19 Which of the following measures brain structure rather than brain function?
- A EEG
- B fMRI
- C MRI
- D PET

20 Which of the following statements about concussion is true?
- A Concussion causes permanent brain dysfunction.
- B Concussion can be diagnosed using a CT scan or MRI scan.
- C Concussion can only be diagnosed by observing its symptoms.
- D The symptoms of concussion appear immediately after the head trauma.

Section B: Short answer

1 A neurosurgeon wants to carry out experimental brain surgery on one of their patients who has an acquired brain injury (ABI). Identify two ethical guidelines the neurosurgeon must follow. Explain what they would have to do to follow these guidelines. [4 marks]

2 Explain why the decline in myelination that occurs as a result of normal ageing results in slower cognition and processing of stimuli. [2 marks]

3 Provide two techniques that could be used to determine whether a person was suffering from Broca's aphasia or Wernicke's aphasia. [4 marks]

Unit 1, Area of Study 2 review

Section A: Multiple-choice

Question 1
From split-brain studies, scientists concluded that

A the two hemispheres of the brain have the same abilities and functions.

B the left hemisphere is dominant in visual motor tasks.

C the corpus callosum enables information to flow between the two hemispheres.

D information travels to the right hemisphere quicker than the left hemisphere.

Question 2
The introduction of neuroimaging techniques enabled researchers to

A electrically and magnetically stimulate the brain to see the effects on behaviour.

B physically stimulate the brain to see the effects on behaviour.

C test the brains of people during autopsy.

D chemically and physically stimulate the brain to see the effects on behaviour.

Question 3
Which of the following is part of the forebrain?

A cerebellum

B RAS

C thalamus

D medulla

Question 4
Which of the following is not a part of the hindbrain?

A cerebellum

B medulla

C pons

D RAS

Question 5
What does the term 'RAS' stand for in brain structure?

A reticular activating system

B radical activated system

C reticular activated system

D radical activating system

Question 6
In which brain region is the medulla located?

A forebrain

B hindbrain

C midbrain

D backbrain

Question 7
Which of the following lobes of the brain is responsible for processing somatosensory information?

A frontal

B temporal

C parietal

D occipital

Use the following information to answer questions **8–12**.

Naomi was driving in her car while she was on her phone and not paying attention to the road. When she looked up, she saw that traffic had stopped but she didn't have time to brake. She crashed into the car in front of her and hit her head on the roof of the car because her seatbelt malfunctioned. As a result, she was unable to move her right arm immediately after the accident.

Question 8
Naomi has most likely suffered damage to which lobe and hemisphere of the brain, leading to her inability to move her right arm?

A frontal, left

B parietal, left

C frontal, right

D parietal, right

Question 9
Which region of the brain most likely sustained damage?

A brain stem

B forebrain

C midbrain

D hindbrain

Question 10

Which of the following would help Naomi to maximise the recovery of movement for her right arm?

A watching videos on how to move your arm
B constraining movement in the left arm
C constraining movement in the right arm
D consuming high volumes of red meat

Question 11

What could have been a social impact of Naomi's brain injury?

A problems with critical thinking
B emotional and personality changes
C no longer being able to play basketball
D struggles with motor movement

Question 12

Based on the region of her brain that sustained damage, what may have been another side effect experienced from this trauma?

A problems with vision
B inability to recognise faces
C issues with emotional regulation
D struggles with understanding sounds

Question 13

At age 5, Kiki learnt to ride her bike. Which of the following is true of the changes that occurred at the neural level due to this learning?

A She experienced developmental plasticity, which occurs in response to general learning.
B She experienced adaptive plasticity, which occurs in response to injury.
C She experienced developmental plasticity, which occurs in response to injury.
D She experienced adaptive plasticity, which occurs in response to general learning.

Question 14

Which of the following is not a process of developmental plasticity?

A pruning
B rerouting
C synaptogenesis
D myelination

Question 15

Which of the following negatively impacts on neuroplasticity?

A good sleep hygiene
B antioxidants
C active lifestyle
D non-stimulating activities

Question 16

Which of the following is a way a person can maximise brain functioning?

A playing strategic video games
B having a sedentary lifestyle
C consuming a diet high in processed foods
D overthinking problems

Question 17

Which of the following is not a psychological impact on functioning that can result from acquired brain injury?

A loss of memory
B emotional dysregulation
C trouble with problem solving
D trouble with sleeping

Question 18

Jackson is experiencing extreme hot and cold temperatures following an accident causing damage to his brain. Which type of effect is this?

A social
B biological
C psychological
D psychosocial

Question 19

Miriam recently suffered brain damage and her mother noticed that she was having difficulty speaking. Her speech was laboured, she only used simple sentences and she often left out the connecting words. Miriam is most likely suffering from

A epilepsy.
B Broca's aphasia.
C a stroke.
D Wernicke's aphasia.

Question 20

What does CTE stand for?

A computerised trauma encephalopathy
B chronic trauma endoscopy
C computerised trauma encephalopathy
D chronic traumatic encephalopathy

Question 21

Layla lost function in her right thumb 2 months ago after an accident that left her concussed. She now has full movement back in her thumb, after weeks of rehabilitation.

Layla's regained function is due to

A developmental plasticity.

B pruning.

C synaptogenesis.

D adaptive plasticity.

Question 22

Which of the following is not a part of the forebrain?

A cerebellum

B cerebrum

C thalamus

D hypothalamus

Question 23

Which of the following is correct in relation to CTE?

A It is caused by excessive consumption of alcohol.

B It is not fatal or progressive.

C It can be diagnosed with an MRI scan.

D There is no treatment for CTE.

Use the following information to answer questions **24–29.**

Jaskaran is a researcher interested in the impacts of traumatic brain injury on memory in people who suffer multiple concussions. He selected seven people aged 50–70 who had each suffered more than three concussions. He conducted a memory task where they were required to learn a series of words and recall them with no cues 1 week later. He also tested six people who had never experienced any head trauma using the same tasks. He counted the number of correctly identified words by each individual.

Question 24

Identify the dependent variable.

A suffering 3 or more concussions

B suffering no concussions

C age range of 50–70

D recall on memory task

Question 25

Identify the control group.

A individuals who suffered no brain damage

B sufferers of brain damage

C individuals aged 50–70

D there was no control

Question 26

Identify the type of experimental design used.

A matched participants

B random sampling

C independent groups

D repeated measures

Question 27

What is one potential limitation of this study?

A too large a sample size

B no control groups

C unrepresentative sample

D placebo effect

Question 28

Using a memory task is an example of what type of impact on functioning?

A social

B physiological

C biological

D psychological

Question 29

It was found that individuals who had experienced trauma had significantly lower memory abilities than those who did not. What is a possible condition these participants may have?

A spatial neglect

B Broca's aphasia

C chronic traumatic encephalopathy

D frontal haemorrhage

Question 30

You are researching current findings of neurological disorders. You have found a website that uses information from 2005. Is this website a good resource to use?

A Yes, a lot of important breakthroughs occurred in the early 2000s.

B No, you only ever want to use websites created in the last 3 years.

C Yes, if the website had more important information it would have been automatically updated.

D No, it is not current enough for this fast-moving topic.

Section B: Short answer

Question 1

Draw and label a diagram of the brain, including all three major brain areas. [4 marks]

Question 2

Jake was playing AFL when he got tackled, hit his head and became concussed. He couldn't speak properly for a few hours after the incident, and it took him 2 months to regain function in his right hand.

a What impact did Jake's injury have on his biological functioning? [2 marks]

b What was responsible for Jake regaining the ability to use his right hand? Describe what was happening at a biological level to explain Jake's recovery. [3 marks]

c If this injury happened repeatedly to Jake, what could he be diagnosed with? Describe what this means. [2 marks]

[Total = 7 marks]

Question 3

Name and describe one region of the cerebral cortex and its role in mental functioning. [2 marks]

Question 4

Luca was in a car accident and sustained an injury to the back of his head, consequently causing damage to that region of his brain.

a What brain region was most likely affected? Why have you chosen this region? [2 marks]

b What are three likely side effects Luca may suffer due to damage in this region? [3 marks]

[Total = 5 marks]

Question 5

Name and explain one biological, one psychological and one social factor that affect functioning after a traumatic brain injury. Present your answer as a table. [6 marks]

9780170465052

Contemporary psychological research

6

Key knowledge

Scientific evidence

- the distinction between primary and secondary data
- the nature of evidence and information: distinction between opinion, anecdote and evidence, and between scientific and non-scientific ideas
- the quality of evidence, including uncertainty, validity and authority of data and sources of possible errors or bias
- methods of organising, analysing, and evaluating secondary data
- the use of a logbook to authenticate collated secondary data

Scientific communication

- psychological concepts specific to the investigation: definitions of key terms; and use of appropriate psychological terminology, conventions and representations
- the characteristics of effective science communication: accuracy of psychological information; clarity of explanation of scientific concepts, ideas and models; contextual clarity with reference to importance and implications of findings; conciseness and coherence; and appropriateness for purpose and audience
- the use of data representations, models, and theories in organising and explaining observed phenomena and psychological concepts, and their limitations
- the influence of sociocultural, economic, legal and political factors, and application of ethical understanding to science as a human endeavour
- conventions for referencing and acknowledging sources of information

Analysis and evaluation of psychological research

- characteristics of repeatable and reproducible psychological research and the consideration of uncertainty
- criteria to evaluate the validity of psychological research

Source: VCE Psychology Study Design (2023–2027) p. 26

6 Contemporary psychological research

Chapter 6 provides help for you to complete Unit 1 Outcome 3. In this outcome you must come up with a research question from Unit 1 and answer it by finding, analysing and evaluating existing research.

iStock.com/gorodenkoff

iStock.com/PeopleImages

p. 229

6.1 How do we conduct research?

As in other scientific fields, psychological research requires a series of steps for results to be valid and communicated effectively. As covered in Chapter 1, all scientific research begins with a question. You will need to be selective when you decide which sources you will use to gather information. You will also need to develop a keen eye for distinguishing between scientific and non-scientific ideas.

p. 258

6.2 Ethical understandings of research and technology

Some of the early experiments you have learned about in previous chapters probably couldn't be conducted in the same way today. Scientists are responsible for ensuring their work complies with strict ethical concepts and guidelines. As you conduct your research, you will need to consider this in order to avoid any issues that may arise. It's also important to think about the ethics concerning the *use* of the findings from scientific research. There are many external factors that can influence how research is used and it's important to take those factors into consideration.

iStock.com/Halfpoint

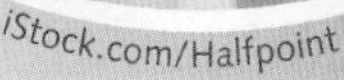

p. 264

6.3 Sample assessment task and rubric

You are now ready to demonstrate your research and science communication skills as you complete your research in a chosen topic related to the field of psychology. Take a look at the sample provided to assist you in doing this.

Adobe Stock/Tryfonov

Scientists have been able to employ well-established research methods to find out just how complex and fascinating our brains are. And yet, researchers in the field of psychology still have so much more to learn about the brain, and continue to make discoveries to this day.

Slideshow
Chapter 6 slideshow

Flashcards
Chapter 6 flashcards

Test
Chapter 6 pre-test

Know your key terms

- Anecdote
- Glossary
- Non-scientific ideas
- Opinion
- Paraphrase
- Peer review
- Plagiarism

In Unit 1 Area of Study 3, you will select and evaluate a recent discovery, innovation, issue or case study linked to a topic you have already studied in Unit 1 Area of Study 1 and/or Area of Study 2. You will incorporate the use of a logbook to organise your work and develop your skills in collecting data from secondary sources. You will apply these skills to identify and explain relevant psychological concepts as well as critically evaluate your evidence for validity, reliability and integrity. Finally, you will communicate your findings to a specific audience in a format that is most appropriate to your purpose. This chapter will guide you through how to do this.

Shutterstock.com/Rawpixel.com

Figure 6.1 Knowing how to communicate your findings is a transferable skill that can be used across a range of occupations. To master this skill, you first need to understand the terms used in research, as well as the relationships between them.

6.1 How do we conduct research?

Literature reviews

As you have already learned, psychologists conduct scientific investigations using a range of different methodologies. One of these methodologies is the **literature review**. A literature review involves the collation and analysis of secondary data related to other people's scientific findings (VCAA, 2022).

As we learned in Chapter 1, **primary data** is first-hand data that is collected by the researcher, whereas **secondary data** is usually obtained from a research article and can be gained from many different sources (Figure 6.2). You would carry out a literature review to present an overview of the validity of the current state of understanding of a topic, and especially when you are beginning a new piece of scientific research into that topic. It identifies and reports common themes between the results of numerous studies. If there are differences, the authors propose reasons for conflicting results. The authors may also report on any gaps in the existing research and propose avenues for further investigation.

Conducting research

In this section, the process of conducting research will be modelled using the topic *Study strategies*. You can apply this process to your own topic when you complete Outcome 3 in both Units 1 & 2.

Over the years you may have participated in study skills sessions that focused on the importance of developing sound study habits (Figure 6.3). What are these? What do they look like, sound like, feel like and, more importantly, how do you develop them? You may have been introduced to the Pomodoro Technique, and told to keep a planning diary of some description, maintain a regular routine, produce summaries and revise notes. Are these strategies effective and based on evidence, or are they based only on personal experience or anecdote?

In earlier chapters you have covered key concepts relating to the brain and its role in mental processes.

iStock.com/Deepak Sethi

Figure 6.3 As a VCE student, you are probably being bombarded by advice about different study techniques.

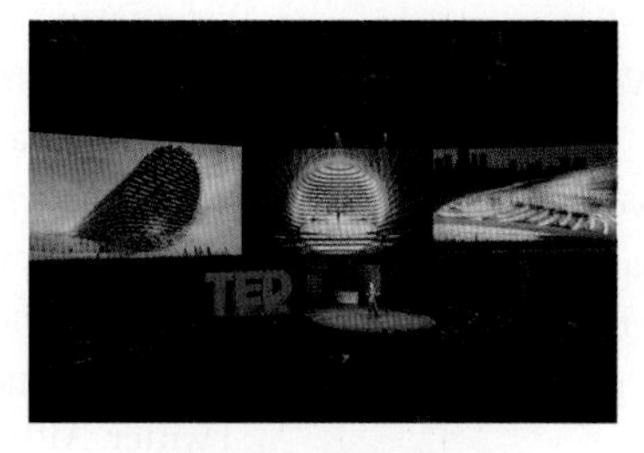

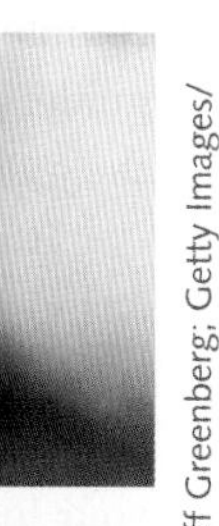

Top row, left to right: Getty Images/Jeff Greenberg; Getty Images/ Lawrence Sumulong; Shutterstock.com/rafapress.
Bottom row, left to right: Shutterstock.com/Cristian Dina; Reproduced with permission of The Royal Institution of Australia; Alamy Stock Photo/MShieldsPhotos

Figure 6.2 Science is communicated using a range of platforms.

Now you will expand your knowledge by exploring the research that psychologists have gained, and the evidence for what makes an effective study strategy. The process for the investigation is summarised in Figure 6.4.

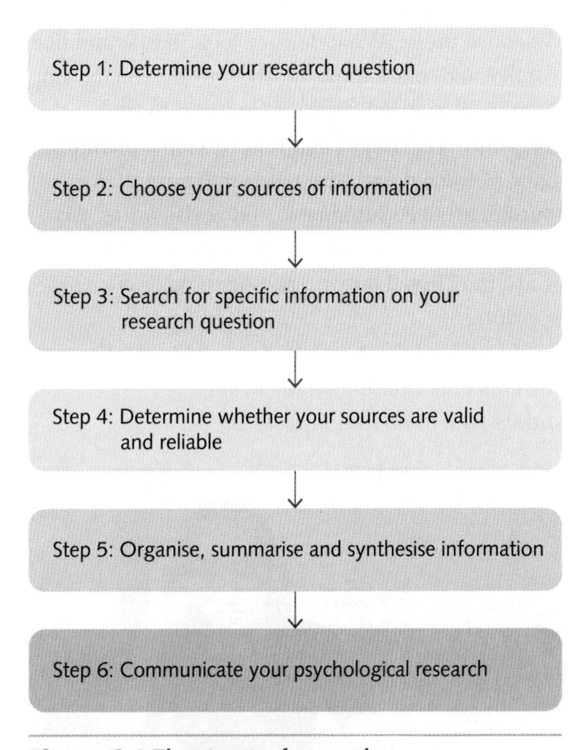

Figure 6.4 The stages of research

Worksheet
A sample research guide

In modelling this process, the skills that you will be required to consider and use for an assessment task on this outcome will be outlined. This will be followed by a worked example for each stage of the process. You will be directed to complete a checkpoint from the online research guide. You can also download a Word version of the guide.

It is recommended that you work through this chapter first and then again for each step when it is time to complete your assessment. Your teacher may suggest you use the research guide as part of your logbook.

Step 1: Determine your research question

As mentioned in Chapter 1, research questions define the exact question you are trying to answer with your research. Being specific about your focus will help guide your research.

When formulating your **research question**, you should:

1. *Brainstorm* potential topics using the list of options provided on page 264.
2. Decide on a *topic*.
3. Decide why you want to research your topic. *What is your purpose?*
4. Consider who might be interested in your topic. *Who is your audience?*
5. Evaluate the most effective way to communicate what you have found out to your specific audience. *What is your format?*
6. Rephrase your topic as a question. Ask an open-ended 'how', 'what' or 'why' question.

Table 6.1 is an example of determining a research question from a broad topic.

Table 6.1 Step 1: Determine your research question from the topic 'Study strategies'.

Topic	Study strategies
Purpose	Provide VCE students with evidence-based strategies they can use
Audience	VCE students
Format options	Review article to be posted on the school's website
Research question	What are some effective, evidenced-based study strategies for VCE students?

Step 2: Choose your sources of information

Before you begin seeking out information, you need to investigate the nature of evidence and information available for the topic. The main sources of information for conducting research are research articles, review articles, journals and news websites. However, other sources of information are emerging from social media, such as Facebook, TikTok and Twitter. Although these sources are an easy way to obtain information, you should take care to carefully evaluate their suitability.

Generally, you are looking for sources of information that have the highest validity and integrity. Figure 6.5 summarises the types of information available from high to low validity and integrity.

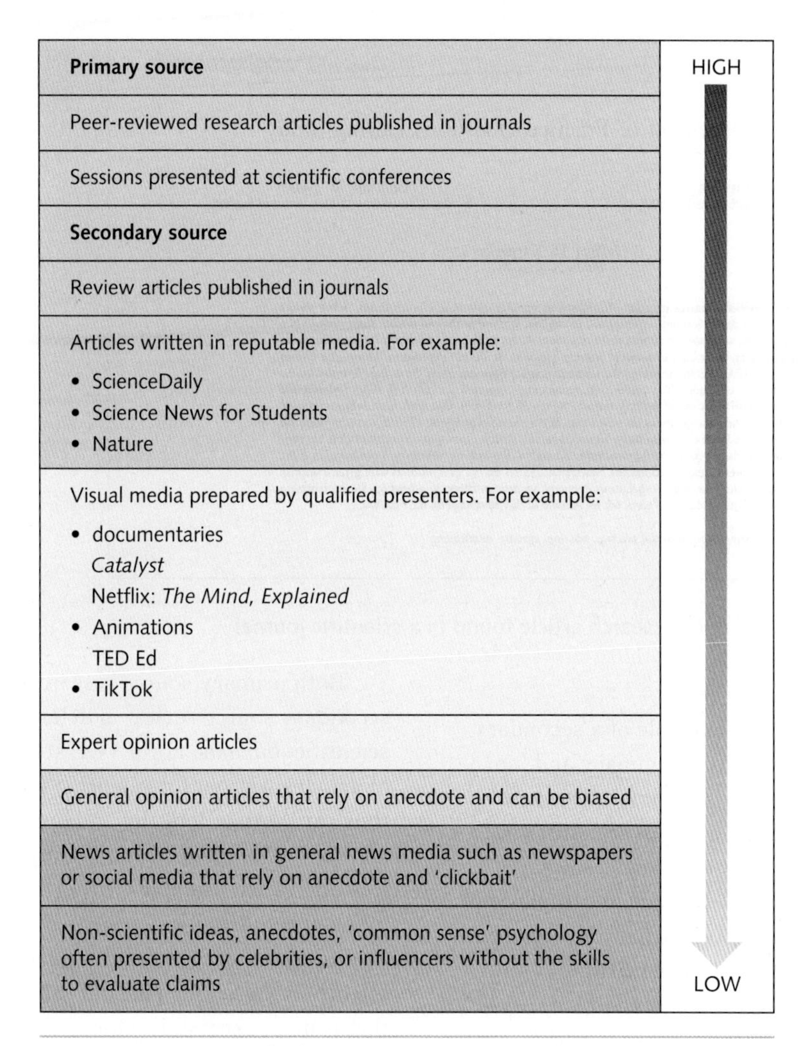

Figure 6.5 Sources of information can be graded from high to low in their validity, reliability and integrity.

Primary sources

Information that has high validity can generally be found in research articles published in science journals. A scientific journal is a periodical publication intended to further the progress of science, usually by reporting new research. Research articles in journals are examples of a primary source (Figure 6.6). A **primary source** in psychological science is a study written by one or more experimenter(s). It generally has an abstract, introduction, methods, results, discussion and references sections. It features figures and tables, and is written in such a way that the study itself could be replicated.

Reports or presentations delivered at conferences are also considered to be primary sources if they are presented by the original researcher(s). An academic conference is a one-day or multi-day event during which researchers present their work to each other. Conferences are an important way researchers stay connected to others in their field and learn about cutting-edge developments.

Secondary sources

A **secondary source** is a written or visual text about a research article or series of articles. Science researchers, science journalists, advertisers, psychologists, governments, companies and social media content developers will take information from primary sources and present it in a way that is accessible to the public or employers.

Journal of Experimental Psychology: Applied
2016, Vol. 22, No. 1, 72–84

1076-898X/16/$12.00 http://dx.doi.org/10.1037/xap0000071

On the Placement of Practice Questions During Study

Yana Weinstein
University of Massachusetts Lowell

Ludmila D. Nunes
Purdue University and University of Lisbon

Jeffrey D. Karpicke
Purdue University

Retrieval practice improves retention of information on later tests. A question remains: When should retrieval occur during learning—interspersed throughout study or at the end of each study period? In a lab experiment, an online experiment, and a classroom study, we aimed to determine the ideal placement (interspersed vs. at-the-end) of retrieval practice questions. In the lab experiment, 64 subjects viewed slides about APA style and answered short-answer practice questions about the content or restudied the slides (restudy condition). The practice questions either appeared 1 every 1–2 slides (interspersed condition), or all at the end of the presentation (at-the-end condition). One week later, subjects returned and answered the same questions on a final test. In the online experiment, 175 subjects completed the same procedure. In the classroom study, 62 undergraduate students took quizzes as part of class lectures. Short-answer practice questions appeared either throughout the lectures (interspersed condition) or at the end of the lectures (at-the-end condition). Nineteen days after the last quiz, students were given a surprise final test. Results from the 3 experiments converge in demonstrating an advantage for interspersing practice questions on the initial tests, but an absence of this advantage on the final test.

Keywords: testing effect, retrieval practice, massing, spacing, interleaving

This research article was published in the *Journal of Experimental Psychology: Applied*

Source: Weinstein, Y., Nunes, L. D., & Karpicke, J. D. (2016). On the placement of practice questions during study. Journal of Experimental Psychology: Applied, 22(1), 72–84. https://doi.org/10.1037/xap0000071

Figure 6.6 An example of a research article found in a scientific journal

Review articles

Review articles are an example of a secondary source that is essentially a summary and/or analysis of one or more primary source articles that cover the same area of research interest (Figure 6.7). A literature review is similar; however, it provides a summary and analysis of a series of articles. Despite being a secondary source of information, a review article is still considered to have a high degree of validity. These articles are well referenced, so they enable you to go to the original sources for further validation.

Both primary source research articles and secondary source review articles are published in scientific journals. For a VCE student, a limitation of articles written for scientific journals is that they can contain lengthy and quite complicated methods. Therefore, when finding information, it is often easier to start with articles written by science journalists or writers who have qualifications in science. They can interpret the science in these journals and report on it in a way that is more accessible. Once you have confidence in your understanding of the key concepts, you can then turn to more academic sources.

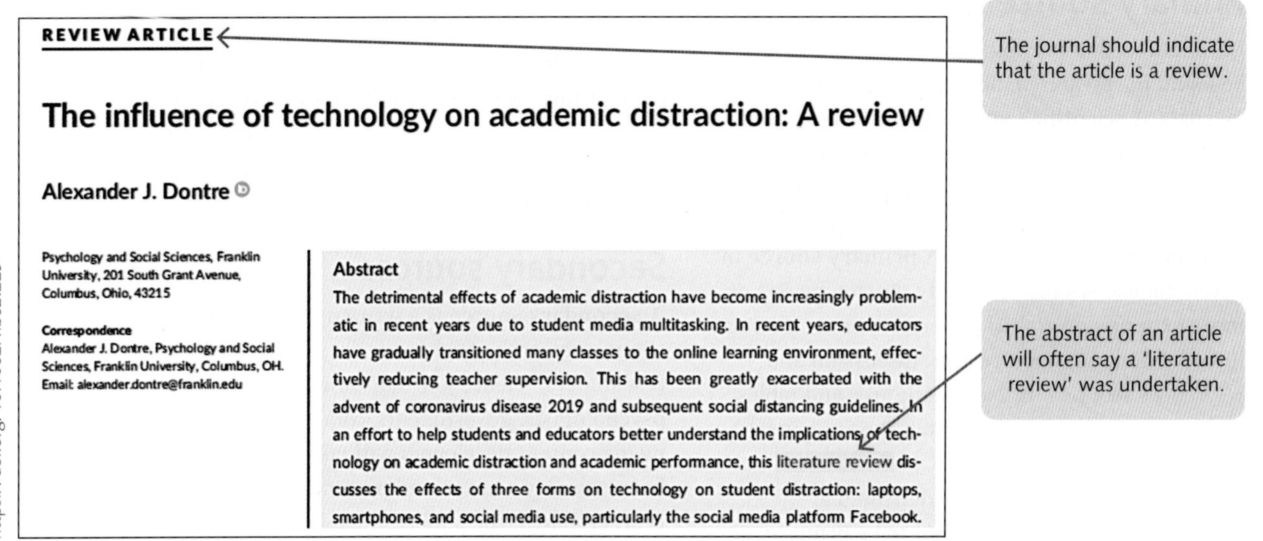
REVIEW ARTICLE

The influence of technology on academic distraction: A review

Alexander J. Dontre

Psychology and Social Sciences, Franklin University, 201 South Grant Avenue, Columbus, Ohio, 43215

Correspondence
Alexander J. Dontre, Psychology and Social Sciences, Franklin University, Columbus, OH.
Email: alexander.dontre@franklin.edu

Abstract
The detrimental effects of academic distraction have become increasingly problematic in recent years due to student media multitasking. In recent years, educators have gradually transitioned many classes to the online learning environment, effectively reducing teacher supervision. This has been greatly exacerbated with the advent of coronavirus disease 2019 and subsequent social distancing guidelines. In an effort to help students and educators better understand the implications of technology on academic distraction and academic performance, this literature review discusses the effects of three forms on technology on student distraction: laptops, smartphones, and social media use, particularly the social media platform Facebook.

Source: Dontre AJ. The influence of technology on academic distraction: A review, Hum Behav & Emerg Tech, 2021, 3:379–390, https://doi.org/10.1002/hbe2.229

Figure 6.7 What to look for when seeking a review article

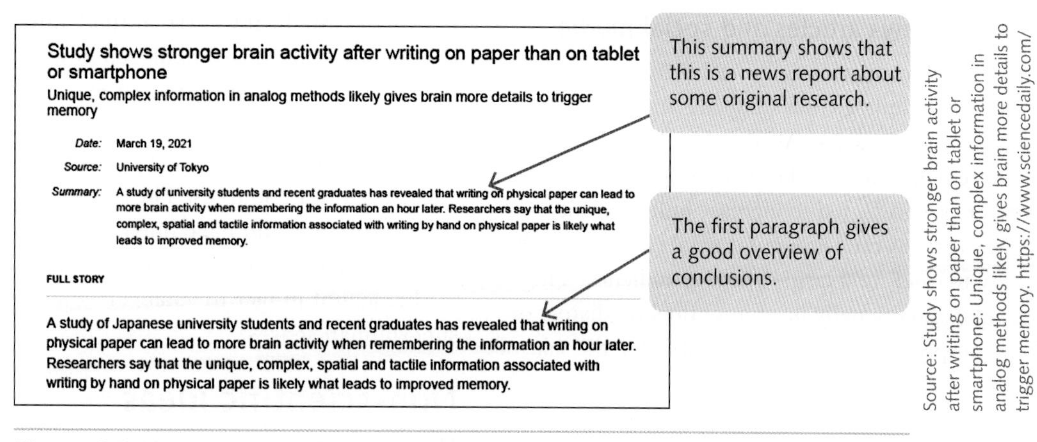

Study shows stronger brain activity after writing on paper than on tablet or smartphone

Unique, complex information in analog methods likely gives brain more details to trigger memory

Date: March 19, 2021

Source: University of Tokyo

Summary: A study of university students and recent graduates has revealed that writing on physical paper can lead to more brain activity when remembering the information an hour later. Researchers say that the unique, complex, spatial and tactile information associated with writing by hand on physical paper is likely what leads to improved memory.

FULL STORY

A study of Japanese university students and recent graduates has revealed that writing on physical paper can lead to more brain activity when remembering the information an hour later. Researchers say that the unique, complex, spatial and tactile information associated with writing by hand on physical paper is likely what leads to improved memory.

Source: Study shows stronger brain activity after writing on paper than on tablet or smartphone: Unique, complex information in analog methods likely gives brain more details to trigger memory. https://www.sciencedaily.com/releases/2021/03/210319080820.htm

Figure 6.8 A science news article is a good place to start your research and gather key terms or concepts that you can use to type into your search engine.

6.1.1 COMMUNICATING SCIENCE

Science media websites

Science media articles from reputable sources such as *Science Daily*, *Science News* or *Scientific American* can be a good place to start your research (Figure 6.8). Authors who write for such publications are required to cite their sources of information. Therefore, it is possible for you to access the original work and confirm what they have written.

TED Talks and YouTube

TED talks are often presented by experts in the field. For example, many of the psychology-based TED talks are presented by psychologists and cognitive scientists and therefore we assume that the information presented is reliable. TED also includes the sources used and speaker qualifications to allow the audience to further investigate the nature of the content.

Expert opinion articles

In an expert **opinion** paper or article, the author focuses on a topic and expresses personal and original thoughts or beliefs that stem from their experience and expertise (Figure 6.9). An opinion piece aims to stimulate debate or new research. For example, sometimes the scientific development of a technology occurs before it is legally approved. This happens especially when the technology is highly experimental. An expert opinion article might speculate on the implications of this recent development. It may also cover similar controversial topics, long-standing problems or a current issue.

The secondary sources of information covered so far have high validity. However, there are other sources that are useful, mainly because they are often easier to understand. If you find that science media articles are also too wordy or complicated, these sources are a useful start. Unfortunately, these sources are rated further down the scale in their validity and integrity (Figure 6.5). Therefore, they need to be used with caution because they can show **bias** or a limited perspective. An author can show bias by leaving out information or by altering facts to force the reader to form a certain opinion.

SHARE

PERSPECTIVES + OPINIONS

For Note Taking, Low-Tech Is Often Best

In college lecture halls, evidence suggests it's time to put down the laptop and pick up a pen

BY: Susan Dynarski

POSTED: August 21, 2017

Source: 'For Note Taking, Low-Tech Is Often Best' by Susan Dynarski, August 21 2017, https://www.gse.harvard.edu/news/uk/17/08/note-taking-low-tech-often-best

Figure 6.9 An example of an expert opinion article

General news articles and social media

General news articles and social media pages that discuss the findings of a study are another example of a secondary source. As a media consumer you, and the public, need to be aware that findings are often misrepresented to increase online traffic. These articles can leave out crucial elements of the study to cater for a larger reading audience. They can also intentionally or unintentionally distort or misinterpret the findings of research for the sake of a good story.

Journalists often use anecdotes to make their stories seem more relevant to people's lives or emphasise why a piece of scientific research matters. An **anecdote** is a usually short narrative of an interesting, amusing or personal story. Anecdotes are usually presented as case studies of individuals' experiences, and they may not include any facts or figures.

Article headlines are commonly designed as 'clickbait', to entice viewers into clicking on and reading the article. At times, they can oversimplify the findings of scientific research; at worse, they sensationalise and misrepresent them.

An example of this can be found on the Mail Online website from 2016 (Figure 6.10). The article suggested that brain implants could soon help humans develop superhero night vision. The research the story was based on had originally been published in the *Journal of Neuroscience* by a team of scientists at Duke University Medical Centre in the United States. Hartmann et al. (2016) explored how easily the sensory processing of adult rats could be changed by implanting them with a brain device to teach them to identify the location of infrared light sources. They suggested that their findings could have important implications for basic neuroscience and rehabilitative medicine, not in fact to 'enable new kinds of superhuman senses'.

Weblink
The learning styles myth

Non-scientific ideas

Non-scientific ideas refer to ideas or practices that are presented to resemble real science, but are not supported by scientific evidence. Sometimes such an idea is referred to as *pseudoscience* (a theory that resembles science but has no scientific basis). For example, there are many suggestions circulating about how to study or learn effectively that are not supported by hard evidence. These include 'sleep learning', which proposes that you learn if you play recordings of new information while you sleep.

You will come across many non-scientific ideas on social media. On such platforms, claims can be put in a way that is very compelling, particularly if they are presented by a celebrity or influencer. Therefore, it is vital that these claims cite original research and that this original research is valid and reliable. If a social media report doesn't have these features, it should be avoided.

Source: 'Could we soon have superhero NIGHT VISION? Brain implants could give us a 'sixth sense' by making us see infrared' By Richard Gray for MailOnline

Could we soon have superhero NIGHT VISION? Brain implants could give us a 'sixth sense' by making us see infrared

- **Infrared sensors that turn light into electrical pulses used as a prosthesis**
- **Electrodes from four of these sensors were implanted into rat brains**
- **The rats learned to detect and reacted to infrared light in under four days**
- **Scientists say it could be used to enable new kinds of superhuman senses**

Sensationalist language

Figure 6.10 Be wary of sensationalised and over-generalised statements.

You can now complete checkpoints 1 and 2 of the research guide.

KEY CONCEPTS 6.1a

- Primary sources of information include research articles published in journals and conference presentations.
- A secondary source is a written or visual text about a research article or series of articles.
- Primary data are generated directly by the investigator analysing them; secondary data are drawn from previous research.
- An expert opinion article focuses on a topic and expresses personal and original thoughts or beliefs, which stem from experience and expertise. It aims to stimulate debate or new research.
- Journalists often use anecdotes or personal stories to make their stories seem more relevant to people's lives or to emphasise why a piece of scientific research matters.
- Be wary of claims that may be exaggerated, or statements such as 'the findings prove' instead of 'the findings suggest'. It takes a great deal of research and review to support any scientific theory.
- Non-scientific ideas refer to ideas or practices that are presented to resemble real science, but are not supported by scientific evidence. They are often supported by celebrities or influencers.

Concept questions 6.1a

Remembering

1. Identify two features of a good research question. r
2. List two primary and two secondary sources of information that you can use to answer a research question. r
3. What is an expert opinion article? r
4. What are non-scientific ideas? Provide an example. r

Understanding

5. Why is it useful to clarify the purpose of your research before you develop your research question? e
6. Identify three differences between a research article and a review article. e

Applying

7. Define the terms opinion, anecdote and bias, and provide an example of each. c
8. Find an example of 'clickbait' on social media and explain how it works. c

HOT Challenge

9. Use the internet to find a primary source journal article on a scientific topic. Then find a secondary source news article on the same topic that exaggerates or misrepresents the original research. Identify how the news article has exaggerated or misrepresented the original research. c

Step 3: Search for specific information

To begin your search for information you can start by using some of the sites suggested in Table 6.2 on the next page.

If you feel daunted by the idea of finding and reading research articles from journals, start with the secondary source information from dedicated psychology websites such as Simply Psychology or Verywell Mind. This will give you an overview of the topic. It will also provide you with themes and key concepts that you can explore or use as search terms when using general science websites or journal websites (Figure 6.11). Another great starting point is The Conversation. The articles are written by researchers for a general audience and often include a primary source that you can further investigate.

iStock.com/Ekaterina Naumenko

Figure 6.11 You will need to find both primary and secondary sources on your topic of choice.

Weblinks
Frontiers in Psychology
Australian Journal of Psychology
American Psychological Association
Psychology Today
Verywell Mind
Simply Psychology
PsyPost
Scientific American, Mind & Brain
The British Psychological Society, Research Digest
PsychNewsDaily
ScienceDaily
Science News for Students
ScienceAlert
Frontiers in Psychology
The Conversation

Table 6.2 Useful websites for both primary and secondary sources

1 Journal articles (open access)	*Frontiers in Psychology* *Australian Journal of Psychology* American Psychological Association publications
2 Science media websites that cover the science of psychology	Psychology Today Verywell Mind Simply Psychology Psypost Scientific American Mind British Psychological Society, *Research Digest* PsychNewsDaily
3 General science media websites	ScienceDaily Science News for Students ScienceAlert

To find specific information about your topic, brainstorm a list of keywords, commonly called 'search terms'. These are the words that you enter into database search boxes. Start with the main concepts of your research and as you progress you will find other keyword and search terms to deepen your investigation. Without the right keywords, you may have difficulty finding the articles that you need.

If you want to narrow your search, you can use Boolean operators. Boolean operators tell databases exactly how you want to combine your search terms. There are five Boolean operators (Table 6.3).

Table 6.3 Functions of Boolean operators

Operator	What does it do?	Example
' '	Holds together the words of a phrase so that the database searches it together instead of separately	'note taking' 'interspersed learning'
*	Shortens a word so that you get alternative endings	educat* for education, educated, educates etc.
AND	Joins two concepts so that the database knows you want both to be in your search results	learning AND 'study techniques'
OR with ( )	Used to join two or more keywords, usually related terms for the same concept, so that the database knows you want either or both of them to be in your search results. Whenever you join keywords with OR, enclose them with parentheses	(techniques OR strategies) AND learning
NOT	Used to tell the database that you do not want this keyword or group of keywords in your search results	habits NOT (habitat OR habituation)

You can now complete checkpoints 3 and 4 of the research guide.

Step 4: Determine whether your sources are valid

We have explored the types of sources available and why you should only use sources that are valid, reliable and have integrity. However, for each source you find it is still useful to do a quick check before you use it.

Validity in primary sources

Remember that validity refers to the extent to which the design of the investigation and what is measured actually assesses the specific concept being investigated. Therefore, if your source presents valid information, the methodology for obtaining the results will accurately measure the variable that is being investigated. For example, a researcher

> *2.1.3. Design and procedure*
> All participants were asked to bring their personal laptops to the experiment. They received an instruction sheet and a consent form. The instruction sheet asked participants to attend to the lecture and use their personal laptop to take notes on the information being presented, just as they might normally do in class. In addition to taking lecture notes, half of the participants (randomly selected) were instructed to complete the 12 online tasks at some point during the lecture.
>
> Evidence of valid method

Source: Weston, Tina & Wiseheart, Melody. (2013). Laptop multitasking hinders classroom learning for both users and nearby peers. Computers & Education. 62. 24–31. https://doi.org/10.1016/j.compedu.2012.10.003 p26

Figure 6.12 Look for terms such as 'control group', 'random sampling', 'randomly assigned', 'controlling for extraneous variables' and 'randomly selected' to help you assess the validity of the research.

interested in measuring the effect of a study technique on brain activity may use brain imaging to measure this. This is because brain imaging actually measures brain activity. There are, however, other variables that will affect brain activity, and to be sure that it is only the study technique that is affecting brain activity, the researcher will need to design the investigation to remove the potential effect of other variables such as time of day tested, blood glucose levels, amount of sleep etc. This ensures that any change to brain activity when compared to a control group is due to the study technique alone.

To find what steps the researcher has taken to ensure validity in a primary source, you will need to read the methods section (Figure 6.12).

As mentioned in Chapter 1, assessing the validity of an investigation will require you to look at factors such as:

» appropriateness of the design
» sampling techniques used
» the impact of extraneous and confounding variables in the results
» whether or not the research can be applied to similar individuals in a different setting.

WEIRD bias

When conducting research, it is advisable to check the method to see if the study had any **WEIRD bias** (Figure 6.13; also see Chapter 1). This reliance on a limited population of participants is problematic for many reasons. Most obviously, it is a threat to the general validity of the findings: if researchers only study a phenomenon in one population, they don't know *if* or *how* it works in others. We will expand on the implications of WEIRD bias later in the chapter.

Weblink
The hidden biases in WEIRD psychology research

Uncertainty in science

As mentioned in Chapter 1, all measurements in science have some degree of **uncertainty** (unless we know the true value). Therefore, in research it's

Worksheet
The limitations of WEIRD samples

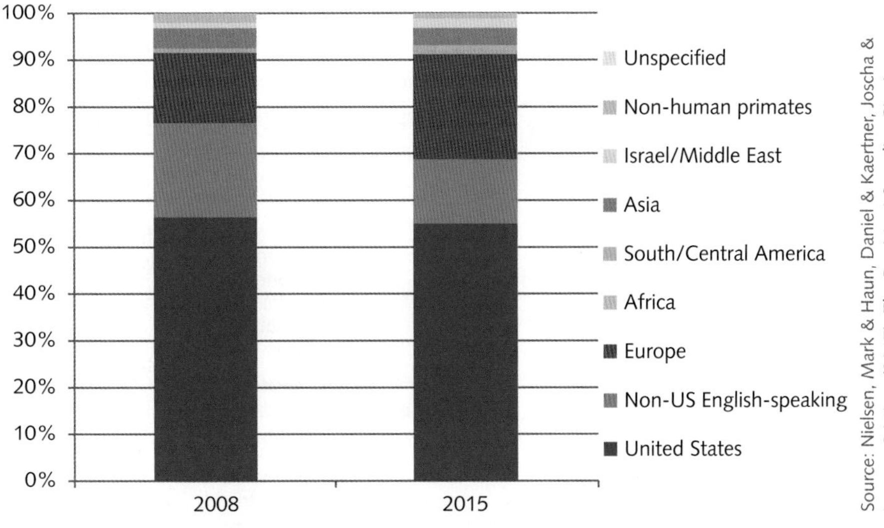

Source: Nielsen, Mark & Haun, Daniel & Kaertner, Joscha & Legare, Cristine. (2017). The Persistent Sampling Bias in Developmental Psychology: A Call to Action. Journal of Experimental Child Psychology. 162. 10.1016/j.jecp.2017.04.017.

Figure 6.13 Many of the generalisations made about human thoughts, feelings and behaviours have been drawn from investigations that used WEIRD samples. This graph shows that more than 50% of investigation subjects in scientific studies in 2008 were from the US, and that Asian, African, Middle Eastern and South American communities were under-represented. By 2015, this had barely changed. The problem with this is that not all populations think, feel or behave in the same way under the same conditions.

important to reduce the amount of uncertainty as much as possible. By acknowledging how much uncertainty is associated with measurements, scientists can communicate their findings more precisely. Sources of uncertainty that can influence the level of uncertainty of any measurement include contradictory data, incomplete data, errors and bias. The more uncertainty there is with any measurement, the less robust the data are and therefore the more uncertain any inference and conclusions are that are dependent on uncertain measurements.

Researchers can assess uncertainty quantitatively and/or qualitatively using professional judgement. This assessment is outlined when they describe the limitations of their research in the discussion section of their report. Indeed, reporting the level of uncertainty strengthens the research results (Figure 6.14). It also provides guidance for the focus of future research projects. Further research on a topic or theory may reduce the level of uncertainty.

It is likely that factors unexplored in our set of experiments may moderate the efficacy of practice quiz questions and their best placement. For instance, for very difficult materials using interspersed practice questions might be better than using at-the-end practice questions because the latter will likely reduce retrieval success to a minimum that might override any context reinstatement and/or retrieval effort benefits. However, for easier materials it is likely that at-the-end practice questions would be more efficient than interspersed practice questions. This possible interaction between materials difficulty and question placement was not investigated in the current set of experiments, and it is a future area that needs clarification. Of course, the role of individual differences between learners and other intrinsic factors such as inclusion and placement of feedback might also have an impact on the ideal placement of practice questions.

The authors outline some of the limitations of their research and suggest further refinement and investigation

Source: Weinstein, Y., Nunes, L. D., & Karpicke, J. D. (2016). On the placement of practice questions during study. Journal of Experimental Psychology: Applied, 22(1), 72–84. https://doi.org/10.1037/xap0000071

Figure 6.14 By acknowledging the limitations and uncertainties of an investigation, further research can be refined and improved.

Uncertainty and scientific models

A **psychological model** is a theory that aims to represent, predict or evaluate how people in real-world situations may think, feel or behave. Each element of a model needs to be tested, and the tests must be repeated and replicated using a range of populations and conditions. If there are gaps or inaccuracies in the model, it needs to be refined or rebuilt. If you encounter a model or theory in your research, you need to able to interpret it and evaluate its strengths and limitations.

Being transparent about uncertainty and limitations helps scientists to review and refine the models that they use to organise and explain concepts and processes. The quality and validity of a model is limited by the research used to construct it. The better a model holds up to testing, the more likely it will become a theory that is used widely by psychologists to explain thoughts, feelings or behaviour (such as the one shown in Figure 6.15).

Let's look at an example of a model of motivation, Maslow's hierarchy of needs. Maslow suggested that human needs could be represented in a pyramid that included physiological, safety, love/belonging, esteem and self-actualisation needs (Figure 6.16). He proposed that higher-order needs, such as accomplishment and achieving one's potential, can only be pursued when the lower needs, such as food, shelter and friendship, are met. Maslow's theory suggests that for you to be motivated enough to want to study for your VCE, you need to be fed, feel safe and have a sense of future. Until these needs are satisfied you won't be motivated to use any study strategies to develop mastery in your learning.

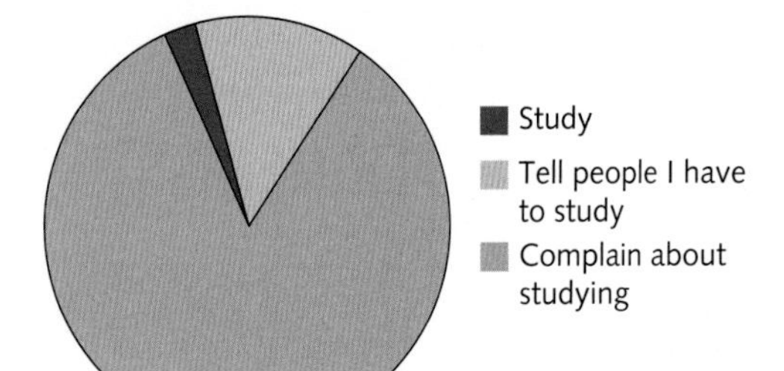

Figure 6.15 Psychological research and models help to explain and provide solutions to common and very relatable problems.

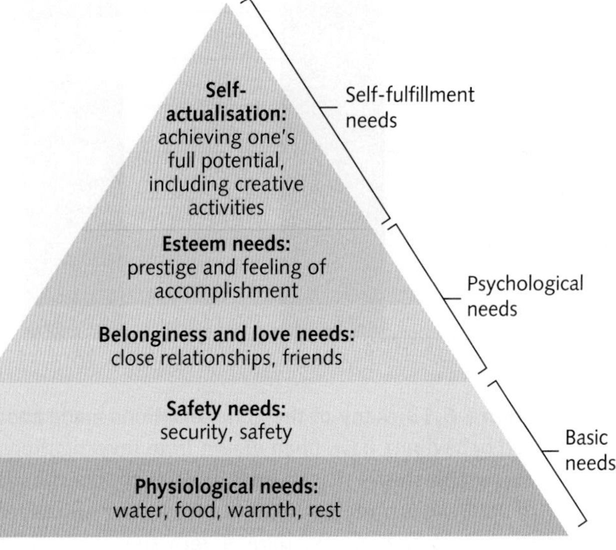

Figure 6.16 Maslow's hierarchy of needs

Table 6.4 An analysis of Maslow's hierarchy of needs. Where there are limitations, the model needs to be refined or rebuilt.

Strengths	Limitations
It is a simple model that is easy to understand and apply. There's an approximate correlation with stages of individual human development. For example, - infancy is primarily about physiological and survival needs, though some social interaction is also crucial - childhood and adolescence focus more explicitly on social needs - the higher drives for self-actualisation and self-transcendence tend to unfold in adulthood.	It is more successfully applied to WEIRD cultures, which focus on the individual, as opposed to non-WEIRD cultures, which are collectivist. Some terms in the hierarchy, such as 'self-esteem' and 'security,' have different definitions in cultures around the globe. Therefore, it is hard for researchers to measure these needs or to generalise them across all human populations (Gambrel & Cianci, 2003). The original study has low validity and reliability as the model was based on a small, biased sample and used self-reporting. It has been found that people can operate at higher levels of the pyramid without having the lower levels met (McLeod, 2020).

What are the strengths and limitations of Maslow's model (Table 6.4)? How should it be tested? How does it need to be refined? When evaluating a model these are the questions that you should ask.

Peer review process

Much of the work involved in testing and evaluating original research, models and theories is conducted by peer review. **Peer review** is an extremely important part of science (Figure 6.17). Before a study can be published in a reputable journal, it must be peer reviewed. This process can last for months. During peer review, the research report is sent to scientists working in the same field. They evaluate whether the methods used were appropriate and the conclusions make sense. These reviewers offer advice on the quality of the paper, whether it should be published and what changes should be made if it is to be published. This process also allows other scientists to offer feedback and, increasingly, to test or reproduce the research to verify its accuracy.

You can find out if an article has been peer reviewed by checking the journal it came from. The journal website usually states that the articles it publishes are peer reviewed. For example, *Frontiers in Psychology* states on its Scope and Mission page that all of its articles are peer reviewed and that it is a member of the Committee of Publication Ethics.

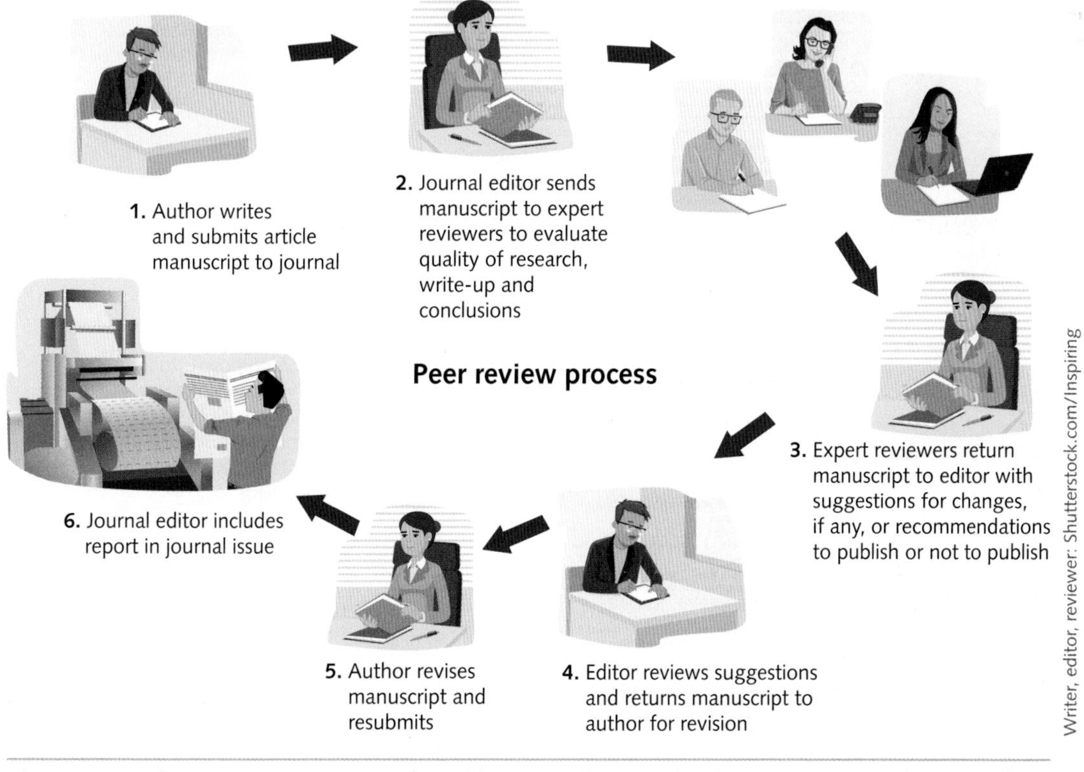

Figure 6.17 The peer review process. If a publication indicates it has been peer reviewed, you can be confident that the information in it is valid, reliable and has integrity.

Worksheet
A detailed checklist of the CRAP test

6.1.2 EVALUATING THE INTEGRITY OF SOURCES (PART 1)

Validity in secondary sources

Determining whether secondary sources are valid requires more work than for primary sources. There are many guides or checklists that you can use to help you make that judgement. One of these guides is the CRAP test.

The CRAP acronym stands for how current the source is, its reliability, its level of authority and its purpose. Figure 6.18 provides a snapshot of how this test works with a more detailed checklist found online. Table 6.5 shows the evaluation of a YouTube secondary source, using the more detailed CRAP test.

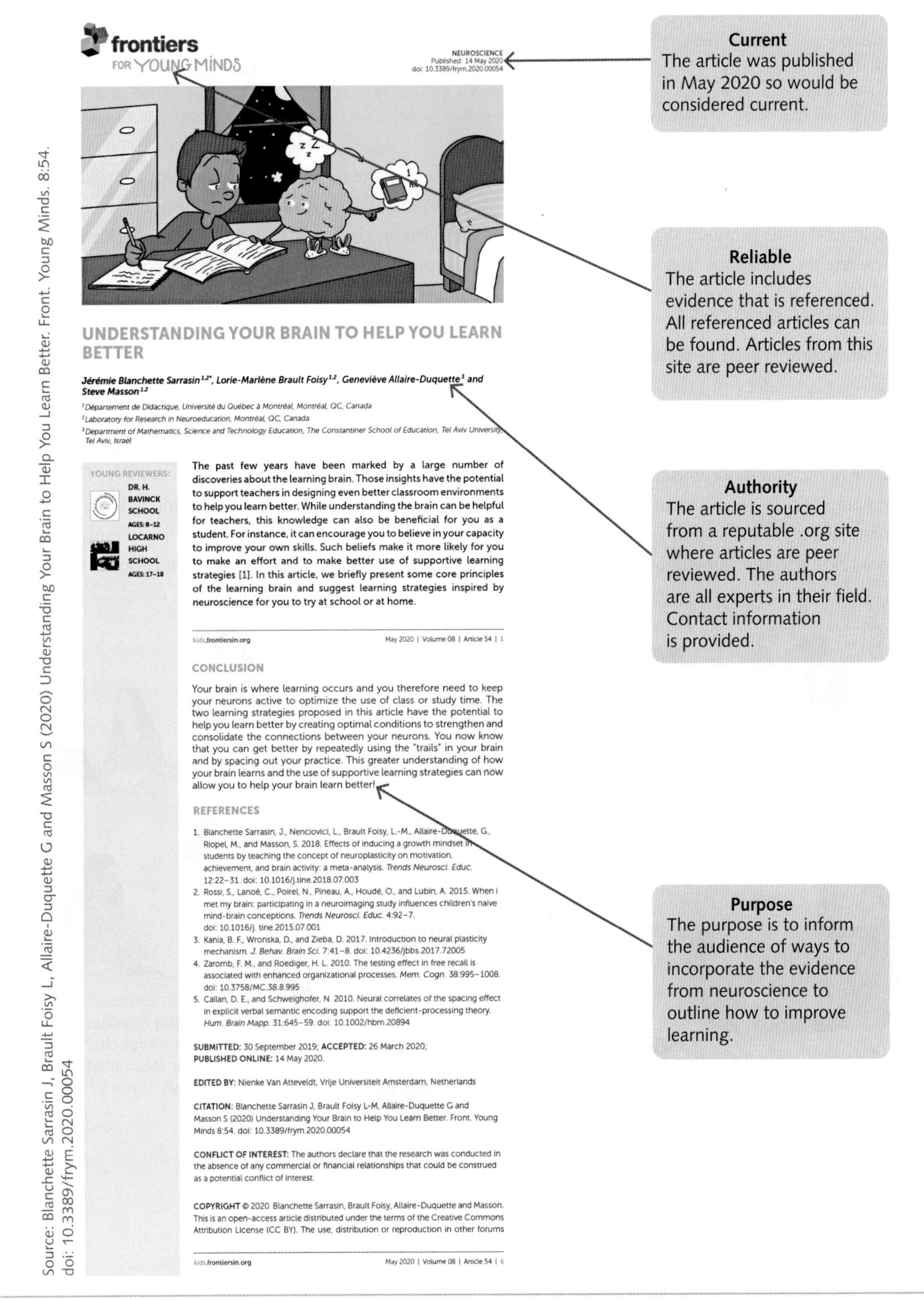

frontiers FOR YOUNG MINDS

NEUROSCIENCE
Published: 14 May 2020
doi: 10.3389/frym.2020.00054

UNDERSTANDING YOUR BRAIN TO HELP YOU LEARN BETTER

Jérémie Blanchette Sarrasin[1,2], Lorie-Marlène Brault Foisy[1,2], Geneviève Allaire-Duquette[3] and Steve Masson[1,2]*

[1] Département de Didactique, Université du Québec à Montréal, Montréal, QC, Canada
[2] Laboratory for Research in Neuroeducation, Montréal, QC, Canada
[3] Department of Mathematics, Science and Technology Education, The Constantiner School of Education, Tel Aviv University, Tel Aviv, Israel

YOUNG REVIEWERS:
DR. H. BAVINCK SCHOOL AGES: 8–12
LOCARNO HIGH SCHOOL AGES: 17–18

The past few years have been marked by a large number of discoveries about the learning brain. Those insights have the potential to support teachers in designing even better classroom environments to help you learn better. While understanding the brain can be helpful for teachers, this knowledge can also be beneficial for you as a student. For instance, it can encourage you to believe in your capacity to improve your own skills. Such beliefs make it more likely for you to make an effort and to make better use of supportive learning strategies [1]. In this article, we briefly present some core principles of the learning brain and suggest learning strategies inspired by neuroscience for you to try at school or at home.

kids.frontiersin.org May 2020 | Volume 08 | Article 54 | 1

CONCLUSION

Your brain is where learning occurs and you therefore need to keep your neurons active to optimize the use of class or study time. The two learning strategies proposed in this article have the potential to help you learn better by creating optimal conditions to strengthen and consolidate the connections between your neurons. You now know that you can get better by repeatedly using the "trails" in your brain and by spacing out your practice. This greater understanding of how your brain learns and the use of supportive learning strategies can now allow you to help your brain learn better!

REFERENCES

1. Blanchette Sarrasin, J., Nenciovici, L., Brault Foisy, L.-M., Allaire-Duquette, G., Riopel, M., and Masson, S. 2018. Effects of inducing a growth mindset in students by teaching the concept of neuroplasticity on motivation, achievement, and brain activity: a meta-analysis. *Trends Neurosci. Educ.* 12:22–31. doi: 10.1016/j.tine.2018.07.003
2. Rossi, S., Lanoë, C., Poirel, N., Pineau, A., Houdé, O., and Lubin, A. 2015. When i met my brain: participating in a neuroimaging study influences children's naive mind-brain conceptions. *Trends Neurosci. Educ.* 4:92–7. doi: 10.1016/j. tine.2015.07.001
3. Kania, B. F., Wronska, D., and Zieba, D. 2017. Introduction to neural plasticity mechanism. *J. Behav. Brain Sci.* 7:41–8. doi: 10.4236/jbbs.2017.72005
4. Zaromb, F. M., and Roediger, H. L. 2010. The testing effect in free recall is associated with enhanced organizational processes. *Mem. Cogn.* 38:995–1008. doi: 10.3758/MC.38.8.995
5. Callan, D. E., and Schweighofer, N. 2010. Neural correlates of the spacing effect in explicit verbal semantic encoding support the deficient-processing theory. *Hum. Brain Mapp.* 31:645–59. doi: 10.1002/hbm.20894

SUBMITTED: 30 September 2019; **ACCEPTED:** 26 March 2020; **PUBLISHED ONLINE:** 14 May 2020.

EDITED BY: Nienke Van Atteveldt, Vrije Universiteit Amsterdam, Netherlands

CITATION: Blanchette Sarrasin J, Brault Foisy L-M, Allaire-Duquette G and Masson S (2020) Understanding Your Brain to Help You Learn Better. Front. Young Minds 8:54. doi: 10.3389/frym.2020.00054

CONFLICT OF INTEREST: The authors declare that the research was conducted in the absence of any commercial or financial relationships that could be construed as a potential conflict of interest.

COPYRIGHT © 2020 Blanchette Sarrasin, Brault Foisy, Allaire-Duquette and Masson. This is an open-access article distributed under the terms of the Creative Commons Attribution License (CC BY). The use, distribution or reproduction in other forums

kids.frontiersin.org May 2020 | Volume 08 | Article 54 | 6

Source: Blanchette Sarrasin J, Brault Foisy L, Allaire-Duquette G and Masson S (2020) Understanding Your Brain to Help You Learn Better. Front. Young Minds. 8:54. doi: 10.3389/frym.2020.00054

Figure 6.18 There are features within your research article that will help you work out whether it is CRAP.

Table 6.5 CRAP test of a YouTube video

Name of Source: How to Study Effectively for School or College [Top 6 Science-Based Study Skills]

Source type: YouTube

URL: https://www.youtube.com/watch?v=CPxSzxylRCI

Date reviewed: 04/05/2022

Currency	Yes/No	Details
Was the information published or updated recently?	No	Published in 2016, which is more than 5 years ago.
If using a website, do the links within the article work?	Yes	Links to the learningscientists.org site, which is current.
Reliability		
Is the information supported by evidence such as data or quotes? Are there references for the evidence?	Yes	There are no references for the evidence in this YouTube, only a link to the learning scientists.org. Further review of this website provides the references of the original scientific investigations reported in reputable journals, but this research is not current.
Does the source make reasonable claims about what the evidence shows?	Yes	Yes, there is no sensationalism or exaggeration about the evidence.
Has the information (or its references) been reviewed?	Yes	The original research has been reviewed but not the YouTube video.
Can you confirm the information using another source?	Yes	Other research on the study strategies has been conducted and peer reviewed.
Does the language or tone seem unbiased and professional?	Yes	The tone isn't traditionally academic, but it does sound professional.
Authority		
Is the author, publisher or sponsor of the information a trustworthy source, such as an educational or government institution?	No	YouTube isn't generally considered trustworthy but the source information for this presentation is.
Is the author qualified to write on the topic?	No	The animator comes from a business background.
Is the author likely to be unbiased about the topic?	No	While there are free videos and resources available from the animator, there are paid courses as well. This could result in a conflict of interest, which could create bias.
Is there any contact information?	Yes	A current website with contact links
Purpose		
Is the purpose of the information to teach or inform rather than to sell, entertain or persuade?	Yes	To inform and possibly to sell memory training
Is the information fact, rather than opinion or anecdote?	Yes	It describes evidence-based strategies.
Overall judgement		Pass

6.1.3 EVALUATING THE INTEGRITY OF SOURCES (PART 2)

Evaluating data representations

It is important that you are aware of some of the ways that data representations can be misleading. This will also help you present your own information accurately.

For example, the graphs shown in Figure 6.19 display identical data. However, in the bar graph on the left, the data appear to show significant differences. In the original bar graph on the right, these differences are hardly visible. Figures 6.20 and 6.21 show further examples of misleading graphs.

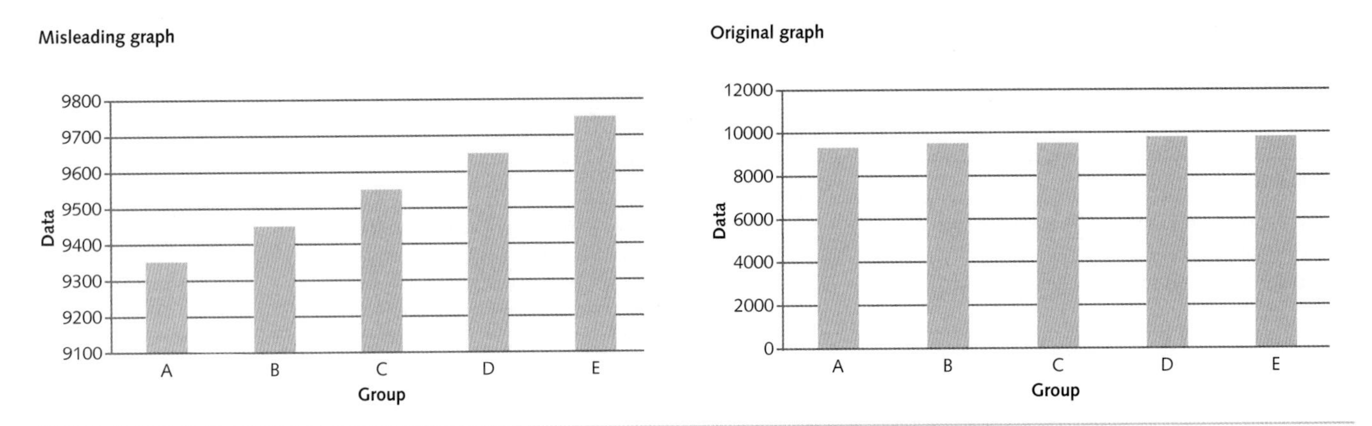

Figure 6.19 The misrepresentation of data can create a perception of differences where, in reality, there are none.

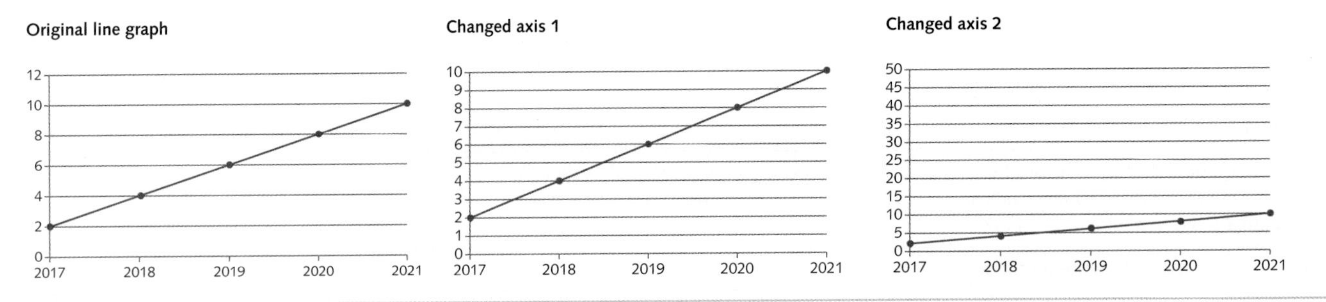

Figure 6.20 By changing the *y*-axis labelling, the slope of the line graph can change, leading the reader to reach very different conclusions.

Weblinks
TED-Ed: How to spot a misleading graph
Misleading graphs: don't get fooled

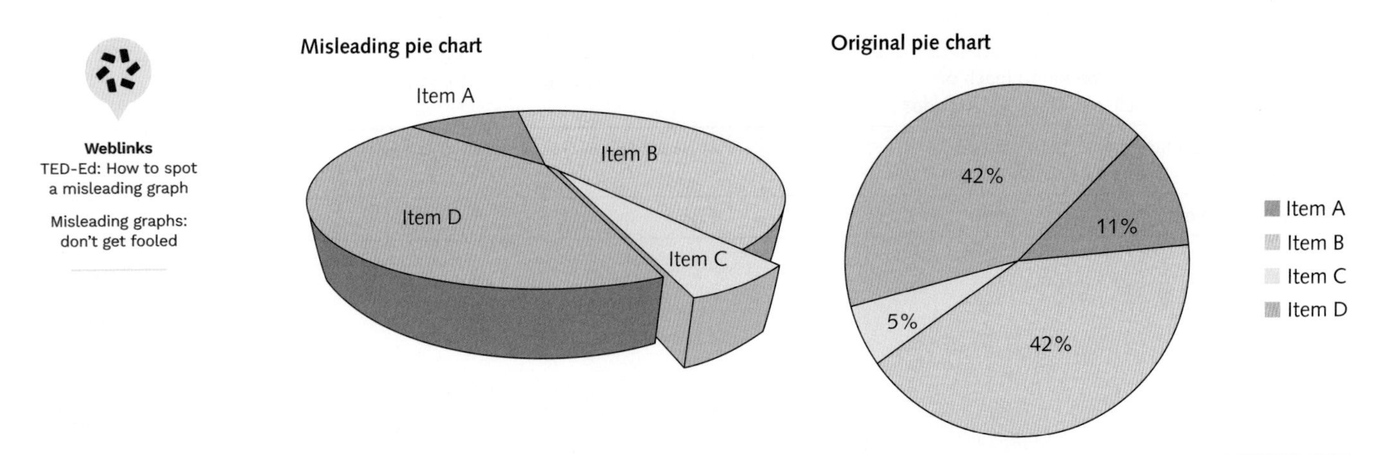

Figure 6.21 Item C appears to be at least as large as Item A in the misleading pie chart, whereas in actuality, it is less than half as large. Item D looks a lot larger than item B, but they are the same size.

KEY CONCEPTS 6.1b

- If you feel daunted by the idea of finding and reading research articles from journals, start with secondary source information from dedicated psychology websites such as Simply Psychology or Verywell Mind.
- When searching for information online, for optimal results you can use Boolean operators to tell databases exactly how you want your search terms combined.
- Many of the generalisations made about human thoughts, feelings and behaviours have been drawn from investigations using WEIRD samples. The problem with this is that not all populations think, feel or behave in the same way under the same conditions.
- Peer review is an extremely important part of the scientific process because it allows other scientists to offer feedback and test the research to verify its accuracy.
- In research reporting, describing the level of uncertainty strengthens the research results and provides guidance for the focus of future research projects.
- Many primary and secondary source articles and alternative platforms describe models or theories to organise and explain psychological concepts.
- The more primary research that the model fits, the more likely it is to become a theory used by psychologists to explain thoughts, feelings and behaviours.

Concept questions 6.1b

Remembering

1 What does each letter of the acronym CRAP represent? r

2 What does each letter of the acronym WEIRD represent? r

Understanding

3 Why is it important to find articles that have been peer reviewed? r

4 Why is it important to find out if authors have a conflict of interest? e

5 Describe how reproducibility strengthens scientific evidence. e

Applying

6 Your search on your topic of interest has landed on an article written for a less reputable media source. It has been written in a way that is easy to read and it passes the CRAP test. Describe how you could use this article. c

HOT Challenge

7 Using the CRAP checklist, evaluate each of the three resources in the weblinks. Rank them from 1 (the most valid source) to 3, justifying your choices. c

Weblinks
Resource 1: Learning and memory tips
Resource 2: Successful study strategies
Resource 3: Why do study skills matter?

You can now complete checkpoint 5 of the research guide.

Step 5: Organise, summarise and synthesise information

Now that you have checked your sources for validity and integrity, it is time to organise these sources in a way that helps you to locate facts, quotes, page numbers, diagrams and other features of interest. This process requires you to **summarise** the main concepts and gather more information from original sources or alternative sources if needed.

Managing sources

As you start to collect sources, a reference manager is vital. All researchers use and maintain some version of a source manager or **logbook** (Figure 6.22). A logbook can be electronic, such as OneNote or Google Docs, or hard copy, such as a standard exercise book.

You can use a logbook to manage your sources as well as record information about the source. However, there are source managers that are specifically designed for this purpose. These include EndNote, Zotero and Mendeley. Regardless of your choice, the purpose of a logbook is to record information from your literature review that you may or may not use in your final presentation. You have been provided with a research guide that you can use as part of your logbook.

Summarising primary sources

Scientific research articles can be difficult to understand unless you have a scientific background. However, an organised approach can make deciphering articles much easier.

Research articles have a specific structure (Table 6.6). Understanding what is included in each section will help you navigate the article to find key ideas and information. Do not try to read the article from start to finish, rather, read parts of certain sections first. As you work through the article, record important information into your logbook. You can use checkpoint 6 of the research guide to help.

Figure 6.22 A logbook is a way to manage and record information about your sources.

Table 6.6 Structure of a primary source research article

Abstract (Summary)	This is a concise summary, including methods, results and their implications, plus the author's interpretations and conclusions.
Introduction (Why)	The topic is introduced and the reason for the investigation is given. Previous research about the topic is presented. A prediction is made.
Methods (How)	This includes the technical details of how the study was conducted; participants, procedures and materials.
Results (What happened)	The data are summarised into tables, graphs and figures.
Discussion (What it means)	The data are analysed and evaluated to determine the level of support for the hypothesis. Limitations and uncertainty are discussed.
Conclusion (What was learned)	How do the results add to existing knowledge? Recommendations for further research are outlined.

1 Read the abstract

Identify the answers to the following questions (Figure 6.23): Why did the researchers conduct the study? Does the study matter? What did they do? What did they find?

2 List the key terms or psychological concepts

Look up any terms you don't know. Add them to your glossary. A **glossary** is a list of words with definitions. For example, after reading the abstract you may need to look up the terms 'concept map' and 'retrieval practice' and find some examples. Add these to your logbook.

3 Read the discussion

The start of the discussion summarises the results and the end outlines the final conclusions. Look for words such as 'found', 'determined', or 'concluded' as they typically signal the results of the study, or the conclusions reached.

4 Identify supporting information from the results or discussion

Find one or two pieces of data/evidence to support the findings. You may also want to find some quotes.

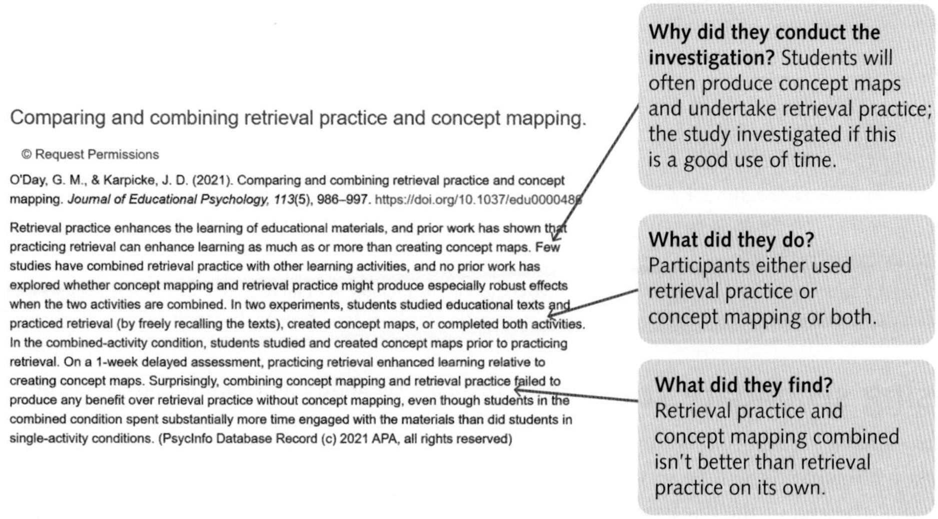

Comparing and combining retrieval practice and concept mapping.

© Request Permissions

O'Day, G. M., & Karpicke, J. D. (2021). Comparing and combining retrieval practice and concept mapping. *Journal of Educational Psychology, 113*(5), 986–997. https://doi.org/10.1037/edu0000486

Retrieval practice enhances the learning of educational materials, and prior work has shown that practicing retrieval can enhance learning as much as or more than creating concept maps. Few studies have combined retrieval practice with other learning activities, and no prior work has explored whether concept mapping and retrieval practice might produce especially robust effects when the two activities are combined. In two experiments, students studied educational texts and practiced retrieval (by freely recalling the texts), created concept maps, or completed both activities. In the combined-activity condition, students studied and created concept maps prior to practicing retrieval. On a 1-week delayed assessment, practicing retrieval enhanced learning relative to creating concept maps. Surprisingly, combining concept mapping and retrieval practice failed to produce any benefit over retrieval practice without concept mapping, even though students in the combined condition spent substantially more time engaged with the materials than did students in single-activity conditions. (PsycInfo Database Record (c) 2021 APA, all rights reserved)

Why did they conduct the investigation? Students will often produce concept maps and undertake retrieval practice; the study investigated if this is a good use of time.

What did they do? Participants either used retrieval practice or concept mapping or both.

What did they find? Retrieval practice and concept mapping combined isn't better than retrieval practice on its own.

Source: O'Day, G. M., & Karpicke, J. D. (2021). Comparing and combining retrieval practice and concept mapping. Journal of Educational Psychology, 113(5), 986–997. https://doi.org/10.1037/edu0000486.

Figure 6.23 What to look for in the abstract section of a research article

5 Evaluate the article

List the article's strengths and weaknesses. These can be from a CRAP test. Alternatively, if you want to evaluate the methodology of an experiment, look for 'sample size', 'sample type', 'repeated trials' and/or 'control group'. Chapter 1 outlines the strengths and limitations of other methodologies, such as correlational studies and case studies, and you can search for these. Table 6.7 also outlines some strengths and weaknesses. Usually, however, the researcher has done this evaluation for you. You can often find a description of the limitations of the research at the end of the discussion. Table 6.8 is a worked example of how to analyse a primary source.

Table 6.7 Some strengths and weaknesses of different methodologies

Type of research	Strength	Limitation
Correlation study	» No manipulation required » Less invasive to study participants » Useful for informing future research	» Can't be used to draw conclusions about causal relationships » Can be time-consuming
Field study	» Researchers can get firsthand knowledge about the participants being studied » Data is detailed » Uncovers social factors that may not have been obvious	» Determines correlation rather than causation » Data collected is specific to the time and place it is collected from
Case study	» Can investigate at the individual level » Very detailed data	» Data gathered is specific to one case, so findings may not be widely applicable
Controlled experiment	» Can determine the cause and effect relationship between variables » Researchers can minimse the effect of extraneous variables	» The environment is somewhat artificial, so it's harder to know if results are applicable in the 'real world'

Table 6.8 A worked example of how to organise and summarise information from a primary source

Name of article	Combining Retrieval Practice and Concept Mapping
Reference citation	O'Day, G. M., & Karpicke, J. D. (2021). Comparing and combining retrieval practice and concept mapping. Journal of Educational Psychology, 113(5), 986–997. https://doi.org/10.1037/edu0000486
Basic summary	Both concept mapping and retrieval practice are effective study strategies. Retrieval practice is more effective. Combining retrieval practice and concept mapping doesn't improve upon retrieval practice used on its own.

Glossary	Concept map – a graphical tool that is used to visualise meaningful relationships among concepts Retrieval practice – the act of trying to recall information without having it in front of you
Quotes	'There is a continuing need for rigorous examinations of learning strategies like concept mapping so that students, teachers, and educators can draw appropriate, evidence-based conclusions about their effectiveness.' p. 996
Evaluation	Article passed the CRAP test. Sample consisted of undergrad students with no indication of neurodiversity or consideration of WEIRD bias. Article was peer reviewed so results are likely valid and reliable.
How can I use this article?	I could use this to suggest further investigations into study techniques that are often recommended OR recommend that students only use retrieval practice for study as this will save them time.

Summarising secondary sources

Secondary sources have a similar structure to an essay; your key information will be in the introduction and at the end of the discussion. You will, however, need to do more work in evaluating them for reliability and validity, so using the CRAP test for secondary sources is essential. You might also need to check the primary source information for further details. Only use those secondary sources that have passed the CRAP test.

Weblinks
Source 1: Retrieval-based learning: Positive effects of retrieval practice in elementary school children

Source 2: Retrieval practice is effective regardless of self-reported need for cognition – behavioral and brain imaging evidence

You can now complete checkpoint 6 of the research guide.

Synthesising sources

The purpose of research is to answer a research question. In this modelled example, the research question is, 'What are effective, evidence-based study strategies?' To examine this, we would need more than one source of evidence for a particular study strategy. This requires synthesising the sources, using a synthesis matrix to track common themes and differences (Table 6.9).

Table 6.9 A synthesis matrix helps you find ideas shared by more than one source. In terms of our research question, the main idea is an example of a study strategy such as retrieval practice. Cells are highlighted where themes and ideas overlap. The more overlap, the more evidence there is for a particular study strategy.

Research question: What are effective, evidence-based study strategies for VCE students?

	Source 1: Karpicke et al. (2016)	Source 2: Wiklund-Hörnqvist et al. (2022)	Source 3: Agarwal (2019)
Main idea 1	Retrieval practice helps students perform better on recall tests than repeated study regardless of their level of skill in reading comprehension or processing speed.	Behavioural and brain-imaging evidence suggests that retrieval practice is also effective for individuals with lower levels of intrinsic motivation.	Building a foundation of knowledge via fact-based retrieval practice may be less effective than engaging in higher-order retrieval practice.
Main idea 2	Participants were of primary school age.	Retrieval practice is more effective than study alone.	Fact-based quizzes are not as effective as quizzes that require students to analyse and evaluate.
Main idea 3		Participants were of high school age.	Mixing quizzes with fact questions and analysis questions is better than analysis questions alone.

Weblink
Source 3: Retrieval practice & Bloom's taxonomy: Do students need fact knowledge before higher-order learning?

Source 1: 'Retrieval practice: beneficial for all students or moderated by individual differences?'
Source 2: 'Retrieval practice is effective regardless of self-reported need for cognition – behavioral and brain imaging evidence'
Source 3: 'Retrieval practice & Bloom's taxonomy: Do students need fact knowledge before higher-order learning?'

You can now complete checkpoint 7 of the research guide.

KEY CONCEPTS 6.1c

» A logbook records information from your literature review that you may or may not use in your final presentation. It keeps track of each source of information.
» When researching journal articles, read the abstract first.
» Summarise the main ideas into your logbook for each source of information, keeping track of data, conclusions, limitations, quotes and page numbers.
» A synthesis matrix can help you to find ideas that are shared by more than one source.

Concept questions 6.1c

Remembering

1 What is the purpose of a logbook? r
2 What part of a journal article should you read first? r
3 What does the start and the end of the discussion section in a journal article tell you? r

Understanding

4 How does using a synthesis matrix help to organise your research so that you find multiple sources of evidence on one idea or concept? e
5 If you are having difficulty unpacking an article of original research even with these guidelines, what step would you take next? Explain. e

Applying

6 Read the following abstract. Why did the researchers do the study? Why does it matter? What did they find? c

Abstract

Retrieval practice improves retention of information on later tests. A question remains: When should retrieval occur during learning – interspersed throughout study or at the end of each study period? In a lab experiment, an online experiment, and a classroom study, we aimed to determine the ideal placement (interspersed vs at-the-end) of retrieval practice questions. In the lab experiment, 64 subjects viewed slides about APA style and answered short-answer practice questions about the content or restudied the slides (restudy condition). The practice questions either appeared 1 every 1–2 slides (interspersed condition), or all at the end of the presentation (at-the-end condition). One week later, subjects returned and answered the same questions on a final test. In the online experiment, 175 subjects completed the same procedure. In the classroom study, 62 undergraduate students took quizzes as part of class lectures. Short-answer practice questions appeared either throughout the lectures (interspersed condition) or at the end of the lectures (at-the-end condition). Nineteen days after the last quiz, students were given a surprise final test. Results from the 3 experiments converge in demonstrating an advantage for interspersing practice questions on the initial tests, but an absence of this advantage on the final test.

Weinstein, Y., Nunes, L. D., & Karpicke, J. D. (2016). On the placement of practice questions during study. Journal of Experimental Psychology: Applied, 22(1), 72–84. https://doi.org/10.1037/xap0000071

HOT Challenge

7 Prepare for your assessment task by setting up your logbook. Your teacher may guide you to use either a hard copy or electronic copy.

Step 6: Communicate your psychological research

Now that you have completed your research, it is time to communicate your findings to others. You must clearly identify your purpose, your audience and what you hope to achieve through this communication. If you want to present information or a position on an ethical issue, you will need to work through Section 6.2 of this chapter first and complete checkpoint 10 of the research guide.

Purpose: What do you want to achieve?

Do you want to:

» inform your audience about the issue by summarising previous research so they can arrive at their own position on it?

» compare, contrast and critically evaluate previous work?
» present an opinion or argument on the issue? (This would also involve discussion of the validity and reliability of the work or even the ethical consequences of the research.)

Audience: Who do you want to inform or influence?

Ask yourself the following questions.

» Who do I want to communicate this information to?
» What expertise does my audience have relating to this issue?
» Why would they want or need the information?
» What do I want to achieve? How does my purpose relate to my audience?

Format: What is the best way to reach your audience?

There is a wide range of formats currently being used to present scientific information. In Unit 2 you will be presenting a formal scientific report or poster based on primary data. In this unit you may be required to present your research as a review article, an opinion piece or one of the newer means of communicating scientific ideas outlined in Table 6.10. Each has a different style and presentation.

Regardless of the presentation format, whether it is a review article, a YouTube animation or an ethical response, your information will still be organised using the same basic structure. However, what is included and the detail that is presented will vary.

Table 6.10 An outline of three social media platforms used by scientists and science communicators

Platform	Features	Example
TikTok Shutterstock.com/XanderSt	Videos should be short (1-, 3- and 10-minute videos are possible). Audience needs to be captured within the first 3 seconds. Data needs to be summarised in a way that the audience can relate to. TikTok has the option of including a *link tree* for presenters to cite their references for evaluation. Limitations: It is easy for misinformation to go viral. It is difficult to debunk the misinformation.	Search for good examples using hashtags such as #sciencecommunications
YouTube vlog or animation Shutterstock.com/PixieMe Reproduced with permission from www.complexly.com	Longer than TikTok. Presenters can list their references for evaluation. Vlog: A video blog or video log, sometimes shortened to vlog, is a form of blog presented as a video. Vlog entries often combine embedded video with supporting text, images and other metadata. Animation: Animated videos are videos created with original designs, drawings, illustrations or computer-generated effects that have been made to move in an eye-catching way, using any number of artistic styles. TEDEd is an example of an education resource that uses animation.	Vlog – Crash Course **Weblinks** The power of motivation How playing an instrument benefits your brain

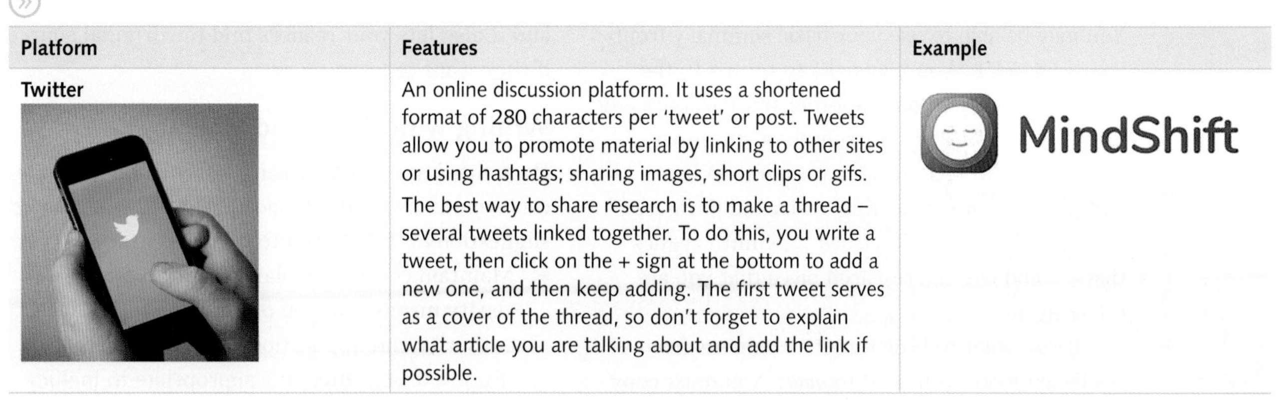

Platform	Features	Example
Twitter	An online discussion platform. It uses a shortened format of 280 characters per 'tweet' or post. Tweets allow you to promote material by linking to other sites or using hashtags; sharing images, short clips or gifs. The best way to share research is to make a thread – several tweets linked together. To do this, you write a tweet, then click on the + sign at the bottom to add a new one, and then keep adding. The first tweet serves as a cover of the thread, so don't forget to explain what article you are talking about and add the link if possible.	MindShift

Alamy Stock Photo/True Images

Table 6.11 provides a checklist of what to include in a YouTube presentation. You can find checklists for a review article and an opinion piece in the Worksheet link.

Constructing your final communication

Your presentation will require several reviews and redrafts before submission as your assessment task. Each type of presentation format has conventions to follow.

Summarising, paraphrasing and the use of quotation(s)

You won't necessarily use all the sources of information from your logbook. Read the basic summaries and evaluations from your logbook and choose which sources you will use that best help you answer your research question for your assessment task.

When reporting you will need to be able to summarise, **paraphrase** and use quotations correctly. Summarising means cutting down information to its essentials.

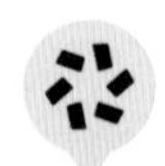

Worksheet
Review article and an opinion piece checklist

Table 6.11 Checklist for a YouTube animation

Introduction	What problem does the research address?	☐
	How does the research relate to the current debate (if presenting a response to an ethical dilemma)?	☐
Method	Not included (but you should still submit your method in your logbook)	☐
Results	Outline the key findings.	☐
	Present secondary data using tables, graphs and models wherever possible. You may take these directly from a study and acknowledge the source or construct your own.	☐
	Text is minimal.	☐
Discussion	Evaluate the social, legal and psychological implications relevant to the research question and/or ethical dilemma.	☐
Conclusion	If presenting a response to an ethical dilemma, what is the position you have taken, if any?	☐
	What future action is required?	☐
References	Not included; however, researchers can be cited within the animation and a *linktree* can be added to the comments section.	☐

Weblink
The nine best scientific study tips

You may be able to use your basic summary from your logbook or you may need to return to the original source and have another try. Paraphrasing means rephrasing text or speech in your own words, without changing its meaning. When paraphrasing, use reporting verbs and phrases, such as 'The author describes…', 'Smith argues that …' and cite the research piece that you are referring to.

If you want to keep direct statements from a published work, you need to *quote*. You must copy the source word for word, enclosing the quoted text in quotation marks. When you are recording a quote, make sure you note the page number it appears in the source.

Quoting and paraphrasing correctly are important if you want to avoid **plagiarism** (presenting someone else's work or ideas as your own, with or without their consent). If you use a source without citing it, you are implying that you created the information on your own. Citing your sources gives proper credit to the original authors, and it also lets your readers find the original source if they want to learn more.

Writing with clarity and accuracy

Paraphrasing or summarising information and data can sometimes result in inaccuracies. The following suggestions will help you to retain accuracy.

» Maintain contextual clarity: Contextual clarity means that you outline the importance, authority and implications of your research (Figure 6.24). Often, it's appropriate to include these in the introduction and the final section of your discussion and conclusion.

» Try to stay on topic: All content must be relevant and necessary. Do not stray off topic, because this distracts your reader.

» Use precise but understandable words: Academic writing requires precise vocabulary and correct terminology. Using unnecessarily sophisticated or complex language makes it harder for readers to understand.

Source: Blanchette Sarrasin J, Brault Foisy L, Allaire-Duquette G and Masson S (2020) Understanding Your Brain to Help You Learn Better. Front. Young Minds. 8:54. doi: 10.3389/frym.2020.00054

CONCLUSION

Your brain is where learning occurs and you therefore need to keep your neurons active to optimize the use of class or study time. The two learning strategies proposed in this article have the potential to help you learn better by creating optimal conditions to strengthen and consolidate the connections between your neurons. You now know that you can get better by repeatedly using the "trails" in your brain and by spacing out your practice. This greater understanding of how your brain learns and the use of supportive learning strategies can now allow you to help your brain learn better!

Contextual clarity: the importance and implications of the research are clearly stated

Figure 6.24 An example of contextual clarity. In this example, context is outlined in the conclusion; however, it is more commonly found in the introduction.

Writing concisely and coherently

Academic writing should be concise; that is, you should use no more words than necessary to convey your meaning. In some cases, the more words you include, the harder it is for the reader to extract meaning. Table 6.12 provides examples of common phrases that can be written more concisely.

Coherence means that information is logically linked, point by point. It enables your reader to understand the context for each sentence. The TEEL approach (topic sentence, explain, evidence, link) is a good way to structure paragraphs (Table 6.13). You can create further structure by using headings, subheadings, numbering and bullet points.

Table 6.12 Commonly used phrases, their shortened forms, and examples

Phrase	Shorter form	Writing example (shortened form in bold font)
A large number of	Many	**Many** students were tested.
Are found to be	Are	The newly developed drugs **are** effective.
Because of the fact that	Because	**Because** he received the treatment, he recovered.
For the purpose of	For	The study was conducted **for** education.
Have the ability to	Can	Students **can** plan their own presentations.
It is possible that	Possibly	**Possibly**, there was bias in participant selection.
In the final analysis	Finally	**Finally**, spaced retrieval was more effective than concept mapping.
In order to	To	Randomisation was performed **to** avoid bias.
Prior to	Before	Patient consent was obtained **before** the study.
The majority of	Most	**Most** of the patients were early adolescents.
With the exception of	Except	All cell types were stained **except** glial cells.

Source: J Korean Med Sci. 2021 Oct;36(40):e275. https://doi.org/10.3346/jkms.2021.36.e275

Short words and phrases can be used to link sentences and paragraphs so that a relationship is established between different ideas. The links usually appear at the beginning of the sentence.

Common examples of these types of transition relationships include:

» addition (Also ...)
» cause–effect (For this reason ...)
» elaboration (Furthermore ...)
» comparison (In contrast ... or In the same way ...)
» time transitions (At the same time ...)
» restatement (In other words ...).

Choice of language

Traditional communication formats, such as review articles, tend to have a serious, neutral and objective tone. In contrast, the voice used for social media can range from trendy, playful or educational, to fun or inspirational. Your choice of language depends on your audience.

Using data and visual representations

In Chapter 1, you were shown how to organise your data in tables and graphs. How you represent your data will depend on the type of data you have collected. When you write your article, presentation or animation, you might want to condense data and information into some form of representation to easily communicate your ideas. You may choose to summarise the data in a table or a visual representation, such as a graph, a chart (Figure 6.25), an image or a diagram (Figure 6.26). These visual representations are collectively referred to as figures. Number tables and figures in sequence as they appear in your report and label them with a concise caption that explains what information they present.

Table 6.13 The TEEL approach to structuring paragraphs

Topic sentence	Main idea
Explain	Further explanation
Evidence	Evidence and or examples
Link	Refer to main idea, sum up or link to next paragraph

For example, you may have researched study techniques and need to explain that learning involves the development of connections between neurons. Presenting this information in diagram form would be an effective way to visually communicate about neural networks.

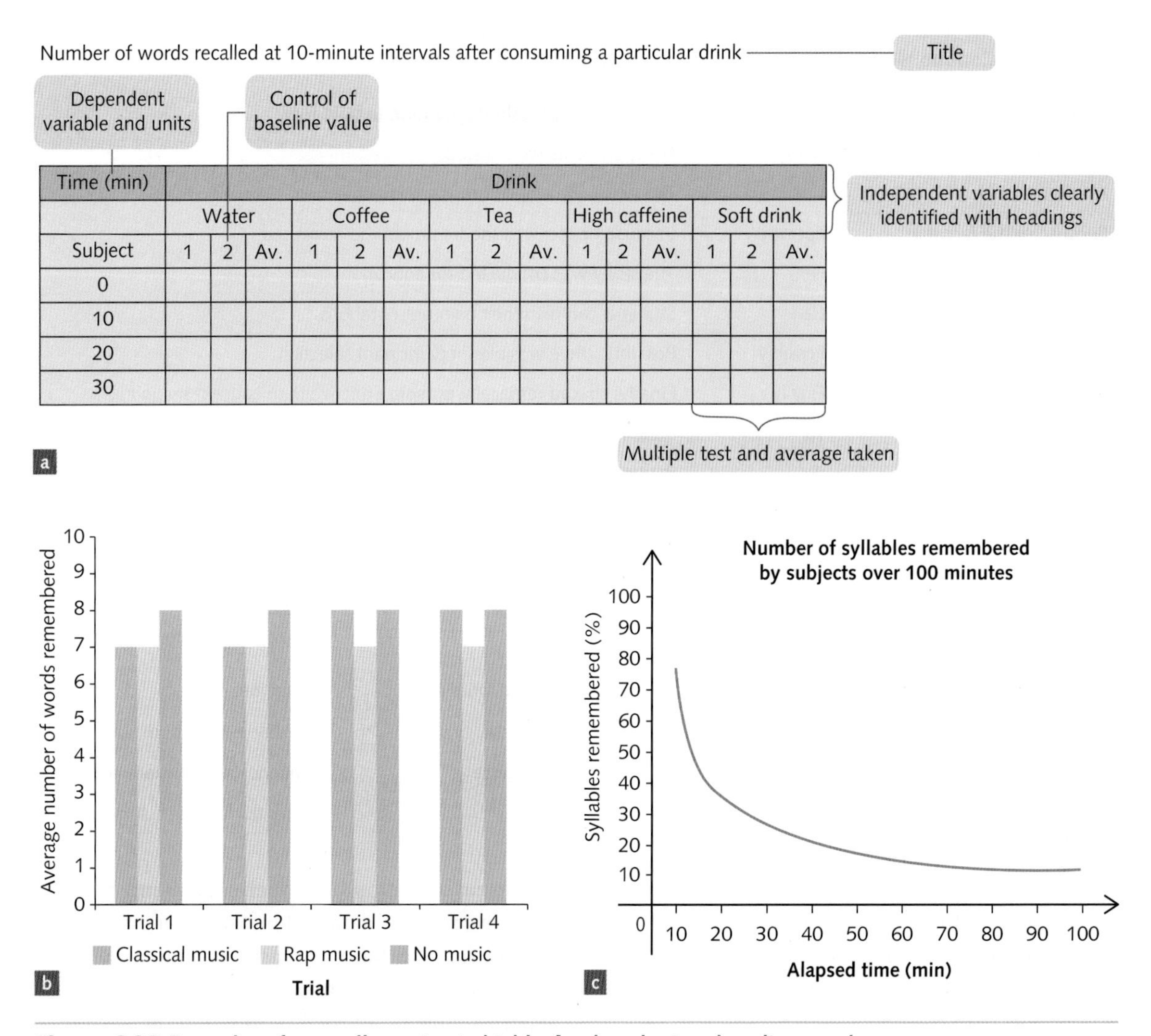

Time (min)	Drink														
	Water			Coffee			Tea			High caffeine			Soft drink		
Subject	1	2	Av.	1	2	Av.	1	2	Av.	1	2	Av.	1	2	Av.
0															
10															
20															
30															

Figure 6.25 Examples of **a** a well-constructed table, **b** a bar chart and **c** a line graph

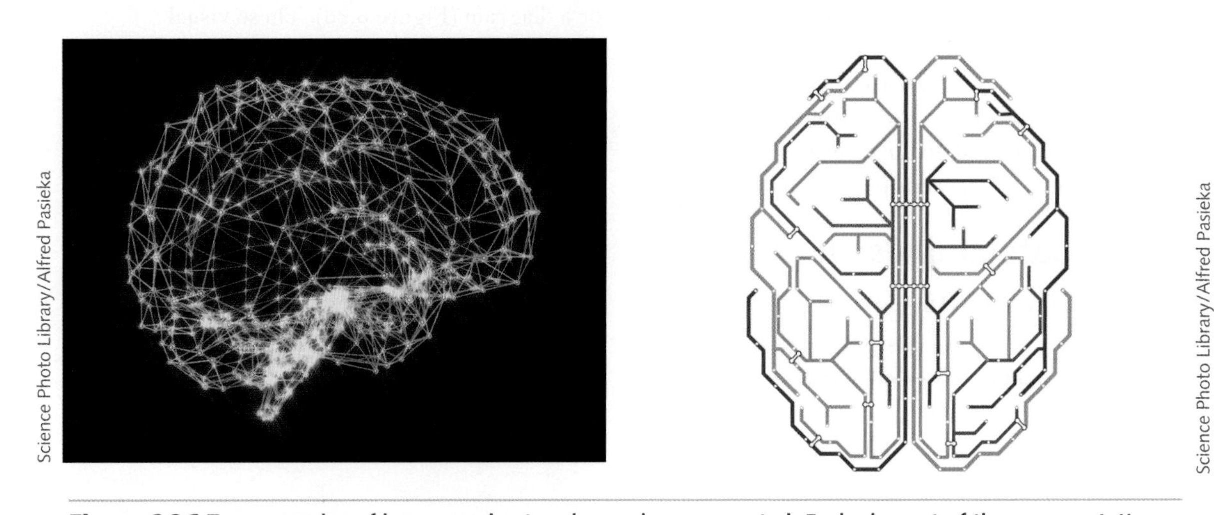

Figure 6.26 Two examples of how neural networks can be represented. Each element of the representation should be supported by scientific evidence.

Referencing and acknowledgements

APA (7th edition) is most commonly used as the source for referencing conventions for psychology research. Examples of how to use this referencing system are listed in Table 6.14. For further information and guidance, visit the weblink provided.

Acknowledgements

Most academic research projects involve many people in the preparation of the written or visual version or the research itself. These include people from a sponsoring institution, a funding body, other researchers, or even family, friends or colleagues who have helped in the preparation. All people involved should be mentioned in the acknowledgements section of your report (Figure 6.27).

Weblink
APA referencing guide

Table 6.14 Citation examples using the APA 7 citation style

Source of information	Citation method	Example
Journal article	Author's Surname, First Initial. Second Initial., & Author's Surname, First Initial. Second Initial. (Year). Article title: Subtitle. *Journal title, Volume*(issue), page range. http://dx.doi.org/xx.xxxxxxxxxx	O' Day, G. M., & Karpicke, J. D. (2021). Comparing and combining retrieval practice and concept mapping. *Journal of educational psychology, 113*(5), 986–997. https://doi.org/10.1037/edu0000486
News article from website	Author Surname, First Initial. Second Initial. (Year, Month Day). Article title: Subtitle. *Newspaper Title*, page range. URL [if viewed online]	Gray, R. (2016, March 18). Could we soon have superhero night vision? Brain implants could give us a 'sixth sense' by making us see infrared. *Daily Mail Australia.* https://www.dailymail.co.uk/sciencetech/article-3496895/Could-soon-superhero-NIGHT-VISION-Brain-implants-rats-sixth-sense-making-infrared.html
YouTube	Author Surname, First Initial. Second Initial. **OR** Author account name. (Year, Month, Day {of video post}). *Title of video* [Video]. Title of website. URL of specific video.	AsapSCIENCE. (2015, Sep 4). *The 9 BEST Scientific Study Tips* [Video]. YouTube. https://www.youtube.com/watch?v=p60rN9JEapg
TikTok	Account name [@TikTok handle]. (Year, Month Date). *First 20 words of caption. #includehashtags* [Video]. TikTok. URL	Dr Ben Rein [@dr.brein]. (2021, Dec 11) *Go for that jog. Your brain will thank you.* **#neuroscience #exercise** [Video]. TikTok. https://www.tiktok.com/@dr.brein/video/7040212135502466351?is_from_webapp=1&sender_device=pc&web_id7038517220075161090
Twitter or Instagram	Post Author Surname, First Initial. Second Initial. or Name of Group [@username]. (Year, Month Date). *Content of the post up to the first 20 words* [Tweet or Instagram Post]. Site Name. URL	MindShift [@MindShiftKQED]. (2019, December 4). *From avoiding distraction to combating procrastination, educators can teach students the necessary skills for overcoming educational challenges.* [Tweet].Twitter. https://twitter.com/MindShiftKQED/status/1484708645535428614

ACKNOWLEDGEMENT

The author thanks Dr Lynne Smelser and Moray Joslyn for their perceptive insights throughout the pursuit of this project.

Source: Dontre AJ. The influence of technology on academic distraction: A review, Hum Behav & Emerg Tech, 2021, 3:379–390, https://doi.org/10.1002/hbe2.229

Figure 6.27 Acknowledging people who have helped a project highlights the collaborative nature of good scientific research.

You can now complete checkpoint 8 of the research guide.

KEY CONCEPTS 6.1d

- When communicating information, you must decide on your purpose and your audience because these will determine the best format for your presentation.
- Your purpose will generally be to present to inform or critique, using a neutral voice.
- If you are responding to an ethical dilemma, your presentation may be designed to persuade or argue.
- Science communication can take traditional forms, such as a review article or news article; or, increasingly, it may take newer forms, such as YouTube, Twitter or TikTok.
- Most traditional formats cite references and structure information using introduction, method, results, discussion and conclusion sections.
- Less traditional formats often condense or leave out one or more sections to highlight key ideas. They should still include links so that the audience can check for validity, reliability and integrity.
- The introduction and/or conclusion must provide contextual clarity, explaining the importance and implications of the research.
- Writing concisely and with coherence is important and takes time and practice.
- Paraphrasing means putting a passage from source material into your own words. Summarising means putting the main idea(s) into your own words, using only the main point(s).
- If you want to keep exact statements from the text, you need to quote. When quoting, match the source word for word, enclosing the text in quotation marks.
- Avoid plagiarism by citing your sources. If you use information from a source without citing it, you are implying that you did the original research behind the information on your own.
- Primary and secondary sources often condense data and information into some form of visual representation, such as a table, graph, chart, photographic image or a diagram, to easily communicate ideas.
- Data representations can be misleading.
- You should follow conventions when using data or visual representations.
- The APA 7th edition referencing system is used in psychology.

Concept questions 6.1d

Remembering

1 List one traditional and two emerging forms of science communication. r

2 What structure is used, regardless of format, in communicating science ideas? r

3 What is the meaning of contextual clarity? r

4 Explain the differences between summarising, paraphrasing and quoting. e

5 What is plagiarism? How do you avoid it? r

Understanding

6 Briefly outline how you would communicate science using an emerging form of communication such as TikTok, while still providing your audience with enough information so that they can review your work for validity and integrity? c

Applying

7 Find the article from ScienceDaily where the following quote appears, and correctly cite it using APA 7 referencing conventions. An APA citation generator such as mybib.com would be helpful. c

'In a large study, scientists have shown that physical fitness is associated with better brain structure and brain functioning in young adults, Sep. 9, 2019.'

HOT Challenge

8 Read the article from Verywell Mind on the Zeigarnik effect (see weblink). Draw a visual representation of this effect, justifying the use of each element in your diagram. Compare your work with that of another student in your class. d

Weblink
The Zeigarnik effect

Sample review article

What are effective evidence-based study strategies for VCE students?

The past few years have been marked by a large number of discoveries about the learning brain. Those insights have the potential to support students to learn better. For instance, it can encourage you to believe in your capacity to improve your own skills. Such beliefs make it more likely for you to make an effort and to make better use of supportive learning strategies (Blanchette Sarrasin et al., 2018).

WHAT HAPPENS IN MY BRAIN WHEN I AM LEARNING?

Your brain is primarily composed of about 85 billion neurons. A neuron is a cell that acts as a messenger, sending information in the form of nerve impulses (like electrical signals) to other neurons (see Figure 1). For example, when you are writing, some neurons in your brain send the 'move fingers' message to other neurons and this message then travels through the nerves (like cables) all the way to your fingers. The electrical signals that are communicated from one neuron to another are therefore what allows you to do everything you do: write, think, see, jump, talk, compute, and so on. Each neuron can be connected with up to 10 000 other neurons, leading to a large number of connections in your brain (Rossi et al., 2015), which looks like a very dense spider web (see Figure 2).

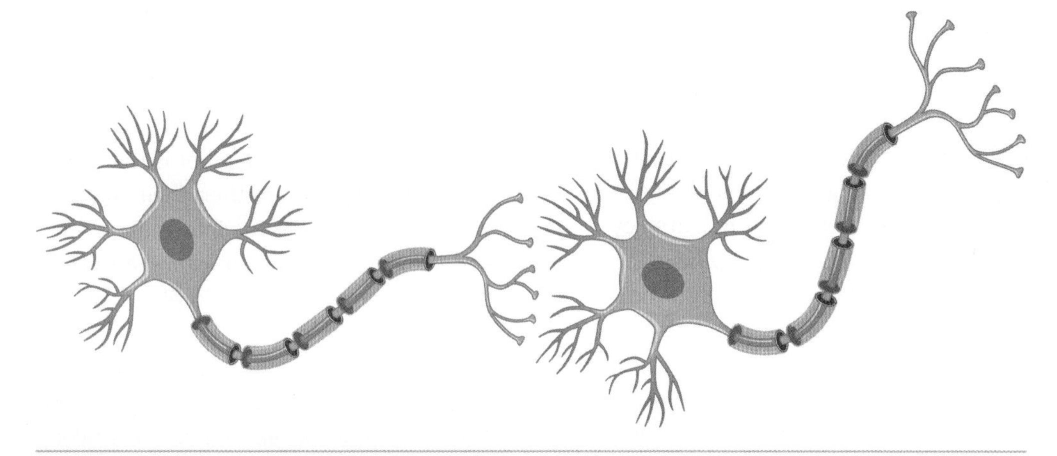

Figure 1 Communicating neuron cells

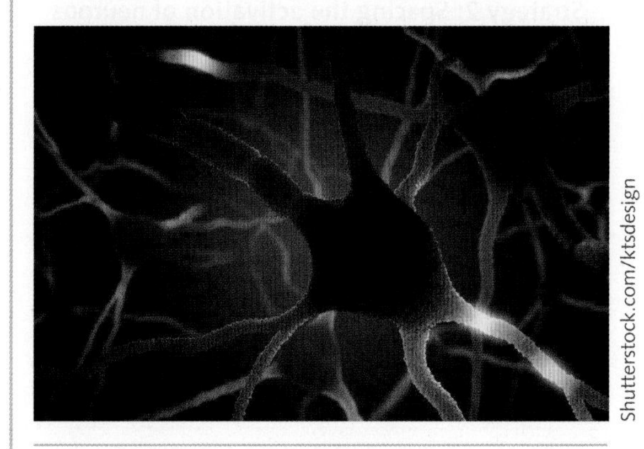

Shutterstock.com/ktsdesign

Figure 2 Neurons form networks

When you are learning, important changes take place in your brain, including the creation of new connections between your neurons. This phenomenon is called neuroplasticity. The more you practice, the stronger these connections become. As your connections strengthen, the messages (nerve impulses) are transmitted increasingly faster, making them more efficient (Kania et al., 2017). That is how you become better at anything you learn whether it is playing football, reading, drawing etc.

When you stop practicing something, the connections between your neurons weaken. That is why it may seem so difficult to start reading again when school starts if you have not read all summer. However, it is possible for some neural networks to become so strong that the trails or connections never completely disappear.

The fact that learning rewires your neurons shows how dynamic (plastic) your brain is – that the brain changes and does not remain fixed. Practicing or rehearsing repeatedly activates your neurons and makes you learn. These changes happen as early as when a baby is in their mother's womb and continues throughout a person's life.

WHICH LEARNING STRATEGIES ARE MORE COMPATIBLE WITH YOUR BRAIN?

Strategy 1: Repeatedly activating your neurons

Because the connections between your neurons need to be activated multiple times to become stronger and more efficient, a first and crucial strategy is to repeatedly activate them. This means that to learn key term definitions, for instance, you have to practice them repeatedly, to establish the 'trail' between your neurons. As a baby, you were not able to speak and walk within 1 day: you practiced a lot. However, it is important to note that only reading or glancing at your summary of definitions will not be that helpful in connecting your neurons. You might also find it quite disengaging and boring. To create the connections between your neurons, you need to retrieve the definitions from your memory. In other words, you have to try recall the answer yourself to activate your connections. We are not saying that this is easy to do! However, scientists think that this 'struggle' improves learning because the challenge is an indication that you are building new connections. When you do try to recall what you have learned and make a mistake, it can help you identify gaps in your learning and give you an indication of which trail still needs to be worked on.

Scientists have also noted that performing tests or exams can help you remember information better than studying alone (Zaromb & Roediger, 2010). For example, if you study your arithmetic tables interspersed with test periods, you will probably perform better on your final test than if you had only studied. Why? The tests require that you retrieve the information from the neurons in which the information is stored, thus activating your connections and contributing to their strengthening. The point is to practice retrieval in an engaging way. There are different strategies that you could try at home; for example, answering practice questions or using flashcards. These should improve learning more than re-reading or listening to lectures. Other strategies include preparing questions to ask to a classmate or a parent as well as redoing tests or exercises. What you need to remember is that, first, for your neurons to strengthen their connections, you need to retrieve the information and avoid just reading or listening to the answer. Second, you should plan a way to get feedback to know whether you got something correct or incorrect. Do not be discouraged if you face challenges; this is a natural step of the learning process taking place in your brain!

Strategy 2: Spacing the activation of neurons

Now that you know that neurons need to be activated repeatedly for learning to occur (and that it means retrieving information), you probably wonder how often you should practice. Scientists who study the learning brain observed that breaks and sleep between learning periods enhance learning and minimise forgetting (Callan & Schweighofer, 2010). It therefore seems better to retrieve often within spaced practice sessions, as opposed to a massed practice (practicing a task continuously without rest). For instance, instead of studying or doing homework for 3 hours, after which you would probably feel exhausted anyway, you could separate this learning period into three 1-hour periods or even into six half-an-hour periods.

When you take a quick break from practicing, let us say a 20 minute recess, you allow for the maintenance or replacement of the receptors on the surface of the neurons. Taking a break helps them work better: your neurons can thus transmit their nerve impulses more easily to other neurons. Finally, when you get a night of sleep between practice sessions, you actually benefit from a free retrieval practice session because, while you sleep, your brain reactivates the connections between the neurons that you activated during the day.

The most obvious change you can make to your study schedule is to break up sessions into smaller sessions. You could also ask your teacher to set daily or weekly review quizzes and other assignments. Finally, spacing can be done by doing interleaved practice. This consists of a set of problems arranged so that consecutive problems cannot be solved by the same strategy. For example, you could mix your math problems so that geometry questions, algebra or inequality problems are randomly sequenced. The added benefit of interleaving is that you engage in different activities between two sessions, making good use of your time. In brief, one thing to keep in mind is that information that was previously learned will require less effort to re-learn because the spacing gives your brain time to consolidate – meaning your brain produces the building blocks required for the connections between your neurons.

CONCLUSION

Your brain is where learning occurs and you therefore need to keep your neurons active to optimise the use of class or study time. The two learning strategies proposed in this article have the potential to help you learn better by creating optimal conditions to strengthen and consolidate the connections between your neurons. You now know that you can get better by repeatedly using the 'trails' in your brain and by spacing out your practice. This greater understanding of how your brain learns and the use of supportive learning strategies can now allow you to help your brain learn better!

REFERENCES

Blanchette Sarrasin, J., Nenciovici, L., Brault Foisy, L.-M., Allaire-Duquette, G., Riopel, M., & Masson, S. (2018). Effects of teaching the concept of neuroplasticity to induce a growth mindset on motivation, achievement, and brain activity: A meta-analysis. *Trends in neuroscience and education, 12,* 22–31. doi: 10.1016/j.tine.2018.07.003

Callan, D. E., & Schweighofer, N. (2010). Neural correlates of the spacing effect in explicit verbal semantic encoding support the deficient-processing theory. *Human brain mapping, 31,* 645–59. doi: 10.1002/hbm.20894

Kania, B. F., Wronska, D., & Zieba, D. (2017). Introduction to neural plasticity mechanism. *Journal of Behavioral and Brain Science, 7,* 41–48. doi: 10.4236/jbbs.2017.72005

Rossi, S., Lanoë, C., Poirel, N., Pineau, A., Houdé, O., & Lubin, A. (2015). When i met my brain: participating in a neuroimaging study influences children's naive mind-brain conceptions. *Trends in neuroscience and education, 4,* 92–97. doi: 10.1016/j. tine.2015.07.001

Zaromb, F. M., & Roediger, H. L. (2010). The testing effect in free recall is associated with enhanced organizational processes. *Memory & cognition, 38,* 995–1008. doi: 10.3758/MC.38.8.995

6.2 Ethical understandings of research and technology

What if your research question touches on current ethical dilemmas in psychology? What process would you go through to unpack and present an opinion on the issue?

You have already examined some current information about neuroscience, and this will be extended in Units 3 and 4. In Chapter 1, you learned about the role of **ethical guidelines** that protect the participants in a study. The broader scientific community also debates the ethics of emerging science and recognises the importance of values, and social, economic, political and legal factors in responsible decision-making about scientific research. Current cutting-edge technologies in neuroscience, such as brain augmentation, implants and brain-to-brain communication, make ethical debates necessary.

Brain implants (often referred to as neural implants) are technological devices that connect directly to a biological subject's brain. They are usually placed on the surface of the brain or attached to the brain's cortex. Neural implants such as those used for deep brain stimulation are becoming increasingly routine for patients with Parkinson's disease and clinical depression. Further, in recent research, Oxley et al. (2021) used cutting edge technology to enable two paralysed patients to wirelessly control, by direct thought, a Surface Book 2 computer running Windows 10 to text, email, shop and bank (Figure 6.28). It is vital that the scientific community explores the ethical implications of new technology such as this.

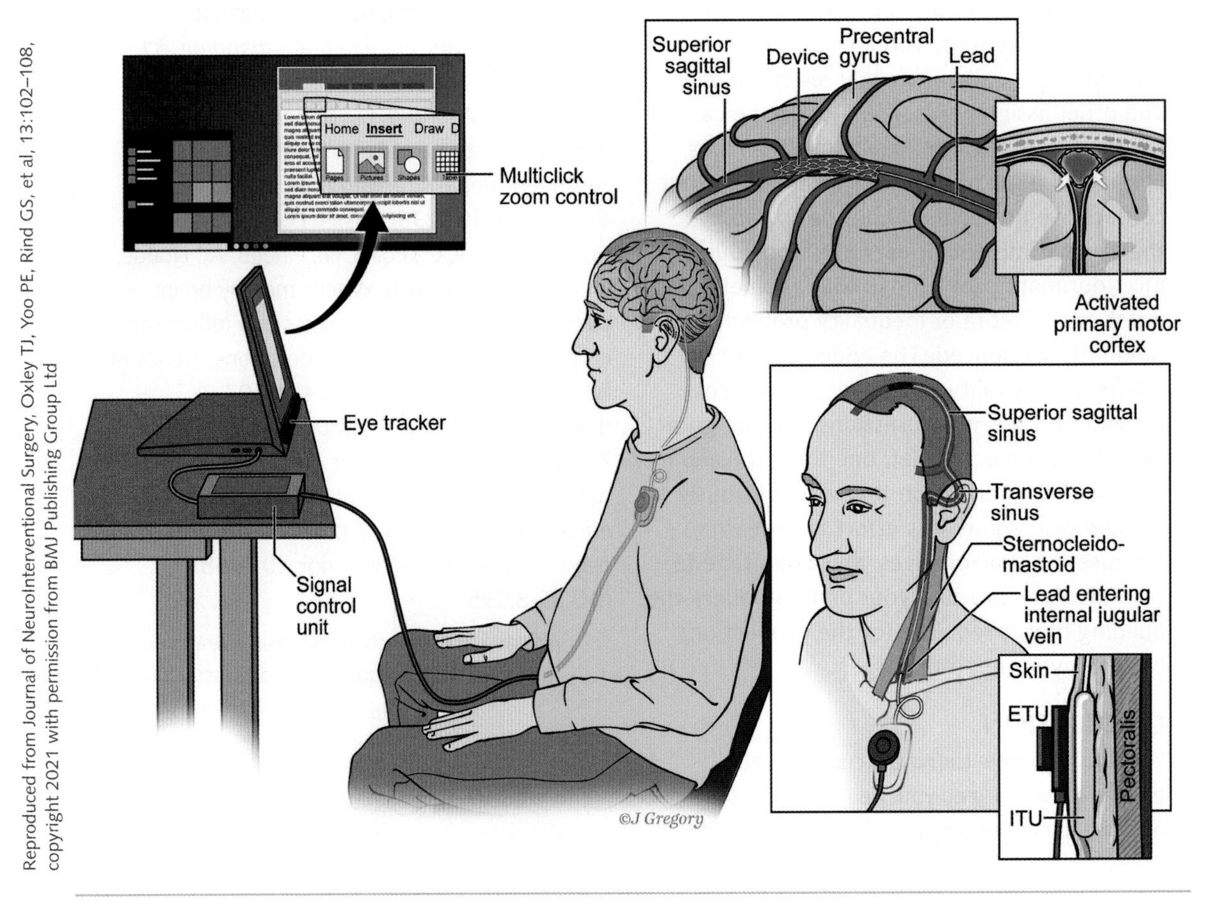

Figure 6.28 Wireless control through a computer interface to text, email, shop and bank using direct thought

Issues to be considered include:

» What laws apply to procedures that involve manipulating brain processing?
» Has the new technology been reviewed by government regulators?
» Who owns the technology once it has been implanted?
» Where is neural data stored?
» Who owns neural data once it is located outside a person's brain?

Ethical concepts

To present an informed position on an issue, writers of opinion pieces should consider whether researchers have applied the following **ethical concepts** when researching an issue: beneficence, integrity, justice, non-maleficence and respect.

Table 6.15 explains how each of these ethical concepts are applied to scientific research.

Researching ethical issues

You need an organised approach when investigating the ethical implications of a new treatment or technology.

Use the research guide to help. Once you have found your sources, evaluated them for CRAP and summarised the main ideas, you then need to evaluate the research using the ethical concepts.

6.2.1 ANALYSING ETHICS IN RESEARCH

Representation in research samples

Another very important aspect of integrity relates to participant representation. This has to do with how representative the research sample is relative to the broader population, as well as how participants' voices are represented.

As we mentioned earlier, WEIRD bias has been found in samples used in psychological research (page 237). An important study by Henrich, Heine and Norenzayan (2010) reviewed the representativeness of samples in published behavioural science research. The researchers found that 96 per cent of participants in such research were from Western industrialised countries. They pointed out that Western industrialised populations only make up 12 per cent of the world population, and yet behavioural scientists often assumed that their findings apply to the general population.

Table 6.15 Key questions to ask when investigating the ethics of a new technology or treatment

Ethical concept	Description	Key questions
Beneficence	The duty to do good	Who will benefit from this technology or treatment? Who needs the most help? How will the technology or treatment benefit people (can include physical, psychological, economic or social benefit)? How many people will benefit?
Integrity	Honesty and transparency	Are the consequences for the technology or treatment supported by scientific evidence? How reliable is the evidence? What has been the conduct of any people Involved or affected (stakeholders)?
Justice	Equal distribution of benefits, risks, costs and resources	Does everyone have equal access to the technology or treatment? Have all stakeholders been recognised? Have some stakeholders been excluded from consideration? How can discrimination be avoided?
Non-maleficence	The duty to minimise harm	Who might be harmed by this technology or treatment? How might they be harmed (can include physical, psychological, economic or social harm)? How many beings could be harmed? Is it possible to minimise the harm?
Respect	Every individual has the right to autonomy and to choose their own course of action	Are the individual rights of all stakeholders considered and respected? Has informed consent been obtained? Is there respect for privacy and confidentiality?

Not only was the typical research sample limited to a narrow range of diversity, but it was largely made up of American undergraduate students. Henrich et al. (2010) reviewed related research to make their case that the WEIRD populations who participate in behavioural science research are actually outliers. Compared with people from other societies, WEIRD populations were shown to be highly individualistic, self-focused, competitive, analytical and open to expressing their ideas and opinions.

One solution to the problem of representativeness is to ensure greater diversity when selecting research samples. Another strategy is to reconsider any assumptions about how generalisable research findings are beyond the immediate sample. These approaches can help, but they do not fully address the issue of representation.

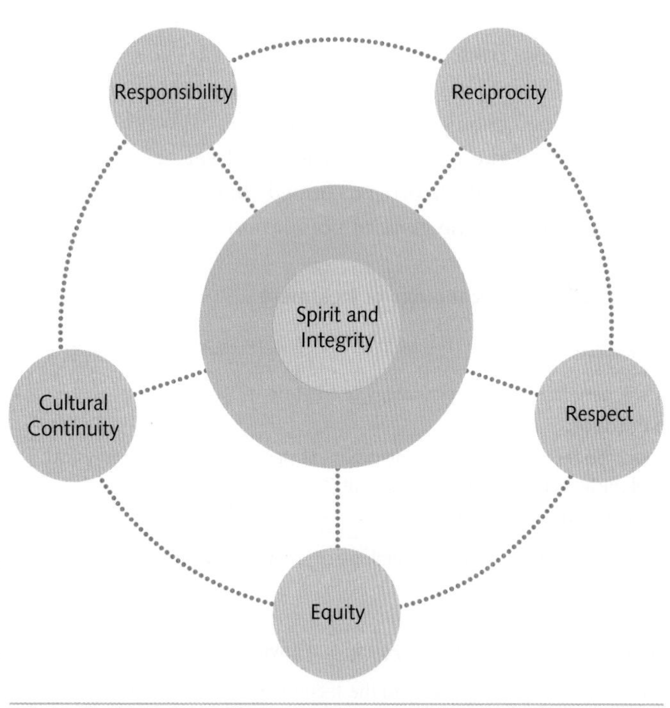

FIGURE 6.29 These ethical principles must be followed when engaging with First Nations communities and research participants

As Henrich et al.'s study shows, pure objectivity in psychological research is rarely achieved. Researchers are inevitably influenced by their own life experiences and by assumptions that derive from the cultural contexts with which they are most familiar. As Shweder (1991) observed, 'When "thinking through cultures" there is no place else, no neutral place, for us to stand' (p. 23). An uncritical 'ethnocentrism', where researchers assume that the cultural norms they accept can be applied to everyone, can lead to systematic bias in the theoretical models they use and in the outcomes of research.

Dudgeon and Walker (2015) argue that Western knowledge systems that insist on objectivity and deductive logic, which are assumed to be superior to the knowledge of other groups, entrench disadvantage and racism. They point out the bias inherent in 'universalising individualistic constructions of human behaviour and … the negation of Aboriginal knowledges and practices' (p. 276).

When we undertake or review psychological research, we need to ask:

» Whose knowledge counts?
» Who is speaking for and about whom?
» Who benefits from the research?

The use of mainstream models and methods in psychological research with First Nations people has been criticised for promoting assimilation into a culture in which individuals are made responsible for difficulties caused by social disadvantage and trauma. A decolonising perspective challenges these assumptions and instead aims to identify the strengths, perspectives and cultural resources that support resilience in First Nations communities.

For researchers, the National Health and Medical Research Council (NHMRC) and the Australian Institute of Aboriginal and Torres Strait Islander Studies (AIATSIS) have outlined important ethical principles that need to be followed when engaging with First Nations communities and research participants (Figure 6.29).

CASE STUDY 6.1

Evaluation of the Aboriginal Girls Circle

Brenda Dobia

Introduction

The Aboriginal Girls Circle (AGC) aimed to increase social connection, participation and self-confidence amongst Aboriginal girls attending a secondary school in regional NSW. The program was initiated by the school, with the goal of re-engaging girls who were disconnected, and reducing challenging behaviours.

Designed and led by educational psychologist Sue Roffey, the AGC focused on building social-emotional skills and connections through a residential camp, weekly circles run by an Aboriginal coordinator, involvement of Elders and student-initiated community projects.

When I was invited to lead the evaluation of the AGC program, I first wanted to know how it involved the community. I asked whether the local Aboriginal community had been consulted about the Girls Circle program and whether they supported its aims. Because the AGC was a program specifically for Aboriginal girls, it was important that the community felt that it was with and for them.

I discussed the project with colleagues and formed a research team that included Aboriginal researchers Shirley Gilbert, Virginia O'Rourke, Gawaian Bodkin Andrews and Annie Daley, as well as Roberto Parada and myself, with each team member contributing their particular expertise and leadership.

Research design

The program was already into its second year, so our options for designing the research were limited. We could not take before and after measures to assess change over time. We also did not want to restrict our approach to only measuring mainstream constructs of psychological resilience. Based on previous experience (Dobia & O'Rourke, 2011), we had identified features of Aboriginal and Torres Strait Islander identities, cultures and life experiences that suggested that resilience was understood differently in Aboriginal communities. We therefore decided to emphasise Aboriginal perspectives on resilience and cultural identity in our research.

Research aims

- To determine the effects of the AGC for participants' resilience, wellbeing and connectedness
- To investigate and develop culturally appropriate tools and methods
- To evaluate the various components of the program and its implementation

Method

We opted for a mixed design with both quantitative and qualitative components.

We conducted qualitative fieldwork that involved interviews and focus groups with community members, school staff and the girls taking the AGC program.

The quantitative data was gathered via a survey that measured different aspects of resilience and cultural identity. We conducted focus groups with community members about what resilience and wellbeing meant to them to inform the design of the new measure.

We investigated resilience, connectedness, self-concept and cultural identity in Aboriginal youth. We gathered quantitative data from the AGC girls, and from other Aboriginal and non-Indigenous youth. This allowed us to compare the relationships between measures and consider the effects of cultural identity differences between the groups.

We consulted Aboriginal Elders and Aboriginal Education Consultative Group from the outset and asked for advice on the most appropriate protocols for conducting the research with their communities. We presented interim and final reports to the local Aboriginal Education Consultative Group and to the research participants.

Key findings

The AGC was very well received within the school and the community, with reported increases in the girls' confidence, social skills and leadership.

»

There were positive impacts for resilience, connectedness, self-concept and cultural identity.

Resilience was strongly related to cultural identity for Aboriginal youth, providing support for cultural models of social-emotional wellbeing.

Limitations

Due to the small sample size and correlational design, the quantitative findings from this research should be regarded as indicative only.

Contribution

The findings from this study have helped to advance theory and research into decolonising psychology (Dobia & Roffey, 2017). The AGC evaluation study was included in a scoping review of research into Indigenous resilience in Australia which provided further support for the importance of the relationship between cultural identity and resilience (Usher et al., 2021).

Analysing sources

When going through your sources, it's a good idea to brainstorm and record your initial thoughts and feelings. This will help to highlight any bias you may have before you investigate the issue. It will make you more aware of articles that might not be compatible with your biases that you may have unconsciously decided not to use. Use the questions listed in Table 6.5 (page 259) as well as the following questions to guide your brainstorm.

- What is the issue?
- What would happen if there was no oversight of ethical considerations?
- Who is affected?
- How does it affect stakeholders, such as patients, families, businesses, medical practitioners, governments and the public? Stakeholders are people or organisations who have an interest in the technology or treatment. Stakeholders include those who are both supportive of the treatment or technology, as well as those who may be less supportive or indeed critical of it.

Use the ethical concepts to organise your thinking. Table 6.16 has modelled this.

Table 6.16 A sample analysis of ethical considerations of neural implants

Technology or treatment	Neural implants
Brainstorm of initial thoughts and feelings	The implant could improve quality of life. Families will need to adjust. Will it change the identity of the person? Excited by what used to seem science fiction. What if the technology is misused? How far has the technology progressed? It could help a lot of people. The research is very expensive so will only benefit the wealthy.
Source of information	Oxley T.J., Yoo P.E., Rind G.S., et al. (2021). J NeuroIntervent Surg, 13, 102–108 http://dx.doi.org/10.1136/neurintsurg-2020-017074
CRAP test	Passed
Summary	Paralysed participants were able to regain some functional independence with the use of a neural implant. They were able to participate in a range of computer tasks, such as online shopping and emails, with high accuracy and independence.

Ethical concept	Analysis
Beneficence	Implants benefit those who are paralysed and improve their quality of life. Deep brain stimulation already benefits those with Parkinson's disease, a neurodegenerative disease that affects movement. It will improve mental health, as affected individuals can increase their independence. Is the cost of developing this technology taking away funds from research that could help more people with other, more prevalent diseases, such as cancer?
Integrity	The research is being published in reputable journals. However, the research is new and there needs to be more trials and refinement before it is used with more people. The trials are positive; however, the sample sizes are small. It is unclear who has been consulted about wider use.

Ethical concept	Analysis
Justice	Only wealthy people can afford this technology.
Non-maleficence	Individuals may be vulnerable to cybercrime or intrusive surveillance. Privacy could be violated by others removing information from the microchips. Will people become more like cyborgs and lose essential human qualities? If participants are part of a trial and the funding is lost, what happens to the implant and the affected individual? Have there been plans developed if issues such as device maintenance arise?
Respect	Participants are informed of the technology and any surgical procedures. Who has legal rights over the implant once it has been implanted? Will neural data that is produced stay with the individual or will there be potential for third parties to access data even without the knowledge of the researchers? Will people in the military inadvertently sign up to trialling neurotechnology? If so, will the military own the data or the individual? Some could argue that we already have so much of our personal information collected through our use of the internet; this technology wouldn't be any different.

You can now complete checkpoint 9 of the research guide.
Note: If you are presenting an ethical issue, you will still need to complete the preceding checkpoints in the research guide.

KEY CONCEPTS 6.2

» Ethical issues often arise when new treatments and technologies are developed.
» Investigating the ethical issues of new psychological research or technology requires identifying the issues, analysing the evidence, analysing the legal, economic and social consequences, and evaluating the ethical considerations for and responsibilities of all stakeholders.
» Ethical concepts are used to unpack the ethical issues involved in research. These concepts include beneficence, integrity, justice, non-maleficence and respect.
» In order for findings to be generalised to a broad population, it is important to consider the participant representation within the study.

Concept questions 6.2

Remembering

1 List the ethical principles that are used to debate emerging treatments and technologies and briefly describe each. r

2 With reference to psychological research, explain what is meant by the term 'stakeholder'. e

Understanding

3 Why is it important to write down your own initial thoughts and feelings about a new treatment and technology before you summarise and evaluate the views of stakeholders? e

4 When you write an opinion article about an ethical issue, do you need to decide your position and outline this in your response, or do you outline the views of stakeholders and leave it to your reader to come to a position? Provide reasons for your answer. e

Applying

5 Analyse the ethical issues arising from the new technology outlined in the weblink article, written by David Tuffley, Senior Lecturer in Applied Ethics and Cybersecurity at Griffith University (see weblink). c

HOT Challenge

6 Create a Menti poll or Google Quiz to find out the views of your peers on the effects of social media on their ability to study effectively. You will need to refine this topic into a research question of your choice. Analyse and present the results of your survey. c

Weblink
Neuralink's monkey can play Pong with its mind

6.3 Sample assessment task and rubric

Below is the assessment task for Unit 1, Area of Study 3: How does contemporary psychology conduct and validate psychological research?

Outcome 3

On completion of this unit the student should be able to identify, analyse and evaluate the evidence available to answer a research question relating to contemporary psychology.

To achieve this outcome the student will draw on key knowledge outlined in Area of Study 3 and related key science skills on pages 12 and 13 of the study design.

Assessment

A response to an investigation into contemporary psychological research and how science can be used to explore and validate psychological research questions.

Source: VCE Psychology Study Design (2023–2027), pp. 26 & 27

Instructions

Produce a piece of science communication for a specific audience and purpose on a research question drawn from the topics below. You will need to submit your logbook as well as your final response.

In your logbook you will need to show evidence of:

1. An evaluation of at least two primary and three secondary sources for validity and reliability.
2. A summary of at least one valid and reliable primary source and two valid and reliable secondary sources.
3. The ability to synthesise ideas from one primary and two secondary sources.

Your final response will follow the conventions of your chosen format. If presenting a video, you will need to submit a transcript.

Your logbook will be worth 75% of your final grade, and your final piece will be worth 25%.

The following is a list of topics to help you brainstorm to find a topic of research interest.

- Experience and what makes us human
- Epigenetics research that shows how gene expression can be affected by environmental influences
- Nature and nurture
- The idea of the human mind as 'tabula rasa'; that is, a blank page that is gradually filled as a result of 'experience'
- Early understandings of human brain functioning compared to modern understandings of human brain functioning
- Case studies into brain trauma
- Challenging human nature by exploring neurodiversity
- Neurodiversity as a competitive advantage
- Twin studies and what they can tell us about psychological development
- The privileging of Western norms and approaches in researching and understanding psychological development and mental wellbeing
- Parenting and its influence on emotional development
- The role of bonds with early caregivers
- The impact of linguistic environments on language development
- Ways individuals construct their sense of self and identity
- The psychology of 'habit formation'
- Psychological growth and change throughout the lifespan
- Impact of intergenerational trauma, including the impact of colonisation on First Nations peoples, on psychological development and/or mental wellbeing.

Table 6.17 is an example of a rubric that you could use to construct your logbook, and that your teacher could use to assess your logbook. Remember, your logbook makes up 75% of the score for this outcome. Check your approach with your teacher before you start.

Table 6.18 is an example of a rubric that you could use to design your report, and your teacher could use to assess it.

Table 6.17 Suggested rubric for a logbook

	Low	Medium	High
Checkpoints 1 and 2 Topic and research question development	Only a few ideas listed Consideration of purpose and audience	Between four and nine ideas listed Consideration of purpose and audience	Comprehensive list of 10 or more ideas Consideration of purpose and audience
Checkpoints 3 and 4 Search strategy	Key terms and concepts listed Some evidence that a search strategy was used Limited sources identified	Key terms and concepts listed Search strategy used Competent use of Boolean operators Two primary sources and three secondary sources identified	Key terms and concepts listed Comprehensive search strategy used Two relevant primary sources and three relevant secondary sources identified
Checkpoint 5 Source evaluation for validity and reliability using the CRAP test	Limited analysis of sources	Brief analysis of two primary and three secondary sources using the CRAP test	Competent and detailed analysis of two primary and three secondary sources using the CRAP test
Checkpoint 6 Source summary	Some relevant ideas outlined	All relevant ideas outlined Paraphrasing and quotations used and correctly formatted	Main ideas accurately outlined Paraphrasing and quotations correctly formatted and appropriate Page reference noted
Checkpoint 7 Source synthesis	Limited ability to synthesise main ideas between sources	Ideas synthesised using the synthesis matrix	Effective use of the synthesis matrix to find common ideas between three sources
Checkpoint 8 Referencing	Referencing of sources using APA 7 style attempted	Complete referencing of five sources using APA 7 style, with minor errors	Complete and accurate referencing of five sources using APA 7 style
Checkpoint 9 (if relevant) Analysis of an ethical issue	Basic analysis of an ethical issue, with limited reflection	Thoughtful use of ethical guidelines to analyse an ethical issue	Comprehensive and considered use of ethical guidelines to analyse an ethical issue
Logbook	Inconsistent use of logbook Most elements completed accurately	Logbook regularly maintained All elements completed accurately	Logbook regularly and well maintained All elements completed accurately and in detail

Table 6.18 Suggested rubric for a report

FINAL RESPONSE	Low	Medium	High
Structure	Some structural conventions followed.	Most structural conventions followed.	All structural conventions followed.
Clarity and coherence of article or transcript	A good attempt at answering the research question has been made. Writing is mostly logical, on topic and provides context. Writing uses precise scientific terminology that can be understood by the audience some of the time. Inconsistent use of voice, tense and tone for chosen format.	Research question is clearly answered. Writing is mostly logical, on topic and provides context. Writing uses precise scientific terminology that can be understood by the audience most of the time. Use of voice, tense and tone mostly consistent and appropriate for chosen format.	Research question is clearly answered. Writing is consistently logical, on topic and provides context. Writing uses precise scientific terminology that can be understood by the audience. Use of voice, tense and tone consistent and appropriate for chosen format.
Visual representations	Tables, graphs, diagrams and/ or animations represent concepts. Labelling conventions are followed.	All tables, graphs, diagrams and/ or animations represent concepts. Labelling conventions are followed.	All tables, graphs, diagrams and/ or animations accurately represent concepts. Labelling conventions are followed.
Referencing and acknowledgements	Embedded within the communication platform.	Fully embedded and accessible within the communication platform.	Fully embedded and accessible within the communication platform. Links to primary sources referenced in secondary sources also included.

6 Chapter summary

KEY CONCEPTS 6.1a

- Primary sources of information include research articles published in journals and conference presentations.
- A secondary source is a written or visual text about a research article or series of articles.
- Primary data are generated directly by the investigator analysing them; secondary data are drawn from previous research.
- An expert opinion article focuses on a topic and expresses personal and original thoughts or beliefs, which stem from experience and expertise. It aims to stimulate debate or new research.
- Journalists often use anecdotes or personal stories to make their stories seem more relevant to people's lives or to emphasise why a piece of scientific research matters.
- Be wary of claims that may be exaggerated, or statements such as 'the findings prove' instead of 'the findings suggest'. It takes a great deal of research and review to support any scientific theory.
- Non-scientific ideas refer to ideas or practices that are presented to resemble real science, but are not supported by scientific evidence. They are often supported by celebrities or influencers.

KEY CONCEPTS 6.1b

- If you feel daunted by the idea of finding and reading research articles from journals, start with secondary source information from dedicated psychology websites such as Simply Psychology or Verywell Mind.
- When searching for information online, for optimal results you can use Boolean operators to tell databases exactly how you want your search terms combined.
- Many of the generalisations made about human thoughts, feelings and behaviours have been drawn from investigations using WEIRD samples. The problem with this is that not all populations think, feel or behave in the same way under the same conditions.
- Peer review is an extremely important part of the scientific process because it allows other scientists to offer feedback and test the research to verify its accuracy.
- In a research reporting, describing the level of uncertainty strengthens the research results and provides guidance for the focus of future research projects.
- Many primary and secondary source articles and alternative platforms describe models or theories to organise and explain psychological concepts.
- The more primary research that the model fits, the more likely it is to become a theory used by psychologists to explain thoughts, feelings and behaviours.

KEY CONCEPTS 6.1c

- A logbook records information from your literature review that you may or may not use in your final presentation. It keeps track of each source of information.
- When researching journal articles, read the abstract first.
- Summarise the main ideas into your logbook for each source of information, keeping track of data, conclusions, limitations, quotes and page numbers.
- A synthesis matrix can help you to find ideas that are shared by more than one source.

KEY CONCEPTS 6.1d

- When communicating information, you must decide on your purpose and your audience because these will determine the best format for your presentation.
- Your purpose will generally be to present to inform or critique, using a neutral voice.
- If you are responding to an ethical dilemma, your presentation may be designed to persuade or argue.
- Science communication can take traditional forms, such as a review article or news article; or, increasingly, it may take newer forms, such as YouTube, Twitter or TikTok.
- Most traditional formats cite references and structure information using introduction, method, results, discussion and conclusion sections.
- Less traditional formats often condense or leave out one or more sections to highlight key ideas. They should still include links so that the audience can check for validity, reliability and integrity.
- The introduction and/or conclusion must provide contextual clarity, explaining the importance and implications of the research.
- Writing concisely and with coherence is important and takes time and practice.
- Paraphrasing means putting a passage from source material into your own words. Summarising means putting the main idea(s) into your own words, using only the main point(s).
- If you want to keep exact statements from the text, you need to quote. When quoting, match the source word for word, enclosing the text in quotation marks.
- Avoid plagiarism by citing your sources. If you use information from a source without citing it, you are implying that you did the original research behind the information on your own.
- Primary and secondary sources often condense data and information into some form of visual representation, such as a table, graph, chart, photographic image or a diagram, to easily communicate ideas.
- Data representations can be misleading.
- You should follow conventions when using data or visual representations.
- The APA 7th edition referencing system is used in psychology.

KEY CONCEPTS 6.2

- Ethical issues often arise when new treatments and technologies are developed.
- Investigating the ethical issues of new psychological research or technology requires identifying the issues, analysing the evidence, analysing the legal, economic and social consequences, and evaluating the ethical considerations for and responsibilities of all stakeholders.
- Ethical concepts are used to unpack the ethical issues involved in research. These concepts include beneficence, integrity, justice, non-maleficence and respect.
- In order for findings to be generalised to a broad population, it is important to consider the participant representation within the study.

9780170465052

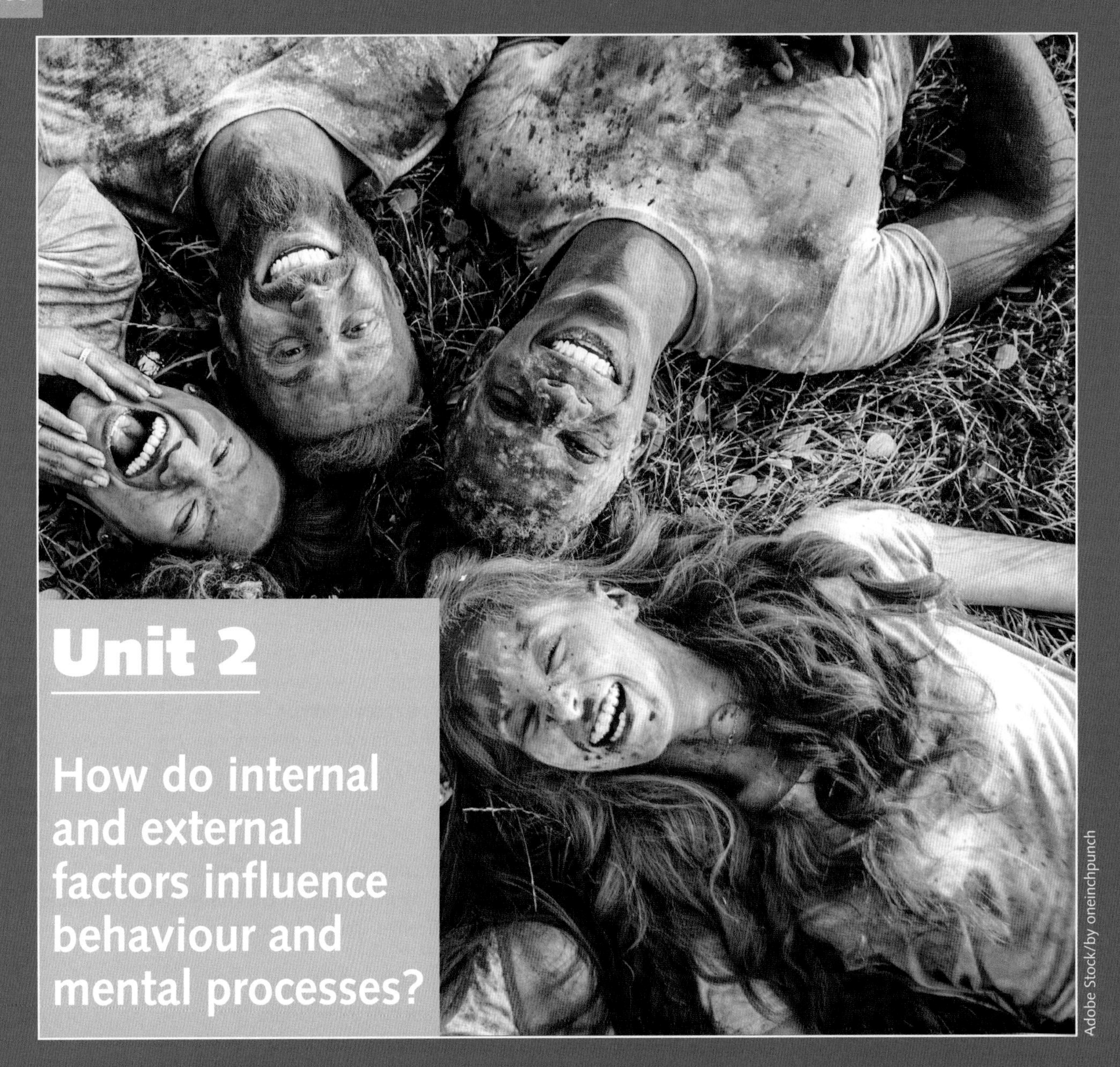
Adobe Stock/by oneinchpunch

Unit 2

How do internal and external factors influence behaviour and mental processes?

Area of study 1: How are people influenced to behave in particular ways?

Area of study 2: What influences a person's perception of the world?

Area of study 3: How do scientific investigations develop understanding of influences on perception and behaviour?

Social cognition

7

Key knowledge

- the role of person perception, attributions, attitudes and stereotypes in interpreting, analysing, remembering, and using information about the social world, including decision-making and interpersonal interactions
- the avoidance of cognitive dissonance using cognitive biases
- the positive and negative influences of heuristics as mechanisms for decision-making and problem-solving
- the influence of prejudice and discrimination within society on a person's and/or group's mental wellbeing and ways to reduce it

Key science skills

Analyse, evaluate and communicate scientific ideas

- discuss relevant psychological information, ideas, concepts, theories and models and the connections between them
- analyse and explain how models and theories are used to organise and understand observed phenomena and concepts related to psychology, identifying limitations of selected models/theories
- critically evaluate and interpret a range of scientific and media texts (including journal articles, mass media communications and opinions in the public domain), processes, claims and conclusions related to psychology by considering the quality of available evidence
- use clear, coherent and concise expression to communicate to specific audiences and for specific purposes in appropriate scientific genres, including scientific reports and posters

Source: VCE Psychology Study Design (2023–2027), pp. 29 & 13

7 Social cognition

So far we've focused on the brain, its physiological processes and how our brains develop as we age. However, very few of us live in an isolated world. Most of us live social lives and we interact with people around us. Our brains work hard to try to interpret the physical and emotional responses of family members, friends and strangers while also considering how to react to these behaviours.

p. 273

7.1 Making sense of the social world

Communication is everything. However, not all communication is verbal. When we interact with someone, we are frequently looking for facial expressions, physical cues and other notable features they may have to figure out what they might *really* be thinking and what they might be like as a person. We are also – whether we admit to it or not – already making judgements about that person, often before they have even had a chance to introduce themselves!

iStock.com/wildpixel

p. 280

7.2 Cognitive dissonance and cognitive bias

Life would be much simpler if everything we did was consistent with our attitudes and beliefs. However, this is rarely the case. Many of us value our health, for example, but continue to eat processed, sugary foods that we know are not good for us. None of us are perfect human beings and often we try to simplify information to make sense of the world, resulting in errors in our decision-making.

Adobe Stock/Kekyalyaynen

p. 286

7.3
Heuristics

No matter what we are doing with our lives, whether we are studying, working, travelling or socialising, we are constantly making decisions – what to do, what to say and how to act. As you learned in Chapter 4, decision-making requires different regions of the brain to work together, and this carries a high cognitive load. To simplify this process, we sometimes make snap judgements and quick decisions based on limited information.

iStock.com/Daisy-Daisy

p. 289

7.4
The effects of prejudice and discrimination

Our brains are always looking for patterns to try to lessen the cognitive load by using shortcuts. This can lead us to hold certain beliefs or attitudes toward another person based on a pattern we've seen before. Unfortunately, this can result in us holding negative preconceived notions about someone based on their membership of a group and treating that person unfairly.

iStock.com/Rowan Jordan

Human beings are social creatures, and the way we think, feel and act is often influenced by our interactions with others. While we may not like to admit it, our brains like to take shortcuts and we need to work hard to avoid making snap judgements and having preconceived notions about people. As the saying goes, don't judge a book by its cover!

Slideshow
Chapter 7 slideshow

Flashcards
Chapter 7 flashcards

Test
Chapter 7 pre-test

Assessment
- Pre-test
- End-of-chapter exam

Revision
- Chapter map
- Key term flashcards
- Key concept summary
- Slideshow

Investigation
- Investigation: Classification and identification of cognitive biases
- Data calculator
- Logbook template: Scientific

Worksheet
- Experiences of racism in Victorian racial and ethnic minorities

To access these resources, visit **cengage.com.au/nelsonmindtap**

Nelson MindTap

Know your key terms

Actor–observer bias	Halo effect
Affect heuristic	Heuristic
Anchoring bias	Hindsight bias
Attentional bias	Internal attributions
Attitude	Misinformation effect
Attribution	Optimism bias
Availability heuristic	Person perception
Cognitive bias	Physical cue
Cognitive dissonance	Prejudice
Confirmation bias	Representative heuristic
Discrimination	Saliency detection
Dunning–Kruger effect	Self-serving bias
External attributions	Social categorisation
False-consensus bias	Social cognition
Functional fixedness	Stereotype

In 1961, African-American student James Meredith applied to study at the University of Mississippi. While this would not raise eyebrows today, at that time the University of Mississippi only accepted white students.

Meredith's application was rejected twice; however, following a successful court battle, he was allowed to enrol. The Mississippi State Government did everything in its powers to prevent Meredith from attending the university, including the Governor appointing himself to a position within the university to personally block Meredith's enrolment.

Following confirmation that Meredith would start attending classes, violent riots broke out when large groups of angry white protesters confronted the police at the university. Soldiers were called in to end the violence. In the violence that followed, two people died and over 200 were injured.

Throughout his time at university, Meredith reported that he was rejected by most of his fellow students and even teaching staff, and needed 24-hour protection from police (Figure 7.1).

Trikosko, M. S., photographer. (1962) Integration, Miss. Univ - James Meredith. , 1962. October 4. [Photograph] Retrieved from the Library of Congress, https://www.loc.gov/item/2016646448/.

Figure 7.1 James Meredith being escorted to class at the University of Mississippi

What can explain the attitudes of those who attempted to prevent an African-American man from attending university? Why do people make judgements about individuals before getting to know them? This chapter will explore the concepts of prejudice and discrimination and their effect on a person's mental health and wellbeing. Also, this chapter will outline several consistent errors that occur in our decision-making.

7.1 Making sense of the social world

Social psychology is a branch of psychology interested in the way individuals' thoughts, feelings and behaviours are influenced by others. This chapter focuses on **social cognition**, which relates to the way that we behave in social settings and also how we interpret the behaviours of others. Social cognition involves:

1 the detection of facial expressions and emotional responses (social perception)
2 understanding other people's cognitive or emotional states (social understanding), and
3 carrying out behaviours that take into consideration the goals and needs of ourselves and others (social decision-making) (Arioli et al., 2018).

A person's social cognition and behaviour influence the way they view themselves and the way they relate to others.

Person perception

When we meet an individual for the first time we make quick judgements, or form an impression, about that person based on their overt (observable) characteristics (Figure 7.2). This is known as person perception. **Person perception** is the processes by which people think about, appraise and evaluate other people. For example, if a person is perceived as friendly, we are more likely to help them when they are in need. Conversely, if we perceive someone as aggressive or dominant, we are less likely to choose them as a group member for a collaborative project (Fang et al., 2018).

Person perception leads us to make judgements about people based on limited information; therefore, it is not always accurate. When making these judgements we use physical cues, saliency detection and social categorisation.

Physical cues

Physical cues such as physical appearance, facial expressions and overall manner serve as signals that allow us to draw conclusions about a person. Consider the following scenario. It's peak hour and you're walking through the city. There are hundreds of people walking the same streets as you, all of whom are providing physical cues. You notice a man in a suit carrying a briefcase walking briskly through the crowd. He is weaving past people with a serious look on his face. Behind him, you notice a woman dressed casually who is walking much slower. She has her hands in her pockets and appears carefree. What judgements might you make about these individuals based on the physical cues they present?

iStock.com/PeskyMonkey

Figure 7.2 What assumptions can we make about a man in a suit, based on his appearance?

We quickly arrive at conclusions regarding people based on the physical cues that they present. For example, there are several physical cues that indicate 'warmth' in a person. To demonstrate their interest and engagement, a person might smile widely, lean forward, nod, and move their body to face the person they are interacting with. On the other hand, 'coldness' can be perceived from a person's tense posture, if they're leaning back, or if they move their body away from the person they are interacting with (Cuddy et al., 2008).

7.1.1 PERSON PERCEPTION

Saliency detection

When we look at the world around us, we are bombarded by a range of visual information that competes for our attention. Salient features, also known as attention-getting features, are those physical features that are prominent or distinctive.

Shutterstock.com/Margy Crane

Figure 7.3 The person's blue hair is considered a salient feature, a distinct characteristic that stands out from the crowded visual scene.

Saliency detection refers to the tendency to notice physical features that are unique, novel or stand out from the norm. People with salient features capture our attention and we tend to spend longer looking at people we perceive to be salient (Figure 7.3). McArthur and Ginsberg (1981) found that when participants watched videos of people with salient and non-salient features, they fixated longer on the salient individuals (in their experiment, a person with a striped shirt, a person with red hair and a person with a leg brace) than those who didn't possess distinctive characteristics.

Social categorisation

Imagine you are walking home late at night and you notice a young man in the distance. Worried about your safety, you cross the road to avoid interacting with him. Now imagine the same scenario, but the young man is replaced by an elderly woman. Presumably, you would be less likely to cross the road, as the characteristics for this person (older and female) are less threatening than those of the first individual (young and male). This process of judging people based on their appearance is known as **social categorisation**.

Social categorisation is the process by which we group individuals based upon their perceived social category. The most common social categories are sex, race and age. We also make judgements about people based on other characteristics. These include occupation and perceived social status, as well as perceived sexual orientation (Stolier & Freeman, 2016). Social categorisation helps individuals simplify their complex social world by making assumptions about an individual's supposed beliefs and behaviours based on their social category (Liberman et al., 2017). When we apply social categories, we quickly identify those who fit the same categories as us. Consequently, we also identify those who belong to different categories. As a result, we may stereotype individuals and judge them incorrectly based purely on their social category. The negative result of this is that stereotyping can lead to prejudice, and prejudice can lead to discrimination. These concepts are discussed later in this chapter.

Table 7.1 gives an overview of the components of person perception.

Table 7.1 The components that contribute to person perception

Component	Explanation
Physical cues	A person's physical appearance, facial expressions and overall demeanour act as signals that allow us to draw conclusions about a person.
Saliency detection	We tend to notice physical features that are unique, novel or stand out from the crowd. These features capture our attention, and we tend to spend longer looking at salient features.
Social categorisation	The process by which we group individuals based upon the perceived social category they belong to; the most overt social categories are sex, race and age.

ACTIVITY 7.1 PERSON PERCEPTION

Try it

1 Figure 7.4 shows a group of people. Copy and complete the table below to record assumptions that might be made about the individuals in the group.

Figure 7.4

Components of person perception	Examples from the group photo
Physical cues	
Salient features	
Social categorisation	

Apply it

2 Explain how physical cues, saliency detection and social categorisation contribute to the process of person perception. In your answer, ensure you briefly explain each of the relevant psychological terms.

Exam ready

3 One benefit of person perception is ________, while one limitation of person perception is ________.

A it allows us to make quick judgements about others without much mental effort; it can lead to inaccurate judgements about stereotyping

B it allows us to stereotype others; it requires a great deal of mental effort

C it allows us to make quick judgements about others without much mental effort; it leads to saliency detection

D it leads to saliency detection; it allows us to make quick judgements about others without much mental effort

Attributions

It is not possible to know everything about another person. In fact, we often form impressions about others' behaviour based on very limited evidence. These impressions are influenced by the cognitive process of attribution. **Attribution** is the process of attaching meaning to behaviour by looking for a cause or causes to explain the behaviour. As we observe others, we tend to make inferences about their actions. Why did Mr Taylor yell at the class? Why did Oscar walk home with Sophie and not Meg? Why did Kim leave the party early? When considering the behaviours of others, we attempt to attribute (explain) their behaviour in relation to various causes. For example, if someone is happy, is it because that is the way they typically are (part of their personality) or because something positive happened to cause the good mood, or elements of both?

Austro-American psychologist Fritz Heider (1958) was the first to describe how people make attributions. Heider believed that people make two types of attributions: internal and external. **Internal attributions** (also known as dispositional attributions) refer to factors within the person that shape their behaviour. These include personality characteristics, motivation, ability and effort. **External attributions** (also known as situational attributions) refer to environmental factors that are external to the individual, such as their location or the people around them (Maio & Augoustinos, 2005). For example, if a Year 10 student accidentally dropped and smashed their tablet, we might consider the cause of their behaviour in terms of internal factors (such as their clumsiness or inability to take care of their property). We may also look at external attributions (such as whether the library was overcrowded or someone was fooling around with the tablet) when evaluating their behaviour. Thus, the causes of the behaviour (dropping and smashing the tablet) might be attributed to both internal and external causes (Figure 7.5).

7.1.2 ATTRIBUTIONS

Heider suggested we often make errors when we attempt to interpret the behaviour of others. We tend to overestimate the role of internal factors and underestimate the role of external factors. This was later called the **fundamental attribution error**.

The error lies in the belief that internal factors are the main cause of their behaviour. Also, we see the behaviour as a reflection of who the person is and ignore the influence of external factors. The fundamental attribution error occurs because observers are not aware of external factors and historical considerations (events that have occurred in the past) that may influence and guide another's behaviour. Often, when we view a person's behaviour, much of the time we do so without knowing what occurred before the behaviour and without knowing many of the external factors. The historical considerations and the situation are often difficult to judge, so we rely heavily on internal attributions.

Dreamstime/Wavebreakmedia Ltd

Figure 7.5 If this person dropped their tablet, is their behaviour caused by internal or external factors?

KEY CONCEPTS 7.1a

- Social psychology is a field of psychology that studies the way in which our behaviour is influenced by others.
- Social cognition refers to the way in which we judge our own behaviours and the behaviours of others in a social setting.
- Person perception refers to the process of making judgements about individuals based on their apparent characteristics. While person perception allows us to make quick judgements about individuals, it can be inaccurate because we are using a small amount of information to make broad judgements.
- Attributions are the process of attaching meaning to our own and other people's behaviours. Attributions can be influenced by internal factors (personality, motivation, mood, ability and effort) and external factors such as the environment (i.e. school, work etc.).
- When explaining a person's behaviour, we tend to overestimate the role of internal factors and downplay the role of external factors. This is known as the fundamental attribution error.

Concept questions 7.1a

Remembering

1 What is social cognition? r

2 What is person perception? r

Understanding

3 Explain how internal attributions differ from external attributions. e

4 Use an example to explain the fundamental attribution error. c

5 Outline one benefit and one limitation of social categorisation. e

6 Our initial judgements of individuals are not always accurate. With reference to person perception, explain why this is the case. e

Applying

7 Explain how physical cues and saliency detection combine to shape person perception. e

HOT Challenge

8 Alex borrowed his brother's jacket to wear to a party. While at the party, a smoker bumped into Alex and put a cigarette burn in the jacket. The next morning Alex showed his brother the cigarette burn and attempted to explain what happened. Alex's brother didn't believe him, saying that he is inconsiderate and that he can't borrow his clothes anymore. c

Explain how the fundamental attribution error accounts for Alex's brother's decision to not let Alex borrow his clothes anymore.

Attitude formation

Attitudes are learned ideas we hold about ourselves, others, objects and experiences. Attitudes are not innate. They are learned through exposure to the environment and can cause a person to respond in a positive or negative way. If a person positively evaluates someone or something, they will have a positive attitude towards that person or object. For example, a person who has a gun licence and goes hunting will have a positive attitude towards guns and hunting. People who oppose guns and hunting and who do not hunt will have a negative attitude towards guns and hunting. If someone has never thought much about guns and hunting, they may have an indifferent attitude.

Many factors contribute to how we learn an attitude. These include experience, personal influence and exposure to the media. Attitudes can come from direct contact (personal experience) with the object of the attitude; for example, you might support special bicycle lanes on roads because you were once knocked off your bike by a car. Attitudes are also learned indirectly through interaction with others. For instance, if three of your friends all volunteer for an environmental organisation, and you talk to them about their beliefs, you may develop more environmentally friendly attitudes, too.

The tri-component model of attitudes

Attitudes influence the way people relate to each other. The tri-component model of attitudes proposes that there are three elements that contribute to attitude formation. These components are known as the ABC of attitudes. They are the:

» *affective component:* feelings or *emotions* towards the object or person
» *behavioural component:* the *actions* towards various people, objects or institutions
» *cognitive component:* the *belief* about an object or person.

The tri-component model is applicable to many situations; for example, examining factors that influence customer behaviour, determining why you choose a travel destination and why you like certain music. Consider, for example, your attitude towards wearing a school uniform. Your emotional response (the affective component) could be that school uniforms are attractive and useful, or ugly and uncomfortable. Therefore, you will either agree or disagree with wearing a uniform and may or may not wear it properly (the behavioural component). Also, you may think about whether wearing a uniform should be compulsory or not (the cognitive component). For other examples, see Figure 7.6 and Table 7.2.

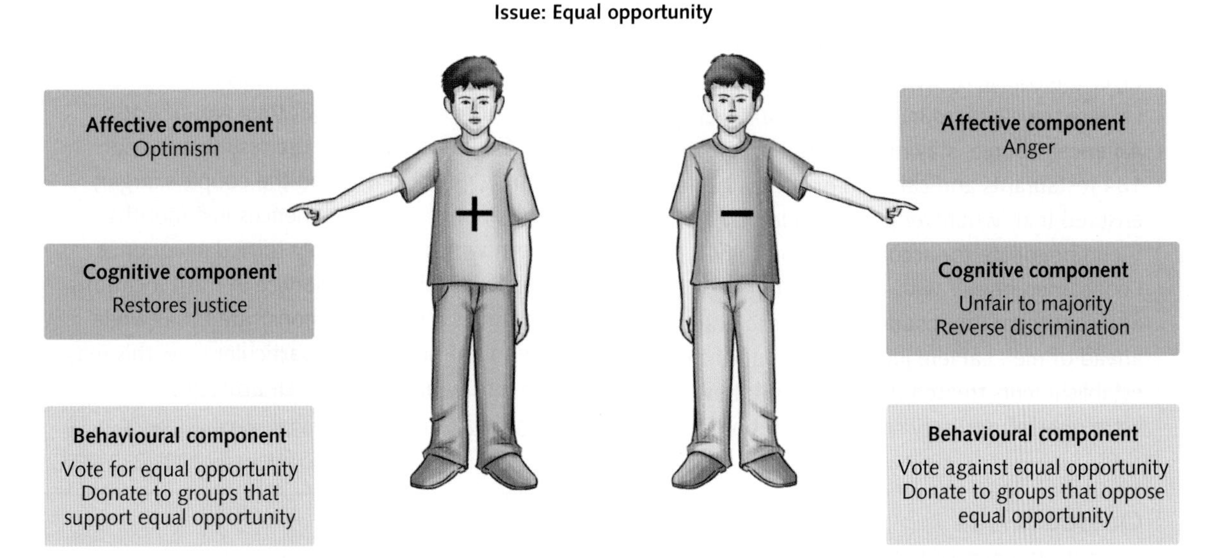

Figure 7.6 Positive and negative attitudes towards equal opportunity in the workplace, showing the parts of the tri-component model of attitudes

Table 7.2 Examples of the tri-component model of attitudes

Example	Affective component	Behavioural component	Cognitive component
Ella really enjoys going to the beach. Ella thinks that being at the beach brings her closer to nature. As such, Ella spends every summer holidays at her parents' beach house.	Ella really enjoys the beach.	Ella spends every summer holidays at her parents' beach house.	Ella believes that being at the beach brings her closer to nature.
Louis goes to the gym three times a week. Louis enjoys exercising at the gym because it's a healthy way to reduce his levels of stress.	Louis enjoys exercising at the gym.	Louis goes to the gym three times a week.	Going to the gym is a healthy way to reduce stress.
Emma is a 'morning person' who likes to wake up early. Emma believes she can make the most of the day if she wakes up early. Emma sets an alarm to get up at 5:45 every morning – even on weekends.	Emma is a 'morning person' who likes to wake up early.	Emma sets an alarm for 5:45 each morning.	Emma believes she can make the most of the day if she gets up early.

7.1.4 COMPONENTS OF ATTITUDE

Limitations of the tri-component model of attitudes

While the tri-component model provides a useful framework to unpack an individual's attitudes, it has limitations. These include:

» It does not indicate the strength of an attitude, and understanding the strength of an attitude is useful. Strong attitudes are generally seen to be firmly held and resistant to change, therefore they impact behaviour (Olson & Maio, 2003).

» Inconsistencies often exist between what a person thinks and feels and their behaviour (cognitive dissonance). For example, an individual might hold negative views towards a particular ethnic or religious group, but not express them in their daily interactions with individuals from that group. For an example of this, see Analysing research 7.1.

ANALYSING RESEARCH 7.1

The inconsistency between behaviours and attitudes

The inconsistency in the attitudes we hold and behaviours we demonstrate was clearly shown in an influential study by Richard LaPiere (1934). Over the course of two years, starting in 1930, LaPiere travelled around America with a Chinese-American couple, staying at 66 hotels and visiting 184 restaurants and cafes. In his study, LaPiere ensured that 'wherever possible, I let my Chinese friend negotiate for accommodation (while I concerned myself with the car or luggage) and sent them [the couple] into a restaurant ahead of me'. LaPiere judged that half of the establishments treated them well and they were only refused service at one location. In the subsequent months, LaPiere sent questionnaires to the venues that he and the couple had visited. The survey included the question: 'Will you accept members of the Chinese race as guests in your establishment?' One hundred and twenty-eight of the restaurants and hotels responded, with 92 per cent saying that they *would not* accept Chinese visitors. These results were not in keeping with LaPiere and the couple's largely positive and pleasant experiences just months earlier, in which he was only turned away once and reported reasonable service in most locations.

LaPiere's research demonstrated that while people might behave in a particular way, this may not be a true indicator of their attitudes.

Source: LaPiere, R.T. (1934). Attitudes vs. Actions. *Social Forces*, *13*(2), 230–237

Weblink
Attitudes vs actions

Questions

1 State the aim of LaPiere's research.
2 Briefly outline the procedure of the study.
3 Explain the findings and conclusion of LaPiere's study.

HOT Challenge

4 Identify and explain one qualitative measurement of attitudes and one quantitative measurement of attitudes present in LaPiere's study.

5 In his study, LaPiere ensured that 'wherever possible, I let my Chinese friend negotiate for accommodation (while I concerned myself with the car or luggage) and sent them [the couple] into a restaurant ahead of me'. How might this approach have encouraged a more 'natural' response from the hotel and restaurant staff?

The effects of stereotypes

Stereotypes are a major contributor to attitude formation. **Stereotypes** are generalised views about the personal attributes or characteristics of a group of people. Stereotyping places people in categories based on their shared characteristics and their membership of a particular group. Common stereotypes include those based on gender, age, ethnicity and occupation; for example, 'People over the age of 65 are old and senile'. Stereotypes tend to simplify people by grouping them into 'us' and 'them' categories. People who share common traits with you are seen as part of the 'in-group' and those who do not share these traits (or share traits that you perceive to be negative) are in the 'out-group'.

Even though stereotypes sometimes include positive traits, they can still lead to prejudice. This is because stereotypes are mainly an expression of a negative attitude towards a group, and are used to maintain control over other people. When people have been stereotyped, the easiest thing for them to do is to abide by others' expectations – even if the expectations are demeaning. That is why no one likes to be stereotyped. Being forced into a small, distorted social 'box' by stereotyping is limiting and insulting. Stereotypes rob people of their individuality (Fiske, 1993).

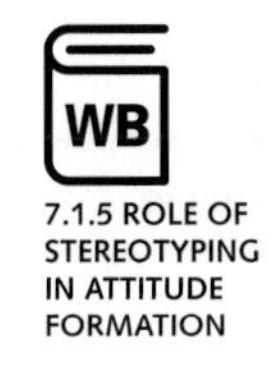

7.1.5 ROLE OF STEREOTYPING IN ATTITUDE FORMATION

KEY CONCEPTS 7.1b

- Attitudes are the learned perspectives we hold about individuals, issues, events and so on.
- The tri-component model of attitudes proposes that attitudes contain an affective (emotional) component, a behavioural (actions) component and a cognitive (reasoning) component.
- Stereotyping involves placing people in categories because of perceived similarities.
- Stereotyping is a quick but often flawed way of making judgements about individuals because it does not take into consideration factors beyond the stereotype.

Concept questions 7.1b

Remembering

1 Explain the affective, behavioural and cognitive components of attitudes. e

2 What are stereotypes and why does stereotyping occur? r

Understanding

3 Outline two limitations of the tri-component model of attitudes. e

Applying

4 Identify the affective, behavioural and cognitive elements of the tri-component model of attitudes present in the scenarios below. c

a Sophie doesn't like going to the city on the weekends. She thinks that too many people go out to get drunk and end up in fights. As a consequence, Sophie avoids going to the city on the weekends.

b Ben prefers cooking dinner as opposed to going out to restaurants. Ben believes that restaurants are over-priced. When his friends suggested going out to dinner for his birthday, Ben insisted on hosting a dinner party instead.

c As he got older, Oscar had to choose between playing cricket and playing basketball. Oscar decided to give up cricket so he could focus on basketball. Oscar prefers playing basketball because the games are shorter and more of his friends play basketball than play cricket.

HOT Challenge

5 Use an example to explain the role of learning in the formation of an attitude. i

7.2 Cognitive dissonance and cognitive bias

Cognitive dissonance

7.2.1 COGNITIVE DISSONANCE

When a person acts in a way that is inconsistent with their attitudes or perception of themselves, the contradiction will generally make them uncomfortable. This discomfort can motivate them to make their thoughts or attitudes align with their actions (Petty et al., 1997). This inconsistency is known as **cognitive dissonance**.

For example, smokers are told on every pack that cigarettes endanger their lives, but they light up and smoke anyway (Figure 7.7). The overwhelming medical advice is that every cigarette is doing you damage (cognitive component) and many people who smoke don't feel happy about doing it (affective component). To reduce their cognitive dissonance, a person who smokes could quit smoking, which would align with their understanding that smoking is dangerous. But it may be easier to convince themselves that smoking is not really so dangerous. To do this, many smokers seek examples of heavy smokers who have lived long lives. They spend their time with other smokers, and they avoid information about the link between smoking and cancer. Additionally, a smoker may highlight the perceived positives of smoking; for example, that it relieves their tension, leading to a calm state (Festinger, 1957). By applying these justifications for their behaviour, the individual eases their discomfort via the use of cognitive biases (flawed ways of thinking).

iStock.com/Nustock

Figure 7.7 Even with images like this one, people continue to smoke. Some people ease the discomfort that this information may produce by changing their cognitive and affective components to be in line with their behaviour.

Cognitive bias

7.2.2 COGNITIVE BIAS

Despite our best attempts to make rational, logical judgements, flaws can emerge that influence our choices. These flaws are called cognitive biases. **Cognitive biases** are systematic errors (consistent mistakes) that occur in our decision-making. Cognitive biases occur when we attempt to simplify the information we are processing, especially if we have to interpret it quickly.

A number of cognitive biases are outlined in Table 7.3.

Table 7.3 Cognitive biases

Cognitive bias	Key components	Example
Actor–observer bias	The tendency to attribute your own behaviours to external (situational/environmental) factors while attributing other peoples' behaviours on internal (dispositional) factors. Reduces cognitive dissonance by attributing our behaviours to environmental factors.	You throw litter on the ground because there are no nearby bins. Other people throw litter on the ground because they are inconsiderate.
Anchoring bias	The tendency to rely on the first piece of information offered (the 'anchor') when making decisions. Judgements are shaped by the anchor.	A car is advertised for $9900 (the anchor). You offer $8000 for it. The car dealer counter-offers with $9500. You both agree on a $9000 sales price. You feel this is a good deal compared to the original price quoted.

9780170465052

Cognitive bias	Key components	Example
Attentional bias	The tendency to focus on particular stimuli while overlooking or ignoring other relevant pieces of information. Individuals prioritise information that captures their attention. Can result in failure to attend to all pieces of information present.	The attention systems of smokers are more responsive to smoking-related stimuli; i.e. lighters, cigarettes, ash-trays.
Confirmation bias	Occurs when a person focuses on and favours information that supports their perspectives while ignoring contradictory information or views. Prejudicial views about certain groups in society can then be reinforced by negative interactions with a member of that group or by unfavourable reporting of that group in the media.	Evidence suggests that people will rate the quality of health information that supports their beliefs more highly than information that contradicts their beliefs (Meppelink et al., 2019).
False-consensus bias	The tendency for people to assume that their attitudes, beliefs and behaviours are relatively common and are more widely shared by others than is the case. By assuming that our behaviours or beliefs are common, cognitive dissonance can be reduced.	People might justify their risk-taking behaviour (e.g. drug use or binge-drinking) by stating that 'everyone does it'.
Functional fixedness	The tendency to believe an object or item can only be used in a particular way (its intended use). Functional fixedness limits the way we approach problems, leading us to try to solve problems or arrive at a decision using tried and true methods that may not be appropriate or applicable for the problem at hand.	Problem-solving tasks that require people to 'think outside the box'; e.g. the candle problem or nine-dot problem.
Halo effect	Occurs when the positive evaluation we hold about one quality of a person influences our beliefs and expectations regarding other qualities of that person. The halo effect is particularly evident when we perceive someone to be physically attractive. Due to their appearance, we might also assume other positive traits about the individual, including higher levels of intelligence or competency.	One study found that lawyers who were considered attractive earned 10–12% more than their less attractive peers (Engemann & Owyang, 2005).
Hindsight bias	Occurs when a person believes an outcome to have been more predictable than they did before it occurred. There are three elements to hindsight bias: 1 foreseeability ('I knew it would happen') 2 inevitability ('it had to happen') 3 memory distortion ('I said it would happen'). One explanation for the hindsight bias is that people have a desire to see the world as orderly and predictable.	A person tipped one team to win the match, but after the game they said that they knew the team they selected would lose.
Misinformation effect	Occurs when a person demonstrates poor recall of events following exposure to additional information after the event took place. Said to occur because the original memory is combined with the new information as part of an updated memory; the memory is incomplete and vulnerable to additional information being added, and recently learned information is easier to retrieve.	According to the Innocence Project (2022), mistaken eye-witness testimony contributed to nearly 70% of the 375 wrongful convictions that were later overturned due to DNA evidence in the United States.
Optimism bias	Our inclination to overestimate our likelihood of experiencing positive events and underestimate our likelihood of experiencing negative events. The optimism bias reduces our cognitive dissonance by drawing attention to examples that challenge the dangers of our behaviours.	Researchers found that individuals with high levels of optimism bias were less likely to follow social distancing rules and lockdown protocols during the COVID-19 pandemic (Fragkaki et al., 2021).
Self-serving bias	A tendency for a person to attribute a positive outcome to internal (dispositional) factors, yet attribute negative outcomes to external (situational) factors. We perceive positive outcomes to be under our control but interpret undesirable outcomes to be beyond our control. In this way, the self-serving bias helps to protect our self-esteem when things don't pan out as we might hope.	A footballer who misses a crucial shot at goal may blame the wind for their inaccuracy and not their own skill deficit. If the same footballer kicked the goal, they might attribute their success to their high skill level.
Dunning–Kruger effect	Occurs when people with low ability at a task overestimate their own skill set, and that people with high ability at a task underestimate their own skill set. When learning new information, we feel as though we have a strong grasp of the content. Once we learn more about the topic, we start to realise how little we know.	You've studied hard throughout the year and have achieved excellent results in English. Following the English exam, you doubt you did well. A friend says that even though they didn't read any of the texts throughout the year, they found the exam easy. Consequently, when the results are released, you find you did well and your friend did poorly.

INVESTIGATION 7.1 CLASSIFICATION AND IDENTIFICATION OF COGNITIVE BIASES

Scientific investigation methodology

Classification and identification

Data calculator

Aim

To identify and classify different types of cognitive bias

Logbook template: Classification and identification

Introduction

Imagine you are the coach of a junior basketball team. You have 12 children who want to join your team. You select the seven tallest children, even though you know nothing about their basketball abilities. This is an example of cognitive bias – just because the children are tall, you think they must be good at playing basketball. This may not be the case.

Cognitive bias is an automatic tendency or preference for processing or interpreting information in a particular way. This can produce systematic errors in thinking when making judgements or decisions. Cognitive biases occur when we try to simplify the information we are processing, especially if we have to interpret it quickly.

Your task

Follow the steps below.

1 Read the nine examples of cognitive bias in the table below.

2 Decide, from the list below, which type of cognitive bias they represent. Not all are represented.

- Actor–observer bias
- Anchoring bias
- Attentional bias
- Confirmation bias
- False-consensus bias
- Functional fixedness
- Halo effect
- Hindsight bias
- Misinformation effect
- Optimism bias
- Self-serving bias
- The Dunning–Kruger effect

3 Provide more information for each type of cognitive bias you have identified.

  a Describe its key components.

  b Find an image of an everyday example portraying the bias.

  c Give an explanation of its influence on behaviour.

4 Prepare your findings as a visual presentation.

Jane tells her friends that Rottweiler dogs are the smartest dogs. Jane does not know anything about dogs apart from owning a Rottweiler.	A policeman asked an eyewitness to a crime 'Did you see the man in the blue shirt break into the clothes shop?'	Luke is organising his 18th birthday party. He tells his friends to come along because it is going to be a great time.
James won the 400 m sprint at the school athletics carnival because he put in hours of training. The next day he failed a maths test because the school bus was late that morning and he could not have a final read of his notes.	Anand and May spent their June holiday in a small mountain town. The weather was wet and cold for the whole of their stay. Anand told May at the end of the holiday that he knew that would happen and that his holiday had been ruined. He said that they should not have come.	Joe's first car was a Holden car and for 30 years he has only bought Holden cars.
Everybody knows that Essendon is the best football club.	Jay does not like Ian being in his Chemistry class, even though everyone else in the class thinks Ian is a really nice person. Jay once saw Ian push in front of someone in the queue for the canteen.	Georgia is shopping for a new pair of shoes. She goes into the first shoe shop that she comes across and sees a 50% off sale. She buys her shoes there because, she tells herself, she is getting a great deal.

ANALYSING RESEARCH 7.2

The false-consensus bias

Consider the following question: do most people share similar attitudes and values to you? While we might expect some variation based on factors such as age, culture or religion, many of us would expect that other people are similar to us. By arriving at this conclusion, we project our beliefs onto others. This is known as the false-consensus bias.

The false-consensus bias is the tendency for people to assume that their attitudes, beliefs and behaviours are relatively common and more widely shared by others than is actually the case. In other words, it's the tendency to overestimate the extent to which people agree with you or behave in the same way as you.

The real-world application of the false-consensus bias was investigated by Monin and Norton (2003). When a sudden tropical storm hit the east coast of America, students at Princeton University were asked to limit their water usage to prevent a water shortage. For example, the students were told that they couldn't shower for 3 days. The tropical storm provided researchers from Princeton and a neighbouring university with the perfect opportunity to investigate the false-consensus bias.

Four hundred and fifteen students were interviewed during the shower ban and were asked (1) if they had showered or not, and (2) to estimate the percentage of students who had showered during the ban. The researchers hypothesised that those who showered would predict that higher numbers of students showered, as opposed to those who didn't shower during the ban. The results showed that students who showered during the ban estimated that 72 per cent of their peers had also showered during this time; whereas those who didn't shower during the ban estimated that only 44 per cent of their peers had showered during this time. The researchers concluded that this difference between the groups could be attributed to the false-consensus bias; that the estimations of who had showered during the ban was influenced by the individual's own behaviour.

Questions

1 Identify the variables of interest in the study and explain how the researchers used them to investigate their research question.
2 A questionnaire was used to gather data in this study. Outline one benefit and one limitation of using questionnaires.
3 Explain the sampling technique that was used in this study.
4 Explain how the false-consensus bias was demonstrated in the findings of the study.

HOT Challenge

5 A controlled experiment is characterised by the manipulation of variables by the researchers, the use of control groups, and, where possible, random sampling and random allocation. Based on these characteristics, would the above study be considered an experiment?

ACTIVITY 7.2 THE NINE-DOT PROBLEM

Try it

1 The nine-dot problem has been used as a means of challenging functional fixedness. Connect all the dots in Figure 7.8 by drawing four straight, continuous lines that pass through each of the nine dots, while never lifting the pen from the page.

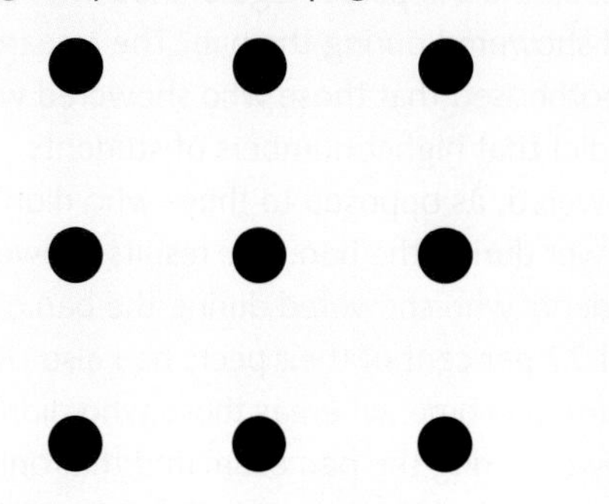

Figure 7.8

Apply it

2 Functional fixedness is one of many cognitive biases that we experience.
   a What is functional fixedness?
   b Did you solve the nine-dot problem?
   c If you were able to solve the nine-dot problem, how did you avoid functional fixedness?

Exam ready

4 Dan needed to change the batteries in the TV remote, but this required a screw-driver. Dan checked the house for a screw-driver but couldn't find one. Dan's wife, Jen, suggested that he use a knife to turn the screws, but Dan said that wouldn't work. That night, Jen changed the batteries using a knife. When Dan saw what Jen was doing, he said, 'I thought a knife would work'.

In this scenario, Dan demonstrated

A functional fixedness and optimism bias.
B functional fixedness and the halo effect.
C functional fixedness and the Dunning–Kruger effect.
D functional fixedness and hindsight bias.

ANALYSING RESEARCH 7.3

The Dunning–Kruger effect

The Dunning–Kruger effect is a cognitive bias that suggests that people with low ability at a task overestimate their own skill set, and that people with high ability at a task underestimate their own skill set (Figure 7.9).

In 1999, researchers David Dunning and Justin Kruger conducted four experiments to test how accurately people estimate their own skill set. Dunning and Kruger hypothesised that incompetent individuals would significantly overestimate their ability and performance compared to their more competent peers. Students from Cornell University were recruited to take part in one of four tests. These tests assessed the following: judgement of humour, grammar, and two logical reasoning tests. After completing the test, participants were also required to estimate their performance on the test.

The researchers found that participants who scored in the bottom 25 per cent for the tests consistently overestimated their performance. Interestingly, Dunning and Kruger also found that participants who placed in the top 25 per cent of test scores significantly underestimated their test performance in relation to their peers.

According to Kruger and Dunning (1999), 'people who lack the knowledge or wisdom to perform well are often unaware of this fact'. The authors suggested that incompetent people lack the required cognitive ability to 'distinguish accuracy from error'; hence their confidence in their abilities often doesn't match the results of their actions (Kruger & Dunning, 1999). Interestingly, the researchers also noted that the high-performing individuals fell prey to the false-consensus bias; that is, they underestimated their performance as they expected that most other people did as well as them, if not better.

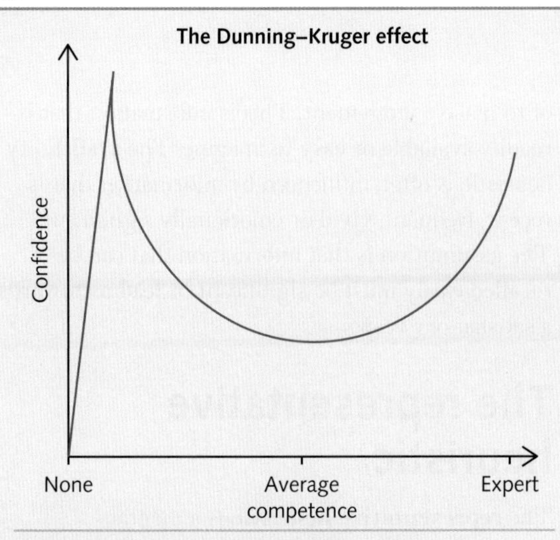

Figure 7.9 The Dunning–Kruger effect is a cognitive bias that suggests that people with low ability at a task overestimate their own skill set, and that people with high ability at a task underestimate their own skill set.

Questions

1 Write an aim for Dunning and Kruger's study.
2 What was the hypothesis for this study?
3 A between-group design was used in this study. Outline one strength and one limitation of the design.
4 What conclusions can be drawn from Dunning and Kruger's study regarding individuals' perceived levels of competency?

HOT Challenge

5 How was the false-consensus effect demonstrated in the estimations of highly competent individuals?

Source: 'Lessons Learned from a Study of the Integration of a Point-of-Care Ultrasound Course into the Undergraduate Medical School Curriculum' by Mateusz Zawadka et al, Medical Science Monitor: International Medical Journal of Experimental and Clinical Research 25:4104–4109, DOI: 10.12659/MSM.914781

KEY CONCEPTS 7.2

- Cognitive dissonance is experienced when there is a misalignment between our behaviours and our attitudes.
- Cognitive biases are systematic errors in judgement that occur when we try to simplify the information we are processing.
- Cognitive biases reduce the experience of cognitive dissonance as we process information or arrive at decisions that justify our behaviours or beliefs.

Concept questions 7.2

Remembering

1 What are cognitive biases and why do they occur? r
2 Explain why the anchoring bias occurs. e
3 What is confirmation bias and why does it occur? r

Understanding

4 Explain how our interpretation of external and internal factors contributes to the actor–observer bias. e

Applying

5 Explain how we would be likely to attribute behaviour in each scenario below if we were the actor (the person completing the behaviour) or the observer (the person witnessing the behaviour). c

a A person driving home from work has a car accident on a sunny day.
b A parent with a screaming child forgets to pay for their groceries at the supermarket.
c A teenager who is late for work jaywalks on a busy road.

HOT Challenge

6 One study investigating functional fixedness (German & Defeyter, 2000) found that children aged 5 solved a problem twice as fast as children aged 6 or 7.

Based on your understanding of functional fixedness, why might these results have occurred? c

7.3 Heuristics

On any given day, we make hundreds if not thousands of decisions. Some require a great deal of mental effort, but many are made without much cognitive effort or awareness. For many of the decisions we make, we use shortcuts. These are known as **heuristics**. Heuristics are mental shortcuts we use to make quicker, more efficient decisions. Although heuristics lead us to a decision without critically evaluating it first, they are helpful because:

» they reduce the cognitive load required in decision-making
» we can make snap judgements that save time and mental effort
» they simplify complex information
» they direct attention towards the question, probable outcomes and plausible answers.

Three types of heuristics are the availability heuristic, the representative heuristic and the affect heuristic.

The availability heuristic

7.3.1 APPLYING HEURISTICS

Consider the following question. Which profession is more dangerous: being a police officer (Figure 7.10) or being a garbage collector? According to the US Bureau of Labor statistics, garbage collectors are twice as likely to die at work than police officers. If your answer to the question was 'police officer', this was probably a result of the availability heuristic. Through exposure to news, television and movies, we've come to expect that police officers are confronted with significant dangers a fair amount of the time.

The **availability heuristic** is a mental shortcut in which we use the information that we first think of to make a judgement. This is information that is readily available or easy to imagine. The availability heuristic is often influenced by information that is recent, frequent, vivid or emotionally significant. The assumption is that information that can be recalled easily must be significant or lead us towards a satisfactory response.

Shutterstock.com/Leonard Zhukovsky

Figure 7.10 Is it more dangerous to be a police officer or a garbage collector?

The representative heuristic

The **representative heuristic** is a mental shortcut in decision-making where we estimate the likelihood of something occurring or being true based on its similarity to our existing understanding and expectations. As the saying goes, 'If it looks like a duck, and quacks like a duck, then it probably is a duck'. The representative heuristic allows us to quickly categorise information. However, it can lead to a biased or inaccurate estimation of probability, often for scenarios that we couldn't know the answer to.

A study by Kahneman and Tversky (1973) illustrates the representative heuristic. The researchers provided participants with the following personality profile of a university student named Tom W. Participants in one condition had to predict what type of course Tom W was completing, based on a personality profile given to them. The profile read as follows:

> *Tom W is of high intelligence, although lacking in true creativity. He has a need for order and clarity, and for neat and tidy systems in which every detail finds its appropriate place. His writing is rather dull and mechanical, occasionally enlivened by somewhat corny puns and by flashes of imagination of the sci-fi type. He has a strong drive for competence. He seems to feel little sympathy for other people and does not enjoy interacting with others. Self-centred, he nonetheless has a deep moral sense.*
>
> Source: Kahneman, D., & Tversky, A. (1973). On the psychology of prediction. *Psychological Review, 80*(4), 237–251

The nine courses given to the participants were business administration, computer science, engineering, humanities and education, law, library science, medicine, physical and life sciences, and social science and social work.

Before reading the profile, participants made a separate estimation: that there were three times the number of people studying humanities and education than computer science. However, after reading the profile 95 per cent of respondents predicted that Tom W was more likely to study computer science than humanities and education. The researchers concluded that the profile of Tom W was more representative of a computer science student and this contributed to the judgement that he was most likely a computer science student.

The affect heuristic

The **affect heuristic** is an approach that is used when decision-making is influenced by an individual's current emotional state or mood. It is similar to 'going with your gut'. Affective responses are useful because they tend to occur quickly and efficiently. However, we make these responses without taking the time to weigh up all relevant information (Bateman et al., 2007). We also make judgements about certain words, images and objects because of the emotional response they prompt.

Finucane et al. (2000) tested the hypothesis that our mood influences our decision-making. Initially, participants were required to rate the risks and benefits of nuclear power on a scale from 0 to 10, with 10 being very beneficial. The participants then read one of four passages of text about the use of nuclear power. The passages suggested that nuclear power was: (1) high risk; (2) low risk; (3) high benefit; or (4) low benefit. The high risk and low benefit conditions were designed to create negative emotional states, whereas the low risk and high benefit conditions were designed to create positive emotional states.

The results showed that 45 per cent of the participants changed their opinion regarding the risks and benefits of nuclear power after reading one of the four passages. The researchers concluded that providing information designed to alter a person's feelings towards nuclear power tended to change their perception of the benefits and risks associated with the topic. In other words, a positive emotional state led people to perceive nuclear power as having higher benefits and lower risks. Additionally, those exposed to the negative passages perceived nuclear power to have higher risks and lower benefits (Figure 7.11).

Figure 7.11 In their study, Finucane et al. (2000) asked participants to rate the perceived benefits and risks of nuclear power. The researchers found that a significant number of participants changed their opinions depending on whether they were in a positive mood or negative mood.

ACTIVITY 7.3 USING HEURISTICS TO MAKE JUDGEMENTS

Try it

1 Answer each of the following questions with the first information that comes into your mind.
   - a What is the most common colour of car in Australia?
   - b Globally, what is the highest selling brand of smartphone?
   - c What percentage of the world's population has blue eyes?
   - d What percentage of people are left-handed?

   Identify and explain the heuristic that was used to answer each question.

Apply it

2 Consider the following question: 'Do more English words start with the letter "s" than the letter "x"?' Almost all English-speakers would select 's' as the answer, which is the correct answer.

9780170465052

Outline the heuristic that was used to arrive at the correct answer.

Exam ready

3 Heuristics are mental shortcuts that we use to arrive at a decision. Which one of the following statements about heuristics is correct?

A Availability, representativeness and functional fixedness are examples of heuristics.

B Heuristics are useful for arriving at a quick decision; however, they increase the mental effort required to make a decision.

C Heuristics are useful for arriving at a quick decision; however, our responses are not always correct.

D Heuristics are useful for arriving at a quick decision to reduce cognitive dissonance.

KEY CONCEPTS 7.3

- Heuristics are mental shortcuts that help us make decisions.
- Heuristics are useful in reducing the cognitive load that is required in decision-making because they allow us to make snap judgements that save us time and mental effort, yet they are susceptible to bias.
- The availability heuristic is a mental shortcut in which we use the information that first comes to mind to make a judgement; that is, information that is readily available or easy to imagine.
- The representative heuristic is a mental shortcut for decision-making in which we estimate the likelihood of something occurring or being true based on its similarity to our existing understanding and expectations.
- The affect heuristic is a heuristic in which decision-making is influenced by an individual's current emotional state or mood.

Concept questions 7.3

Remembering

1 Define the term 'heuristics'. r

Understanding

2 Using an example, explain the representative heuristic. c

3 Using an example, explain the availability heuristic. c

4 Using an example, explain the affect heuristic. c

5 Outline one benefit and one limitation of using heuristics in decision-making. e

Applying

6 Tess was watching the news when she saw a story about a shark attack. The following weekend, Tess and her friends went to the beach. Though Tess' friends happily went swimming in the ocean, Tess refused to go in. Explain the heuristic that best accounts for Tess' decision to not swim in the ocean? c

HOT Challenge

7 In regard to decision-making, outline a similarity between person perception and heuristics. e

7.4 The effects of prejudice and discrimination

The positive and negative attitudes we hold about ourselves, others, objects and experiences play a significant role in the way people relate to each other.

Prejudice

Prejudice is a negative preconceived notion that we hold towards individuals due to their membership in a particular group. Prejudice leads to discrimination or the unequal treatment of people who should have the same rights as others.

Like other attitudes, prejudicial attitudes can include the ABC components of the tri-component model. For example, 'I dislike old people' (affective component), 'I wouldn't hire an older worker' (behavioural component) and 'Old people are always ill and are not as strong as young people' (cognitive component).

Discrimination

Discrimination is the action of being prejudiced or treating others in an unfair manner based on the negative attitude held about that person or the group to which they belong. Discrimination frequently prevents people from doing things they should have the opportunity to do. These include buying a house, getting a job, travelling on a bus or attending a particular school. Discrimination can be broadly categorised as direct or indirect. Discrimination can occur directly or indirectly.

» *Direct discrimination:* When a person or a group is treated less favourably than another person or group because of their background or certain personal characteristics. For example, the 1901 White Australia Policy was a federal government policy that placed heavy restrictions on people of non-white backgrounds migrating to Australia (Figure 7.12). The policy was gradually dismantled by the 1975 *Racial Discrimination Act*, making it illegal for people to be denied to the right to migrate based on racial criteria.

» *Indirect discrimination:* An unreasonable rule or policy that is the same for everyone but has an unfair effect on a particular group.

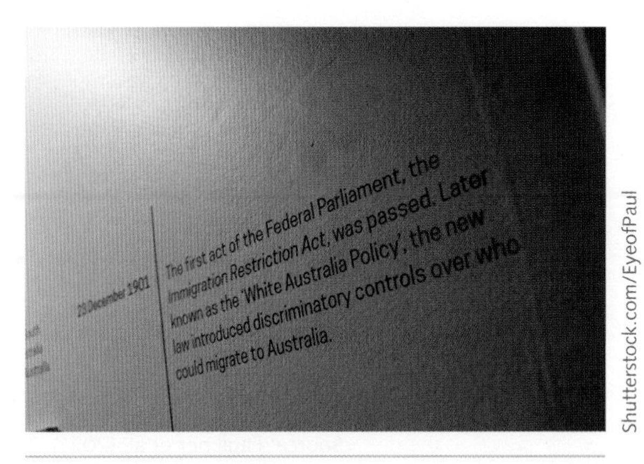

Shutterstock.com/EyeofPaul

Figure 7.12 The White Australia Policy was a clear example of direct discrimination.

For example, if a workplace policy says that managers must work full-time, this disadvantages women with family responsibilities because they are more likely to work part-time (Australian Human Rights Commission, 2021).

Forms of discrimination

The unequal treatment of people based on group association can occur in several forms. In all cases the individual discriminated against becomes disempowered and is afforded fewer rights and opportunities than others. Sexism, racism and ageism are all forms of discrimination based on prejudice.

» Sexism is an attitude that is a mixture of negative thoughts and stereotypes, and feelings of fear, envy or hostility, that results in discrimination based on gender. For example, a woman who is the best-qualified and most experienced candidate is not given the job because she is pregnant or may become pregnant.

» Racism is an attitude that is a mixture of negative thoughts and stereotypes, and feelings of fear, envy or hostility, that results in discrimination based on race. For example, a person might be called names or teased about their cultural heritage by their peers.

» Ageism is an attitude that is a mixture of negative thoughts and stereotypes, and feelings of fear, envy or hostility, that results in discrimination based on age. For example, a person may be denied a promotion at work because they are perceived to be 'close to retirement age'.

Table 7.4 Prejudicial attitudes towards particular groups in Australian society

Social group	Respondents who 'moderately' or 'strongly' agreed with prejudicial statements
Racial minorities	21%
Aboriginal and Torres Strait Islanders	21%
LGBTIQ+ people	20%
Religious minorities	19%
Young people (18–24 years)	16%
Women	14%
People with a disability	7%
Older people	7%

Source: Faulkner et al. (2021). *The inclusive Australia social inclusion index: 2021 report.* Monash University. https://inclusive-australia.s3.amazonaws.com/files/Inclusive-Australia-2020-21-Social-Inclusion-Index-min.pdf

Prejudice and discrimination in Australian society

While much work has been done to reduce prejudice and discrimination, these attitudes are still present in Australian society.

The 2021 Social Inclusion Index, conducted by researchers at Monash University, found that many Australians still harbour highly prejudicial views. On average, the respondents 'moderately' or 'strongly' agreed with statements that indicated prejudice. Table 7.4 shows the percentage of people who demonstrated high levels of prejudice to particular social groups in Australia.

The survey also found that almost 50 per cent of First Nations peoples reported experiencing major forms of discrimination in the last year (Figure 7.13). This included being unfairly stopped by police, being denied employment opportunities or promotions, and being discouraged from continuing their education.

Historical research has shown that one way of reducing prejudice is to increase contact between social groups. Yet, staggeringly, approximately 40 per cent of Australians reported that they have not had contact with a First Nations Australian in the past year.

Shutterstock.com/simez78

Figure 7.13 In a 2021 study, nearly 50 per cent of First Nations peoples reported experiencing major forms of discrimination in the past year.

ACTIVITY 7.4 DISCRIMINATION IN AUSTRALIAN SOCIETY

Prejudice and discrimination are still widely experienced in Australia. Figure 7.14 outlines the experiences of discrimination encountered by marginalised groups in Australian society.

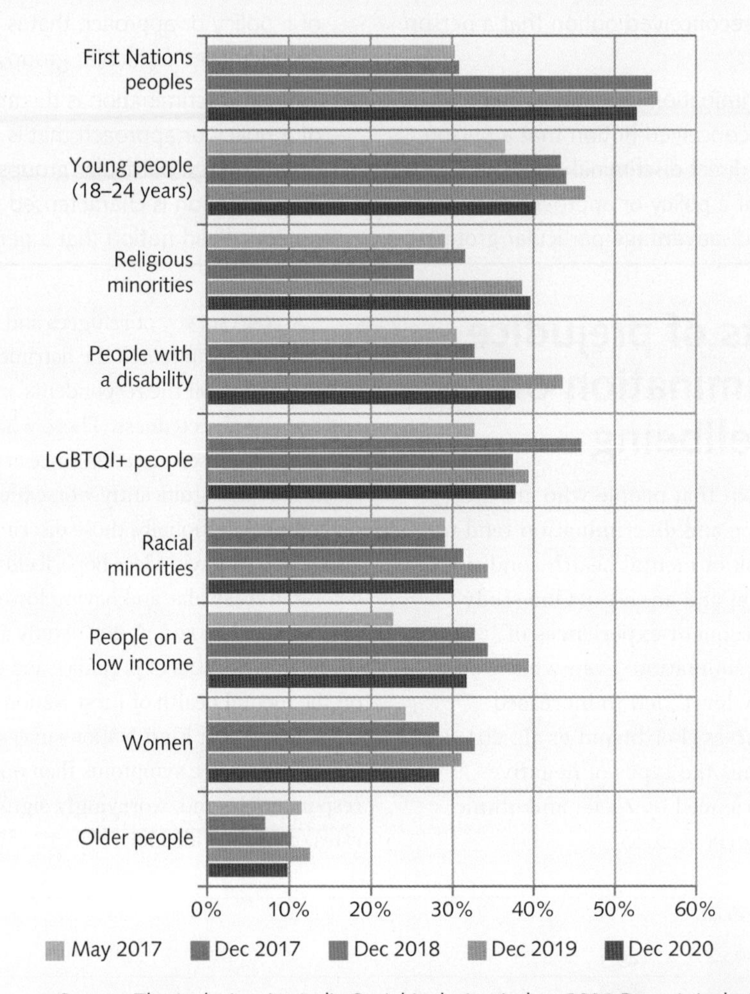

Source: The Inclusive Australia Social Inclusion Index: 2021 Report, Inclusive Australia, 2021

Figure 7.14 **Percentage of people in groups who experience one or more forms of everyday discrimination at least weekly**

Try it

1 Use the data in Figure 7.14 to complete the questions below.

a As of December 2020, which group in Australian society reported the highest levels of discrimination?

b As of December 2020, which group in Australian society reported the lowest levels of discrimination?

c Which groups reported an increase in discrimination between May 2017 and December 2020?

d Are there any pieces of data from the survey that you find interesting or surprising? Discuss any points of interest with a class mate.

Apply it

2 Explain how the use of stereotyping can lead to prejudice and discrimination.

Exam ready

3 Which statement best describes how indirect discrimination differs from direct discrimination?

7.4.3 RELATIONSHIPS BETWEEN ATTITUDES, PREJUDICE AND DISCRIMINATION

A Indirect discrimination is characterised by the application of a blanket rule that disadvantages one group more than another, while direct discrimination is characterised by a negative preconceived notion that a person holds.

B Indirect discrimination is characterised by a negative preconceived notion that a person holds, while direct discrimination is the application of a policy or approach that is designed to disadvantage particular groups.

C Indirect discrimination is characterised by the application of a blanket rule that disadvantages one group more than another, while direct discrimination is the application of a policy or approach that is designed to disadvantage particular groups.

D Indirect discrimination is the application of a policy or approach that is designed to disadvantage particular groups, while direct discrimination is characterised by a negative preconceived notion that a person holds.

The effects of prejudice and discrimination on mental wellbeing

Research has shown that people who are the targets of prejudice and discrimination tend to be at increased risk of mental health conditions (namely, depression and anxiety). One study found that more frequent experiences of prejudice and discrimination (even what would be considered 'low level') led to increased psychological distress (Ferdinand et al., 2015). Figure 7.15 outlines the types of negative interactions experienced by racial and ethnic minorities in Victoria in 2015.

A 2020 survey of refugees and asylum seekers in Australia documented the detrimental effects of discrimination on the respondents' mental health and social connectedness. Those who reported experiencing discrimination since arriving in Australia had significantly worse mental health outcomes. Additionally, those discriminated against also reported having less hope, feeling like they don't belong in Australia, and having lower levels of trust (Ziersch et al., 2020). A 2021 study also highlighted the negative effects of prejudice and discrimination on the mental health of First Nations males. The study found that First Nations males reported more severe depressive symptoms than non-First Nations respondents – and, worryingly, significantly higher rates of suicidal ideation (Haregu et al., 2021).

Worksheet
Experiences of racism in Victorian racial and ethnic minorities

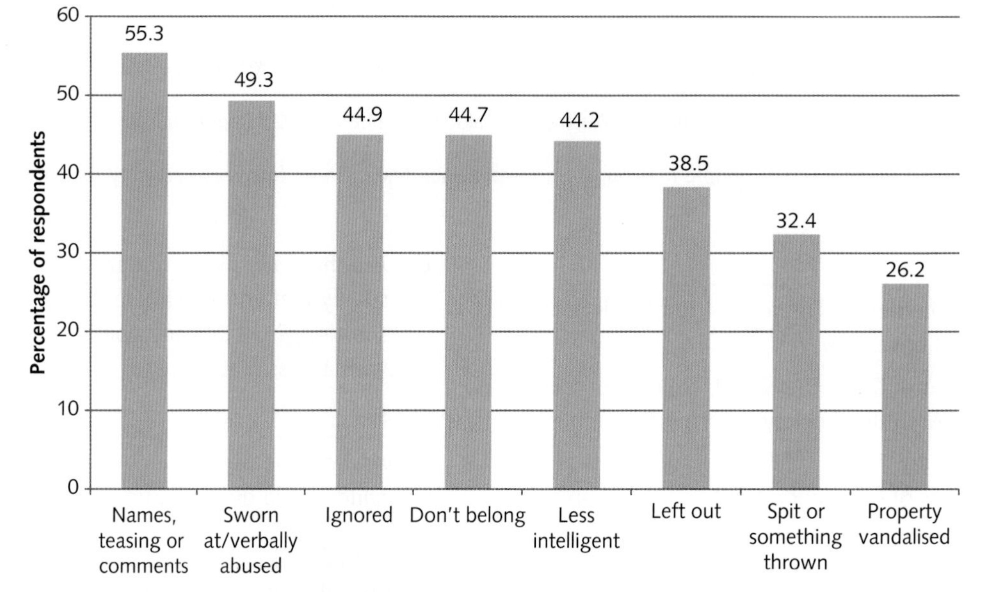

Figure, 'Experiences of racism for Victorians from racial/ethnic minority backgrounds', from Ferdinand, A.S., Paradies, Y. & Kelaher, M. Mental health impacts of racial discrimination in Australian culturally and linguistically diverse communities: a cross-sectional survey. BMC Public Health 15, 401 (2015). https://doi.org/10.1186/s12889-015-1661-1. Licensed under Creative Commons 2.0 https://creativecommons.org/licenses/by/2.0/

Figure 7.15 The experiences of racism of Victorians from racial and ethnic minorities, from a 2015 study

KEY CONCEPTS 7.4a

- Prejudice is a negative preconceived notion held towards individuals due to their membership of a particular group.
- Discrimination is the action of being prejudiced or treating others in an unfair manner, based on the negative attitude held about that person or group.
- Direct discrimination happens when a person, or a group of people, is treated less favourably than another person or group because of their background or certain personal characteristics.
- Indirect discrimination occurs when there is an unreasonable rule or policy that is the same for everyone but has an unfair effect on people who share a particular attribute.
- People who are the targets of prejudice and discrimination tend to have poorer mental health outcomes.

Concept questions 7.4a

Remembering

1 Explain the following forms of discrimination. **r**

  a Racism

  b Sexism

  c Ageism

Understanding

2 Using examples, explain how direct discrimination differs from indirect discrimination. **c**

3 Explain how prejudice and discrimination are similar yet different. **e**

Applying

4 A vision-impaired person attempts to buy tickets to a sporting match online but finds it difficult. They call the venue to ask whether they can purchase the tickets over the phone, but they are told that it is company policy that tickets can only be purchased online. Would the venue's policy be considered indirect discrimination? Justify your answer. **c**

HOT Challenge

5 A recent study in the UK found ageism to be the most common form of discrimination. Why might this be the case when compared to other forms of discrimination, including racism and sexism? **c**

CASE STUDY 7.1

Respecting cultural practices

In 2020, a 5-year-old boy of Cook Island descent started prep at a co-educational Christian school attended by his older sister. It is a Cook Island custom to not cut the firstborn son's hair until a hair-cutting ceremony is held. The boy's parents planned to hold a ceremony the following year when the boy would be aged seven. In the meantime, for school, the boy wore his hair in a neat bun. The school uniform policy requires boys' hair to be cut above the collar. The boy's parents were given notice that the boy would be unenrolled if his hair was not cut to meet the school uniform policy.

The Queensland Civil and Administrative Tribunal found that the hair-cutting ceremony for the eldest son is a cultural tradition associated with the Cook Island culture, and the ceremony should take place at a time chosen by the parents. The Tribunal held that the threat to unenrol the boy from school amounted to unfavourable treatment. It said that excluding a child from school is a serious step to take, and has the potential to lead to dislocation, and cause emotional distress and embarrassment, and long-lasting effects on the child's perception of their place in a community.

»

The threat to unenrol the boy had not been carried out, and he had been allowed to attend school without having his hair cut until the matter was decided by the tribunal. The Tribunal found there had been unlawful discrimination of the boy on the basis of race and there was no obligation to cut the boy's hair. The school had to write an apology to the boy's mother. A 2022 appeal decision confirmed that there had been discrimination.

Adapted with permission from Recent case studies, Queensland Human Rights Commission, https://www.qhrc.qld.gov.au/resources/case-studies/race-case-studies

Questions

1 Explain the difference between prejudice and discrimination.
2 In this case, would the school's uniform policy be considered a form of direct or indirect discrimination?
3 According to the Tribunal, what are two potential negative effects on the wellbeing of the child if he was not allowed to attend school?
4 Racism, sexism and ageism are still major issues in Australian society. Based on your knowledge of these concepts, write down five key points about stereotyping, prejudice and discrimination that could help improve people's understanding of these concepts as a whole.
5 Are there any policies or rules at your school or in the broader community that you consider to be discriminatory?

ANALYSING RESEARCH 7.4

A classroom experiment on discrimination

Eye colour seems like an unrealistic basis for creating prejudices, yet people already use skin colour to make decisions about the race of another person (Brown et al., 1998). In 1968, American primary school teacher Jane Elliot sought to give her all-white class of students direct experience with prejudice, using eye colour.

On the first day of the experiment, Elliot announced that brown-eyed children were to sit in the back of the room and that they could not use the drinking fountain (Figure 7.16). Blue-eyed children were given extra recess time and got to leave first for lunch. At lunch, brown-eyed children were prevented from taking second helpings because they would 'just waste it'. Brown-eyed and blue-eyed children were kept from mingling, and the blue-eyed children were told they were 'cleaner' and 'smarter' (Peters, 1971).

At first, Elliot had to maintain these imposed conditions of prejudice. She also made an effort to constantly criticise and belittle the brown-eyed children. To her surprise, the blue-eyed children rapidly joined in and were soon outdoing her in the viciousness of their attacks. The blue-eyed children began to feel superior,

Shutterstock.com/Rob Byron

Figure 7.16 A brown-eyed girl looks on while her blue-eyed classmate uses the drinking fountain, a 'privilege' denied to her in Jane Elliot's experiments.

and the brown-eyed children felt awful. Fights broke out. Test results of the brown-eyed children fell.

The effects were short-lived, because two days later the roles of the children were reversed. Before long, the same destructive effects occurred, but this time in reverse.

The implications of this study are unmistakable. In less than one day, it was possible to get children to look down on each other because of eye colour and status inequalities (differences in power, prestige or privileges).

Questions

1 What was the aim of the study?
2 Briefly explain the procedure of the study.
3 Explain how Elliot created division between the members of the class.

HOT Challenge

4 What are the independent and dependent variables in this study?
5 Outline two ethical guidelines that would need to be followed to ensure that this experiment was ethical by today's standards.

Weblink
PBS documentary: A Class Divided

Reducing prejudice

Becoming aware of the factors that work towards reducing prejudice is a good starting point for reducing prejudice. These factors include intergroup contact, sustained contact, superordinate goals, mutual interdependence, equality of status and changing social norms (see Table 7.5 for a summary).

Intergroup contact

One way to reduce prejudice is to increase the intergroup contact between the people who hold the stereotype and those who are the target of the stereotype. Stereotypes can be broken down and replaced by more positive attitudes when the intergroup contact is maintained over a period of time (known as sustained contact, discussed shortly), when meeting the goals for each group requires mutual interdependence, and when equality exists through equal-status participation.

Intergroup contact makes people aware that members of various racial and ethnic groups share the same goals, ambitions, feelings and frustrations as they do. This has a positive effect on attitudes and can reduce prejudice. This improvement in attitude can be observed when employees of different ethnic groups work side by side in the workplace.

Sustained contact

For maximum benefit, intergroup contact should be sustained (Figure 7.17). This means there should be prolonged and involved cooperative activity, rather than casual and purposeless contact (Allport, 1954). Individuals who work together for extended periods of time to achieve a shared goal have an opportunity to get to know the reality of the other groups. They may then arrive at an informed attitude about these groups rather than accept a stereotype. For example, a junior sporting team may be made of individuals from different religious and ethnic backgrounds, different

iStock.com/Leontura

Figure 7.17 Sustained contact between members of different groups may reduce the incidence of prejudice.

Table 7.5 Methods to reduce prejudice

Method	Definition
Intergroup contact	Members of opposing groups should be encouraged to spend time with each other, which can assist in breaking down stereotypes and recognising shared beliefs and values.
Sustained contact	Members of opposing groups should spend prolonged periods of time cooperating to achieve a goal.
Superordinate goals	Members of opposing groups should have shared goals that individuals or groups can only achieve with the cooperation of others.
Mutual interdependence	Members of the group are reliant on each other to achieve their own goals.
Equality of status	Members of the in-group and out-group interact with each other, but without one group exerting power over the other.
Changing social norms	Prejudice can be reduced by changing the negative or harmful standards or behaviours that might be accepted within society.

economic levels, and a variety of neighbouring schools. The stereotypes the members hold in relation to any of the above characteristics may be broken down by their sustained contact with each other at training and on match days.

7.4.4 PREJUDICE AND WAYS TO REDUCE PREJUDICE

Superordinate goals

Superordinate goals are shared goals, which individuals or groups can't achieve without the cooperation of others. These goals exceed or override other, lesser goals that the individual or group may have. As members of the groups are forced to cooperate, this can lead to hostilities and prejudices lessening or even disappearing. Cooperation and shared goals seem to help reduce conflict by encouraging people in opposing groups to see themselves as members of a single, larger group (Gaertner et al., 1990).

Mutual interdependence

Mutual interdependence is another element of intergroup contact that is effective in reducing prejudice between groups. When people are mutually interdependent, they must rely on one another to meet each person's goals. When each person's needs are linked to those of others in the group, cooperation is encouraged (Deutsch, 1993). An example of this is an AFL football team that is composed of several racial and age groups. For the team to achieve its aim of winning a football game, all the players are dependent on each other. This mutual interdependence helps to reduce prejudices between players of different races and of different ages.

Equality of status

A final condition that assists in reducing prejudice is equality of status amongst members of the in-group and out-group. Equality of status involves social interaction that occurs at the same level, without obvious differences in power or status. In other words, members of the in-group and out-group interact with each other, but without one group exerting power over the other.

Changing social norms

Many of our behaviours are governed by social norms; these are rules, standards or behaviours that are generally understood or accepted within society. While many social norms help to maintain order, some social norms reinforce prejudicial ideas and discriminatory behaviour. For example, until the mid-1970s, homosexuality was illegal in all regions of Australia. Following decades of advocacy and attempts to change social norms and negative attitudes towards members of the LGBTQI+ (lesbian, gay, bisexual, transgender, queer, intersex) community, meaningful progress has been made. Homosexuality is no longer illegal, and in 2017 the Australian Parliament legalised same-sex marriage. To change social norms, efforts are made to change the way people think about particular groups, and consequently how they act in the presence of members of that group. Norms can be changed through the passing of legislation, awareness campaigns, listening to people's experiences and spending time with members of out-groups.

ANALYSING RESEARCH 7.5

The Robbers Cave experiment

One of the most well-known studies regarding the reduction of prejudice was the 1954 Robbers Cave experiment conducted by American psychologist Muzafer Sherif. Sherif recruited 22 young boys aged 11 to 12 years, who were randomly allocated to one of two groups. The boys were from similar backgrounds, with similar academic abilities, and all had no history of psychological or behavioural issues. Over the course of a three-week summer camp at the Robbers Cave State Park, the boys were placed in a number of situations designed to form distinct in-groups and out-groups. Finally, the boys were required to complete a series of activities that could only be completed if they cooperated with each other, known as superordinate goals. Sherif hypothesised that, following the establishment of the in-group, the introduction of superordinate goals would reduce tension between the groups.

Stage 1: In-group formation

To assist in the formation of in-groups, the boys were allocated to one of two groups. The boys were collected and taken to the campsite on separate buses and were placed at different ends of the campsite, under the impression that they were the only group there. For the next week, the individual groups completed team-building activities such as organising campfires and preparing meals, which strengthened the bond between the boys.

During this stage of the experiment, the boys gave themselves distinct group names: one group was called the 'Rattlers' (Figure 7.18) and the other was called the 'Eagles'. The boys made flags with their group names on them and also stencilled their shirts and hats with their group's name.

Figure 7.18 Members of the Rattlers, one of the groups in the Robbers Cave experiment, proudly display their flags.

Stage 2: The friction phase

By the beginning of Stage 2, the boys in each group had become aware of each other's existence. Camp staff announced that a tournament of activities would be taking place over the coming days. During this time, the boys played basketball, baseball and tug-of-war against each other, all while trading insults at the opposing team. Tensions reached a tipping point when the Eagles set the Rattlers' flag on fire. In an attempted act of retribution, the Rattlers hatched a plan to steal the Eagles' flag, which culminated in a fist-fight between members of the two groups.

Stage 3: The integration phase

Now that strong in-groups had been formed and negative attitudes established, Sherif planned to use the third stage of the experiment to reduce hostilities by creating scenarios where the Rattlers and Eagles had to work together.

Initially, the boys attended a movie together and ate meals in the same location; however, their attitudes did not change. At mealtime, the boys would sit on separate sides of the food hall, often culminating in food-fights. Simply placing the boys in the same location for extended periods of time did not reduce the tensions between the Rattlers and the Eagles. Sherif and his associates now introduced superordinate goals: worthwhile goals that can only be achieved via cooperation. Working together, the boys repaired the camp's water supply; pulled a broken down truck carrying food for the camp; and also pooled their money to hold a movie night.

At the end of the three weeks at the camp, the boys suggested that they head back home on the same bus. Sherif noted that on the ride home, the boys intermingled, sang songs and many of the boys exchanged details to meet again in the future. The introduction of superordinate goals greatly reduced the tensions between the two groups. The reliance on each other encouraged the boys to put their differences aside. Additionally, this helped to reduce the prejudice the boys held about the opposing group.

Source: Adapted from Sherif et al. (1961). *Intergroup conflict and cooperation: The Robbers Cave experiment.* Norman: University Book Exchange.

7.4.5 FORMATION AND CHANGING OF ATTITUDES

Weblink
The Robbers Cave experiment

Questions

1 What was the aim of the study?
2 What was the hypothesis for the study?
3 Explain how in-groups and out-groups were formed in the Robbers Cave experiment.
4 What are superordinate goals and how were they utilised in the Robbers Cave experiment?

HOT Challenge

5 Explain how Sherif attempted to control for the unwanted effect of individual participant differences in the study.
6 Explain why sustained contact alone did not effectively reduce prejudice amongst the boys.

KEY CONCEPTS 7.4b

- Intergroup contact can be used to reduce prejudice; however, it must be sustained contact, accompanied by mutual interdependence, equality of status and the pursuit of superordinate goals.
- Superordinate goals are shared goals that individuals or groups can't achieve without the cooperation of others. Superordinate goals reduce prejudice because members of the in-group become dependent on members of the out-group to achieve a common goal.
- Equality of status involves social interaction that occurs at the same level, without obvious differences in power or status. As members of the in-group and out-group are seen as 'equals', one group doesn't exert power or dominance over the other.

Concept questions 7.4b

Remembering

1 What is sustained contact? r

Understanding

2 What are superordinate goals and how can they be used to reduce prejudice? e

3 What is mutual interdependence and how can it be used to reduce prejudice? e

4 What is equality of status and how can it be used to reduce prejudice? e

5 Explain how intergroup contact reduces the prejudicial attitudes between groups. e

Applying

6 Using an example, explain how changing social norms can reduce prejudice. c

HOT Challenge

7 Using the example of a sporting team, explain how sustained contact, mutual interdependence, equality of status and the pursuit of superordinate goals can reduce prejudice and strengthen bonds between members of the team. c

7 Chapter summary

KEY CONCEPTS 7.1a

- Social psychology is a field of psychology that studies the way in which our behaviour is influenced by others.
- Social cognition refers to the way in which we judge our own behaviours and the behaviours of others in a social setting.
- Person perception refers to the process of making judgements about individuals based on their apparent characteristics. While person perception allows us to make quick judgements about individuals, it can be inaccurate because we are using a small amount of information to make broad judgements.
- Attributions are the process of attaching meaning to our own and other people's behaviours. Attributions can be influenced by internal factors (personality, motivation, mood, ability and effort) and external factors such as the environment (i.e. school, work etc.).
- When explaining a person's behaviour, we tend to overestimate the role of internal factors and downplay the role of external factors. This is known as the fundamental attribution error.

KEY CONCEPTS 7.1b

- Attitudes are the learned perspectives we hold about individuals, issues, events and so on.
- The tri-component model of attitudes proposes that attitudes contain an affective (emotional) component, a behavioural (actions) component and a cognitive (reasoning) component.
- Stereotyping involves placing people in categories because of perceived similarities.
- Stereotyping is a quick but often flawed way of making judgements about individuals because it does not take into consideration factors beyond the stereotype.

KEY CONCEPTS 7.2

- Cognitive dissonance is experienced when there is a misalignment between our behaviours and our attitudes.
- Cognitive biases are systematic errors in judgement that occur when we try to simplify the information we are processing.
- Cognitive biases reduce the experience of cognitive dissonance as we process information or arrive at decisions that justify our behaviours or beliefs.

KEY CONCEPTS 7.3

- Heuristics are mental shortcuts that help us make decisions.
- Heuristics are useful in reducing the cognitive load that is required in decision-making because they allow us to make snap judgements that save us time and mental effort, yet they are susceptible to bias.
- The availability heuristic is a mental shortcut in which we use the information that first comes to mind to make a judgement; that is, information that is readily available or easy to imagine.
- The representative heuristic is a mental shortcut for decision-making in which we estimate the likelihood of something occurring or being true based on its similarity to our existing understanding and expectations.
- The affect heuristic is a heuristic in which decision-making is influenced by an individual's current emotional state or mood.

9780170465052

KEY CONCEPTS 7.4a

- Prejudice is a negative preconceived notion held towards individuals due to their membership of a particular group.
- Discrimination is the action of being prejudiced or treating others in an unfair manner, based on the negative attitude held about that person or group.
- Direct discrimination happens when a person, or a group of people, is treated less favourably than another person or group because of their background or certain personal characteristics.
- Indirect discrimination occurs when there is an unreasonable rule or policy that is the same for everyone but has an unfair effect on people who share a particular attribute.
- People who are the targets of prejudice and discrimination tend to have poorer mental health outcomes.

KEY CONCEPTS 7.4b

- Intergroup contact can be used to reduce prejudice; however, it must be sustained contact, accompanied by mutual interdependence, equality of status and the pursuit of superordinate goals.
- Superordinate goals are shared goals that individuals or groups can't achieve without the cooperation of others. Superordinate goals reduce prejudice because members of the in-group become dependent on members of the out-group to achieve a common goal.
- Equality of status involves social interaction that occurs at the same level, without obvious differences in power or status. As members of the in-group and out-group are seen as 'equals', one group doesn't exert power or dominance over the other.

7 End-of-chapter exam

Section A: Multiple-choice

1 Which of the following is not a component of social cognition?
   A understanding other people's cognitive or emotional states
   B carrying out behaviours that take into consideration the goals and needs of ourselves and others
   C understanding our behaviour when we're alone
   D the detection of facial expressions and emotional responses

2 According to psychologists, attitudes are
   A positive evaluations we hold about people, objects, events or issues.
   B statements of factual knowledge about something or someone.
   C general evaluations we make about people, objects, events or issues.
   D negative evaluations we hold about people, objects, events or issues.

3 Liam has driven a car for 20 years. One night while he was driving, torrential rain began falling, causing Liam to skid and smash into a pole. Liam's behaviour is likely to be attributed to
   A internal attributions.
   B cognitive dissonance.
   C external attributions.
   D the fundamental attribution error.

4 When a student receives a poor result on their exam, they attribute the mark to their teacher's lack of ability instead of their own, even though before the exam they were impressed with the way the teacher was performing. Blaming the situation instead of accepting personal responsibility is called
   A the behavioural component of an attitude.
   B self-serving bias.
   C hindsight bias.
   D confirmation bias.

5 Ben believes that 'all P-plate drivers are hoons'. Ben's statement is an example of
   A false-consensus bias.
   B affect heuristic.
   C stereotyping.
   D the behavioural component of an attitude.

6 Which of the following statements regarding the tri-component model of attitudes is not correct?
   A The affective component attempts to explain how we feel about a person, object, event or issue.
   B The cognitive component attempts to explain how we feel about a person, object, event or issue.
   C The behavioural component attempts to explain the way we act due to our attitudes about a person, object, event or issue.
   D A limitation of the tri-component model is that people's behaviours do not always reflect their true attitude about a person, object, event or issue.

7 What are the two types of attributions that can be used to explain a person's behaviour?
   A cognitive and internal
   B external and affective
   C internal and external
   D cognitive and behavioural

8 Which of the following statements about stereotypes is most correct?
   A A stereotype is a positive evaluation about a person or object.
   B A stereotype is a negative evaluation about a person or object.
   C Stereotypes reduce the tension between in-groups and out-groups.
   D Stereotypes can be limiting and insulting.

9 The three components of an attitude are
   A cognitive, internal and external.
   B behavioural, affective and external.
   C affective, behavioural and cognitive.
   D affective, behavioural and internal.

10 Prejudice is ________, while discrimination is ________.

- **A** an attitude; a behaviour
- **B** external; internal
- **C** an opinion; a judgement
- **D** social; personal

11 Superordinate goals are

- **A** reached if one group uses their power to drive the groups.
- **B** aimed at reducing the intergroup contact between members of in-groups and out-groups.
- **C** designed to reduce the mutual interdependence between an in-group and out-group.
- **D** shared goals that can only be achieved by groups working together.

12 The psychological discomfort that comes from an inconsistency in our attitudes and our behaviours is known as

- **A** cognitive bias.
- **B** the cognitive component of our attitude.
- **C** cognitive dissonance.
- **D** confirmation bias.

13 The tendency for people to assume that their attitudes, beliefs, and behaviours are relatively common in others is known as

- **A** confirmation bias.
- **B** false-consensus bias.
- **C** the actor–observer bias.
- **D** attentional bias.

14 An explanation for why a person might use hindsight bias is

- **A** our abilities to make accurate estimations decreases over time.
- **B** by using hindsight bias, we increase our level of cognitive dissonance.
- **C** people have a desire to see the world as orderly and predictable.
- **D** once a memory has been formed, we can't change it.

15 Cognitive biases are ____________, whereas heuristics are ____________.

- **A** mental shortcuts we use to arrive at a decision; systematic errors in thinking
- **B** the psychological discomfort we experience when our beliefs don't align with our behaviours; mental shortcuts we use to arrive at a decision
- **C** systematic errors in thinking; mental shortcuts we use to arrive at a decision
- **D** systematic errors in thinking; techniques used to reduce cognitive dissonance

16 The Dunning–Kruger effect is a cognitive bias that suggests that people with low ability at a task ___________ their own skill set, and that people with high ability at a task ___________ their own skill set.

- **A** underestimate; overestimate
- **B** overestimate; underestimate
- **C** overestimate; overestimate
- **D** underestimate; correctly estimate

17 A researcher wanted to test for the halo effect. To do this, they showed participants 40 photos of individuals, half of whom were considered 'attractive looking', the other half of whom were considered 'average looking'. After seeing each photo, the participants had to estimate the person's level of academic achievement. The independent variable and dependent variable, respectively, for this study are

- **A** the perceived academic performance of the individuals; the level of attractiveness of the people in the photos
- **B** the attractiveness of the participants who took part in the study; the level of attractiveness of the people in the photos
- **C** the level of attractiveness of the people in the photos; the perceived academic performance of the individuals
- **D** the perceived academic performance of the individuals; the attractiveness of the participants who took part in the study

18 A similarity between optimism bias and confirmation bias is that

- **A** both biases encourage us to think critically about the world.
- **B** individuals tend to focus their attention towards particular information that eases their cognitive dissonance while excluding information that challenges their beliefs.
- **C** optimism bias and confirmation bias both increase our cognitive dissonance.
- **D** both are examples of heuristics we use to reduce the cognitive dissonance caused by information that challenges our existing beliefs.

19 Which statement best describes the informed consent procedures that Jane Elliot would have to abide by for the eye-colour experiment to be considered ethical by today's standards?

- **A** Written consent must be obtained from the child's legal guardian or the school's principal.
- **B** Written consent must be obtained from the child.
- **C** Written consent must be obtained from the child and the school's principal.
- **D** Written consent must be obtained from the parent or legal guardian of the child.

20 A number of studies explored in this chapter utilise deception to gather natural responses from participants. What ethical guideline should be followed at the conclusion of a study that utilises deception?

- **A** withdrawal rights
- **B** informed consent
- **C** debriefing
- **D** voluntary participation

Section B: Short answer

1 Explain how prejudice and discrimination differ. [2 marks]

2 Rashad is opposed to the detention of refugees and asylum seekers. Rashad believes that refugees and asylum seekers should be allowed to live in Australia, because the government has the resources to support them. Consequently, Rashad donates money to charities that support refugees and asylum seekers.

Explain the affective, behavioural and cognitive components of Rashad's attitude towards asylum seekers and refugees. [3 marks]

3 Explain why cognitive dissonance occurs and how it can be reduced through the use of cognitive biases. In your response, include an example that relates to the use of one cognitive bias. [3 marks]

Factors that influence individual and group behaviour

8

Key knowledge

- the influence of social groups and culture on individual behaviour
- the concepts of obedience and conformity and their relative influence on individual behaviour
- positive and negative influences of different media sources on individual and group behaviour, such as changing nature of social connections, social comparison, addictive behaviours and information access
- the development of independence and anti-conformity to empower individual decision-making when in groups

Key science skills

Plan and conduct investigations

- design and conduct investigations; select and use methods appropriate to the investigation, including consideration of sampling technique and size, equipment and procedures, taking into account potential sources of error and uncertainty; determine the type and amount of qualitative and/or quantitative data to be generated or collated
- work independently and collaboratively as appropriate and within identified research constraints, adapting or extending processes as required and recording such modifications

Comply with safety and ethical guidelines

- demonstrate ethical conduct and apply ethical principles when undertaking and reporting investigations

Generate, collate and record data

- systematically generate and record primary data, and collate secondary data, appropriate to the investigation

Analyse and evaluate data and investigation methods

- identify and analyse experimental data qualitatively, applying where appropriate concepts of: accuracy, precision, repeatability, reproducibility and validity of measurements; errors (random and systematic); and certainty in data, including effects of sample size on the quality of data obtained

Analyse, evaluate and communicate scientific ideas

- discuss relevant psychological information, ideas, concepts, theories and models and the connections between them
- use clear, coherent and concise expression to communicate to specific audiences and for specific purposes in appropriate scientific genres, including scientific reports and posters

Source: VCE Psychology Study Design (2023–2027), pp. 29 & 12–13

9780170465052

8 Factors that influence individual and group behaviour

You've already learned that human beings are social creatures. How we act when we are alone compared with when we're in a group can be very different. It gets even more interesting when we look at how people act in different types of groups and how behaviour can change based on a person's social status within the group.

p. 308

8.1 Social groups and culture

Think about all the different groups you interact with regularly: social groups, family, sports teams, your Psychology class ... Chances are that your social status or place in the hierarchy within each group varies. This can affect how much influence you hold within each group.

iStock.com/arsenisspyros

iStock.com/SolStock

p. 315

8.2 Obedience and conformity

Obedience might be something you would normally associate with puppy training, but it applies to humans too. We tend to obey someone who has a higher status and social power than us. Peer pressure can also change the way we think and behave.

p. 328

8.3
Media and mental wellbeing

Advertisers often take advantage of our need to belong. Have you ever wanted to buy a new product – not because you needed it, but because others in your social group had already bought it? Social media can also affect our self-esteem and mental health because we compare ourselves with the thousands of online connections we have.

iStock.com/Valeriy_G

p. 333

8.4
Independence and anti-conformity

People aren't always swayed by group pressure. Perhaps you feel too strongly about something to change your stance; at other times, you might be influenced by the group, but not in the way you were expecting. You might choose to do the exact opposite or believe something different purely because it's different from the rest of the group.

iStock.com/Rawpixel

If films were anything to go by, you would think social hierarchies, peer pressure and conformity only affected teenagers; but in most societies, this is not the case, and we are all influenced heavily by those around us.

Slideshow
Chapter 8 slideshow

Flashcards
Chapter 8 flashcards

Test
Chapter 8 pre-test

Assessment
- Pre-test
- End-of-chapter exam

Revision
- Chapter map
- Key term flashcards
- Key concept summary
- Slideshow

Investigation
- Investigation: Conformity at school
- Data calculator
- Logbook template: Fieldwork

Worksheet
- Miligram's Studies of Obedience

To access these resources, visit **cengage.com.au/nelsonmindtap**

Nelson MindTap

Know your key terms

Anti-conformity
Collectivist culture
Conformity
Deindividuation
Group
Group norms
Group shift
Groupthink
Independence
Individualist culture
Informational influence
Legitimacy of authority
Normative influence
Obedience
Social comparison theory
Social connection
Social power
Status
Unanimity

Between 2003 and 2004, American soldiers at Abu Ghraib prison in Iraq committed severe human rights abuses against Iraqi detainees (Figure 8.1). Many of the detainees were humiliated, and physically and psychologically tortured – with one prisoner dying as a result. The United States government authorised the use of torture, with orders coming from senior members of the military being carried out by the lower ranked officers at the prison. Many of the soldiers charged with the atrocities at Abu Ghraib claimed that were 'just following orders'; therefore, they shouldn't be held accountable for their actions.

Time and again, it has been shown that people behave quite differently in the presence of a group as opposed to when alone. Our behaviour is also heavily shaped by our position within the group, and the amount of power that we believe we have. This chapter will explain how groups, and society more broadly, shape our behaviour.

Alamy Stock Photo/Pictorial Press Ltd

Figure 8.1 An Iraqi detainee at Abu Ghraib prison was told that he would be electrocuted if he stepped off the box

8.1 Social groups and culture

A **group** is formed when two or more people interact, influence each other and share a common objective. Most people belong to a number of groups. These range from structured, permanent groups such as sporting teams, to fluid, temporary groups, such as study groups. Within these groups, individuals differ in terms of the status and social power that each member has. Although status and social power are related, they are not the same thing. **Status** refers to a person's position in the hierarchy of a group. **Social power** refers to the amount of influence that an individual can exert over another person. In general, the higher a person's status, the more power they exert.

Status

All societies have people of unequal status (Brown, 1965). Status may be based on lineage (ancestry), occupation or wealth. People recognise the different statuses and have accepted norms of treatment for people of high or low status.

A person's status can determine their degree of influence over others. Imagine a traffic signal brightly flashing the word 'WAIT'. As you and other pedestrians wait for it to change, a well-dressed man in a suit crosses against the light. How many people will follow him? Do you think the answer would be different if the man had an overgrown beard and was dressed in faded, old clothes (Figure 8.2)? This street-corner scenario was used in an early experiment on social influence. As you might have guessed, more people followed the well-dressed man than the man dressed in faded, old clothes (Lefkowitz et al., 1955).

The status given to an individual because of their role within a group can greatly affect the degree of influence they exert. Psychologist Phillip Zimbardo explored this issue in a now-infamous experiment that he set in a mock prison (Zimbardo et al., 1973). We will examine this experiment later in this chapter.

Social power

While the concepts of status and social power are related, they are based on separate ideas. Status refers to a person's standing in the community or group. Social power is about a person's capacity to exert influence over others. Often a person with high status also has social power. For example, a teacher has a high level of status in a class and also has power, in that they can influence the behaviour of class members.

iStock.com/Holger Mettem

Getty Images/RubberBall Productions

Figure 8.2 Dress sends a clear message about a person's status

8.1.1 SOCIAL POWER

A person could have high status and a lot of social power in one situation or group, but not in another. Consider the oldest child in a family who has high status among the family's children. That same child may have low status at school and little social power in that context.

Sometimes a person's social power can be subtle and they may not realise that they are influencing others. A person who is perceived by their peers as 'popular' may not be aware of the social power they possess. In this instance, the influence they have on the behaviour of others may not be intentional. For example, an older child who has high status may not intend to influence their younger siblings' choices in music and clothing, but as a result of their status they may exert influence anyway.

Psychologists Raven and French (1958) identified five different bases of social power: reward, coercive, legitimate, referent and expert power (Table 8.1). They suggested that, depending on the situation or group, the nature of the influence a person has differs. According to Raven and French, some individuals may possess more than one type of power. For example, a boss who is well respected by their employees could exert both reward power and referent power. Their employees are influenced because the boss pays them, but also because they aspire to be like them.

8.1.2 SOCIAL INFLUENCE

Table 8.1 Social power types, descriptions and examples

Social power	Description	Examples
Reward	This power is based on the ability to reward a person who complies with the desired behaviour.	Teachers reward students with results. Employers reward employees with bonuses.
Coercive	This power is based on the ability to punish a person for failure to comply.	Fines and imprisonment are used to control behaviour.
Legitimate	This power is based on our acceptance of a person as being part of an established social order.	Elected leaders Teachers in a classroom Parents at home
Referent	We refer to the person for direction. We want to be like the person.	Celebrities Sporting heroes
Expert	We recognise a person has knowledge or expertise in a specific field because of their training and experience.	Doctors Lawyers Computer technicians

Culture and behaviour

In addition to the immediate social groups we identify with (our school, our sporting teams etc.) the culture in which we live also greatly shapes our behaviour. Broadly speaking, cultures can be categorised as either collectivist or individualist.

Collectivist cultures (for example, many Asian countries; Figure 8.3):

- value group needs or interests over the interests of individuals (Triandis, 2001)
- prioritise loyalty to the group (as the belief is that the group will in turn support the individual), as well as interdependence, and understanding your role within the group (Darwish & Huber, 2003).

Individualist cultures (for example, many Western countries):

- value individual interests over the interests of groups
- view people as independent and accept that individuals' actions are focused on the attainment of their goals and the meeting of their needs over any broad group goals (Triandis, 2001)
- promote the interests of the individual and encourage the development of independence and personal identity (Darwish & Huber, 2003).

The influence of culture can be seen in the way we are raised, the way we tend to behave as teenagers, and even in who we choose to marry, as seen in the prevalence of arranged marriages in collectivist cultures such as India, China and Israel (Enzmann, 2012). Broadly speaking, child-rearing in individualist cultures encourages young children to explore their environment so they develop independence and self-reliance. Collectivist cultures emphasise conformity (following the group), obedience (respecting authority and following the rules) and security (Triandis, 2003). Cultural differences also exist in our likelihood of committing deviant and delinquent acts as a teenager. Studies show that children in individualist cultures with low levels of parental attachment and low parental supervision were more likely to commit delinquent acts than their collectivist peers who also had low levels of parental attachment and low parental supervision (Kotlaja, 2020).

Alamy Stock Photo/RYU SEUNG IL

Figure 8.3 In South Korea, all men aged between 18 and 28 must complete 18 months of compulsory military service. Many have to postpone their studies or careers to do this, for the 'greater good' of their society

Individualist and collectivist cultures are also classified as either 'horizontal' or 'vertical'. Cultures that operate horizontally prioritise equality, hence the term horizontal to represent the apparent evenness of the society. Cultures that operate vertically accept that inequalities exist and stress the importance of status and hierarchies (Singelis et al., 1995). The basic principles of horizontal-individualist, vertical-individualist, horizontal-collectivist and vertical-collectivist are outlined in Table 8.2.

8.1.3 COMPARING CULTURES

Table 8.2 Different types of cultural structure

	Individualist	Collectivist
Horizontal	» Individuals are encouraged to be autonomous but see others as relatively equal. » Equality is encouraged and embedded within the culture. » Examples of horizontal-individualist cultures include Australia and Sweden.	» The individual is part of a larger in-group, where all members are seen as similar to each other. » All members of the in-group are considered equals. » While least common, examples of horizontal-collectivist cultures include an Israeli kibbutz and rural cultures in Central America.
Vertical	» Individuals are encouraged to be autonomous, with others seen as different. » Inequality is accepted as not all members of the culture have the same degree of status. » Competition is encouraged within this cultural system. » Examples of vertical-individualist cultures include the USA and France.	» The individual is part of a larger in-group, but all members are different. » Inequality is accepted as not all members have the same degree of status. » It is generally accepted that some members of society will have to serve and some will have to make sacrifices, all of which benefit the in-group. » Examples of vertical-collectivist cultures include Japan and South Korea.

Source: Adapted from Singelis et al. (1995). Horizontal and vertical dimensions of individualism and collectivism: A theoretical and measurement refinement. *Cross-cultural research, 29*(3), 240–275; and Torelli & Stoner (2015). Managing cultural equity: A theoretical framework for building iconic brands in globalized markets. *Review of marketing research, 12*, 83–120.

ACTIVITY 8.1 COLLECTIVIST OR INDIVIDUALIST CULTURE

Try it

1 Broadly speaking, societies can be categorised as individualist or collectivist. Make a table like the one below to explain how these cultures shape individual and group behaviour.

The influence of **individualist** cultures on individual and group behaviour:	The influence of **collectivist** cultures on individual and group behaviour:
•	•
•	•
•	•
Overall, I would rather live in a ____________ culture because:	

Apply it

2 Kim and Markus (1999) conducted a study to investigate differences between members of individualist and collectivist cultures. Travellers at an airport were asked to fill out a survey (with the researchers choosing European American travellers to represent individualists and East Asian travellers to represent members of collectivist cultures). At the conclusion of the survey, the respondents were offered a pen as a token of appreciation. The travellers were given five pens to choose from (four that were one colour and one that was a different colour: 'the unique pen'). American travellers picked the unique-coloured pen 77 per cent of the time, whereas the East Asian travellers selected the unique pen less than 30 per cent of the time. Outline how culture might have contributed to the results of the study.

Exam ready

3 A similarity between vertical-individualist cultures and vertical-collectivist cultures is that

A equality is encouraged and is clearly embedded in the culture.

B not all members share the same level of status and inequality is accepted.

C competition is discouraged.

D individual expression and uniqueness are encouraged.

KEY CONCEPTS 8.1a

- A group is formed when two or more people interact and influence each other and share a common objective.
- Status refers to a person's position in the hierarchy of a group. The higher a person's status, the more power they exert.
- Social power refers to the amount of influence that an individual can exert over another person.
- There are five different bases of social power: reward, coercive, legitimate, referent and expert power.
- Cultures can be categorised as either collectivist or individualist.
- Collectivist cultures value group needs or interests over the interests of the individual. Individualist cultures, on the other hand, place the needs of the individual above the interests of groups.

9780170465052

Concept questions 8.1a

Remembering

1 Define the term 'group'. r

Understanding

2 Use an example to explain how a person's status in a group influences their social power. c

3 Identify three differences between individualist cultures and collectivist cultures. e

4 If there are five people in line at the supermarket checkout, are they classified as a group? Explain your answer with reference to the characteristics of a group. e

Applying

5 A teacher keeps a group of students back at recess for misbehaving. Identify and explain the type of power the teacher is using in this scenario. c

6 An apprentice carpenter on a work site asks their boss to show them how to correctly install a door. Identify and explain the type of power possessed by the boss in this scenario. c

HOT Challenge

7 Grant is running late for work, so he speeds on the freeway. In his rear-view mirror he notices the flashing lights of a police car. Grant pulls over to the side of the road and is subsequently given a fine by the police officer. Explain how legitimate power and coercive power are present in the scenario above. c

Stanford prison experiment

American psychologist Phillip Zimbardo was interested in how perceptions of status in a group influence the way people relate to one another. More specifically, he was interested in finding out whether prison guards and prisoners behave the way they do because of their personal characteristics or because of the roles they are given (Zimbardo et al., 1971).

Zimbardo placed an advertisement in the local newspaper seeking male university students to take part in a study of prison life. Of the 75 people who responded to the ad, Zimbardo chose 24 male participants. He selected only young men who showed stable personalities.

Using a coin toss, Zimbardo randomly allocated participants to the role of prisoner or guard (Figure 8.4) so that any individual differences between participants were spread equally across the experimental and the control groups.

To make the situation realistic, prisoners were collected from their homes by police in police cars. They were then driven to the simulated prison located in the basement of a building at Stanford University in California. Once they arrived at the 'prison', the prisoners were stripped of any individuality. They were ordered to undress, sprayed for lice, required to wear prison uniforms, assigned a number and ordered to answer to their numbers.

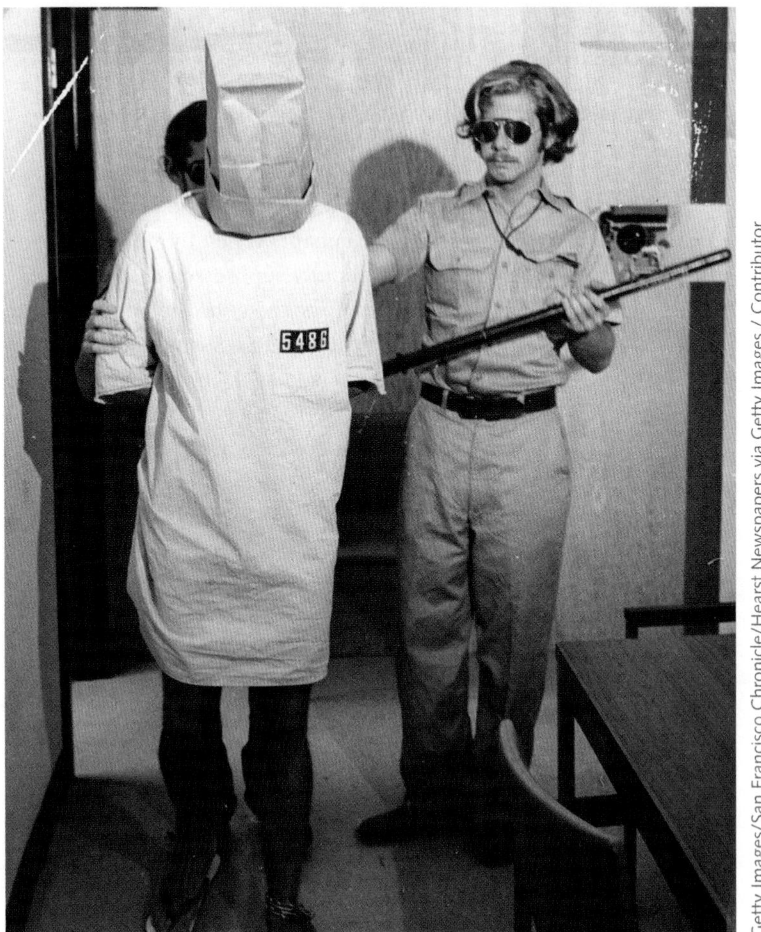

Getty Images/San Francisco Chronicle/Hearst Newspapers via Getty Images / Contributor

Figure 8.4 A 'prisoner' and 'guard' in Zimbardo's experiment

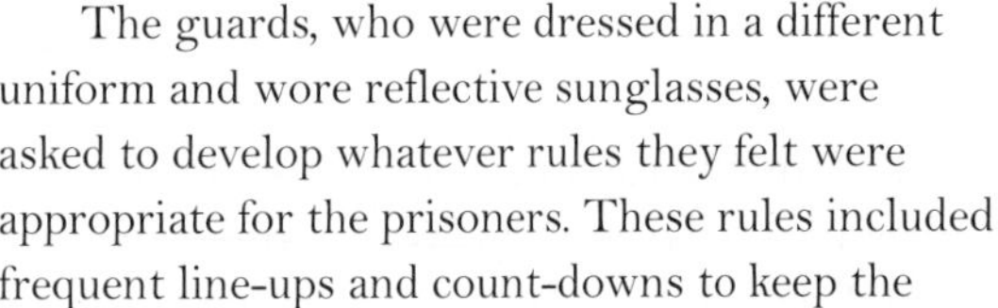

The guards, who were dressed in a different uniform and wore reflective sunglasses, were asked to develop whatever rules they felt were appropriate for the prisoners. These rules included frequent line-ups and count-downs to keep the prisoners in their place (Figure 8.5).

Figure 8.5 **Participants in the Stanford prison experiment**

After just 2 days in 'prison', the prisoners grew restless and defiant. When they staged a riot, the guards unmercifully suppressed the rebellion. Over the next few days, the guards clamped down with increasing brutality. In a surprisingly short time, the fake convicts looked, behaved and felt like real prisoners: they were dejected, traumatised, passive and dehumanised. Four of them had to be released because of hysterical crying and a confused state. Each day, the guards tormented the prisoners with more frequent commands, insults and demeaning tasks. After 6 days, the experiment was halted.

WB

8.1.4 ZIMBARDO'S EXPERIMENT ON THE EFFECTS OF PERCEIVED POWER ON BEHAVIOUR

According to Zimbardo, the assigned roles – prisoner and guard – were so powerful that in just a few days the experiment became 'reality' for those involved. Afterwards, it was difficult for many of the guards to believe their own behaviour. As one recalls, 'I was surprised at myself. I made them call each other names and clean toilets out with their bare hands. I practically considered the prisoners cattle' (Zimbardo et al., 1973).

Rethinking the Stanford prison experiment

Since the publication of the Stanford prison experiment, Zimbardo and his associates have faced criticism regarding whether the study was truly scientific, specifically with regard to its validity. Letters and interviews from the Stanford prison experiment archives showed that:

- Many of the extreme behaviours of both the guards and prisoners were not genuine. Guards received clear instructions from the researchers as to how they should treat the prisoners. Additionally, they may not have been told that they were participants in the study, with some assuming that they were serving as research assistants (Le Texier, 2019).
- Many of the guards felt it was their responsibility to behave as oppressive prison guards to benefit the study. One guard stated that 'most of the time I was conscious it was an experiment'. He also stated that he had to behave in certain ways 'or else the experiment won't come off right' (Le Texier, 2019, p. 830).
- Two of the more aggressive guards claimed to be acting the entire time. One stated that he was 'hamming it up'. Another wrote that he 'was always acting' (Le Texier, 2019, p. 830).
- Zimbardo was aware that at least two of the guards were acting. One guard wrote, 'I was always very conscious of the responsibility involved in the guards' and the experimenters' positions. I mentioned this to various people at various times, including to you [Zimbardo] during the debriefing' (Le Texier, 2019, p. 830).
- The study was scheduled to last 2 weeks and there were only three ways in which a prisoner could leave early: if they fell ill, if they had a nervous breakdown, or if they obtained special authorisation from the experimenters. One prisoner attempted to leave by faking a stomach ache. When his leave was rejected by Zimbardo, he successfully left the experiment by pretending to have a nervous breakdown.

In 2019, the American Psychological Association (APA) acknowledged criticism of the authenticity and validity of the Stanford prison experiment. The APA stated that the recent information regarding the Stanford prison

9780170465052

experiment provides useful opportunities to investigate researcher bias objective scientific inquiry, and acts as a reminder of the 'illusory truth effect': a mental error that occurs when we associate repetition of information with truth, in this case an experiment that has been studied for decades by psychology students of all ages (Bartels & Griggs, 2019).

KEY CONCEPTS 8.1b

- » The Stanford prison experiment investigated the influence of assigned roles (prisoner or prison guard) on the behaviour of individuals.
- » While considered ground-breaking at the time, the scientific validity of the study is now being debated.

Concept questions 8.1b

Remembering

1 State the aim of Zimbardo's experiment. i

Understanding

2 Identify the independent and dependent variables in the Stanford prison experiment. i

3 What is random allocation and how was it used in the Stanford prison experiment? i

4 Identify and explain the type of power that the guards displayed in the Stanford prison experiment. e

5 What methods did Zimbardo use to indicate to the participants that the guards had a higher status and greater social power than the prisoners? e

6 Outline three criticisms of the Stanford prison experiment. e

Applying

7 In 2004, photos emerged from Abu Ghraib prison in Iraq. US soldiers had forced detainees in the prison to wear leashes, pile themselves on top of each other, and stand in degrading poses. Many of the prisoners were tortured and one died in custody.
Outline how the concepts of status and power can explain the actions of US soldiers at Abu Ghraib prison. c

HOT Challenge

8 Recent revelations suggest that some of the guards in the Stanford prison experiment might have been unaware they were actually participants in the study, rather than assistants to the researchers. How could this have influenced the results of the study? i

8.2 Obedience and conformity

The presence of individuals and groups can have a significant influence on individual behaviour. This can result in a person behaving in ways that are not in keeping with how they would behave when not in the presence of others. Two effects that can happen are obedience and conformity.

Obedience

Obedience occurs in situations in which people change their behaviour in response to direct commands from others. A person with higher status and social power is more likely to be obeyed (Birchnell, 1993). This does not mean that you necessarily agree with the command, only that you obey and do what you are told to do. For instance, you may pick up some rubbish in the school yard if a teacher asks you to do so (obeying a direct command), even if you think that you should not have to do it because you didn't drop the rubbish. In some situations, we may ignore the norms that usually govern our lives, and obey direct orders that are inconsistent with those norms. Under direction from others, we may even carry out acts that would normally be unthinkable to us and that we may even feel are morally wrong.

8.2.1 OBEDIENCE

Obedience requires the issuing of a direct command; therefore, it is more direct and less subtle than social power. However, whether the individual obeys a command might depend on the social power of the person issuing the command.

Milgram's studies of obedience

The classic studies of obedience were conducted in the 1960s by Stanley Milgram. Milgram wanted to see at what point people would disobey a direct command issued by an authority figure, in this case, an experimenter.

In the initial study in 1963, Milgram studied 40 males aged 20 to 50 years. These men were drawn from a range of occupational backgrounds and included clerks, engineers and teachers (Figure 8.6).

Participants arrived at a laboratory and were asked to draw slips of paper to determine if they would be the 'teacher' or 'learner' in the experiment. A confederate (research actor) was always assigned to the role of 'learner', with the legitimate participants serving as 'teachers'. The teachers were instructed to read a list of word pairs to the learner. The learner was then tested on their ability to recall the list. An experimenter (authority figure) was positioned next to the teacher to ensure the experiment ran correctly.

Public Announcement

WE WILL PAY YOU $4.00 FOR ONE HOUR OF YOUR TIME

Persons Needed for a Study of Memory

*We will pay five hundred New Haven men to help us complete a scientific study of memory and learning. The study is being done at Yale University.

*Each person who participates will be paid $4.00 (plus 50c carfare) for approximately 1 hour's time. We need you for only one hour: there are no further obligations. You may choose the time you would like to come (evenings, weekdays, or weekends).

*No special training, education, or experience is needed. We want:

Factory workers	Businessmen	Construction workers
City employees	Clerks	Salespeople
Laborers	Professional people	White-collar workers
Barbers	Telephone workers	Others

All persons must be between the ages of 20 and 50. High school and college students cannot be used.

*If you meet these qualifications, fill out the coupon below and mail it now to Professor Stanley Milgram, Department of Psychology, Yale University, New Haven. You will be notified later of the specific time and place of the study. We reserve the right to decline any application.

*You will be paid $4.00 (plus 50c carfare) as soon as you arrive at the laboratory.

TO:
PROF. STANLEY MILGRAM, DEPARTMENT OF PSYCHOLOGY, YALE UNIVERSITY, NEW HAVEN, CONN. I want to take part in this study of memory and learning. I am between the ages of 20 and 50. I will be paid $4.00 (plus 50c carfare) if I participate.

NAME (Please Print)....................................

ADDRESS....................................

TELEPHONE NO. .... Best time to call you ....

AGE........ OCCUPATION.................... SEX......
CAN YOU COME:

WEEKDAYS .... EVENINGS ...... WEEKENDS.........

Alamy Stock Photo/GRANGER - Historical Picture Archvie

Figure 8.6 The advertisement used to recruit participants for Milgram experiment

Each time the learner made a mistake, the teacher was required to administer an electric shock by pressing a lever on a shock generator. The shocks ranged from 15 volts (slight shock) to 450 volts (danger: severe shock). The voltage increased by 15 volts for each incorrect response. Of course, the 'learner' didn't actually receive any shocks, but they were instructed to respond in particular ways to each 'shock'.

As the intensity of shocks increased, the learner would make audible protests, complaining of pain, asking the teacher to stop and so on. If the teacher demonstrated hesitation or wanted to withdraw from the experiment, the experimenter would instruct the teacher to continue, using prompts such as 'the experiment requires you to continue' and 'it is absolutely essential that you continue' (Milgram, 1963). Eventually, the learner would stop responding. The experimenter would then instruct the teacher to continue to administer shocks, saying that no response should be counted as an incorrect response.

Once the experiment ended, the teacher was interviewed by the experimenter. They were told that the learner did not receive any shocks, but the dilemma of the teacher was quite real. 'Teachers' protested, sweated, trembled, stuttered, bit their lips and laughed nervously. Clearly, they were disturbed by what they were doing, but most obeyed the experimenter's orders.

The experiment was designed to test at what point people would disobey. The results surprised everyone, even Milgram. Few participants withdrew from the experiment before delivering shocks of 300 volts, and 65 per cent of participants delivered the maximum voltage of 450 volts (Figure 8.7).

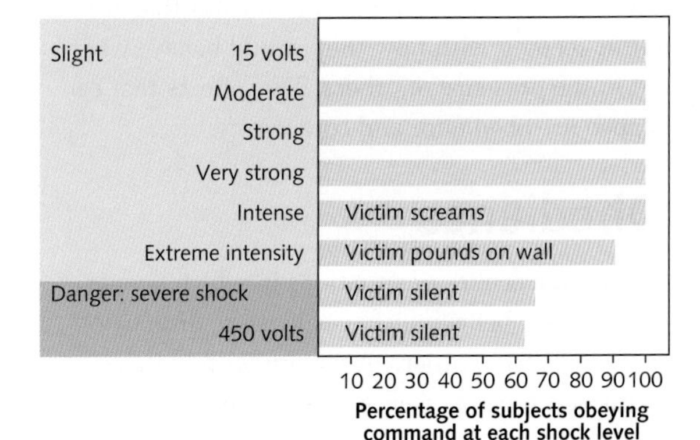

Figure 8.7 The level of shock shown on the shock generator, and the response of the 'learner'

Milgram's conclusions

Milgram concluded that the participants were influenced by the gradual nature of the situation. Participants took on a task and gradually found that it was more and more inconsistent with their beliefs about what they should and should not do. However, there was no clear cut-off point at which to switch from obeying what seemed to be a reasonable request to disobeying what seemed to be an unreasonable command. The participants became locked into the situation. By obeying the first small command, they felt they had to keep going and obey larger and larger commands.

Factors that affect obedience

Milgram's research suggests that a number of environmental factors have an effect on obedience, including social proximity and the legitimacy of authority figures.

Social proximity

Whether a person will or won't obey orders is partly dependent on **social proximity**, or the social distance between two parties. In the case of Milgram's experiments, the social proximity was the distance between the person inflicting the harm (the teacher) and the person being harmed (the learner), and this influenced the level of obedience. The greatest level of obedience occurred in the situations in which the learner was more remote from the teacher.

When the learner was in a different room from the teacher and the teacher could only see the learner's responses on a display panel in front of him (Figure 8.8), the proportion of participants obeying the experimenter's directions was highest. When the learner was able to be seen or touched by the teacher, the proportion of participants administering the high level of shock decreased (Figure 8.9).

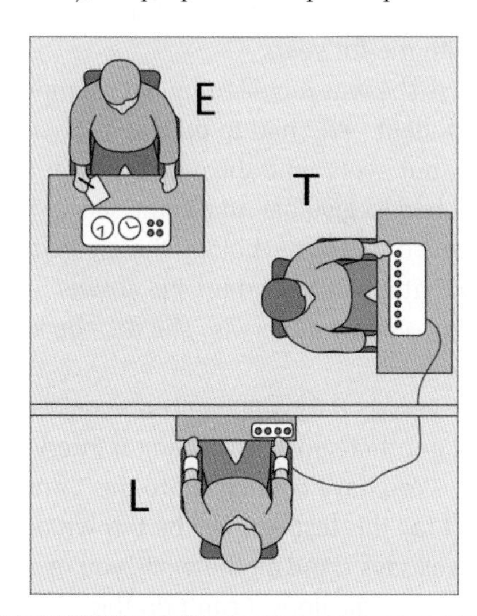

Figure 8.8 'Experimenter', 'teacher' and 'learner' in Milgram's experiment

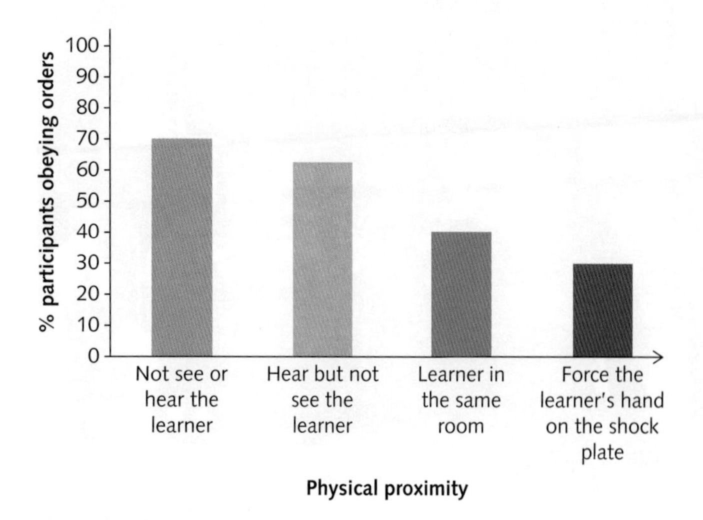

Figure 8.9 The physical distance from the 'learner' had a significant effect on the percentage of 'teachers' obeying orders

Legitimacy of authority

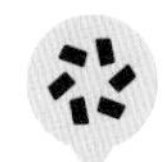

Worksheet
Miligram's Studies of Obedience

Several factors influence participants' obedience to authority in this type of situation. **Legitimacy of authority** is when people assign authority to an individual and are obedient to their instructions. Milgram's study was conducted at a well-known university by a high-status figure. The higher status attributed to a person who issues an instruction or command can lead participants to the conclusion that they are not responsible for their actions. They may conclude that the person who issues the instruction or command is responsible for their behaviour and they were only following orders. Milgram's participants believed that important scientists must know more about these things than they do, and that it was the scientists who were responsible for the participants' behaviour. To test this idea Milgram replicated the experiment on another site, away from the university. Similar results were obtained, but fewer participants fully obeyed the command to keep going.

8.2.2 MILGRAM'S EXPERIMENT ON OBEDIENCE

The implications of Milgram's studies

Milgram initially conducted his experiments about obedience because of his concern about the killing of European Jews from 1933 to 1945. He was interested in when people obey orders issued by others and when they obey the dictates of their own conscience (Figure 8.10).

8.2.3 REPRODUCING MILGRAM'S FINDINGS

Getty Images/Keystone-France/Gamma-Keystone

Figure 8.10 The persecution of Jews and the establishment of concentration camps in Europe during WWII were one result of people obeying orders. This image shows prisoners being liberated from a concentration camp in Germany in 1945

His research raises questions about our willingness to commit anti-social or inhumane acts commanded by a 'legitimate authority'. The excuse so often given by war criminals – 'I was only following orders' – takes on new meaning in this light. Milgram suggested that when directions come from authority, people rationalise that they are not personally responsible for their actions. Others have pointed out that 'crimes of obedience' may be committed by ordinary people responding to normal social psychological processes. In locales as diverse as Vietnam, Rwanda, Bosnia, South Africa, Nicaragua, Sri Lanka and Laos, for example, the tragic result has been 'sanctioned massacres' of chilling proportions (Kelman & Hamilton, 1989).

Even in everyday life, 'crimes of obedience' are common. To keep their jobs, many people obey orders to do things that they personally regard as dishonest, unethical or harmful (Hamilton & Sanders, 1995). These range from a salesperson stretching the truth to sell a product, to encouraging someone to take out a loan that they probably cannot afford.

CASE STUDY 8.1

La Trobe 'torture' study anguish

Tim Elliott

In 1973, arts student Dianne Backwell tortured her roommate to death. Or so she thought.

Ms Backwell, then a 19-year-old student at La Trobe University, believed she was taking part in research into the effect of punishment on learning. But the friend whose screams she heard from another room every time she pushed a button was only pretending to receive electric shocks.

Nonetheless, the experiment, record of which has only now come to light, traumatised Ms Backwell for years. According to the book, *Behind the Shock Machine*, by Melbourne psychologist Gina Perry, Ms Backwell was one of about 200 La Trobe students who took part in 1973 and 1974 in controversial experiments conducted by the university's psychology department.

The experiments were modelled on the notorious 'obedience tests' carried out by US psychologist Stanley Milgram at Yale University in 1961, in which participants were ordered to shock students in another room, even when they believed it would kill them. Dr Milgram's study caused an uproar when it was published in 1963, not least because he linked his findings to the trial of prominent Nazi Adolf Eichmann.

Several students at La Trobe University in the 1970s say they suffered long-term damage from participating in the Australian experiments.

'I was humiliated, ashamed, angry,' says Ms Backwell, who claims the effect of the experiment 'stayed with me for years'.

She says she was recruited by her roommate, a fellow student. 'All I had to do was ask her questions, but every time she didn't answer correctly I had to give her an electric shock. She also said she'd failed psychology the previous year, and that it was important she answer everything correctly, otherwise she'd … be kicked out of uni.'

Whenever Ms Backwell baulked at delivering her friend another shock, the monitor intervened. 'He'd say, "You have committed to this", and "She can't fail this test or she'll be thrown out",' Ms Backwell says. 'And all the while you're thinking this is ridiculous, I can't do this.'

»

Finally, Ms Backwell's friend let out a 'distressingly loud scream, like she was in real pain', then went silent. 'I remember thinking: "I killed her." And then this ridiculous thought went through my head, that if I give her one more shock it won't matter because she's dead.'

Just then, her friend emerged from the other room. 'She had this big grin on her face, and she said: "We were just testing to see how far you'd go, and you went all the way! You would have killed me!"'

The man in charge of the La Trobe experiments was Dr Bob Montgomery, now a clinical and forensic psychologist on the Gold Coast and, until 2010, president of the Australian Psychological Society. Dr Montgomery stands by the experiments. He also says everyone was debriefed afterwards, 'probably in groups'.

But Ms Backwell and others deny this. Former psychology student Maurie Hasen, who took part in 1974, says: 'I don't think any of us were debriefed by Bob Montgomery, nor … by any of his staff.'

Mr Hasen was left feeling 'shame and remorse' after deceiving a friend into taking part. (The two men never spoke again.) 'We were being quite obedient ourselves: we had to pass our course, and so we just did the experiment. In the end, I don't know that we had the maturity to argue about the ethical situation we were in.'

Source: Elliot, T. (2012, April 26). La Trobe 'torture' study anguish. The Age https://www.theage.com.au/national/victoria/la-trobe-torture-study-anguish-20120425-1xlmf.html

Weblink
ABC report: Psychology study leaves Australians in shock

Questions

1 What was the aim of the experiment?
2 What sampling technique was used to recruit the participants?
3 Explain how deception was used in the study.
4 Participants in the study have suggested debriefing didn't occur. Outline the negative psychological effects experienced by Ms Backwell following the study and suggest how debriefing might have reduced them.
5 At lunchtime, Kim and Emma left rubbish at their table and walked away. The teacher on yard duty instructed them to clean up their mess, but they ignored her. When the school principal asked them to clean up their mess, they did.

Which factor that affects obedience best explains the students' behaviour in this situation? Provide a reason for your answer.

KEY CONCEPTS 8.2a

» Obedience occurs in situations in which people change their behaviour in response to direct commands from others.
» Milgram's study demonstrated that individuals are willing to inflict serious physical harm if instructed to by an authority figure.
» Obedience is influenced by several factors, including social proximity and the legitimacy of the authority figure.

Concept questions 8.2a

Remembering

1 Define the term 'obedience' as it relates to psychology. i

Understanding

2 Explain how social proximity and the legitimacy of authority figures influences levels of obedience. e
3 Complete a flowchart for Milgram's study of obedience. Your flowchart should include the aim, hypothesis, procedure, results and conclusion of the study. e

Applying

4 Provide an example of a scenario where it would be necessary or beneficial to obey an authority figure. c

HOT Challenge

5 Identify two ethical guidelines that were breached in Milgram's studies of obedience and explain how they were breached. Do you believe Milgram would have achieved the same results for his study if these ethical guidelines weren't breached? i

Conformity

Conformity occurs in situations in which individuals change their behaviour as the result of real or implied pressure from others. The process of conformity was first studied experimentally by Solomon Asch in the 1950s through a series of studies that are now seen as classics in the psychological literature (Asch 1952, 1955).

Asch's studies

As a participant in one of Asch's experiments, you would walk into a room with about four other participants, who were confederates of the experimenter. You would sit at the side of a long table and assist the experimenter in a perception task. You would look at a vertical line (called the standard line) and then look at three vertical lines (called the comparison lines). Then you would indicate which line is the same length as the standard line (Figure 8.11).

In the first few trials, all participants give the same, obviously correct answer. Then, in the next trial, the four people before you all give the same answer, an answer which to you appears to be obviously incorrect. What do you do? Do you give the correct answer, or do you agree with the preceding four participants and give the answer they have already given?

Data presented by Asch (1955) suggested that 75 per cent of participants 'made a mistake' and went along with the group on at least one occasion, although 24 per cent of participants never conformed.

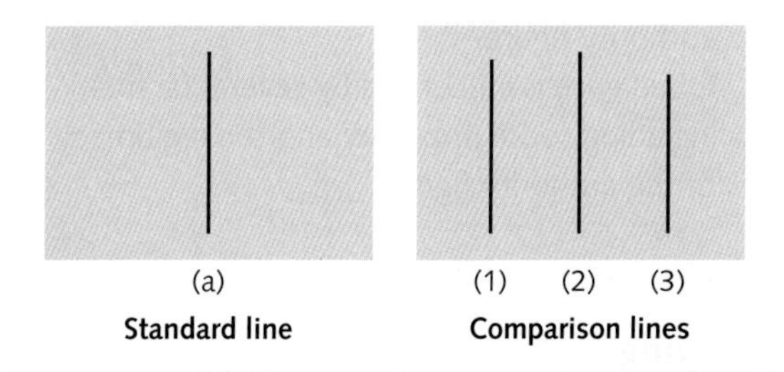

Figure 8.11 An example of the judgement task in an Asch conformity experiment

Weblink
Asch conformity experiment

In contrast, people tested alone erred in less than 1 per cent of their judgements. Those who yielded to group pressures were clearly denying what their eyes told them, bowing to the pressure of the group. The other participants gave the wrong answer because they were confederates. Asch's study showed that participants were strongly influenced by the actions of those around them.

Factors that affect conformity

A number of factors affect conformity, including normative influence, informational influence, unanimity, group size and deindividuation. These are summarised in Table 8.3 (page 322).

Normative influence

Normative influence refers to the impact of the established behaviour of the group, termed group norms, on the likelihood of a person conforming. If a group you belong to, such as your family, has an established norm (for example, that shoes are not worn inside the house), then it is likely that you will conform to this. To avoid being ridiculed by our friends and family, we have learned to conform to group norms.

Group norms are standards that are set by the groups to which we belong. They may not be explicitly stated but they are established patterns of behaviour. This may include the way you dress (Figure 8.12), the time you arrive at a party, or the way you speak.

The influence of others can lead us to conform to differing degrees. If we conform only in the presence of group members, this is referred to as public conformity. In this situation your beliefs are not consistent with your behaviour. Asch's study was an excellent example of this. In the case of private conformity, your beliefs and behaviour are consistent. You are therefore likely to exhibit similar behaviour when you are alone to what you would in a group context.

Shutterstock.com/Yuriy Rudyy

Figure 8.12 Peer groups often dress and act similarly

Informational influence

When we visit a foreign country, or even when we move from primary school to secondary school, or from school to work, we observe the behaviour of others to find out how we should behave. When we are influenced by the behaviours of people around us and these behaviours shape our actions, it is called **informational influence**. Informational influence also impacts on conformity, as we look at others' behaviour to guide us in whether we should behave the same way. For instance, if you were to visit an unfamiliar place of worship, such as a church or a mosque, you would look at the behaviour of others to provide an example of how to act, such as where to sit or stand (Figure 8.13).

Alamy Stock Photo/Cultura RM

Figure 8.13 People who are not Buddhist and visit a Buddhist temple would look at the behaviour of locals for guidance on how to act

Unanimity

When everyone in the group agrees or does the same thing, this is referred to as **unanimity**. A unanimous group is a very powerful influence for conformity. In Asch's (1952) studies, there was unanimous agreement from the group about which line was the correct line and this had a strong influence on the participants' behaviour. If we have one supporter (or ally) then we are unlikely to be influenced by the group. In terms of how much influence a unanimous group can have, a unanimous group of three is more powerful than a group of eight that has one dissenting person.

Group size

The extent of group influence also depends on the size of the group. Studies have shown that the number of conforming participants increases dramatically as the size of the group grows from two to four. This means that if you were in a group with just one other person you would be less likely to conform than if there was a group of four people where the other three were all doing the same thing. However, a majority of four produces about as much conformity as a majority of eight. The next time you want to talk someone into (or out of) something, take two friends along and see what a difference it makes.

8.2.4 CONFORMITY

Deindividuation

Sometimes, particularly when we are in a large group, our behaviour can be influenced by a sense of anonymity, and this can lead us to behave in a way that we never would when alone. For instance, crowds of students in a cafeteria may throw food around and crowds at the football will scream at the umpire. Psychologists explain this behaviour in terms of deindividuation. To be deindividuated is to be less self-conscious and less inhibited, becoming excited by the current situation and ignoring any possible consequences of our actions. **Deindividuation** is the loss of social identity and inhibition when in a group, which results in a person engaging in behaviours with little regard for potential consequences (Figure 8.14). The group behaviour acts as a guide for behaviour: the individual feels that 'if everyone else is doing it, it must be okay'.

Shutterstock.com/Ververidis Vasilis

Figure 8.14 Fans at a soccer match in Europe clash with police. The behaviour of many of the fans could be partly attributed to deindividuation

Data calculator

Logbook template: Fieldwork

Table 8.3 Factors that affect conformity

Factors that affect conformity	Explanation
Normative influence	Our likelihood to conform is influenced by the established behaviour of the group. The group can be friends, family or society.
Informational influence	We look at the behaviour of others to guide us in whether we should behave in a similar way.
Unanimity	We are more likely to conform when other members of the group behave in the same way.
Group size	Our conformity increases as the group size expands from two to four, then stays at similar levels of conformity in a group of eight.
Deindividuation	When we are in a large group, our behaviour can be influenced by a sense of anonymity, and this can lead us to behave in a way that we never would when alone.

INVESTIGATION 8.1 CONFORMITY AT SCHOOL

Scientific investigation methodology

Fieldwork

Aim

To investigate examples of conformity within the school environment

Introduction

Conformity is adjusting one's beliefs, judgements or actions to bring them into agreement with those of a social group or situation. Conformity occurs when individuals change their behaviour as the result of real or implied pressure from others in a range of social settings including schools and professional settings. Research suggests that conformity is influenced both by a need to fit in and a belief that other people are smarter or better informed than you are. Factors that affect conformity include normative influence, informational influence, unanimity, group size and deindividuation. Revise information about these factors using your textbook or other resources.

Procedure

In this investigation you will develop, conduct and analyse a questionnaire that investigates conformity in your school environment. You are to work with your classmates to develop a questionnaire that helps fellow students to identify examples of when they have conformed and why. You may choose to conduct the survey on a range of students across different year levels to identify differences in conformity across age groups. Once you have carried out the survey, analyse the questionnaire responses to identify the range of factors affecting conformity.

Results

Record, organise and present data using an appropriate table and/or graph. Process any quantitative data using appropriate mathematical relationships and units.

Discussion

1 What was the size and make-up of your sample?
2 Do your data show any patterns or trends?
3 Were the trends different across the different year levels that you surveyed?
4 Were there any questions in your survey that produced a large range of responses? If yes, why might this have been?
5 Identify any limitations of your survey.
6 What would you do differently if you were to repeat this investigation?
7 Draw an evidence-based conclusion that is consistent with your results.

ANALYSING RESEARCH 8.1

Deindividuation and Halloween behaviour

A naturalistic study by Diener et al. (1976) observed American children during Halloween to see whether the presence of a group and deindividuation would influence anti-social behaviours.

Either alone or as part of a group, children who were trick-or-treating entered the front of a house where there was a bowl of candy and a bowl of coins. A female research assistant greeted each child or group and told them that they could take *one* piece of candy. If the children asked about the money, they were again told that they could take *one* piece of candy. The experimenter then said that they had work to do and left the children with the bowls of candy and coins. The research assistant then observed the behaviours of the children through a peep-hole.

Deindividuation was created in the study by the children either being 'anonymous' or 'identified'. The researcher would comment on each child's costumes and then asked approximately half of them their name and where they lived ('identified'), while the other children weren't asked their names or addresses ('anonymous').

Results

Study condition	Percentage who stole money or took extra candy
Identified and alone	7.5%
Identified and in a group	20.7%
Anonymous and alone	21.4%
Anonymous and in a group	57.2%

The results showed that children were significantly more likely to take money or extra candy when in a group as opposed to alone. Additionally, the children were more likely to take money or extra candy when they were anonymous as opposed to identified.

Questions

1 Write an aim for the study.
2 Outline the sampling procedure that was used in this study.
3 Outline two ethical guidelines that would need to be adhered to for this study to be considered ethical.
4 Write a conclusion for the study.

HOT Challenge

5 Two individual variables were investigated in this study. What were they?

Source: Diener et al. (1976). Effects of deindividuation variables on stealing among Halloween trick-or-treaters. *Journal of Personality and Social Psychology*, *33*(2), 178–183.

KEY CONCEPTS 8.2b

- Conformity occurs when individuals change their behaviour as the result of real or implied pressure from others.
- Normative influence refers to the established norms (rules) a group follows.
- Informational influence occurs when we look to others' behaviour to guide our own behaviour.
- People are more likely to conform when the group is unanimous in its decision.
- Deindividuation is the loss of social identity and inhibition when in a group, which results in a person engaging in behaviours with little regard for potential consequences.

Concept questions 8.2b

Remembering

1 Define the term 'conformity' as it relates to the study of psychology. r

Understanding

2 Create a flowchart for Asch's studies of conformity. Ensure you include the aim, hypothesis, procedure, results and conclusion of the study. e

Applying

3 In July 2021, thousands of ticketless English soccer fans attempted to break into a London stadium to watch the Euro final between England and Italy. The crowd clashed with police, threw beer bottles and damaged property in surrounding areas. Explain how deindividuation could account for the behaviour of some of these soccer fans. c

HOT Challenge

4 In Asch's study, participants gave their responses regarding the length of the lines after the confederates gave incorrect responses. If you were to replicate Asch's study, why would it be advisable to use this approach in a study of conformity? i

Decision-making in groups

As seen throughout this chapter, a group can bring about clear changes in the behaviour of its members. These range from selecting clearly incorrect responses to fit in with the group, to making decisions to maintain group harmony. Research has also shown that groups tend to make more drastic decisions than individuals.

Groupthink

In May 1996, two teams of climbers readied themselves to reach the summit of Mount Everest. Both teams were led by experienced and internationally renowned climbers; however, by the following day, five people were dead – including the leaders of the two groups. How did a small expedition, led by experienced climbers, end in such tragedy? The answer can be partly attributed to groupthink.

Groupthink occurs when the members of a group prioritise the strong bonds of the group over clear decision-making. When groupthink occurs within a group, there is a tendency to evaluate every decision or judgement in relation to how it could influence the cohesion of the group, which leads to close-mindedness. In the scenario above, the climbers had become a reasonably close-knit team over their journey up the mountain. Additionally, despite having strict safety protocols in place in the case of bad weather, the leaders chose to disregard them. The pursuit of the goal and the unity of the group prevented clear decision-making. The symptoms or 'warning signs' of groupthink are set out in Table 8.4.

8.2.5 SOCIAL FACTORS THAT AFFECT GROUP DECISION-MAKING

Table 8.4 The eight symptoms of groupthink

Symptom of groupthink	Effect on the group
1 Invulnerability (Feeling unable to be harmed)	Members of the group develop an illusion of invulnerability, which encourages excessive levels of optimism and leads to risk-taking behaviours.
2 Rationale (Rationalising reasons for a course of action)	Members of the group discount warnings and downplay negative feedback that could challenge the assumptions of the group.
3 Morality (Distinguishing between right and wrong)	Members of the group believe that they are morally justified in their decisions, consequently ignoring the possible ethical consequences of their actions.
4 Stereotypes (Applying widely held, over-simplified views about a particular group)	Members of the group apply negative stereotypical views to people in an opposing group. For example, they may perceive them as inherently bad, weak, untrustworthy or stupid.

Symptom of groupthink	Effect on the group
5 Pressure (Using persuasion and coercion)	Members of the group apply direct pressure to any member who expresses doubt, or opposing views or opinions.
6 Self-censorship (Remaining quiet to avoid criticism)	Members of the group actively avoid challenging the group consensus. They may do this by remaining silent or by downplaying their doubts.
7 Unanimity (Expecting agreement by all people involved)	Members of the group believe that the decisions of the group are shared by all. By extension, remaining silent during a discussion is taken to mean that the individual is in full support of the group.
8 Mindguards (Filtering out any negative information)	Members of the group adopt the role of 'mindguards' to prevent the leader and members of the group from exposure to information that challenges the decision-making of the group.

iStock.com/SDI Productions

Figure 8.15 When members of a group prioritise maintaining high levels of cohesion over achieving their objectives, they become susceptible to groupthink

Close-knit groups are likely to show the symptoms and experience the consequences of groupthink (Figure 8.15). However, there are a number of ways in which groupthink can be prevented (Janis, 1971).

These include:

- each member being given the role of 'critical evaluator', with priority given to sharing objections and doubts
- the leader not sharing their opinion about the topic before the group starts work
- at specific points during the decision-making process, members discussing the group's progress with someone they trust
- the group inviting experts who are not part of the group to attend meetings and critically evaluate the group's work
- at least one member playing 'devil's advocate' at meetings that require the group to evaluate its progress
- when a decision affects others, the group spending time considering the consequences of its decision and the likely response from those affected.

WB 8.2.6 TRUE OR FALSE

ACTIVITY 8.2 OBEDIENCE AND CONFORMITY

Try it

1 Your Mission: Lost on the Moon

Your spaceship has just crash-landed on the lighted side of the Moon. You were scheduled to rendezvous with the mother ship 200 miles away, also on the lighted surface of the moon, but the rough landing has ruined your ship and destroyed all equipment on board, except for the 15 items listed on the next page.

Your crew's survival depends on reaching the mother ship, so you must choose the most crucial items available for the 200-mile trip. Your task is to rank the 15 items in terms of their importance for survival. Place number 1 by the most important item, number 2 by the second most important item, and so on through to number 15 (the least important item).

My own rank		Our group's ranks
______	Box of matches	_____
______	Food concentrate	_____
______	Fifty feet of nylon rope	_____
______	Solar-powered portable heating unit	_____
______	Two. 45 caliber pistols	_____
______	One case of dehydrated milk	_____
______	Two 100-pound tanks of oxygen	_____
______	Stellar map of the moon's constellations	_____
______	Self-inflating life raft	_____
______	Magnetic compass	_____
______	Five gallons of water	_____
______	Battery-operated signal flares (with batteries)	_____
______	First-aid kit containing needles, vitamins, medicines etc.	_____
______	Solar-powered FM receiver-transmitter	_____

Now that you've decided your ranks, do this again as a group. Share your individual solutions and reach a consensus – one ranking for each of the 15 items that best satisfies all group members. Put the group's ranks in the 'Our group's ranks' section.

Source: Mueller, J. (2000–2022). Resources for the teaching of social psychology. http://jfmueller.faculty.noctrl.edu/crow/group.htm

Apply it

2 In 1986, the space-shuttle Challenger exploded 73 seconds after lift-off. Senior staff at NASA had been warned by the manufacturing company, Morton Thiokol, that the rubber seals on the shuttle could not handle the low morning temperatures expected before lift-off. Senior staff at NASA dismissed the concerns and urged members of the manufacturing company to change their evaluation so that the mission could continue, which they did – ultimately resulting in disaster.

Explain how the groupthink symptoms of 'rationale' and 'pressure' apply to the decision-making of NASA and the company Morton Thiokol.

Exam ready

3 Asch (1955) investigated the influence of conformity in a group setting. Asch's experiment utilised deception by telling the participants that they were taking part in a visual perception test.

Given that Asch used deception, he should have ensured that he

A obtained informed consent from the participants before starting the study.

B debriefed the participants before starting the study.

C debriefed the participants following the study.

D obtained informed consent from the participants following the study.

Adapted with permission from Jon Mueller, Professor of Psychology, North Central College, http://jfmueller.faculty.noctrl.edu/crow/lotmquestions.pdf

Group shift

Group shift, also known as group polarisation, is the tendency for group members discussing an issue or dilemma to adopt a more extreme position than their individual position before the discussion. This shift in decision-making can go in either direction: towards a more dangerous approach or towards a more cautious or safe approach. For example, a person may be opposed to eating animals. However, following a discussion with like-minded individuals, their opposition to eating meat becomes even stronger.

When considering dilemmas (where a difficult decision must be made), group shift can occur in one of two ways (Figure 8.16). It can move towards riskier behaviour (risky shift) or away from riskier behaviour (cautious shift). If a group is made up

iStock.com/AlexSava

Figure 8.16 When group shift occurs, the group adopts a more extreme approach than the individual members would have adopted, which can lead to taking greater risk (risky shift) or taking more care (cautious shift)

largely of individuals who don't like risk, it tends to favour an even more cautious approach. If it is made up largely of individuals who aren't strongly opposed to risk, it tends to favour an even riskier approach.

Causes of group shift

Research has shown that groups tend to make more extreme decisions than individuals, but what accounts for this disparity in decision-making? Group shift has been attributed to the social comparison theory, the persuasive argument theory and the use of social decision schemes.

- **Social comparison theory** proposes that the tendency to evaluate or compare ourselves to others influences our self-image and our wellbeing. To be accepted we behave in a way we think fits in with the group norm. Social comparison theory is discussed in greater detail later in the chapter.
- The persuasive argument theory proposes that when a group discusses a moral dilemma, members will offer a number of arguments for and against. If they favour a certain approach, the number of arguments they offer that support that approach will outweigh the number of argument against that approach. As the members continue to generate more arguments in favour of their overall approach, the position of the group tends to become more extreme (Sieber & Zeigler, 2019; Burnstein & Vinokur, 1977).
- Social decision schemes are strategies used by groups to select one approach from various approaches. They can be explicit, such as voting and going with the majority, or implicit, where the decision tends to favour the most powerful members of the group.

KEY CONCEPTS 8.2c

- Groupthink is a phenomenon that occurs when a members of a group prioritise the cohesion of the group over objective decision-making.
- Groupthink is characterised by the use of stereotypes, discouraging differing opinions, unanimity in decision-making, and downplaying of risk.
- Group shift is the tendency for group members discussing an issue or dilemma to adopt a more extreme position than their individual position prior to the discussion.
- Cautious shift occurs when the group suggests an even safer approach than an individual would have taken if acting alone.
- Risky shift occurs when a group demonstrates less caution than the individual members would have if they were acting alone.
- Group shift can be attributed to social comparison theory, the persuasive argument theory and social decision schemes.

Concept questions 8.2c

Remembering

1 What is groupthink and how does it influence group decision-making? r

2 What is group shift and how does it influence group decision-making? r

Understanding

3 Identify and explain the three causes of group shift. e

4 Using an example, explain how risky shift differs to cautious shift. c

Applying

5 Imagine you are part of the student leadership group that is organising the school formal. Outline three techniques that could be used to reduce groupthink in this situation. c

HOT Challenge

6 Explain how groupthink and group shift demonstrate the concept of conformity. e

8.3 Media and mental wellbeing

Recently we've seen a shift in information access from traditional sources, such as print newspapers, TV news and magazines, to digital information sources, such as social media, streaming services and online news sites. This change in how information is accessed has had both positive and negative effects on people's mental health and wellbeing.

Social connections

Social connection (also known as **social connectedness**) refers to the belief that we belong to a group and generally feel close to other people. Strong social connections have a beneficial effect on our mental health because they help foster empathy, higher self-esteem and the development of trusting relationships. Social connection begins from birth and continues throughout our lifespan. The bond between a child and its caregivers serves as our first link to others. Throughout childhood, adolescence, adulthood and old age, we continue to form links to others. However, these links may weaken due to loss of friendship, changes in work environment and potential exclusion from groups (Lee et al., 2001).

Research shows that people with high levels of social connection tend to easily identify with others, feel more connected to others, and participate in social activities. Those who report lower levels of social connection (potentially linked to abandonment, isolation, rejection etc.) tend to feel misunderstood, have difficulty relating to others, and see themselves as 'outsiders' (Lee et al., 2001). These individuals also tend to experience higher levels of anger, anxiety and loneliness (Baumeister & Leary, 1995). An extensive New Zealand study demonstrated that a person's level of social connection is also a strong predictor of their mental health. The 21 227 participants were surveyed at five different points from 2010 to 2013. The results of the study indicated those with higher levels of social connection tended to have better mental health, while those with lower levels of social connection tended to have poorer mental health (Saeri et al., 2018).

Social media and mental wellbeing

Social media has radically changed the way we communicate (Figure 8.17). Recent estimates suggest that four billion people use some form of social media (Cheng et al., 2021). Studies indicate that social networking has a positive

iStock.com/bokan76

Figure 8.17 The rise of smart devices has changed the nature of social connections impact on our social connections (Valkenburg & Peter, 2009). However, prolonged exposure to social media can lead to low self-esteem, because of comparisons we make between ourselves and those we see online.

Social media provides us with immediate exposure to current events, the ability to maintain connections and communicate across geographic barriers, and social inclusion for those who may be otherwise excluded in their day-to-day lives; for example, sexually diverse and gender-diverse individuals (Riehm et al., 2019).

Social media and social comparison theory

First outlined by American psychology Leon Festinger (1954), social comparison theory suggests that to gain an accurate understanding of who we are, we have a drive to make decisions, evaluations and judgements about ourselves in relation to those around us.

We make three types of social comparisons. These are:

- *Upward social comparison*: we compare ourselves to those we view as superior or possessing positive characteristics. This:
  - helps us to make decisions, regulate our emotions and improve our wellbeing, and inspires us.
  - can lead to feelings of inadequacy, negative mood and low self-esteem (Vogel et al., 2014).
- *Downward social comparison:* we compare ourselves to those we view as inferior or possessing negative characteristics This.
  - improves our mood and our evaluations of ourselves (Vogel, et al., 2014).
- *Lateral social comparison:* we compare ourselves to those we view more or less as our equals.

Traditionally, social comparison occurred in relation to those physically around us (friends, family, co-workers etc.). During the course of a normal day, we might make upward comparisons, downward comparisons and lateral comparisons. Now, with the rise of social media, users are faced with hundreds – if not thousands – of sources of comparison in the form of profiles, images and videos. Also, much of the content on social media falls in the category of upward social comparison (Figure 8.18). Users take time to manipulate their online image to present themselves in

Shutterstock.com/Odua Images

Figure 8.18 Many of the images and videos we're exposed to on social media lead to upward social comparison and the pursuit of unattainable beauty standards

8.3.1 SOCIAL COMPARISON SCALE

an interesting light. For example, they share posts that highlight their successes and upload flattering photos.

Research suggests that frequent social media use leads to people having lower self-esteem than is seen in people who only have casual social media use. The researchers concluded this difference in self-esteem levels could be attributed to chronic upward social comparison that occurs via social media use (Vogel et al., 2014). A 2019 longitudinal study tracked the mental health and social media use of 6595 American teenagers over the course of 3 years. The results of the study indicated that teenagers who use social media for more than 3 hours a day are at an increased risk of developing depression and anxiety (Riehm et al., 2019).

In the wake of the COVID-19 pandemic, researchers Meredith David and James Roberts set out to determine the effects of social distancing and smartphone use on the wellbeing of American university students.

The results of the study indicated that those who used their phone more frequently during lockdowns felt greater social connection than those who used their phone less frequently. Those who used their phones more frequently were more likely to use this time to connect or reconnect with others. While the researchers conceded that connections made via smartphones aren't as effective as face-to-face interactions, they did serve as a viable alternative to maintain social connection in light of social distancing requirements (David & Roberts, 2021). This demonstrates that during times of isolation, technology can serve to maintain our social connections.

8.3.2 IDENTIFYING AND APPLYING SOCIAL COMPARISON THEORY

ANALYSING RESEARCH 8.2

Social media and mental health

A team of researchers in the US set out to determine whether greater time spent using social media by US adolescents is associated with higher rates of mental health concerns and behavioural problems. Between 2013 and 2016, 6595 American adolescents aged 12–15 took part in a longitudinal study. The researchers completed three rounds of interviews with the participants over the course of the study. In the final round of interviews, the researchers gathered information about internalising problems (including symptoms of depression and anxiety) and externalising problems (such as bullying behaviour and attention issues).

The adolescents were divided into five groups based on their social media consumption:

1 no social media use
2 less than or equal to 30 minutes per day
3 more than 30 minutes to 3 hours
4 3 hours to 6 hours
5 more than 6 hours.

The findings of the study are outlined in Table 8.5 below.

Table 8.5 The percentage of respondents reporting internalising problems (depression and anxiety), or both internalising problems (depression and anxiety) and externalising problems (behavioural issues), in the five groups with different levels of social media usage

Time spent on social media each day	Internalising problems	Internalising problems and externalising problems
1 No social media use	6.6%	10.7%
2 Less than or equal to 30 minutes	7.8%	13.6%
3 More than 30 minutes to 3 hours	9.8%	19.6%
4 3 hours to 6 hours	11.9%	24.6%
5 More than 6 hours	12.1%	29.7%

Based on their findings, the researchers concluded that adolescents who spend more than 3 hours per day using social media may be at heightened risk of mental health problems, particularly depression and anxiety. Moreover, the greater time spent on social media, the greater the chance that an adolescent will experience both internalising problems and externalising problems.

Source: Riehm, K. E. et al. (2019). Associations between time spent using social media and internalizing and externalizing problems among US youth. *JAMA psychiatry, 76*(12), 1266–1273. https://jamanetwork.com/journals/jamapsychiatry/fullarticle/2749480

Questions

1 Write an aim for this study.
2 Who were the population and sample in this study?
3 Write a conclusion for this study.

HOT Challenge

4 What was the purpose of including a group of adolescents who did not use social media at all during the course of the study?

Advertising and mental wellbeing

Advertising aims to persuade an individual or group to behave in a certain way via the use of learning theories (namely, classical conditioning, which is explored in Unit 3). Most forms of advertising use persuasive techniques that are very subtle. They encourage us to adopt a new behaviour or change an old behaviour with minimal conscious effort (Figure 8.19).

Shutterstock.com/Liv friis-larsen

Figure 8.19 Soft drink companies try to create an association between their product and feeling refreshed

Advertisements often portray the models as having the 'ideal' body size and shape. However, these models are not representative of the general public. The portrayal of the 'ideal' body or the 'thin ideal' can have negative impacts on all genders.

Eating disorders are caused by a range of factors, including the influence of family and peers, and the media's promotion of the 'ideal' body. Advertising can exacerbate this issue by using models with 'ideal' body size and shape to promote products in the hope that it will enhance the products and create body dissatisfaction. This is also true of products targeting young males that promote a muscular and strong body shape as the 'ideal' shape that all males should aspire to. Body dissatisfaction and emotions such as depression, stress, guilt, shame, unhappiness and low self-esteem can lead to a negative change in eating behaviour (Stice & Shaw, 1994; Almond, 2008).

Advertising as a means of influencing behaviour can also have benefits for mental wellbeing. For instance, in the field of health promotion, advertising is used to target individual and group behaviour that may be unhealthy or pose a risk. The same advertising techniques can be used to stop individuals behaving in a particular way or to promote a healthier lifestyle. For example, advertising aimed at smokers, people who have a poor diet and people who drive when they are tired or are alcohol- or drug-affected aims to decrease negative behaviour and replace it with positive behaviour. These healthier choices can lead to improved mood and self-esteem. Additionally, mental health support networks can use conventional advertising platforms to promote their services.

Media and addictive behaviours

The increased access to smart devices and online platforms has also raised concerns regarding the prevalence of addictive disorders, specifically problem gambling and video game addiction.

Internet gambling

Internet gambling refers to betting or wagering through any device that is connected to the internet; for example, the use of betting applications or the purchasing of lottery tickets. Internet gambling is the fastest growing form of gambling and has also changed how people engage in gambling behaviour. Unlike traditional methods of gambling, which involve physical cash transactions, internet gambling can take place anywhere that an individual has internet access. Surveys of gamblers have also shown that approximately a quarter reported that it was easier to spend more when gambling online, because it felt as though they weren't spending 'real money' (Gainsbury, 2015). Additionally, 15 per cent of gamblers reported that online gambling was more addictive than traditional gambling (McCormack & Griffiths, 2012).

Gambling disorders have clear negative effects on a person's mental health. Current estimates state that people with gambling disorders are twice as likely to be depressed than someone without a gambling disorder (Beyond Blue, 2022).

Gaming disorder

Over the last few decades, the video game industry has adapted to the shifting nature of media and entertainment. Once confined to a large machine or console, video games are now readily available online and on smart devices. This increased access to video games has generated debate as to whether excessive gaming can be considered an addictive behaviour or merely a hobby (Figure 8.20).

While the *Diagnostic and Statistical Manual of Mental Disorders* (5th ed.; *DSM-5-TR*™) does not list excessive gaming as an addictive disorder, the *International Classification of Diseases* (11th revision; *ICD-11*) included gaming disorder as a form of addictive disorder.

According to the *ICD-11* (World Health Organization, 2018), a diagnosis for gaming disorder can be made if:

» persistent and repetitive gaming occurs over a period of at least 12 months
» the following three symptoms are demonstrated: (i) impaired control over gaming (impaired self-regulation); (ii) increased priority given to gaming (impaired functioning); and (iii) continuation or escalation of gaming despite the occurrence of negative consequences (inability to stop gaming) (Jo et al., 2019).
» gaming habits negatively affect relationships with friends and family, and impair educational or occupational functioning.

However, the inclusion of gaming disorder has been met with criticism from a number of experts in the field of psychology and addictive disorders. These include:

» playing video games is a common and safe recreational activity
» the classification of gaming disorder will lead to a moral panic and the subsequent stigmatising of gamers
» the inclusion of gaming disorder might lead to an over-diagnosis of video game addiction
» the criteria used to classify gaming disorder rely too heavily on the criteria used to classify substance abuse and gambling disorders (Aarseth et al., 2017).

Figure 8.20 Whether excessive gaming is a form of addiction is still up for debate

ACTIVITY 8.3 UPWARD SOCIAL COMPARISON

Try it

1 Choose one person that you admire (they can be someone you know, a celebrity, athlete, historical figure and so on). Jot down as many characteristics, behaviours and traits you can think of that you admire about this person.

Apply it

2 The 'Try it' task required you to complete upward social comparison. Upward social comparison has benefits; however, it also has downsides, particularly when we are comparing ourselves to people we see online. Outline two benefits and two downsides of upward social comparison.

Exam ready

3 Which of the following statements regarding social comparison theory is not correct?

A Downward social comparison occurs when we compare ourselves to someone we perceive to be inferior or lesser than us.
B Lateral social comparison occurs when we compare ourselves to someone we perceive as superior and a person we perceive as inferior.
C Upward social comparison occurs when we compare ourselves to someone we perceive to be superior to us.
D Lateral social comparison occurs when we compare ourselves to someone we perceive to be relatively equal to us.

KEY CONCEPTS 8.3

- Advertisers use subtle persuasive techniques that encourage us to adopt a new behaviour or change an old behaviour with minimal conscious effort, which can have a positive or negative effect.
- We make three types of social comparisons: upward social comparison, downward social comparison and lateral social comparison.
- Many of the social comparisons that we make online are upward and they can have a negative effect on our self-esteem and mental health.
- Increased access to smart devices and the internet has resulted in new forms of addiction, namely gaming disorder and online gambling addiction.

Concept questions 8.3

Remembering

1 Define the term 'social connection'. r

Understanding

2 Explain the concept of social comparison. e

3 Outline the positive and negative effects of advertising on a person's wellbeing. e

4 Explain the possible influence of prolonged social media use on our mental wellbeing. e

Applying

5 James received a gaming console for Christmas. Each day after school, James spends 2–3 hours playing video games.

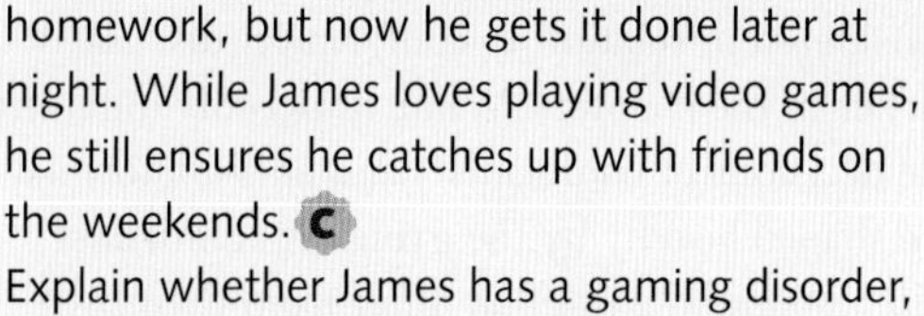
At first, James struggled to complete his homework, but now he gets it done later at night. While James loves playing video games, he still ensures he catches up with friends on the weekends. c

Explain whether James has a gaming disorder, according to the ICD-11 classification.

HOT Challenge

6 Explain how the nature of online social comparisons differs from those we make offline – that is, in the physical world. e

8.4 Independence and anti-conformity

It has been estimated that each day we are confronted by approximately 200 attempts to influence our opinions (Martin & Hewstone, 2001). These include advertisements we see on television, posts on social media, news reports, and conversations we hold with friends and loved ones. At times we will conform to these information sources, but at other times we might not conform. In social psychology, it is widely suggested that there are two types of non-conformity. These are independence and anti-conformity (Willis, 1963).

Independence

Independence occurs when an individual is aware of how the group expects them to behave or respond, but their decision-making is not swayed in any way by the expectations of the group. In other words, independence is 'resisting influence' (Padalia, 2014). In this context, independence does not mean that the individual fails to recognise or consider the views of the group. Instead, they reject the views or norms of the group in order to arrive at their own objective decision.

At times, group pressure makes it difficult to act independently. By acting independently, you may alienate yourself or share opinions that may be unpopular with the group. However, when people act independently within a group setting, there is consistency in what they believe to be correct privately and the stance that they take publicly; as a consequence, they are true to themselves.

Anti-conformity

Anti-conformity (also known as **counter-conformity**) is a deliberate refusal to comply with social norms as demonstrated by ideas, beliefs or judgements that challenge these social norms. Anti-conformity is not a genuine attempt to express one's ideas or perspectives. Instead, anti-conformity comes from a desire to be stubborn or rebellious. Anti-conformity within a group setting occurs when a person who is trying to be swayed purposefully

8.4.1 TECHNIQUES TO EMPOWER INDIVIDUAL DECISION-MAKING

says or does the opposite of what the group suggests. As a result, the person ends up publicly disagreeing with the group, even if the individual agrees with the group's position privately (Nail et al., 2013). In other words, anti-conformity is 'rebelling against influence' (Padalia, 2014).

Quite often, behaviour considered anti-conformist to one group would be seen as conformist to another group. For example, many of the behaviours and choices of teenagers that conform to the norms of their friendship group can be perceived as anti-conformist by the parents (Krueger et al., 2017).

Anti-conformity and independence must be considered in relation to conformity (discussed earlier in the chapter). While noticeably different, the decisions of conformists and anti-conformists are shaped heavily by the group. A conformist will respond in ways that differ from their own opinion in order to match the decision of the group. Similarly, an anti-conformist will provide answers that are incorrect or don't align with their beliefs, but with the intention of being different from the group. In both situations, the norms of the group influence the behaviour of the individual, with the conformist holding positive views towards the group and the anti-conformist holding negative views towards the group (Frager, 1970).

Playing 'devil's advocate'

As discussed earlier, a technique that can be used to prevent the occurrence of groupthink is to assign one group member to play the 'devil's advocate'. The role of the devil's advocate is to actively argue against the points of group members, even if they share the same beliefs as the group members. The devil's advocate takes the opposite position of the group members to encourage them to reflect on their positions and prevent them from hastily arriving at a solution without considering potential consequences (Akhmad et al., 2021). In doing so, the devil's advocate is actively encouraged to be anti-conformist, to ensure that counterpoints are considered by the group, including the potential risks that could outweigh potential rewards (Figure 8.21).

There are many benefits of using a devil's advocate. Individuals who play devil's advocate are not constrained by the bias of the group, because they are free to challenge the group's assumptions. Moreover, the devil's advocate will encourage other group members to consider alternatives and allow them to challenge proposed solutions (Akhmad et al., 2021).

Shutterstock.com/fizkes

Figure 8.21 Playing 'devil's advocate' is a form of anti-conformity that can be useful in preventing groupthink

ACTIVITY 8.4 DE BONO'S THINKING HATS

Try it

1 Psychologist Edward de Bono devised the 'six thinking hats' as a framework for groups to use to arrive at a well-thought-out decision by encouraging group members to evaluate the problem and solution from a range of perspectives.

One of the thinking hats is the Black Hat or the cautious hat. The member wearing the Black Hat plays the role of 'devil's advocate'. They think of worst-case scenarios, potential errors in thinking and possible dangers by pointing out any risks.

Risks, difficulties, problems: The risk management hat, probably the most powerful hat; a problem, however, if overused; used to spot difficulties where things might go wrong, why something may not work; inherently an action hat used to point out issues of risk with the aim of overcoming them.

Source: Six Thinking Hats, The de Bono Group, https://www.debonogroup.com/services/core-programs/six-thinking-hats/

Figure 8.22 Edward de Bono's Black Hat

In groups of three, discuss the scenario below. Allocate the role of Black Hat to one of the members of the group.

Scenario: A 15-year-old girl leaves her phone unlocked on the bench. The child's mother decides to quickly scroll through her daughter's messages and notices that her daughter's best friend has asked her to bring alcohol to a party on the weekend.

How should the mother respond to this situation?

Apply it

2 Sienna is a 15-year-old girl who wants to get her nose pierced, but her parents have said no on numerous occasions, saying that she'll regret it when she's older. All of Sienna's friendship group have their piercings, and she tells her parents that 'everyone has their nose pierced these days' and that she 'feels like the odd one out'.

On the weekend, some of Sienna's friends go to get more piercings. While at the piercing studio, Sienna's friends encourage her to get her nose pierced, which she does.

Explain how Sienna's behaviour can be described as both anti-conformist and conformist.

Exam ready

3 When someone is making a decision in a group setting, anti-conformity is ___________; whereas independence is ___________.

A rebelling against the influence of the group; responding to fit in with the group
B rebelling against the influence of the group; resisting the influence of the group
C responding to fit in with the group; resisting the influence of the group
D resisting the influence of the group; rebelling against the influence of the group

ANALYSING RESEARCH 8.3

An early study of conformity, anti-conformity and independence

Using a similar approach to the early research of Asch, Richard Willis and his associate E. P. Hollander (1964) created conditions that they believed would lead to participants demonstrating behaviour that was conformist, independent or anti-conformist.

Willis and Hollander told 36 university students (12 male; 24 female) that they would estimate the length of lines with a partner, whom they wouldn't meet during the study. Unbeknownst to them, participants had been randomly allocated to one of three conditions: the conformity condition, the independence condition, and the anti-conformity condition. To create these conditions, all participants completed a pre-test where they had to determine if the measurements of 20 lines were longer or shorter than listed. The participants were told that they would be judging the lengths of the lines with a partner; however, there was no partner. The participants in the conformity group were told that during the pre-test their partner correctly judged 18 out of 20 lines correctly; participants in the independence group were told that they themselves had judged 18 out of 20 lines correctly; and participants allocated to the anti-conformity group were told that their partner had only judged 3 out of the 20 lines correctly during the pre-test.

During the experimental phase of the study, the participants would have to judge if 100 lines were longer or shorter than their stated length. Four lines were drawn on each page, with the length of the line written under each one (with each line ranging in length from approximately 8 centimetres to 23 centimetres). The participants had to state if the line was longer or shorter than the written length. Prior to the experiment commencing the researchers informed the participants that half the lines were longer than the number listed and half were shorter than the number listed; therefore, half of their estimations should be higher and half should be lower. In actuality, the lengths listed below each line were correct.

9780170465052

They were told that they would have a partner who would complete the same test after them. They then would be able to see their responses and their partner's responses and decide if they wanted to change their estimation.

Willis found significant differences in all three groups. Those in the anti-conformity group became more anti-conformist as the experiment continued – with the same being true for the independence group, and the conformity group. Members of the independence group tended to disagree with their 'partner's' estimations, favouring their own choices; members of the anti-conformity group also showed a tendency to go against the choices of their 'partner'; and finally, those in the conformity group demonstrated a tendency to change their choice when presented with their 'partner's' response.

Source: Adapted from Willis, R. H., & Hollander, E. P. (1964). An experimental study of three response modes in social influence situations. *The Journal of Abnormal and Social Psychology*, *69*(2), 150–156. https://apps.dtic.mil/sti/pdfs/AD0402791.pdf

Questions

1 Write an aim for this study.
2 Write a likely hypothesis for this study.
3 Write a conclusion for this study.

HOT Challenge

4 Random allocation was used in this study. What is random allocation and why is it used?
5 Explain how deception was used to create the conformity, independence and anti-conformity condition.

KEY CONCEPTS 8.4

» Independence occurs when the individual is aware of how the group expects them to behave or respond, but their decision-making is not swayed in any way by the expectations of the group.
» Anti-conformity is a deliberate refusal to comply with social norms as demonstrated by ideas, beliefs or judgements that challenge those norms.
» Independence involves resisting the influence of the group, whereas anti-conformity involves rebelling against the influence of the group.
» Anti-conformity in the form of playing devil's advocate can help groups to make more objective decisions.

Concept questions 8.4

Remembering

1 Define the term 'independence'. r
2 Define the term 'anti-conformity'. r

Understanding

3 Explain how playing the role of devil's advocate relates to the concept of anti-conformity. e

Applying

4 Explain the influence of group norms on the behaviours of someone who is conformist and someone who is anti-conformist. e

HOT Challenge

5 Using an example not mentioned in the text, explain how the behaviour of an individual can be perceived as anti-conformist by one group and conformist by another. c

8 Chapter summary

CHAPTER 8 SUMMARY

KEY CONCEPTS 8.1a

- A group is formed when two or more people interact and influence each other and share a common objective.
- Status refers to a person's position in the hierarchy of a group. The higher a person's status, the more power they exert.
- Social power refers to the amount of influence that an individual can exert over another person.
- There are five different bases of social power: reward, coercive, legitimate, referent and expert power.
- Cultures can be categorised as either collectivist or individualist.
- Collectivist cultures value group needs or interests over the interests of the individual. Individualist cultures, on the other hand, place the needs of the individual above the interests of groups.

KEY CONCEPTS 8.1b

- The Stanford prison experiment investigated the influence of assigned roles (prisoner or prison guard) on the behaviour of individuals.
- While considered ground-breaking at the time, the scientific validity of the study is now being debated.

KEY CONCEPTS 8.2a

- Obedience occurs in situations in which people change their behaviour in response to direct commands from others.
- Milgram's study demonstrated that individuals are willing to inflict serious physical harm if instructed to by an authority figure.
- Obedience is influenced by several factors, including social proximity and the legitimacy of the authority figure.

KEY CONCEPTS 8.2b

- Conformity occurs when individuals change their behaviour as the result of real or implied pressure from others.
- Normative influence refers to the established norms (rules) a group follows.
- Informational influence occurs when we look to others' behaviour to guide our own behaviour.
- People are more likely to conform when the group is unanimous in its decision.
- Deindividuation is the loss of social identity and inhibition when in a group, which results in a person engaging in behaviours with little regard for potential consequences.

KEY CONCEPTS 8.2c

- Groupthink is a phenomenon that occurs when a members of a group prioritise the cohesion of the group over objective decision-making.
- Groupthink is characterised by the use of stereotypes, discouraging differing opinions, unanimity in decision-making, and downplaying of risk.
- Group shift is the tendency for group members discussing an issue or dilemma to adopt a more extreme position than their individual position prior to the discussion.
- Cautious shift occurs when the group suggests an even safer approach than an individual would have taken if acting alone.
- Risky shift occurs when a group demonstrates less caution than the individual members would have if they were acting alone.
- Group shift can be attributed to social comparison theory, the persuasive argument theory, and social decision schemes.

KEY CONCEPTS 8.3

- Advertisers use subtle persuasive techniques that encourage us to adopt a new behaviour or change an old behaviour with minimal conscious effort, which can have a positive or negative effect.
- We make three types of social comparisons: upward social comparison, downward social comparison, and lateral social comparison.
- Many of the social comparisons that we make online are upward and they can have a negative effect on our self-esteem and mental health.
- Increased access to smart devices and the internet has resulted in new forms of addiction, namely gaming disorder and online gambling addiction.

KEY CONCEPTS 8.4

- Independence occurs when the individual is aware of how the group expects them to behave or respond, but their decision-making is not swayed in any way by the expectations of the group.
- Anti-conformity is a deliberate refusal to comply with social norms as demonstrated by ideas, beliefs or judgements that challenge those norms.
- Independence involves resisting the influence of the group, whereas anti-conformity involves rebelling against the influence of the group.
- Anti-conformity in the form of playing devil's advocate can help groups to make more objective decisions.

8 End-of-chapter exam

Section A: Multiple-choice

1 What are the three main characteristics of a group?
  A Four or more people; interaction between members; share a common goal
  B Two or more people; share a common goal; share all interests
  C People who share a common goal; see each other at least once a week; influence each other
  D Two or more people; influence and interaction between members; share a common goal.

2 Status refers to
  A a person's position in the hierarchy of a group.
  B the shared purpose of a group.
  C how obedient a person is.
  D whether the person possesses coercive power or not.

3 An individual with reward power refers to a person
  A who has the ability to punish another for failing to comply.
  B with particular knowledge or expertise in an area because of their training and experience.
  C we want to be like or who inspires us to do what they do.
  D who has been given the power to reinforce others in the group who comply with the desired behaviour.

4 Studies have shown that obedience is more likely if
  A the person giving the command has higher status.
  B the person giving the command is further away.
  C the person giving the command has no authority.
  D other people refused to obey the command.

5 Generally speaking, the _______ a person's status, the ____ power they exert.
  A higher; less
  B lower; more
  C higher; more
  D lower; more coercive

6 Asch's (1952) experiments on conformity demonstrated that people would conform to the group consensus even when the response was clearly incorrect. However, the experiments also showed that the degree to which people conformed was affected by
  A group size and the participants' intelligence.
  B the group leader's personality and presence of an ally.
  C task difficulty and group size.
  D group size and the presence of an ally.

7 Informational influence allows people to
  A look at the behaviour of others in order to make judgements about how to behave themselves.
  B look at the behaviour of others in order to feel better about themselves.
  C behave in ways they usually would not because of the commands of an authority figure.
  D behave in negative ways due to a loss of identity.

8 Culture can affect conformity because some cultures
  A are categorised as 'horizontal'; therefore, they are more likely to conform.
  B are individualist, so there is more conformity.
  C are collectivist, so there is less conformity.
  D are collectivist, so there is more conformity.

9 One of the main lessons to be learned from Milgram's studies is that
  A people are naturally aggressive.
  B ordinary people can act aggressively in some situations.
  C being accepted by others is a strong human motive.
  D people do the opposite of what they are told.

10 Playing 'devil's advocate' is a form of ________, which can reduce ________.
  A anti-conformity; group shift
  B groupthink; anti-conformity
  C anti-conformity; groupthink
  D group shift; anti-conformity

11 What is the process whereby individuals lose social identity, ignore norms and social responsibility, and engage in potentially negative or destructive behaviours?

- **A** group shift
- **B** social status
- **C** anti-conformity
- **D** deindividuation

12 In the Stanford prison experiment, the independent variable and dependent variable, respectively, were

- **A** the behaviour of the participants that was indicative of their role; whether the participants were assigned to the role of guard or prisoner.
- **B** the behaviour of the participants that was indicative of their role; how often the prisoners were referred to by their number.
- **C** whether the participants were assigned to the role of guard or prisoner; how many days it took for participants to quit the experiment.
- **D** whether the participants were assigned to the role of guard or prisoner; the behaviour of the participants that was indicative of their role.

13 Downward social comparison occurs when we compare ourselves to

- **A** individuals we perceive to be inferior.
- **B** individuals we perceive to be more or less equal to us.
- **C** individuals we perceive to be superior to us.
- **D** people who are younger than us.

14 Which one of the following statements is true for both anti-conformity and conformity?

- **A** Both are influenced by the need to demonstrate independence.
- **B** Conformist and anti-conformist behaviours are both direct responses to the influence or wishes of the group.
- **C** Anti-conformity and conformity can both reduce groupthink.
- **D** Both involve resisting the influence of the group when making decisions.

15 Risky shift often occurs when

- **A** prior to group discussion, the individual opinions of group members favoured a more cautious approach.
- **B** prior to group discussion, one individual was assigned the role of 'devil's advocate'.
- **C** prior to group discussion, the individual opinions of group members favoured a riskier approach.
- **D** the individual opinions of group members are split between caution and risk.

16 A potential negative psychological effect of advertising is

- **A** we are informed of the newest forms of technology.
- **B** some companies portray unrealistic or unattainable beauty standards, leading to lower self-esteem.
- **C** many advertisements lead to lateral social comparison, which can lead to feelings of inadequacy.
- **D** some companies or charities use advertising to draw attention to social issues.

17 Which of the following is a not a criticism of the Stanford prison experiment?

- **A** Some of the guards may not have been aware that they were also participants in the experiment.
- **B** Some of the prisoners stated that they pretended to have breakdowns so they could leave the experiment early.
- **C** Zimbardo didn't use random allocation, meaning that individual participant differences may have influenced the results.
- **D** The guards were given explicit instructions as to how to treat the prisoners.

18 The term social connection refers to

- **A** a person's position within a group.
- **B** a measurement of how influential social media is on a person's life.
- **C** the number of cultural influences that shape a person's behaviour.
- **D** the belief that we belong to a group and generally feel close to other people.

19 A similarity between the studies of Asch and Milgram is that

- **A** both investigated the influence of an authority figure on behaviour, with Asch investigating conformity and Milgram investigating obedience.
- **B** both investigated the influence of groups on conformity.
- **C** both ensured that withdrawal rights were adhered to.
- **D** both used deception.

20 Vertical-collectivist cultures and vertical-individualist cultures are similar in that

- **A** both strive to promote equality within the society.
- **B** both encourage the pursuit of individual goals.
- **C** both accept that inequalities exist and stress the need for hierarchies.
- **D** both stress the importance of pursuing goals that benefit the wider group.

Section B: Short answer

1 Outline three techniques used in the Stanford prison experiment to create differences in status between the prisoners and the guards. [3 marks]

2 Outline two differences between independence and anti-conformity in relation to how individuals make decisions when in a group. [2 marks]

3 Identify one positive psychological effect and one negative psychological effect of using social media. [2 marks]

Unit 2, Area of Study 1 review

Section A: Multiple-choice

Question 1

According to psychologists, attitudes are

A personal characteristics we have.

B statements of factual knowledge about something or someone.

C judgements about how much we like or dislike something or someone.

D learned ideas we have about people, objects, events or issues.

Use the following information to answer questions 2 and 3.

Jack is a very skilled football player. He hates it when his teachers assume that, just because he is good at football, he is not intelligent.

Question 2

Jack's reaction is an example of the ____________________ component of the tri-component model of attitudes.

A affective

B behavioural

C cognitive

D social

Question 3

Jack's teachers' attitudes of assuming footballers are not intelligent is an example of which component of the tri-component model of attitudes?

A affective

B behavioural

C cognitive

D all of the above

Question 4

A researcher wants to see whether an advertising campaign has affected how people select soap powders in a supermarket. They arrange to sit in the store manager's office and use the store's closed-circuit television to watch how people behave. This method of data collection is called

A a survey.

B self-reporting.

C experimentation.

D observation.

Question 5

The three components that contribute to attitude formation are

A feelings, beliefs and actions.

B affective, behavioural and consistency.

C affective, opinions and values.

D cognition, learning and behaviour.

Question 6

Which of the following is false regarding attitudes

A they are short term judgements made before meeting a person

B often derived based on experiences

C they impact the way we interact with others and behave

D involve an interaction between feelings, cognitions and behaviour

Question 7

Often, stereotyping people can lead to prejudice. What is a stereotype?

A an oversimplified opinion of people who belong to a particular group

B a true statement about every member of a group

C a behaviour all members of a group share

D all of the above

Question 8

While walking home from school, Mark sees some of his friends and stops to have a chat. His friends are spray-painting graffiti on a wall. Even though Mark believes that vandalism is wrong, he joins in and paints graffiti with his friends. Such inconsistency between a person's attitude and their behaviour is known as

A willpower.

B cognitive dissonance.

C social conformity.

D obedience.

Question 9

Which of the following is true of cognitive bias?

A It occurs when we hold two conflicting beliefs.

B It causes unease or discomfort.

C It is a coping mechanism for cognitive dissonance.

D It is the prejudices we hold against people.

Question 10

An educated guess is an example of which of the following?

A cognitive dissonance

B cognitive bias

C heuristics

D willpower

Question 11

Which of the following is a disadvantage in using heuristics in decision-making?

A They can lead to inaccurate judgements.

B They are hard to understand and implement.

C They are too time consuming.

D The information obtained is too complex.

Question 12

Judging a person negatively because of their membership of a group is an example of

A prejudice.

B deindividuation.

C unanimity.

D groupthink.

Question 13

Superordinate goals are ____________________. If the goal is not met, the result can __________________ prejudice.

A goals that take a long time to achieve; have no impact on

B shared goals that individuals cannot achieve on their own; reduce

C shared goals that individuals cannot achieve on their own; increase

D shared goals where only a few members of a group need to do any work; have no impact on

Question 14

One way to give groups equal status in society is to promote multiculturalism and to learn more about individuals' cultures and backgrounds. This technique seeks to abolish

A sexism.

B ageism.

C culturism.

D racism.

Question 15

Which of the following is not an example of a cognitive intervention that may be used to reduce prejudice?

A seeking information about an individual

B looking for characteristics that we share with others

C becoming more aware of different cultures

D working over an extended period of time with others

Question 16

Shehani hates the fact that when she goes to the football she always has to wait for the old people to climb up the stairs and find their seats. She thinks all old people are slow. This belief is an example of

A sexism.

B ageism.

C racism.

D discrimination.

Question 17

Reward power involves an individual

A giving negative consequences to others.

B removing positive consequences from others.

C giving positive consequences to others.

D All of the above

Question 18

As part of a study for her Social Psychology class, Bronte left some coins in a phone booth. As each new person used the booth, Bronte approached them and asked for her coins back. When Bronte was well-dressed, people generally returned the coins, yet when she was dressed poorly, the people generally refused her. This experiment most likely demonstrated the

A difference between ascribed and achieved roles.

B effect of status on our behaviour towards others.

C effect of coercive power on others.

D impact of roles on our behaviour towards others.

Question 19

Pro-social behaviour is best defined as

A being helpful; having expected rewards; positively valued.

B being harmful; having expected rewards; positively valued.

C being helpful, having no expected rewards; positively valued.

D being helpful, having expected rewards; negatively valued.

Question 20

Which of the following is a way in which we can empower individualistic thinking and reduce conformity?

A increasing the size of a group
B demonstrations anti-conformity by peers
C both A and B are correct
D none of the above

Question 21

Which of the following is NOT an example of a heuristic

A a person stuck in traffic decides to go another route without knowing the way
B trying to guess the price of a property based on past trends
C doing research to come to a conclusion on what subjects you will pick in year 12
D making an educated guess on how many people are in your school based on how many you see daily

Question 22

Josie was so excited to go to the opening of a new clothing store in her city due to her favourite singer promoting it on her Instagram page.

The type of power exerted by the singer is

A expert.
B coercive.
C reward.
D referent.

Question 23

In his obedience studies, Milgram instructed subjects to deliver electric shocks to an unseen 'learner' in a fake learning study. Milgram found that the majority of the 'teachers'

A reluctantly administered shocks up to the highest, presumably most dangerous, levels.
B obeyed up to the point where the 'learner' screamed in pain, and then refused to go further when instructed to do so.
C administered weak shocks only and refused to go further when instructed to do so.
D obeyed readily and did not protest or become visibly upset.

Question 24

In Asch's experiment on the effect of group pressure on conformity, participants were asked to compare line segments and say which one matched a standard.

These studies showed that the most influential factor in reducing conformity was

A the gender composition of the group.
B the strength of conviction of group members.
C the presence of an ally in the group.
D the level of authority of the leader of the group.

Question 25

Milgram's electric shock experiment is considered by many psychologists to be unethical because Milgram

A allowed the participants to suffer physical harm.
B used university students as part of his research.
C deceived the participants.
D did not debrief the participants sufficiently.

Question 26

In his classic studies of conformity, Asch demonstrated that the degree of conformity was influenced by group size. Conformity reaches its peak when a group contains

A one person.
B two people.
C three people.
D four people.

Question 27

How do gambling advertisements contribute to addictive behaviours?

A They trigger addictive behaviours.
B They make gambling more accessible.
C They encourage gambling behaviours.
D All of the above

Question 28

How does social media impact on social interactions?

A It encourages more in-person social gatherings.
B It provides ease of contact between individuals.
C It potentially encourages anti-social behaviour in person.
D Both B and C

Question 29

How does anti-conformity empower decision-making in groups?

A Individuals feel more inclined to dispute authority when supported by peers.

B People are less likely to rebel against authority when in a group.

C Individuals will make decisions to rebel without regard for the pros or cons.

D Only certain personality types will rebel against authority.

Question 30

Jo refuses to follow the social norms of school and decides to dye her hair pink despite school rules. This is an example of

A cognitive bias.

B cognitive dissonance.

C anti-conformity.

D obedience.

Section B: Short answer

Question 1

Jerimiah doesn't like his boss because he believes he plays favourites and does a poor job. Yet Jerimiah continues to work at his job and be nice to his boss.

a According to the ABC model of attitudes, how is Jerimiah perceiving this situation? [3 marks]

b Is Jerimiah's attitude congruent or does it display cognitive dissonance? Explain your answer. [2 marks]

[Total = 5 marks]

Question 2

Asch did an experiment on conformity. Use your knowledge of this study to answer the following questions.

a What was the aim of the Asch study? [2 marks]

b What conclusions did Asch come to? [2 marks]

[Total = 4 marks]

Question 3

Differentiate between prejudice and discrimination. [2 marks]

Question 4

How can an individual use cognitive bias to reduce cognitive dissonance? [2 marks]

Question 5

How can obedience be useful within society? Provide an example. [2 marks]

Perception 9

Key knowledge

- the role of attention (sustained, divided, selective) in making sense of the world around us
- the role of perception in the processing and interpretation of sensory information, as demonstrated through top-down and bottom-up processing
- the influence of biological, psychological and social factors on visual perception and gustatory perception

Key science skills

Develop aims and questions, formulate hypotheses and make predictions

- identify independent, dependent and controlled variables in controlled experiments
- formulate hypotheses to focus investigation

Plan and conduct investigations

- design and conduct investigations; select and use methods appropriate to the investigation, including consideration of sampling technique and size, equipment and procedures, taking into account potential sources of error and uncertainty; determine the type and amount of qualitative and/or quantitative data to be generated or collated

Comply with safety and ethical guidelines

- demonstrate ethical conduct and apply ethical principles when undertaking and reporting investigations

Generate, collate and record data

- record and summarise both qualitative and quantitative data, including use of a logbook as an authentication of generated or collated data

Analyse and evaluate data and investigation methods

- process quantitative data using appropriate mathematical relationships and units, including calculations of percentages, percentage change and measures of central tendencies, and demonstrate an understanding of standard deviation as a measure of variability
- identify and analyse experimental data qualitatively, applying where appropriate concepts of: accuracy, precision, repeatability, reproducibility and validity of measurements; errors (random and systematic); and certainty in data, including effects of sample size on the quality of data obtained

Construct evidence-based arguments and draw conclusions

- use reasoning to construct scientific arguments, and to draw and justify conclusions consistent with the evidence and relevant to the question under investigation
- identify, describe and explain the limitations of conclusions, including identification of further evidence required

Analyse, evaluate and communicate scientific ideas

- discuss relevant psychological information, ideas, concepts, theories and models and the connections between them

Source: VCE Psychology Study Design (2023–2027), pp. 30 & 12–13

9 Perception

We all perceive the world around us via sensory stimuli: the things we see, smell hear, taste and touch. However, because we all have experienced different lives, emotions and hold different beliefs and social status, how we perceive the stimuli in our environments can be very different.

p. 351

9.1 Making sense of the world

Making sense of the world begins with a sensation such as a sweet scent. This sensation is converted into electrochemical energy and sent to different areas of the brain to be processed. Your brain can then 'perceive' the sensation and this is when you realise that the smell means, for example, that the cake is ready to come out of the oven.

Shutterstock.com/Bolkins

iStock.com/Imgorthand

p. 359

9.2 The role of attention in perception

Our attention and what or who we focus it on can make all the difference when it comes to what you perceive. Have you ever been so caught up in watching media that you have not noticed someone repeatedly calling your name? There are different types of attention depending on the task you have at hand and what your brain needs to focus on to achieve it.

p. 364

9.3
Factors influencing visual perception

Look around you, carefully. Do you notice things that you might not have noticed before? We are usually surrounded by such a great number of stimuli that if we were to pay attention to all of them, our brains would never have time to process it all. This is why we are selective in what we pay attention to.

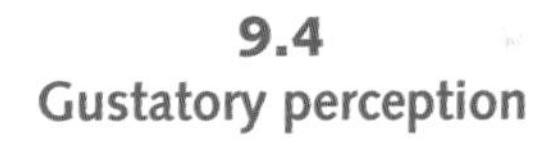
Shutterstock.com/dodotone

p. 376

9.4
Gustatory perception

Like all the other sensations, your perception of taste sensations can be influenced by a range of biological, psychological and social factors. To some people, for example, coriander tastes like soap, whereas other people love it. This is because of the genes that they have inherited (a biological factor).

iStock.com/triloks

Our bodies are constantly experiencing a range of sensations, but everyone's individual perception can be influenced by a range of biological, psychological and social factors. Don't believe us? Note what you see, feel, hear, taste and smell at lunchtime, and compare your perceptions with your friends who you spend lunchtime with. What differences are there?

Slideshow
Chapter 9 slideshow

Flashcards
Chapter 9 flashcards

Test
Chapter 9 pre-test

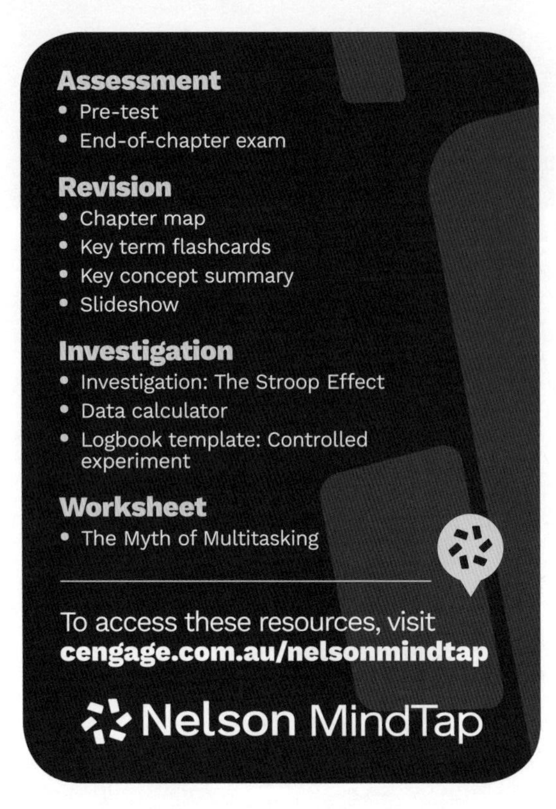

Know your key terms

Accommodation
Attention
Automatic processes
Binocular cues
Bottom-up processing
Closure
Context
Controlled processes
Convergence
Depth cue
Depth perception
Divided attention
Figure–ground
Gestalt principles
Gustation
Gustatory receptors
Height in the visual field
Interposition
Linear perspective
Monocular cues
Motivation
Past experience
Perception
Perceptual hypothesis
Perceptual organisation
Pictorial cues
Proximity
Reception
Relative size
Retinal disparity
Selection
Selective attention
Sensation
Sensory receptors
Similarity
Sustained attention
Taste
Taste buds
Texture gradient
Top-down processing
Transduction
Transmission
Visual constancies
Visual perception principles

Most people who broke an ankle would feel intense pain when they tried to walk. However, some rare individuals are born with an inability to sense any kind of physical pain. Ashlyn Blocker, an American 16-year-old, is one of these people (Figure 9.1). When she was younger she ran around on a broken ankle for two days before her parents noticed it was broken. As a baby, she almost chewed off part of her tongue when her teeth started to come through, she dipped her hands in boiling water, and she was bitten by fire ants hundreds of times without sensing any pain. Needless to say, childhood was a hazardous time for Ashlyn.

Being able to perceive (understand) sensory stimuli correctly has an important evolutionary purpose. It improves our chance of survival because it alerts us to potentially life-threatening injuries. However, to perceive sensory stimuli, we must first sense it. In this chapter we will explore the role of perception in processing and interpreting sensory information.

Getty Images/Jeff Riedel

Figure 9.1 Ashlyn Blocker was born with insensitivity to pain.

9.1 Making sense of the world

Each day our environment constantly bombards us with a changing array of light, sound and chemical energy. We detect this energy through our sense organs and experience it as sights, sounds, tastes and smells. Table 9.1 shows the types of stimulus energy our sense organs take in.

Initially, our response is an automatic physical (bodily) response that does not involve the brain. When **sensory receptors** located in the body's sense organs detect and respond to stimuli, they transmit electrochemical signals to specific sensory areas of the brain. This generates a sensory experience known as **sensation**. To understand the world around us, we must give meaning to these sensations.

Table 9.1 Types of stimulus energy, and the senses and sensory organs used to detect them

Stimulus energy	Sense	Sensory organ
Electromagnetic radiation (visible light)	Sight (vision)	Eye
Sound waves	Hearing (audition)	Ear
Chemical energy (molecules)	Smell (olfaction)	Nose
Chemical energy (molecules)	Taste (gustation)	Tongue
Mechanical and thermal energy	Skin (touch, pressure, pain, cold and warmth)	Skin

Sensation: our connection to the environment

Sensation changes various kinds of external energy into electrochemical neural impulses. Sensation is an automatic physical reaction to a stimulus that is the same for everyone. It occurs in the cells in the sense organs and neural pathways and does not involve the brain. Sensation is a process that involves three stages: reception, transduction and transmission.

Sensation begins with **reception** when the stimulus, or change to a stimulus, is detected by sensory receptor cells located in our various sense organs. These cells are specialised to detect and respond to a specific type and level of energy from the environment. For example, sensory receptors in your eye's retina only respond to a light stimulus, and sensory receptors in your tongue only respond to specific chemical molecules. If the stimulus is strong enough to activate a response, the second stage of sensation – transduction – begins.

9.1.1 SENSATION

Transduction involves sensory receptors converting stimulus energy into impulses of electrochemical energy. For example, when light hits the eye's retina, receptor cells (photoreceptors) are activated. These cells convert the light energy into individual impulses of electrochemical energy, which are then transmitted to the brain for further processing. Transduction is necessary because the nervous system can only transmit and process energy in electrochemical form.

Transmission begins when electrochemically charged neural impulses leave their sensory receptor site and travel along specific nerve fibres (neural pathways). These nerve fibres connect to specific sensory areas in the brain specialised to receive them. For example, auditory information detected by sensory receptors in our ears is registered in the brain's temporal lobe, but visual information detected by our eyes is registered in the occipital lobe.

When transmission finishes, so does the physical process of sensation. However, to understand what the energy means about the outside world our brain must assign meaning to the stimulus. This psychological response to a sensation is referred to as perception.

KEY CONCEPTS 9.1a

- Experience of the external environment begins with sensation.
- Sensation is an automatic physical process that involves the body's sensory receptors detecting and responding to external energy.
- Reception is the first stage in sensation. It involves sensory receptors taking in and being activated by the raw energy.
- Transduction is the second stage in sensation. It involves sensory receptors converting stimulus energy into individual impulses of electrochemical energy so it can be transmitted by neural pathways to specific areas of the brain.
- Information about specific features of the stimulus is coded into the neural impulses.
- Transmission is the final stage of the sensation process. It involves electrochemical impulses travelling along neural pathways that connect sensory receptors to brain areas specialised to receive them. Perception then occurs.

Concept questions 9.1a

Remembering

1 Name the five types of sensory stimuli we experience. **r**

2 What is sensation? **r**

3 What are the three stages in sensation? **r**

Understanding

4 Explain why raw stimuli needs to be transduced into individual impulses of electrochemical energy. **e**

Applying

5 Using bullet points, explain what each stage of the sensation process contributes to your experience of knowing that Figure 9.2 is a cat. **c**

Figure 9.2

HOT Challenge

6 Each of your eyes is physically connected to your brain by an optic nerve. The optic nerve is a bundle of more than 1 million nerve fibres that carry visual messages from the eye to the brain. Draw on your understanding of the sensation process to explain how damage to your optic nerves would affect your ability to read a book. **c**

Perception: how we create meaning

Perception is a psychological activity that gives meaning to the stimuli our sense organs detect (Figure 9.3). It allows us to understand our world. Perception is a process that occurs in the brain and involves a number of cognitive processes, such as thinking, learning, memory and emotions. It involves three stages: selection, organisation and interpretation (Figure 9.4).

Figure 9.3 Although we all sense stimuli in the same way, perceptions of a single stimulus can vary between individuals.

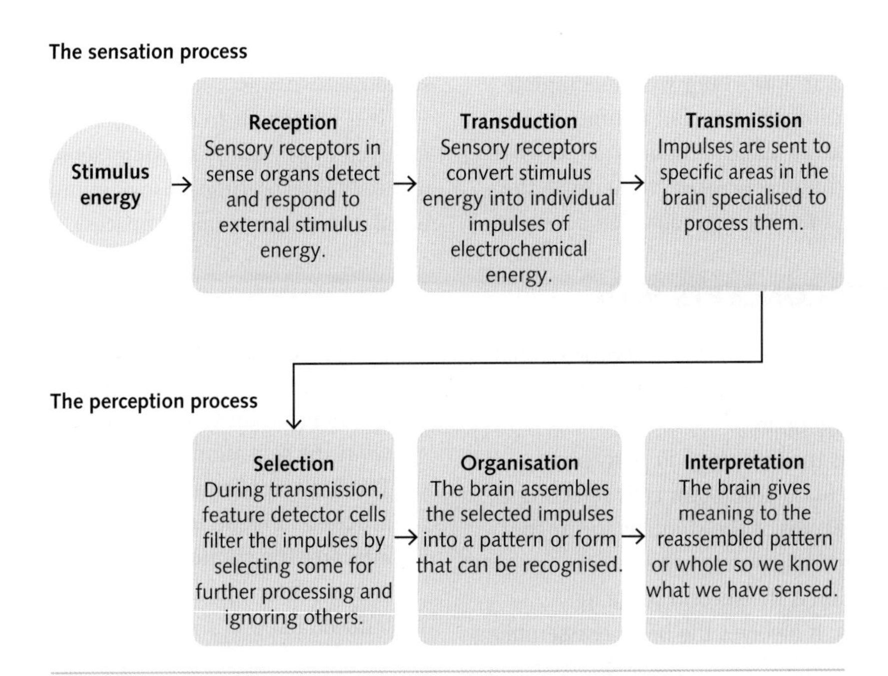

Figure 9.4 Sensation and perception are parts of a continuous process that enables us to understand our environment.

WB
9.1.2
SENSATION AND PERCEPTION

Information detected by sense organs moves along neurons in the form of electrochemical impulses to the brain. Specialised neurons, called feature detectors, select specific features of these impulses so they can be organised into meaningful patterns or wholes that can be interpreted. The brain cannot possibly process all the varied and constantly changing information it receives. So, it pays attention to only the most important pieces and ignores the rest. This is known as **selection**. Stimuli that are ignored receive no further processing. The selected stimuli contain important information about the original stimulus. Because the information now exists in the form of separate impulses, and also because some features have been ignored, individually they have no meaning. Before the brain can interpret the stimuli it receives, it must organise (or reassemble) them into a meaningful pattern or whole.

Perceptual organisation refers to the reassembling of features of sensory stimuli to form a whole or pattern that can be given meaning. How these features are organised depends on a range of mental processes. Some processes are shared by us all; others (for example, memory and past experience) may vary between individuals.

Once the information has been organised, we need to know what it represents. In other words, we need to interpret it. Interpretation involves giving meaning to stimuli so we understand what they represent about the external world (Figure 9.5).

Figure 9.5 What do you see? During sensation, the light image of the picture is taken in by your eyes, converted to neural impulses and sent to your brain. During perception, your brain automatically organises the stimuli into a whole that can be given meaning. When you interpret this whole you recognise some of the randomly arranged blotches as a dog sniffing the ground. You have formed a mental representation of the information you sensed (Plotnik & Mollenauer, 1986).

When the brain is attempting to organise and interpret stimuli, it is working with incomplete and inconsistent information. So, at times the brain has to make an 'educated guess' as to what the pattern of stimuli represent. Sometimes the guess is correct, but this approach also accounts for why our first impression of a stimulus is sometimes wrong.

It is important to remember that the stages involved in physically sensing (sensation) and mentally interpreting sensory stimuli (perception) blend into one continuous process.

KEY CONCEPTS 9.1b

- Perception is the psychological process that gives meaning to sensations. Perception occurs in the brain.
- Selection is the first stage in perception. It involves feature detector cells selecting specific features of the stimulus for further processing and ignoring others.
- Organisation is the second stage in perception. It involves the brain reassembling the individual features of the stimulus into a whole or pattern that can be given meaning.
- Interpretation is the third and final stage in perception. It involves the brain giving meaning to the stimuli so we can understand what the sensation represents.
- The brain applies a number of psychological processes unique to the individual to help it interpret the stimulus. This is why perceptions of a common stimulus may vary between individuals.

Concept questions 9.1b

Remembering

1 Define the term 'perception'. r

2 What are the three stages in perception? r

Understanding

3 Explain why sensation is described as a physical process, but perception is described as a psychological process. e

4 Explain the role of feature detector cells in perception. e

Applying

5 Using bullet points, explain how each stage of the perception process contributes to your experience of knowing that Figure 9.6 is a dog. c

Figure 9.6

HOT Challenge

6 When you look at Figure 9.7, what do you see? Show the figure to someone else and ask them what they see. Do your answers differ? Use your knowledge of the perception process to explain why your perceptions differ. c

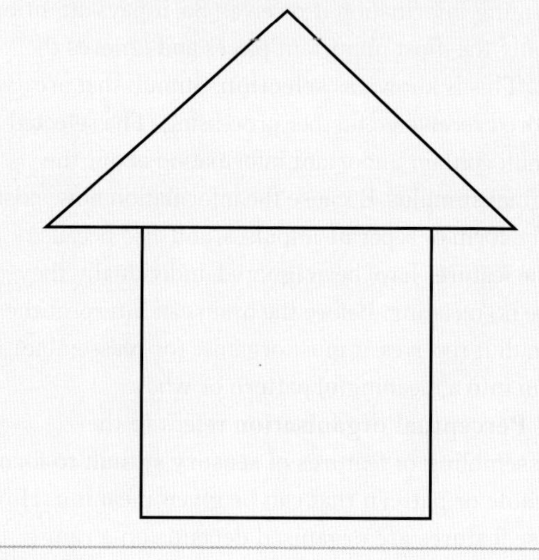

Figure 9.7

Processing and interpreting sensory information

There are two basic approaches to understanding how we process and interpret sensory stimuli and form a perception. They are bottom-up processing and top-down processing.

Bottom-up processing: from specific to general

Bottom-up processing is when our sensory receptors, such as our eyes, receive sensory information and we do not rely on prior experience in order to interpret it. In 1966, James J. Gibson proposed the theory of bottom-up processing and suggested these principles.

- Perceptions are built upwards, starting from the sensory stimuli that sensory receptors take in.
- Perception starts with the processing of low-level features (tiny details) of sensory stimuli (such as lines, edges, shape and colour).
- More and more analysis occurs as we build upwards towards a complete understanding, or perception, of what the original stimulus represents.
- Bottom-up processing requires no previous knowledge or learning (however, research now suggests that other factors, such as expectations and motivation, can impact our perceptions; Rousay, 2021).
- We use bottom-up processing when we are presented with unfamiliar or highly complex stimuli.

Processing of visual stimuli is carried out in one direction from the eye's retinas to the visual cortex. Initially, light bounces off an object and hits the retina. It is then transduced into individual electrochemical impulses that contain information about the object, such as colour and shape. As these impulses move from the retina to the brain's visual cortex, we move deeper and deeper into an analysis of what we are seeing. The visual features are then combined into objects, and the objects are combined into scenes. Our interpretation of this information is based only on the sensory information available.

So, with bottom-up processing, we experience the visual stimulus first and analyse it later. For example, if you are lying on a couch reading a book and suddenly you hear some crackling noises and you smell smoke, you quickly understand that there is a fire nearby. You can't see the fire, but bottom-up processing helps you use your other senses to determine what is happening.

9.1.3 BOTTOM-UP AND TOP-DOWN PROCESSING

Top-down processing: from general to specific

Top-down processing is when your brain starts with an overall hypothesis about a stimulus and uses context and general knowledge to fill in any blanks. In 1970, Richard Gregory introduced the theory of top-down processing because he did not agree with James J. Gibson. Gregory argued that these principles apply.

- Perception is driven by cognition because we perceive the world around us by drawing from what we already know in order to interpret new information (Gregory, 1970).
- Perception starts with perceiving the whole object. You begin with the most general features and move downwards towards the more specific or smaller details.
- Because we constantly take in an endless stream of stimuli, we are not able to attend to each sensation and most of it is lost by the time it reaches the brain. In order to fill in the blanks, the brain formulates a **perceptual hypothesis** (an 'educated guess' about how to interpret a particular pattern of sensory stimulation) and anticipates what comes next.
- Top-down processing relies heavily on higher level cognitive processes. These include expectations, motivation, memories, context, and prior knowledge and experiences. These frameworks are known as **schemas** (concepts that help us recognise patterns in new stimuli so we can rapidly organise sensory information to form a meaningful perception).

Processing information from the top down allows us to make sense of information that has already been brought in by the senses, working downward, from initial impressions down to particular details (Gregory, 1970).

PURPLE YELLOW RED
BLACK RED GREEN
RED YELLOW ORANGE
BLUE PURPLE BLACK
RED GREEN ORANGE

Figure 9.8 **The Stroop effect demonstrates that our past experience and knowledge of words and our expectations about how to spell them allows us to identify the word quicker than the colour it is printed in. This is an example of top-down processing.**

A classic example of top-down processing is a phenomenon known as the Stroop effect (Figure 9.8). In this task, people are shown a list of words printed in different colours. They're asked to say the colour of the word instead of the word itself. People make more mistakes and take longer to do this when the meaning of the word and the ink colour doesn't match; for example, when the word 'purple' is printed in red ink. According to the top-down processing theory, this task is difficult because people automatically recognise the word before they think about its specific features (like what colour it is written in). This makes it easier to read the word aloud rather than to say the colour of the word.

Weblink
Top-down versus bottom-up processing

It is thought that as much as 90 per cent of visual information sensed is not selected for further processing (Gregory, 1970). When the brain receives this incomplete information, it combines it to construct a hypothesis about what the information represents based on factors such as memory, past experience and context. For instance, you will perceive a man with a height of 180 centimetres as looking 'tall' when surrounded by others of average height, and 'short' among a group of professional basketball players. You are using top-down processing to interpret the visual information based on the surrounding visual cues.

Expectation of what we assume a stimulus to be also influences perception (Bugelski & Alampay, 1961). For example, look at the image in Figure 9.9. What do you see? If you were asked if you saw a duck in the image, you would probably recognise a duck's face right away. But if you were asked whether you had seen a rabbit in the image, your perception of the drawing would have been totally different.

Figure 9.10 illustrates how bottom-up processing and top-down processing influence perception, and Table 9.2 summarises their key features.

Alamy Stock Photo/Heritage Image Partnership Ltd

Figure 9.9 **Is this an image of a rabbit or a duck?**

Source: 'Cognitive biases and mindfulness' by Philip Z. Maymin and Ellen J. Lange, Humanities and Social Sciences Communications (2021) 8:40 | https://doi.org/10.1057/s41599-021-00712-1. licensed under a Creative Commons Attribution 4.0 International License, https://creativecommons.org/licenses/by/4.0/

Figure 9.10 **A beginning reader would use bottom-up processing when reading the quote aloud. They would carefully attend to each letter and word and realise that having a second 'the' doesn't make sense. An experienced reader doesn't expect there to be a second 'the', so they would probably skip over it. They used top-down processing because their past experience with written language changed the way they perceived the writing in the triangle.**

Table 9.2 Top-down and bottom-up processing

Bottom-up processing	Top-down processing
Definition: Information processing in which incoming stimulus data initiate and determine the higher level processes involved in their organisation and interpretation	Definition: Information processing in which an overall hypothesis about a stimulus is applied to and influences the analysis of incoming stimulus information into a meaningful perception
Key features: » analyses separate features of the stimulus then integrates them to build upwards towards a perception » uses pre-existing schemas to interpret sensory information » begins when we sense a stimulus and process its low-level features » involves data-driven analysis of sensory stimuli » has no previous learning required » typically used for processing new or complex stimuli	Key features: » starts with perception of the general features of a stimulus and moves downwards towards the specific, smaller details » involves forming a perceptual hypothesis » relies on higher cognitive processes or schemas (such as prior knowledge, experiences, memory, motivations, context and expectations) » typically used for processing familiar and relatively simple stimuli

ANALYSING RESEARCH 9.1

Blind food-taste challenge

A blind taste-testing challenge focuses on isolating all but one of the five senses – taste. Participants are asked to place a blindfold on and determine different characteristics about the food and/or drink items that are presented in front of them.

In a study by Lowengart (2012), researchers interested in whether branding made a difference in consumer purchases conducted a taste-test challenge. For two days, students, visitors and staff members at a large university were approached and asked to participate in the taste test. The challenge involved two taste tests by the same participants – one a blind taste-test and the other a non-blind test. They were presented with bottles of wine to taste which were a mixture of brand name wines with a good reputation and generic brand wines. Then they were asked to assess the colour intensity, aroma, taste, tartness, harmony and aftertaste of several wines. These features are consistent with how wine is judged in the wider community. Participants were asked to use a 10-point interval scale with 1 representing a very low level of the attribute and 10 representing a very high level of the attribute.

In the blind test, participants tasted a variety of wines while blindfolded. Then they were asked which wine was better. They were given no information about the wine they were drinking. The blindfold was intended to eliminate the possibility of participants being influenced by any senses other than taste. In the non-blind test, the same participants followed the same procedure, but in this case, they could see the labels of the different wines.

The results indicated that when participants could see the brands, the ratings of the more reputable brands increased.

Sources: Adapted from Lowengart, O. (2012). The effect of branding on consumer choice through blind and non-blind taste tests. *Innovative Marketing, 8*(4).
Rousay, V. (2021, January 21). Bottom-up processing. *Simply Psychology*. www.simplypsychology.org/bottom-up-processing.html

Questions

1 Identify the sampling procedure used by the researchers.
2 Identify the independent and dependent variables in this study.
3 What were the results of this study?
4 What conclusion(s) can be drawn from this study?
5 Explain whether top-down or bottom-up processing was demonstrated in the blind test. Give a reason for your answer.

ACTIVITY 9.1 BLIND TASTE-TESTS

Try it

1 In this activity you are going to conduct your own blind taste-test 'experiment'. You will need a blindfold and two brands of soft drink, crisps or chocolate, one plain label (or supermarket brand) and the other a known brand. You will also need a willing participant. There are two stages to this experiment.
Stage 1: Blindfold your participant and tell them they will be completing a blind taste-test. Give them a small sample of each brand with a drink of water in between. Ask them which one they prefer. Record their response. This is the bottom-up phase of the experiment
Stage 2: Show the participant both products and ask them which one they prefer. Record their response. This is the top-down phase of the experiment.

Apply it

2 Identify the example of top-down and bottom-up processing in the following scenario. You take a sip of a cup of coffee and the coffee is so hot that it burns your tongue. Before you take the next sip, you blow gently on the coffee to cool it down.

Exam ready

3 Which is true about bottom-up processing?
A It relies on information from the senses.
B It is schema driven.
C It relies on high level cognition.
D It involves the interpretation of information.

KEY CONCEPTS 9.1c

» We use top-down processing and bottom-up processing to help us quickly organise and interpret sensory stimuli.
» Bottom-up processing involves the perception of the details of sensory stimuli first, and builds upwards towards a perception of the whole stimulus.
» Bottom-up processing is used when we are presented with unfamiliar stimuli.
» Bottom-up processing does not rely on schemas, because it only processes the stimuli itself.
» Top-down processing involves perceiving the general features of a whole stimulus first, then its specific details.
» Top-down processing is influenced by schemas, including previous experience, memory and motivation, expectations, context and previous knowledge.

Concept questions 9.1c

Remembering

1 Define the term 'top-down processing'. r
2 Define the term 'bottom-up processing'. r
3 Identify three factors that influence top-down processing. e
4 What are schemas? Provide an example. c
5 What role do schemas play in top-down processing? e

Understanding

6 Explain the difference between top-down processing and bottom-up processing. e

Applying

7 How does prior knowledge influence perception? e
8 How does top-down processing help you to read the following sentence? Pysolchogy is a facsiantnig subject. c

HOT Challenge

9 Identify whether you would use top-down processing or bottom-up processing in the following situations. c
a You are trying to solve a maze by looking at it from above.
b You are trying to find your way out of a dark maze using a torch.

9.2 The role of attention in perception

Our perception of a sensory stimulus largely depends on whether we select it for attention and how much attention we direct towards it. **Attention** is a voluntary (conscious) or involuntary (unconscious) tendency to focus awareness on a specific stimulus and ignore other stimuli. Imagine you were standing in the crowd at a noisy football match. You get so involved in a conversation with your friend that you block out all the background noise. Suddenly you notice that people around you are starting to leave the stands. You had been paying so much attention to your conversation that you didn't hear the half-time siren!

Key features of attention

Attention is part of our executive (higher level) thinking and decision-making processes. It allows us to 'tune out' irrelevant information so we can focus on important information. The focus of attention may be an external or an internal stimulus; for example, it could be a person, a beach or a television (external), or a thought, a pain or recalling a person's name (internal).

We tend to select stimuli for attention if:

» *it is unusual, intense, unexpected or stands out from its surroundings.* For example, if you are listening to your favourite music and suddenly hear a loud crash outside, your attention will switch from the music to the sound of the crash
» *we are motivated to or expect to encounter a particular stimulus.* For example, if you are hungry when you are shopping, you are more likely to notice the aromas from the food court more than the people around you
» *it is personally significant.* For example: parents can hear their baby crying in the next room while other adults may fail to notice
» *it is moving or changing.* For example, if you were riding your bike down a busy street, you are more likely to notice a car pulling out in front of you than one parked on the side of the street
» *it becomes repetitious.* For example, you might notice a dripping tap or flickering light.

Automatic and controlled processes

When we are perceptually processing sensory information, we rely on two cognitive processes: **automatic processes** and **controlled processes**. The key features of automatic and controlled processes are outlined in Table 9.3.

Sustained, divided and selective attention

Attention is limited, both in terms of how much information we can focus on at once (capacity) and how long we can maintain this focus (duration).

Table 9.3 Controlled processes and automatic processes

Controlled processes	Automatic processes
Definition: Conscious, voluntary actions or cognitive processes that require a high level of attention and monitoring	Definition: Actions that require little conscious awareness or mental effort, and do not interfere with performance on other activities
Key features: » require a high level of attention, so involves conscious, deliberate, voluntary effort to think about and perform tasks » can be stopped and resumed later » requires more mental energy » part of conscious awareness » takes time and practice to develop » difficult to engage in more than one task relying on controlled processes at the same time; for example, when you learn to drive a car you must pay close attention to each step of the driving process. This is why it's hard to hold a focused conversation while learning to drive a car.	Key features: » does not require attention or a deliberate effort – are often unintentional and typically have an environmental trigger » can't be stopped once started » require little mental energy » after acquisition, can be performed with little conscious awareness » acquired quicker than automatic processes » can engage in more than one task dependent on automatic processes at the same time » with a great deal of practice, many mental processes may become automatic. For example, typing, riding a bike, driving a car and identifying the meaning of words all require attention at first. However, after regular practice these tasks may become automatic.

There are three types of attention that help us effectively manage our attentional resources so we can understand the environment around us. They are sustained attention, divided attention and selective attention.

When we focus attention on an activity or stimulus over a prolonged period of time without being distracted by other stimuli, we are using **sustained attention**. The key features of sustained attention are as follows.

» Sustained attention deeply focuses your attention, so it is most beneficial when learning something. For example, if you are learning to play chess, you need to focus, and hold, your attention on the placement of the chess pieces in front of you to successfully learn winning moves.

WB 9.2.1 UNDERSTANDING ATTENTION

» Visually, sustained attention is critical for assessing motion information, such as how fast cars are moving when you want to cross the road.
» It can be challenging to maintain sustained attention for long periods of time without becoming distracted. Therefore, your level of sustained attention often varies, from being focused one minute to fading the next. However, you are usually able to re-focus on the task after a distraction has passed.
» Research suggests that sustained attention peaks during the early 40s and then gradually declines as people age (Fortenbaugh et al., 2015).

Divided attention (often referred to as multi-tasking) refers to rapidly switching the focus of your awareness between two (or more) sources of information so you can perform two (or more) tasks at the same time. The key features of divided attention are as follows.

Worksheet The Myth of Multitasking

» It is dependent on the types of tasks being performed, how difficult they are, how similar they are and whether they involve automatic or controlled processes.
» You do not completely change from one task to another entirely different task. Instead, you try to perform both tasks at the same time. You divide your attention; for example, like when you are buttering your toast while holding a conversation, or when you are able to take in what the teacher is saying and write notes at the same time.
» It is necessary for responding to important daily events, therefore distraction can be useful. For example, if you start to pass a car when you are driving on the freeway, you may suddenly notice a sign for your exit ahead. If you aren't able to safely pass the car and pay attention to traffic signs at the same time, you may lose important information or risk causing an accident.
» Tasks requiring a high level of attention and controlled processes are difficult to successfully complete using divided attention. This is why it is both difficult and dangerous to attempt to send a text message at the same time as driving a car.
» Tasks requiring automatic processes, such as watching TV and eating a meal at the same time, require low levels of attention, so they are suited to divided attention.
» Two similar tasks are much more difficult to complete at the same time than two different ones.
» Your responses to changing stimuli will be slower and you may miss attending to crucial stimuli. For example, if you are talking on a mobile phone while crossing a street, you may fail to notice a bus coming around the corner.

Using divided attention when you are trying to learn something requires you to share cognitive resources between the tasks. This causes your memory and understanding of the information to not be as efficient as if you used sustained attention (Figure 9.11). For instance, if you are listening to the TV while trying to solve a new mathematics problem, your ability to focus attention on problem solving and remembering suffers because you are devoting cognitive resources to listening to the TV. Research suggests that using divided attention, or multi-tasking, often does not work well because our attention is, in reality, limited (Srna et al., 2018).

Shutterstock.com/Ollyy

Figure 9.11 Dividing your attention when you are trying to study diverts cognitive resources away from learning. This may result in an inability to successfully learn new information.

Selective attention involves being able to choose to focus your awareness on a specific or limited range of stimuli while ignoring others. The key features of selective attention are as follows.

» Selective attention allows you to focus on the most important stimuli and ignore less important stimuli. This helps you to survive. For example, when crossing a busy street, you need to focus on moving cars or when the traffic lights change while you ignore other sights and sounds.

» Selective attention can be intentional. For example, you can consciously choose to focus attention on one person who is talking in a group and ignore others who are present.

» Selective attention can be automatic. For instance, if your mobile phone rings, your attention automatically shifts from what you were doing to focus on answering the phone call.

» Selective attention makes it difficult to pay full attention to more than one thing at a time.

9.2.2 MEDIA ANALYSIS

ACTIVITY 9.2 ATTENTION

Try it

1 Have music playing during this activity! Study the pattern sets in Figure 9.12 by comparing the left and right columns. Circle where the pattern on the right is different from the pattern on the left. If both sets match, move to the next set of patterns.
Your ability to complete the task while blocking out the music is an example of selective attention.
If you can sing along to the music *while* completing the task, this is an example of divided attention.
The pattern task is new and a little challenging so you might find you need to use selective attention to complete it successfully or quickly.

HappyNeuron Pro

Figure 9.12

Apply it

2 Sometimes, when you are in class, you need to listen to what the teacher is saying, read the notes on the board and write down the important information in your book. When you are at home studying, you need to read and reread your notes, and stay focused on studying your notes for hours. Identify the different types of attention being described in these scenarios.

Exam ready

3 You are working at a clothing outlet and a customer asks you about the store's opening hours when paying for some clothes they have just purchased. You need to scan and bag the items while answering the question about opening hours.

This would be an example of

A selective attention.

B divided attention.

C sustained attention.

D focused attention.

ANALYSING RESEARCH 9.2

Checking phones in lectures can cost students half a grade in exams

According to a study undertaken by researchers from Rutgers University in the USA, students who access electronic devices, such as a phone or tablet, for personal use when they are in a lecture do not perform as well as they could in end-of-term exams. Students who don't use such devices but attend lectures where their use is permitted also do worse. This suggests that phone/tablet use damages the group learning environment.

The research team performed an in-class experiment to test whether dividing attention between electronic devices and the lecturer during the class affected students' performance in within-lecture tests and an end-of-term exam. One hundred and eighteen psychology students at Rutgers University participated in the experiment during one term of their course. Laptops, phones and tablets were banned in half of the lectures and permitted in the other half. When devices were allowed, students were asked to record whether they had used them for non-academic purposes during the lecture.

The study found that having a device didn't lower students' scores in comprehension tests of the lectures, but it did lower scores in the end-of-term exam by at least 5 per cent, or half a grade.

This suggests that the main effect of divided attention in the classroom is on long-term retention of material presented, which negatively affected unit exam and final exam performance.

Also, when the use of electronic devices was allowed in class, performance was poorer for students who did not use devices as well as for those who did.

The study's lead author, Professor Arnold Glass, suggested that these findings should alert students and instructors that dividing attention has a harmful effect on exam performance and final grades.

Adapted from 'Dividing attention in the classroom reduces exam performance' by Arnold L. Glass and Mengxue Kang, Educational Psychology, Volume 39, 2019 - Issue 3, Pages 395–408, Taylor & Francis, https://doi.org/10.1080/01443410.2018.1489046

Questions

1 What was the aim of this study?
2 Identify the independent and dependent variables in this study.
3 What sampling method did the researchers use?
4 Identify one limitation of this study. How could this limitation be overcome?
5 What were the results of this study?
6 What conclusion(s) can be drawn from these results?

HOT Challenge

7 Mary is very concerned that her son Jad, a first-year university student, repeatedly looks at his mobile phone when studying. Jad tells her not to worry because multi-tasking won't affect his ability to learn and it won't affect his grades. What would you say to Mary in response to Jad's statement?

CASE STUDY 9.1

Celeste Barber

Australian comedian Celeste Barber (Figure 9.13) has opened up about being diagnosed with attention deficit disorder (now more commonly referred to as attention deficit hyperactivity disorder, or ADHD) when she was 16 years old. Barber spoke about the myths that surround the neurodevelopmental condition, which affects the behaviour and learning of around one in every 20 Australians. [Individuals with ADHD find it difficult to self-regulate and control their behaviour, thoughts, emotions and impulses. This often makes it difficult for them to plan and problem solve and achieve long-term goals.]

'People just think ADD is like "You're just loud" and "She's got ADD, she's nuts",' she explains. 'It's not what it is. It's the inability to focus. It's the inability to sit, pick up a pen and do your homework.'

Shutterstock.com/Dfree

Figure 9.13 Celeste Barber

The 38-year-old says she remembers overhearing her mother discussing her medical condition and medication with her doctor. 'We know she is loud, we know she is full on, but we like that about her,' she recalls her mother saying. 'We like her drama. She's funny. She's all these things and we don't want that to go away.' The doctor replied: 'It won't. She'll just be able to sit down, have a minute and sit through a class at school.'

The actress and writer added she is grateful for her late diagnosis, because it made her focus on what she was good at from a young age. 'It made me realise my currency and made me work harder and be like, "Well, I'm really loud and I can't sit down, can't sit still in class, can't read a book and can't do anything. But what am I good at? Oh, I'm funny and I can act and I'm dramatic".'

Barber went on to share how ADHD affects her work life now, admitting she can be forgetful. 'With ADD you're quite scattered, which isn't always a bad thing. I can multitask and I can have a lot going on at once. With the medication, it helps me still do all that, but when I throw all the balls in the air, I can now catch most of them as opposed to just going "They're all in the air, let's see what happens" and then running away. It's kind of great because I follow through with things a lot more.'

Adapted with permission from Fitzsimons, B. (2021). 'I'm grateful that I was diagnosed quite late.' Celeste Barber on her ADHD. Mamamia, 21 January <https://www.mamamia.com.au/celeste-barber-adhd>

Questions

1 When Celeste Barber was a child and teenager, what were the indicators that she had ADHD?

2 Based on Celeste Barber's experience, identify one strength and one limitation of having ADHD.

3 Outline one way ADHD could affect a person's sustained attention, divided attention and selective attention.

4 Imagine you are going to conduct a research study on the impact of attention on the academic performance of adolescents. Identify the following information.
   - a Who is your population of interest?
   - b Describe the sampling method you could use to provide the most representative sample of the population.

HOT Challenge

5 Explain whether it would be easy or difficult to hold a conversation with your friend while you were proofreading your English essay.

KEY CONCEPTS 9.2

» Sensory stimuli that is selected for attention receives further processing by the brain.
» Attention can be voluntary (conscious) or involuntary (unconscious).
» Attention allows you to focus on important stimuli and ignore unimportant stimuli.
» Attention can focus on external (environmental) or internal (bodily or psychological) stimuli.
» Attention can be influenced by a range of factors related to the stimuli or to the person.
» Attention can be an automatic process or a controlled process.
» Three major types of attention are sustained attention, divided attention and selective attention.

Concept questions 9.2

Remembering

1 Define the term 'attention'. r

2 Name four factors that influence attention. r

3 Name three different types of attention. r

Understanding

4 With reference to attention, explain the difference between a controlled process and an automatic process. e

5 Explain the difference between sustained attention and selective attention. e

Applying

6 Tash had always wanted to learn the guitar. For his birthday he received a voucher for 10 guitar lessons. When Tash began these lessons, what type of attention was required? Give a reason for your answer. c

7 After Tash had learned to play the guitar, what type of attention did he need to use when playing? Give a reason for your answer. c

HOT Challenge

8 Identify whether sustained, divided or selective attention is being demonstrated in the following scenarios. c

Scenario 1: Jamie was listening to music while making sandwiches.

Scenario 2: At a noisy party, Zoula was deep in conversation with her friend.

Scenario 3: Ineke safely drove her car home from work.

Scenario 4: Lachlan put a random list of 20 names into alphabetical order.

Scenario 5: Martine practiced playing her piano for 30 minutes each day.

9.3 Factors influencing visual perception

When we experience vision, we only take in a limited area of our visual environment. Because there is so much visual stimuli and it is so varied, we are selective in what we pay attention to. So, selective attention is one reason why our perception of visual information is unique. However, other biological, psychological and social factors also cause perceptions of visual stimuli to vary.

Biological factors

'Biological factors' is a broad term that includes physical, physiological, neurological or genetic conditions that affect an individual. Biological factors are entirely internal. Biological factors not only impact our physiological functioning, they also influence the way we perceive the world around us.

Depth cues: judging distance to create a third dimension

One of our most remarkable perceptual abilities is our capacity to create the appearance of three dimensions – commonly called height (depth), length and width – from the flat, two-dimensional images that reach our retinas. This is known as **depth perception** – the ability to see three-dimensional space and to accurately judge

Figure 9.14 Depth perception allows us to see the world in three dimensions and to judge the distance and movement between objects and ourselves.

distances using environmental cues (Figure 9.14). Without depth perception you would be unable to effectively drive a car, ride a bicycle, or simply walk around a room.

To accurately perceive depth in our environment, our visual system needs to be intact. The main components of the visual system are the eyes' retinas, the optic nerves and the brain (Figure 9.15). When light rays land on the eye's retina, they form a two-dimensional image. This image is converted into electrical impulses and transmitted, via the optic nerve, to the brain. The brain organises these impulses into a form that can be given meaning and interprets the two-dimensional retinal image as a three-dimensional representation of the world around us.

Depth perception relies on a number of cues known as depth cues. **Depth cues** are a variety of internal and external stimuli or processes that inform the visual system about the depth of an object or its distance from the observer. Some depth cues are binocular (involve using two eyes), while others are monocular (depend on one eye only).

Binocular cues

Binocular cues are a group of depth cues that require both eyes to work together and that provide the brain with information about depth and distance. Binocular cues include retinal disparity and convergence. They are summarised in Table 9.4.

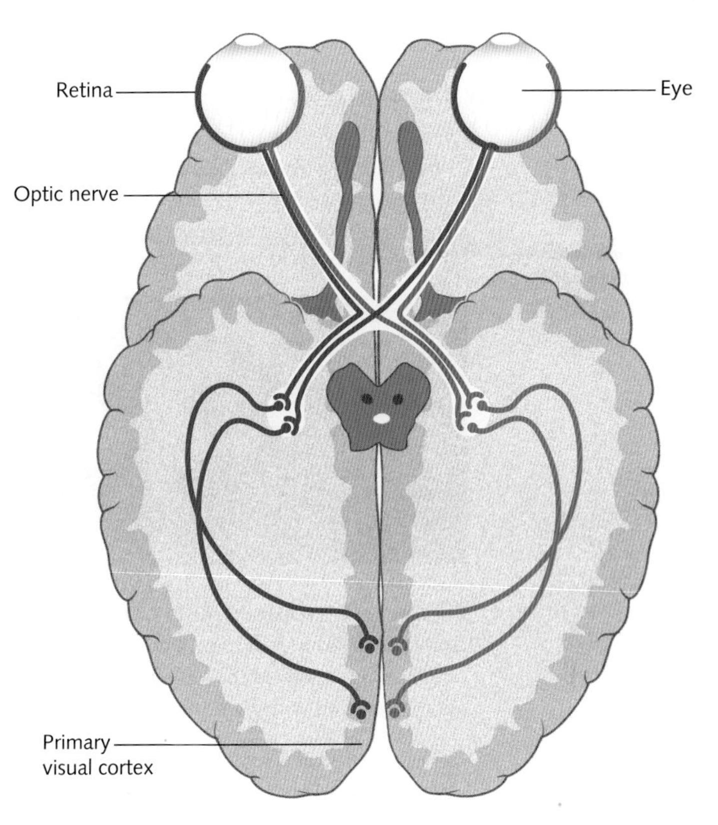

Figure 9.15 The visual system consists of the eyes, the retinas, the optic nerves and the brain.

Monocular cues

9.3.1 SENSATION AND PERCEPTION OF A VISUAL STIMULUS

Monocular cues are the depth perception cues that rely on information from only one eye. Monocular cues include accommodation and pictorial cues. They are summarised in Table 9.5.

If someone has a malformation in their visual system during foetal development, they may experience some form of visual impairment. Common physiological problems that can occur in the visual system and result in loss of vision include:

» abnormalities associated with the shape and size of the eye, retinal damage and damage to the pupil, iris and lens
» damage to or malformation of the optic nerve, which interrupts transmission of visual information
» damage to brain areas involved in visual perception, which may mean the person can sense the visual information and transmit it to the brain efficiently, but their ability to analyse and interpret it may be impaired.

Weblink
Monocular and binocular depth cues

Table 9.4 Binocular depth cues

Cue	Key features and example
Retinal disparity is a depth cue created by small differences between the image that reaches the right eye and the image that reaches the left eye (see example).	» Because our eyes are approximately 6.5 cm apart, each retina receives a slightly different image. » When the brain receives the separate images, it fuses (overlaps) them into one overall image and compares them. » The fused image results in three-dimensional sight (Figure 9.16). » Any disparity (difference) between the two images provides the brain with information about the depth and distance of objects in the environment.   iStock.com/TimAbramowitz **Figure 9.16** Retinal disparity produces stereoscopic vision. Hold the page about 15–20 centimetres from your eyes. Focus your eyes between the two photographs, where an overlapping image of the ivy should appear. Try to merge the leaves into one image. If you are successful, you will see a third dimension appear.
Convergence involves both eyes simultaneously turning inwards as an object moves closer (within approximately 7 metres) in order to maintain focus on the object.	» When you look at a distant object, the lines of vision from your eyes are parallel. However, when you look at something less than 7 metres away, your eyes must converge (turn in) to focus on the object (Figure 9.17). » This simultaneous turning inward allows the image of the object to fall on corresponding points of each eye's retina. This allows two slightly different views of an object seen by each eye to form a single image. » Convergence is controlled by muscles attached to the eyeball. When these muscles tense or relax, they feed information on eye position to the brain to help it judge distance. 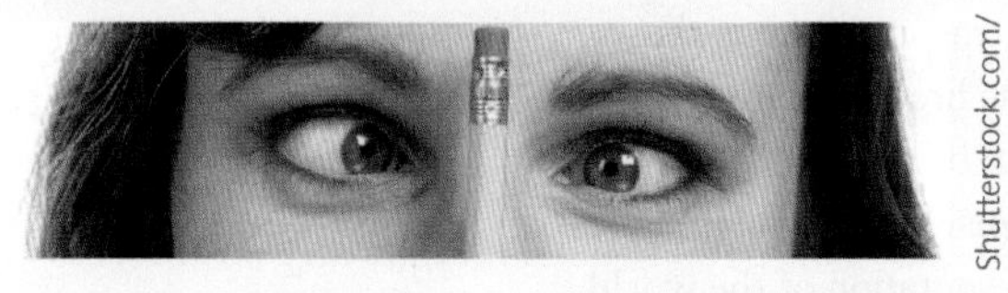 Shutterstock.com/Darrin Henry **Figure 9.17** Convergence: when you look at something close up, your eyes converge.

Table 9.5 Monocular depth cues

Cue	Key features	Example
Accommodation involves the ciliary muscles attached to the lens in each eye voluntarily or involuntarily relaxing or contracting to alter the shape of each lens in each eye so it can focus on objects at varying distances.	» When an object is close (within approximately 3 metres), the ciliary muscles contract so your lens bulges and stays focused on the object. » When the object is further away, the ciliary muscles relax so your lens flattens in order to stay focused on the object. » Sensations from these moving muscles are transmitted to the brain and interpreted.	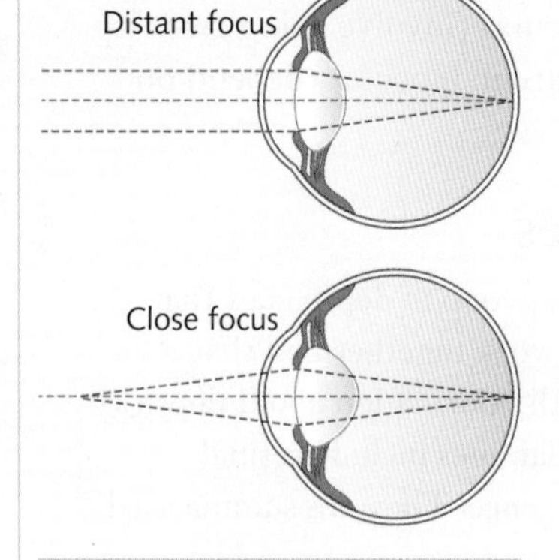 **Figure 9.18** How the eye focuses on distant and close objects

9.3.2 VISUAL PERCEPTION

Cue	Key features	Example
Pictorial cues are monocular cues present in two-dimensional (pictorial) images that allow the brain to perceive apparent three-dimensional depth.	» Pictorial cues create a convincing sense of depth where none exists, and explain why a person with sight in only one eye can learn to estimate depth.	
	Types of pictorial cues » **linear perspective** – The apparent convergence of parallel lines creates the impression of increasing distance. If you look at Figure 9.19a you will see two railway tracks. If you imagine yourself standing between the tracks, they appear to meet near the horizon. Because you know they are parallel, their convergence suggests great distance.	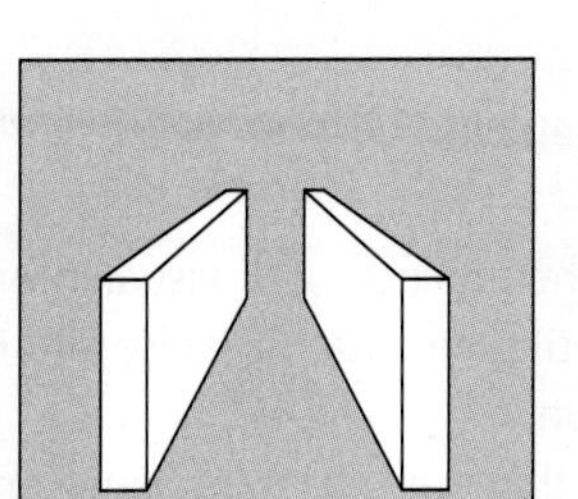 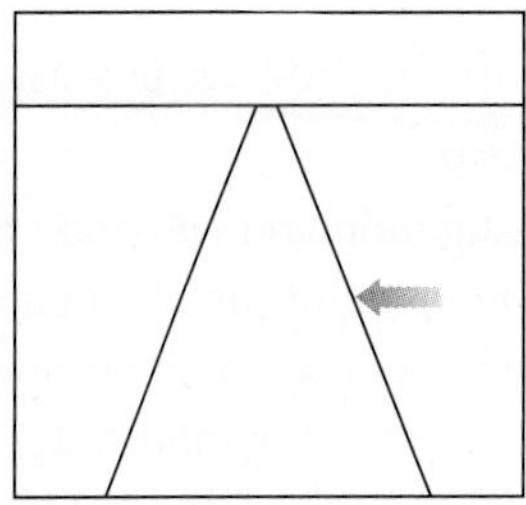 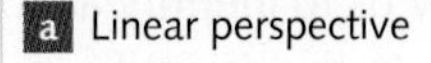 a Linear perspective
	» **relative size** – The smaller retinal image of two objects is perceived as further away, and the larger retinal image of two objects is perceived as being closer. The effect of relative size is strongest when we are familiar with the actual size of the object (Figure 9.19b).	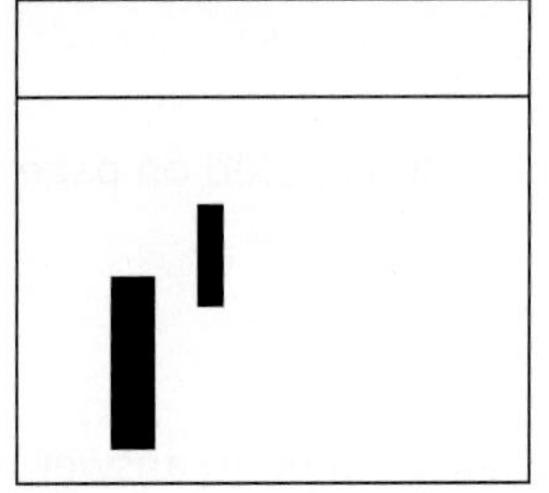  b Relative size
	» **interposition** (also known as overlap) – When one object partially blocks another object it is perceived as being in front of and, therefore, closer than the object it covers (Figure 9.19c).	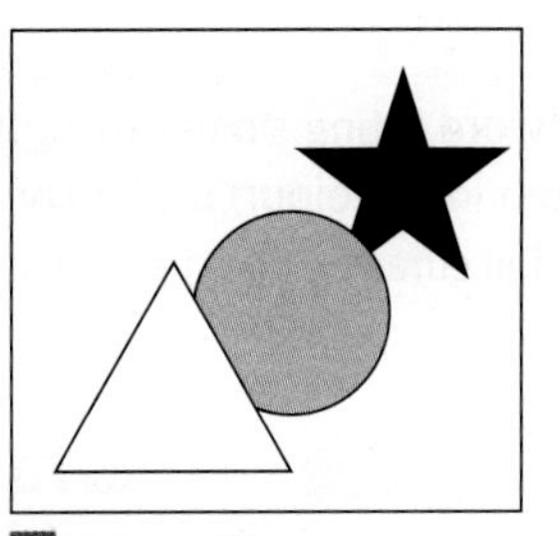 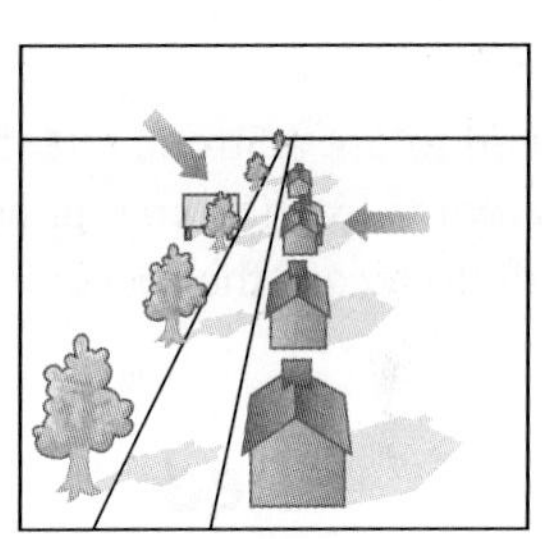 c Interposition
	» **texture gradient** – the surface features of an object are perceived as smaller and less detailed the more distant an object becomes (Figure 9.19d).	 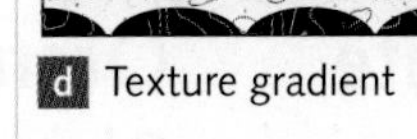 d Texture gradient
	» **height in the visual field** – objects close to the horizon line appear further away; objects below the horizon line placed higher in a picture appear more distant than objects placed further below the horizon line; objects above the horizon line placed lower in a picture appear more distant than objects below, but close to, the horizon line (Figure 9.19e).	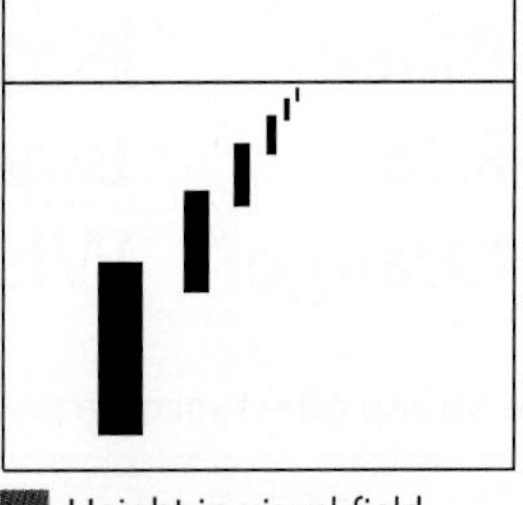  e Height in visual field **Figure 9.19 Pictorial cues**

INVESTIGATION 9.1 THE STROOP EFFECT

Scientific investigation methodology

Controlled experiment

Aim

To investigate into how we process and interpret sensory stimuli and form perception

Data calculator

Introduction

In this investigation you will study the processing of visual information. You will need to generate appropriate qualitative and/or quantitative data, organise and interpret your data, and research a conclusion in response to your research question.

Logbook template: Controlled experiment

The Stroop effect demonstrates that our past experience and knowledge of words and our expectations about how to spell them allow us to identify a word quicker than the colour it is printed in. This is an example of top-down processing. In this scientific investigation you will test the influence of the Stroop effect on perception.

Pre-activity preparation

You are required to apply the key science skills (listed on page 347) to:

- develop a research question
- state an aim
- formulate a hypothesis
- plan an appropriate methodology and method to answer the question, while complying with safety and ethical guidelines.

Procedure

You will need to use a Stroop effect word listing similar to Figures 9.20 and 9.21 below. This word listing has a number of **congruent** words (each word is printed in its colour) and **incongruent** words (each word is printed in a different colour). Students are to record the speed and accuracy of participants reading aloud the colours of the text (not the words as printed).

Red	White	Orange	Red	Blue
Green	Blue	Blue	Black	Orange
Blue	Black	Purple	White	Red
Purple	Red	Black	Orange	Green
Black	Orange	White	Green	Purple

Figure 9.20 The Stroop effect congruent words

Red	White	Orange	Red	Blue
Green	Blue	Blue	Black	Orange
Blue	Black	Purple	White	Red
Purple	Red	Black	Orange	Green
Black	Orange	White	Green	Purple

Figure 9.21 The Stroop effect incongruent words

Results

Collate your results for all participants. Process the quantitative data you recorded, using appropriate mathematical calculations and units (for example, descriptive statistics such as a mean). Organise, present and interpret your data using an appropriate table and/or graph.

Discussion

1. State whether your hypothesis was supported or not.
2. Describe the difference in results (speed and accuracy) for congruent and incongruent words.
3. Discuss any implications of your results.
4. Draw a conclusion in response to your research question.
5. Is it possible to generalise any of your results to the wider population? Why or why not?
6. Discuss any potential extraneous variables and how they could have affected your data.
7. Suggest improvements you could make to address the extraneous variables if the investigation was to be repeated.

Adapted from VCAA VCE Psychology Unit 4 Advice for Teachers https://www.vcaa.vic.edu.au/curriculum/vce/vce-study-designs/Psychology/advice-for-teachers/Pages/Unit4.aspx

KEY CONCEPTS 9.3a

- » Each person's perception of visual stimuli is unique and can be influenced by a range of biological, psychological and social influences.
- » Biological factors that influence visual perception include any physical, neurological, chemical and genetic conditions that affect an individual's physiological functioning and perceptual abilities.
- » We use internal and external depth cues to accurately judge distances.
- » Depth cues allow us to perceive three dimensions from two-dimensional retinal images.
- » There are two types of depth cues: binocular cues and monocular cues.
- » Binocular depth cues include retinal disparity and convergence.
- » Monocular depth cues include accommodation and pictorial cues (linear perspective, relative size, interposition, texture gradient and height in the visual field).
- » Visual perception can be affected by malformations of or damage to any of the components of the visual system, or by ageing or genetics.

Concept questions 9.3a

Remembering

1. What are depth cues? r
2. Name the two categories of depth cues. r

Understanding

3. Explain the difference between binocular depth cues and monocular depth cues. e
4. Name three pictorial cues used by artists to create three-dimensional pictures on two-dimensional surfaces. Explain how each cue contributes to developing a three-dimensional image. c

Applying

5. If you suddenly lost sight in one eye and attempted to kick a soccer ball, you probably would miss because you would not be able to perceive how far away you are from the ball. Based on your knowledge of depth cues, explain one reason why you would miss the kick. c

HOT Challenge

6 Study Figure 9.22. Copy the image and draw arrows on your copy to identify the following pictorial cues: linear perspective, texture gradient, interposition, height in the visual field and relative size. Explain how each cue operates in the photo to create the impression of depth and distance. C

Figure 9.22

Shutterstock.com/PIXEL to the PEOPLE

Psychological factors

During the interpretation stage of perception, a number of psychological factors unique to the individual affect the final perception of sensory stimuli. The term 'psychological factors' means processes that operate at the individual level that impact the mental state and cognition of the individual, thereby influencing behaviours. These factors include Gestalt principles (of organisation), context, motivation, past experiences and memory.

Before the visual elements selected by feature detector cells can be interpreted, the brain must reassemble them into a pattern or form that can be interpreted. Our brain must do this quickly so we can adapt our behaviour to suit the constantly changing external conditions. This increases our chance of survival. The instant nature of this organisation relies heavily on **visual perception principles**. These are rules our brains apply automatically to organise and interpret visual stimuli in a consistent and meaningful way. Gestalt principles and visual constancies are examples of visual perception principles.

Gestalt principles: grouping and separating

The Gestalt psychologists suggest we apply a number of rules, known as **Gestalt principles**, when we organise separate stimuli into meaningful patterns or whole forms so we have some order to our perceptions. Gestalt principles include four main categories (Table 9.6).

The simplest Gestalt principle, **figure–ground**, involves the viewer using an imaginary contour line to perceptually group and separate some features of a stimulus so that a part of the stimulus appears to stand out as an object (the **figure**) against a plainer background (the **ground**) (Figure 9.23).

Closure refers to the viewer's tendency to perceptually complete an incomplete figure by filling in an imaginary contour line so that the figure has a consistent overall form. For example, each drawing in row b of

Shutterstock.com/kooanan007

Figure 9.23 A challenging example of figure–ground organisation. Once the insect becomes visible, it is impossible to view the picture again without seeing the insect.

Table 9.6 Gestalt principles group and separate features of a visual stimulus

Principle	Description	Examples
a Figure–ground	We see these objects as figures against the background.	
b Closure	Here, we see gaps being closed to make a complete figure.	
c Similarity	In these examples, organisation depends on similarity of colour and/or shape.	
d Proximity	Notice how differently a group of six or nine objects can be perceptually organised, depending on their spacing.	

Table 9.6 contains one or more gaps, yet each is perceived as indicating a recognisable and whole shape.

Similarity helps us to perceive stimuli that have similar visual features (such as size, shape, colour or form) as belonging together and forming a meaningful single unit or group.

Proximity allows stimuli close together in space to be perceived as belonging together and forming a meaningful single unit or group. For example, if three people stand near one another and a fourth person stands 5 metres away, the three people standing near each other will be seen as a group and the distant person as an outsider.

Visual constancies: managing change

The sensory information that reaches us is constantly changing, even when it comes from the same object. This means the retinal image of the object changes constantly, yet our brain perceives the external world as stable and unchanging. This is because of **visual constancies** – the group of perception principles that allow us to view objects as unchanging in terms of their actual size, shape, brightness and orientation, even when there are changes to the image that the object casts on the retina. These constancies are summarised in Table 9.7.

Weblink
Using Gestalt principles in design

Table 9.7 Visual constancies

Visual constancy	Definition	Example
Size constancy	The perceived size of an object remains constant despite changes in the size of the object's retinal image.	You do not perceive your hand to be shrinking as you move it away from your face, even though its retinal image is getting smaller. Because you have previously seen your hand at various distances, your perception is that your hand is still its normal size, but that it is moving further away.
Shape constancy	The perceived actual shape of an object remains constant despite changes to the shape of the object's retinal image.	When you look at the page of a book from directly overhead it appears rectangular. When you view it from an angle, the shape of the retinal image changes. In reality the page is rectangular and your perception of its shape remains constant.
Brightness constancy	An object's perceived level of brightness relative to its surroundings stays the same under changing light conditions.	If you look at a white sheet hanging on a clothes line on a sunny day, it will appear bright white. If you view it early in the evening, the retinal image it casts will be duller but your perception is that it is still white in reality.
Orientation constancy	An object's true orientation (position) is perceived as being unchanged despite changes in the orientation of the object's image on our retina.	Suppose you are sitting on a couch or the floor watching TV. To get more comfortable, you lie on your side. Despite your head having rotated 90 degrees, causing a rotation in the retinal image of the TV, you still perceive the TV to be upright.

Context

Context refers to information (conditions or circumstances) that surrounds a stimulus that influences the perception of the stimulus. In Figure 9.24a the centre circle is the same size in both designs, but context alters the circle's apparent size. The importance of context is also shown in Figure 9.24b. If you read across, context causes the central figure to appear as 13. Reading down makes it look like a B.

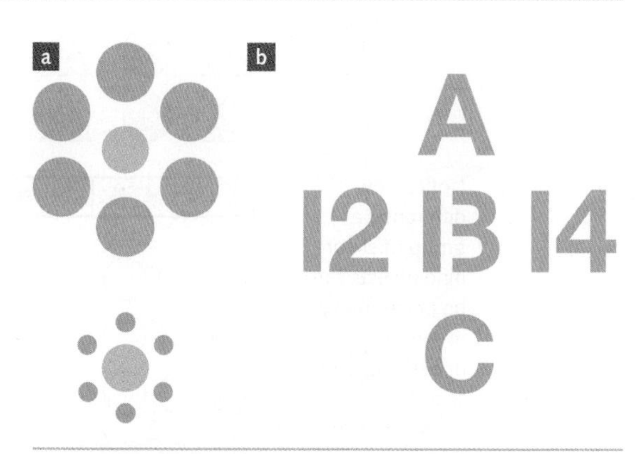

Figure 9.24 a Are the centre dots in both figures the same size? **b** Context alters the meaning of the middle figure (adapted from Bruner & Minturn, 1955).

ACTIVITY 9.3 PERCEPTUAL SETS

Try it

1. Here is a list of words to do with size:
   Large Small Tiny Huge B_g
   What is the last word?
2. Here is a list of words to do with ways to carry groceries:
   Backpack Sack Trolley Basket B_g
   What is the last word?

   Did you write Big for Question 1 and Bag for Question 2?

If you did then you have experienced a perceptual set. Even though your eyes received the same information in each question, 'B_g', when you expected to see words about size or carrying groceries, your brain was predisposed to search your memory for words that matched each category. This is a psychological factor based on context and your past experiences.

»

Apply it

3 In an experiment by Siipola (1935), participants were presented with different non-words, such as sael. Those who were told that they would be reading words related to boats read the word as 'sail,' while those who were told to expect words related to animals read it as 'seal.'
Explain these results with reference to psychological influences on perception.

Exam ready

4 Which of the following does not correctly identify a psychological influence on visual perception?

A changes in vision through ageing

B visual constancies such as shape constancy

C pictorial cues such as linear perspective

D perceptual set experienced through motivation

Motivation

Motivation is an internal state that activates, directs and sustains behaviour in relation to achieving a specific goal. Motivation is a psychological state that can be either conscious or unconscious. Motivation can influence how we interpret the world around us. In turn, motivation can be influenced by a number of factors. These include:

» physiological factors (for example, pain, hunger and body temperature)

» psychological and emotional factors (for example, your interests, priorities and mood).

Imagine you are driving across outback Australia for the first time. You are feeling anxious because you are very low on fuel. Finally, you see a sign and you read the words 'FUEL AHEAD'. But as you draw nearer, you realise the words on the sign are actually 'FOOD AHEAD'. You have never driven in outback Australia before, so why did you misperceive the sign? As shown by this example, and probably by some of your own experiences, motivation can influence perception.

Past experience

Past experience refers to our prior exposure to stimuli and previous life experiences. Past experience creates a tendency, or expectation, to interpret stimuli in the same way we have interpreted similar stimuli in the past. For example, if you are shown a picture of a square with a larger triangle on the top, you might interpret it as an outline of a house, based on your previous experiences of the broad shape of houses. Current research suggests that past experiences shape what we see more than what we are looking at now (González-García et al., 2018).

Memory

As we noted earlier in this chapter, the stimuli we pay attention to (and ignore) determines what our brain receives for further processing. Which features of visual stimuli you attend to are influenced by a range of factors, including memories of your past experiences. When you sense a visual stimulus, it is first encoded, and the memory of its image is held in your short-term memory. It then passes into your long-term memory, where it is stored indefinitely. Later, when you need to interpret a new stimulus, these memories are retrieved and compared to the new stimulus to help. As a result, you can recognise the objects that you see and distinguish them from other objects. Humans recognise what they are looking at by combining current sensory stimuli with comparisons to images stored in memory (NYU Langone Health, 2018).

Social factors

Social factors are the aspects of our interactions with other people, groups, society and culture that influence how we think and behave.

Culture

Culture involves all the distinctive beliefs, values, customs, knowledge, art and language traditions that provide the basis of everyday social behaviours and that are handed down from one generation to the next. Therefore, culture influences how a person behaves, speaks and interacts with others. It also influences how we perceive the environment surrounding us.

Researchers once thought that because all humans have the same physical apparatus for eyesight, visual perception didn't differ between individuals or across cultures. Now, they suggest that cultural or experiential differences cause differences in visual processing. For instance, looking at the same series of images, people from East Asian cultures perceive visual stimuli very differently from people in Western cultures. East Asians view stimuli more holistically. They are more likely to attend to the relationship between the object and the context in which it is located. Westerners tend to focus on a central object (for example, something that is fast moving, large or colourful) and analyse its features (Nisbett & Miyamoto, 2005). These differences have been shown in children as young as 3 years old (Kuwabara & Smith, 2016).

Culture creates differences in contexts and experiences that individuals draw from when perceiving new information. For instance, some cultures use only selected pictorial cues to represent depth, meaning that people from such cultures may not easily recognise other pictorial cues. In 1960, William Hudson tested members of remote South African tribes who did not use the pictorial depth cue of relative size to show depth in their drawings. Hudson showed participants in Western cultures and South Africa a drawing consisting of a man holding a spear, a gazelle and an elephant (Figure 9.25). Various pictorial cues for depth that people in Western cultures are familiar with were also included, such as height in the visual field (the elephant was higher up in the visual field), relative size (the elephant was smaller than the gazelle) and interposition (the man and the gazelle blocked the hills) (Hudson, 1960).

Participants were asked 'What is the man doing?' and 'What is closest to the man?' Westerners answered as if the picture was a three-dimensional image, so they took into account depth cues. For example, they said the gazelle was closer to the man than the elephant. The South Africans did not perceive depth in the picture and said the elephant was closer to the man than the gazelle was. Hudson concluded that the South Africans perceived simplified drawings as flat, two-dimensional designs. He suggested that people in Western societies are trained from a very young age to perceive pictures in three dimensions. Hudson's findings suggest that to perceive depth in a drawing, some familiarity with conventions for representing depth in pictures and photographs is necessary.

Figure 9.25 A Hudson test picture

ANALYSING RESEARCH 9.3

Visual perception may depend on birthplace and environment

In a recent multinational study, a research team led by Kyoto University shows that an ability to perceive differences between similar images depends on the cultural background of the viewer.

Scientists have long recognised that the mental processes behind thinking and reasoning differ between people raised in Western and Eastern cultures. Those in the West tend to use 'analytical' processing – analysing objects independently of context – while those in the East see situations and objects as a whole, which is known as 'holistic' processing.

Volunteers from Canada, the United States and Japan were asked to look at groups of objects such as straight lines with varying properties. They were asked to identify simple differences between them: angle and length, for example. In looking for the one odd line out of a group, North Americans took more time when the line was shorter, rather than if it was longer. No such differences were seen in Japanese volunteers, who in contrast had a significantly harder time identifying a straight line among tilted ones.

The researchers believe such a stimulus-dependent cultural difference cannot be explained simply by analytic-holistic theory.

'There are likely other differences in perceptional mechanisms that caused this discrepancy in visual processing,' continues senior researcher Jun Saiki of Kyoto University. 'Our next step is to find the cause of this discrepancy. One such reason may be the orthographical systems [the way a language is represented visually] the subjects see regularly. In East Asian writing, many characters are distinguished by slight differences in stroke length. In Western alphabets, slight angular alterations in letters result in remarkable changes in the reading of words.'

The researchers eventually hope to gain insight into the role of visual experience in the development – from an early age – of the visual processing system.

Source: Adapted from Kyoto University (2017, May 23). Visual perception may depend on birthplace and environment. *ScienceDaily*. www.sciencedaily.com/releases/2017/05/170523095013.htm

Questions

1 What was the aim of this study?
2 Write a hypothesis for this study.
3 Identify the independent variable and the dependent variable in this study.
4 Identify the sampling method used.
5 Identify the result(s) of this study.
6 What conclusion did the researchers draw from these results?

KEY CONCEPTS 9.3b

» Psychological factors influence our perceptions because they affect the way we organise and interpret sensory stimuli.
» Our brain applies a set of visual perception principles to quickly organise visual stimuli in a consistent and meaningful way.
» These principles include Gestalt principles (figure–ground, closure, similarity and proximity), which are used to group and separate elements of the stimuli into meaningful patterns or wholes.
» Visual constancies help us understand that an object remains unchanged in terms of its actual size, shape, brightness and orientation, regardless of changes in the retinal image of the object.
» The context in which we view a stimulus affects our perception of that stimulus.
» Our past experience with stimuli influences how we perceive a new stimulus because it creates an expectation of what we think that stimulus will be.
» Information stored in our memory influences perception. When we attempt to interpret a new stimulus we compare it with memories of a similar stimulus.
» Social factors such as culture influence perception.

Concept questions 9.3b

Remembering

1 What are Gestalt principles? r
2 At what stage in perception do we apply Gestalt principles? r
3 What is culture? r

Understanding

4 When you watch an aeroplane leave the tarmac and fly into the distance, the image it casts on your retina grows smaller and smaller until it finally disappears in the distance. Explain how you know that the plane is still the same size, in reality. c

Applying

5 When Angela and her friend were lying in shallow water at the beach Angela suddenly jumped up and shrieked 'Jellyfish!' Her friend pointed out it was only some seaweed draped across her leg, and Angela was most relieved. She explained to her friend that she had been stung by a jellyfish at the beach when she was much younger and she became very sick. How can you explain Angela's reaction to the seaweed touching her leg? c

HOT Challenge

6 Figure 9.26 shows an owl sitting in a hollow tree trunk. Explain how the Gestalt principle of figure–ground is manipulated in this photo. C

iStock.com/lavin photography

Figure 9.26

9.4 Gustatory perception

Taste, also known as **gustation**, refers to the sensory experience of a food or drink that is put into the mouth and perceived as flavour. For many years it was generally agreed that humans have four primary tastes: sweet, sour, bitter and salty. However, in 1908, Kikunae Ikeda, a professor at the Imperial University of Tokyo, proposed the existence of a fifth taste. He named this taste *umami* (Japanese for delicious savoury taste). Umami is a taste sensation that is experienced as meaty or savoury (Figure 9.27).

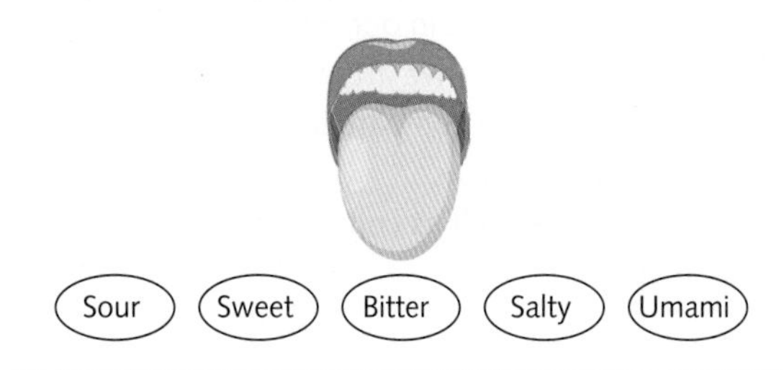

Figure 9.27 The five tastes to which humans are sensitive

Taste sensation and perception

Soluble chemical molecules found in food and liquid are the physical stimuli for taste, and different types of molecules activate different sensory receptors. When food is placed in the mouth, these molecules are dissolved by saliva and come into contact with taste buds. When this happens, the taste sensation begins.

Taste buds are tiny structures contained in goblet-shaped papillae, the small bumps on the surface of the tongue (Figure 9.28). The tongue is the main sensory organ of the gustatory system and most taste buds are found on the tip, sides and back of the tongue. Some are also located in the soft palate, the larynx (the voice box that contains vocal cords) and the pharynx (the section of throat between the mouth and the larynx).

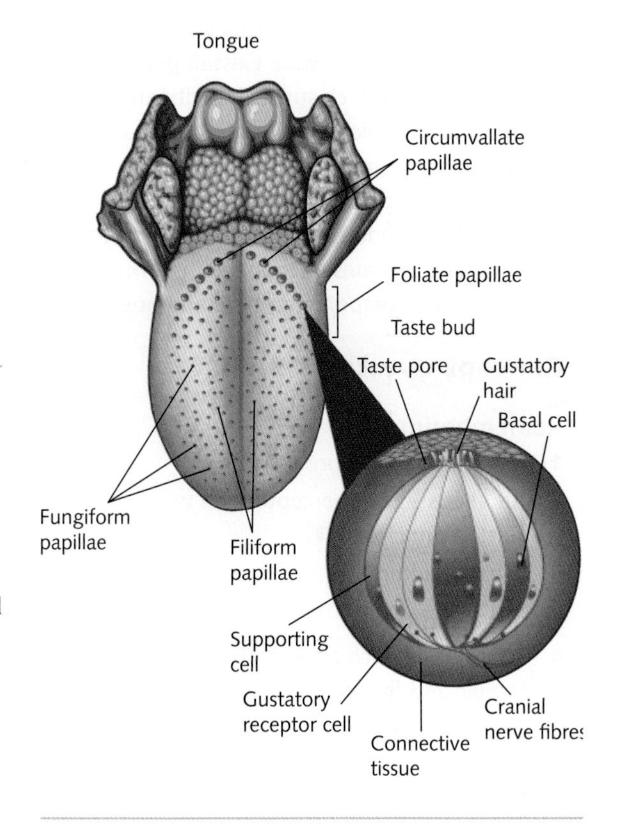

Figure 9.28 The structure of the human tongue and taste buds

Taste buds contain clusters of receptor cells that have in their membranes special proteins, known as **gustatory receptors**, that absorb and identify chemicals within the mouth and throat as they dissolve in saliva. Each taste bud has approximately 100 receptor cells. Taste buds have a life span of approximately 10 days and they are continuously being replaced. This is because they are constantly exposed to hazards, such as heat, infection and toxins, and they are easily damaged. Every receptor cell has a spindly protrusion called a gustatory hair that reaches the outside environment through an opening called a taste pore (Figure 9.28). The chemical molecules in food and drink mix with the saliva, enter the taste pore, interact with the gustatory hairs and bind to the surface of the receptor. This helps chemical molecules of food and drink dissolved in saliva bind to the gustatory receptor protein. As a result, a taste sensation is detected and the neural processes that send information to the taste processing areas of the brain are initiated.

The neural pathway for taste begins in the tongue and ends in the brain, and damage to any section of the neural pathway can cause taste problems.

People vary considerably in their sensitivity to and perception of certain tastes. These differences can be influenced by a range of biological, psychological and social factors.

Biological factors influencing taste perception

Differences in taste perception can be affected by a range of biological factors.

Our genetic make-up influences the amount of and composition of gustatory receptors on taste buds. Some people have as few as 500 taste buds: most of us have approximately 10 000. Research suggests that taste and our general eating behaviour, including meal size and calorie intake, are controlled by our genes. Studies on families and twins have found links between genetic make-up, taste perception and preference to proteins, fat and carbohydrates. These studies have found that preference for sweetness, meat and carbohydrates are strongly determined by genetics (Pallister et al., 2015).

9.4.1 SENSATION AND PERCEPTION OF TASTE

Age also influences our taste perception (Figure 9.29). Experience with taste begins in the womb. Flavours are transmitted from the maternal diet to the amniotic fluid, which is swallowed by the foetus. After birth, the infant experiences flavour through their mother's breast milk. This influences the infant's taste preferences and dietary choices made later in childhood (Forestell, 2017). Evolutionary theory suggests that children are born with a natural preference for sweet and fatty foods and an aversion to bitter tastes. Consumption of high calorie foods when food is scarce aids survival and helps children grow faster. They have a natural aversion to bitter foods because they might be poisonous.

Children have more taste buds than adults; they are more sensitive to flavour and generally prefer bland foods to those with stronger flavours. As we age, the number of taste buds and their sensitivity decreases because taste buds do not regenerate as quickly or they cease regenerating. This impacts a person's flavour perception because it weakens their ability to discriminate between tastes.

iStock.com/Willowpix

Figure 9.29 As we age, our taste buds do not regenerate as quickly. This may explain why older people often find the taste of food bland and lose interest in food. They may also use more salt and spices because their taste buds require more stimulation to achieve the same taste sensations they had when they were younger.

ANALYSING RESEARCH 9.4

Some like it sweet, others not so much: It's partly in the genes

A recent study from Philadelphia's Monell Center suggests that a single set of genes affects a person's perception of sweet taste, regardless of whether the sweetener is a natural sugar or a non-caloric sugar substitute.

In the study, researchers tested 243 pairs of monozygotic (MZ, or identical) twins and 452 pairs of dizygotic (DZ, or fraternal) twins. Studying twin pairs allowed the researchers to determine how much influence the twins' shared genetics contributed to their perception of sweet taste intensity. A further 511 unpaired individuals were also tested. Each participant tasted and then rated the intensity of four sweet solutions: fructose, glucose, aspartame and neohesperidine dihydrochalcone (NHDC). The first two are natural sugars, while the latter two are artificial, non-caloric sweeteners.

The results indicated that genetic factors account for approximately 30 per cent of person-to-person variance in sweet taste perception. In addition, those who perceived the natural sugars as weakly sweet experienced the sugar substitutes as similarly weak. This suggests that there may be a shared pathway in the perception of natural sugar and high-potency sweetener intensity.

The current study also found little evidence for a shared environmental influence on sweet perception. Assuming twin pairs took part in communal meals during childhood, this result challenges the common belief that access to foods high in sugar may make children insensitive to sweetness.

Source: Adapted from Monell Chemical Senses Center. (2015, July 17). Some like it sweet, others not so much: It's partly in the genes: Twin study suggests a common genetic pathway underlies sweet taste perception of natural, non-caloric sweeteners. *ScienceDaily*. https://www.sciencedaily.com/releases/2015/07/150717091937.htm

Questions

1 What was the aim of this research?
2 Construct a possible hypothesis for this study and state a prediction that could be derived from the hypothesis.
3 Explain why this twin study is a correlational study and not a controlled experiment, and identify the variables of interest.
4 What conclusion(s) did the researchers draw?
5 Was the hypothesis you constructed supported or refuted by the results?

Psychological factors influencing taste perception

When we see, smell or taste food, a number of factors influence the perception we form of the food. These factors include memory, food packaging and appearance.

Memory

Our memory of past food experiences helps create an expectation of what the food will taste like. This expectation influences the perception we form. The cerebral cortex does not have a specific lobe for taste or odour. When we see, smell or taste a new food, information is sent to a number of brain areas, including:

» the **gustatory cortex**: the brain area responsible for storing memories of new tastes
» the **hippocampus**: formulates and stores memories of the time and place of the taste experience
» the **amygdala**: adds the emotional element to the taste experience (University of Haifa, 2014).

This is why the smell or sight of a food triggers a memory that influences your perception of the food, often before you have even tasted it. An unpleasant experience with a food (such as vomiting after eating a specific food) can lead to a negative memory of the experience. This may lead to a long-lasting taste aversion to that food, so you will avoid these foods. When you see this food in the future (or perhaps even think about it), the taste cue becomes an aversive signal that triggers your negative memory (Figure 9.30).

The smell, texture and appearance of food can instantly evoke memories of eating the food and the emotions we felt at the time we tasted it. For example, the smell of muffins baking might elicit fond memories of spending time at your grandmother's house when you were a small child (Figure 9.31).

www.CartoonStock.com

Figure 9.30 Negative memories of food experiences influence our perception of the taste of foods.

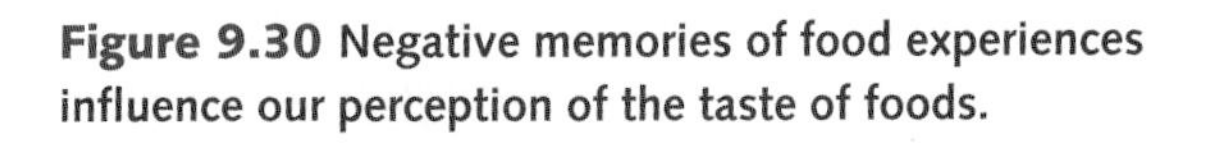

Shutterstock.com/Daxiao Productions

Figure 9.31 The smell of a food can evoke strong memories and emotions associated with the food that influence how we perceive its flavour.

Food packaging

How a product looks (its colour and packaging) sets up expectations for the product's flavour. This was clearly shown in 2011 when Coca-Cola switched the red Coke can colours to a white-coloured can to raise funds for endangered polar bears (Figure 9.32). People complained that the company had changed their recipe, even though they had not. Because of this perception, the brand eventually changed back to red again, although they kept the image of the polar

Alamy Stock Photo/Anthony Pleva

Figure 9.32 The colours used in packaging set up an expectation of the product's flavour. Customers complained that Coke tasted different in a white can.

bears (see weblink). Other characteristics of packaging such as how it feels (its texture and weight), and how it sounds (the noise of a can or bottle being opened) also influence taste perception.

Charles Spence at Oxford University conducted various studies on taste perception where he served food in a variety of coloured bowls. The results of these studies showed that participants perceived:

» salty popcorn to taste sweeter when served in a red bowl (red is usually associated with sweetness)
» a strawberry-flavoured mousse to taste 10 per cent sweeter when served from a white container rather than a black one
» that coffee tastes nearly twice as intense but only two-thirds as sweet when it is drunk from a white mug rather than a clear glass one.

He also discovered that Colombian and British shoppers are twice as willing to choose a juice whose label features a concave, smile-like line rather than a convex, frown-like one (Twilley, 2015).

9.4.2 APPLYING THE BIOPSYCHOSOCIAL MODEL TO GUSTATORY PERCEPTION

Weblink
Can packaging affect taste?

Appearance

Humans rely on vision more than their other senses to provide knowledge of the outside world. Visual appeal is a key element in food acceptance. Foods that may taste great lose appeal if their appearance is not pleasing. The visual cues our brain receives are some of the most powerful cues for taste. Before we eat or drink, we generally look at what we are eating or drinking. So, our brain often receives visual information about the food or drink before it receives taste information.

Generally, we learn to associate visual cues, such as the food's colour, with tastes (Figure 9.33). These learned associations help create expectations about how a food should smell and taste. For example, we may expect a yellow drink to have a banana or lemon flavour and red jellybeans to have a cherry, strawberry or raspberry flavour. In fresh foods, such as fruits and vegetables, we rely on colour to determine how ripe and fresh they are. If the colour does not match our expectations, we may perceive the food's taste and flavour differently. More intensely coloured foods, such as red chilis, and beverages are expected to have a more intense flavour (Carlsmith & Aronson, 1963).

In one study (see weblink), people were served a meal of steak and French fries under dull lighting conditions, and they all said it tasted fine. However, when the lights were turned up, the steak appeared to have been dyed blue, and the fries were dyed green. Many of the participants would not eat any more of the food, and some even became ill.

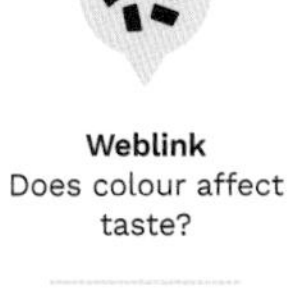

Weblink
Does colour affect taste?

Social factors influencing taste perception

Human beings are social animals. We live in groups and we eat together. So, in many ways, our perception of the taste of food and beverages is a learned experience shaped by our social experience. One of the strongest social influences on taste perception is the culture we were raised in.

Culture

Some basic aspects of taste perception may be innate; however, research suggests that taste preferences are largely learned and are heavily influenced by social processes (Rozin, 1990). This social influence contributes greatly to the striking ethnic and cultural differences found in taste preferences. People from different cultures eat different foods and have different preferences for particular tastes (Figures 9.34 and 9.35). For example, foods that are not generally eaten in many Western cultures – such as worms, insects, fish eyes and blood – are considered delicacies in other cultures. People from Thailand, Congo, Cameroon and Nigeria will happily eat termites and crickets, and these form valuable sources of protein in their diet. In Western countries, we are unlikely to ever develop a taste for, or acceptance of, dog meat or snake blood.

In many cultures, particular flavours and unfamiliar foods are deemed unacceptable for humans to eat. For example, eating meat from a cow is acceptable in Western cultures (and considered tasty) but prohibited by those who practice Hinduism.

Shutterstock.com/chuchiko17

Figure 9.33 Visual cues are associated with the taste of the food. Would you eat blue cabbage?

Shutterstock.com/Benny Marty

Figure 9.34 Culture shapes our attitudes to what is acceptable as a food source and our perception of taste.

Shutterstock.com/Herbert Heinsche

Figure 9.35 Many non-Australians find the yeasty taste of vegemite unpalatable.

Where families live, where their ancestors came from, and their socio-economic background can also influence people's perception of food. In the 1970s, a group of Indian medical students and a group of Indian labourers from the Karnataka province were asked to rate the pleasantness of a number of compounds. They found that Indian labourers favoured sour foods containing citric acid and quinine, while Indian medical students did not. The labourers had been raised on a sparse diet (1200 to 1500 calories a day) that contained a lot of sour food, particularly the sour tamarind fruit. This suggested that the labourers' partiality to sour and bitter tastes was because that was what was available to them on low incomes, and so they learned to like it (Moskowitz et al., 1975).

The values or beliefs a society attaches to food items also defines what families within a cultural group will eat. For example, both plant and animal sources may contribute to meeting nutritional requirements for protein. Soybeans, beef, horsemeat and dog meat are all adequate protein sources. Yet, due to the symbolism attached to these protein sources, they are not equally available or acceptable in all societies. In India, for example, cows are sacred in the Hindu religion; they can provide milk to be used for butter and cheese, but cannot be eaten as beef.

ANALYSING RESEARCH 9.5

Don't like the food? Try paying more

Researchers in nutrition, economics and consumer behaviour often assume that taste is a given – a person naturally either likes or dislikes a food. But a new study suggests taste perception, as well as feelings of overeating and guilt, can be manipulated by price alone. 'We were fascinated to find that pricing has little impact on how much one eats, but a huge impact on how you interpret the experience,' said Brian Wansink, a professor at the Dyson School of Applied Economics and Management at Cornell University who oversaw the research. 'Simply cutting the price of food at a restaurant dramatically affects how customers evaluate and appreciate the food.'

The researchers teamed up with a high-quality Italian buffet in New York to study how pricing affects customers' perceptions. They presented 139 diners with a menu that offered an all-you-can-eat buffet priced at either $4 or $8. Customers were then asked to evaluate the food and the restaurant and rate their first, middle and last taste of the food on a nine-point scale.

Those who paid $8 for the buffet reported enjoying their food on average 11 per cent more than those who paid $4, though the two groups ate the same amount of food overall. People who paid the lower price also more often reported feeling like they had overeaten, felt more guilt about the meal, and reported liking the food less and less throughout the course of the meal.

'We were surprised by the striking pattern we saw,' said one of the researchers. 'If the food is there, you are going to eat it, but the pricing very much affects how you are going to feel about your meal and how you will evaluate the restaurant.'

The researchers said the results could offer lessons about how to optimise a restaurant experience. 'If you're a consumer and want to eat at a buffet, the best thing to do is eat at the most expensive buffet you can afford. You won't eat more, but you'll have a better experience overall,' said Wansink.

The study fits within a constellation of other work by Wansink and others offering insights about how health behaviours can be manipulated by small changes, such as putting the healthiest foods first in a display or using a smaller dinner plate.

Source: Adapted from Federation of American Societies for Experimental Biology (FASEB) (2014, April 29). Don't like the food? Try paying more. *ScienceDaily*.

Questions

1 Identify the aim of the study.
2 Identify the independent and dependent variables in the study.
3 Create a hypothesis for the study.
4 Identify two extraneous variables that may have affected the results.
5 Describe how these extraneous variables could be controlled.

KEY CONCEPTS 9.4

- Humans sense five different tastes: sweet, sour, salty, bitter and umami (savoury).
- The perception that different individuals can have of the same food stimulus can vary.
- Taste involves the gustatory receptor proteins on taste buds detecting chemicals in food, transducing them, and then sending them to the brain for interpretation.
- There is no specific cortical lobe for taste.
- Each person's taste perception is unique and can be influenced by a range of biological, psychological and social influences.
- Biological influences on taste perception include the effects of ageing and genetic factors.
- Psychological factors that influence taste perception include memory, food packaging and appearance.
- Cultural training also influences our taste perception.

Concept questions 9.4

Remembering

1 What is gustation? r
2 Name three factors other than taste that influences our perception of food and beverages. r
3 When does taste perception begin? r
4 How do our taste buds change as we age? r

Understanding

5 Every Sunday Kali cooks lunch for her 80-year-old grandmother. Often, her grandmother complains that the food has no taste and she sprinkles a large portion of salt over her meat and vegetables. Kali is concerned by this. Use your knowledge of how taste perception changes over the lifespan to explain Kali's grandmother's behaviour. c

Applying

6 Adriano walked into a bakery to buy bread. The aromas in the bakery triggered pleasant memories of the delicious pastries his uncle used to bake when he was a child. Adriano bought two apple pies as well as bread. Explain how Adriano's hippocampus and amygdala contributed to this decision. c

HOT Challenge

7 In an attempt to prevent obesity and lifestyle diseases, a number of countries, including Australia, are considering introducing mandatory plain packaging laws for alcohol, confectionary, savoury snacks and sugary drinks. Based on your knowledge of the impact of food packaging and appearance on taste perception, explain why this is being considered. e
8 How can you explain the variations in taste preferences, including vegetarians, vegans and junk food addicts, who exist within the same culture? e

9 Chapter summary

KEY CONCEPTS 9.1a

- Experience of the external environment begins with sensation.
- Sensation is an automatic physical process that involves the body's sensory receptors detecting and responding to external energy.
- Reception is the first stage in sensation. It involves sensory receptors taking in and being activated by the raw energy.
- Transduction is the second stage in sensation. It involves sensory receptors converting stimulus energy into individual impulses of electrochemical energy so it can be transmitted by neural pathways to specific areas of the brain.
- Information about specific features of the stimulus is coded into the neural impulses.
- Transmission is the final stage of sensation process. It involves electrochemical impulses travelling along neural pathways that connect sensory receptors to brain areas specialised to receive them. Perception then occurs.

KEY CONCEPTS 9.1b

- Perception is the psychological process that gives meaning to sensations. Perception occurs in the brain.
- Selection is the first stage in perception. It involves feature detector cells selecting specific features of the stimulus for further processing and ignoring others.
- Organisation is the second stage in perception. It involves the brain reassembling the individual features of the stimulus into a whole or pattern that can be given meaning.
- Interpretation is the third and final stage in perception. It involves the brain giving meaning to the stimuli so we can understand what the sensation represents.
- The brain applies a number of psychological processes unique to the individual to help it interpret the stimulus. This is why perceptions of a common stimulus may vary between individuals.

KEY CONCEPTS 9.1c

- We use top-down processing and bottom-up processing to help us quickly organise and interpret sensory stimuli.
- Bottom-up processing involves the perception of the details of sensory stimuli first, and builds upwards towards a perception of the whole stimulus.
- Bottom-up processing is used when we are presented with unfamiliar stimuli.
- Bottom-up processing does not rely on schemas, because it only processes the stimuli itself.
- Top-down processing involves perceiving the general features of a whole stimulus first, then its specific details.
- Top-down processing is influenced by schemas, including previous experience, memory and motivation, expectations, context and previous knowledge.

KEY CONCEPTS 9.2

- Sensory stimuli that is selected for attention receives further processing by the brain.
- Attention can be voluntary (conscious) or involuntary (unconscious).
- Attention allows you to focus on important stimuli and ignore unimportant stimuli.
- Attention can focus on external (environmental) or internal (bodily or psychological) stimuli.
- Attention can be influenced by a range of factors related to the stimuli or to the person.
- Attention can be an automatic process or a controlled process.
- Three major types of attention are sustained attention, divided attention and selective attention.

KEY CONCEPTS 9.3a

- Each person's perception of visual stimuli is unique and can be influenced by a range of biological, psychological and social influences.
- Biological factors that influence visual perception include any physical, neurological, chemical and genetic conditions that affect an individual's physiological functioning and perceptual abilities.
- We use internal and external depth cues to accurately judge distances.
- Depth cues allow us to perceive three dimensions from two-dimensional retinal images.
- There are two types of depth cues: binocular cues and monocular cues.
- Binocular depth cues include retinal disparity and convergence.
- Monocular depth cues include accommodation and pictorial cues (linear perspective, relative size, interposition, texture gradient and height in the visual field).
- Visual perception can be affected by malformations of or damage to any of the components of the visual system, or by ageing or genetics.

KEY CONCEPTS 9.3b

- Psychological factors influence our perceptions because they affect the way we organise and interpret sensory stimuli.
- Our brain applies a set of visual perception principles to quickly organise visual stimuli in a consistent and meaningful way.
- These principles include Gestalt principles (figure–ground, closure, similarity and proximity), which are used to group and separate elements of the stimuli into meaningful patterns or wholes.
- Visual constancies help us understand that an object remains unchanged in terms of its actual size, shape, brightness and orientation, regardless of changes in the retinal image of the object.
- The context in which we view a stimulus affects our perception of that stimulus.
- Our past experience with stimuli influences how we perceive a new stimulus because it creates an expectation of what we think that stimulus will be.
- Information stored in our memory influences perception. When we attempt to interpret a new stimulus we compare it with memories of a similar stimulus.
- Social factors such as culture influence perception.

KEY CONCEPTS 9.4

- Humans sense five different tastes: sweet, sour, salty, bitter and umami (savoury).
- The perception that different individuals can have of the same food stimulus can vary.
- Taste involves the gustatory receptor proteins on taste buds detecting chemicals in food, transducing them, and then sending them to the brain for interpretation.
- There is no specific cortical lobe for taste.
- Each person's taste perception is unique and can be influenced by a range of biological, psychological and social influences.
- Biological influences on taste perception include the effects of ageing and genetic factors.
- Psychological factors that influence taste perception include memory, food packaging and appearance.
- Cultural training also influences our taste perception.

9 End-of-chapter exam

Section A: Multiple-choice

1 Where does bottom-up processing begin?
- A with emotion
- B with the cognition
- C with bodily sensations
- D during selection, when feature detector cells filter specific elements of the stimuli

2 Which of the following shows the correct sequence for stages in perception?
- A selection, interpretation, organisation
- B reception, transduction, transmission
- C selection, organisation, interpretation
- D transmission, selection, interpretation

3 The conversion of raw energy into electrochemical energy is called
- A transmission.
- B transduction.
- C transformation.
- D reception.

4 The sensation of taste is processed by the
- A frontal lobe.
- B gustatory cortex.
- C somatosensory cortex.
- D olfactory cortex.

5 The little bumps visible on the top of your tongue are called
- A taste buds.
- B gustatory receptors.
- C papillae.
- D retinas.

6 The organisation of sensory information into meaningful experiences is called
- A sensation.
- B perception
- C interpretation.
- D cognition.

7 The process that allows you to become aware of stimuli is
- A sensation.
- B organisation.
- C transduction.
- D perception.

8 Where would you find tastebuds?
- A on the tongue
- B on the walls of the mouth
- C at the back of the throat
- D all of the above

9 Sensation is to ______________ as perception is to ______________.
- A interpretation; organisation
- B detection; interpretation
- C transduction; transmission
- D reception, transduction

10 Binocular depth cues consist of
- A linear perspective and convergence.
- B retinal disparity and accommodation.
- C figure–ground and similarity.
- D convergence and retinal disparity.

11 Which statement about depth cues is incorrect?
- A Depth cues can be environmental (external) or biological (internal).
- B Retinal disparity is caused by our eyes being approximately 6.5 centimetres apart.
- C Accommodation involves the ciliary muscles changing the shape of the lens.
- D Retinal disparity is a monocular depth cue.

12 Which of the following pairs are not Gestalt principles?
- A closure and proximity
- B similarity and proximity
- C figure–ground and closure
- D convergence and retinal disparity

13 Pilots use the converging lines on the tarmac as a depth cue to help them safely land planes. This is an example of
- A texture gradient.
- B relative size.
- C linear perspective.
- D interposition.

14 Which of the following structures is responsible for transduction in the gustatory system?

- **A** papillae
- **B** taste buds
- **C** gustatory receptors
- **D** taste pores

15 When the ciliary muscles change the shape of the lens of the eye, this acts as a cue to the brain to help determine how far away an object is. This is known as

- **A** retinal disparity.
- **B** accommodation.
- **C** convergence.
- **D** interposition.

16 You attend an end-of-year revision session in preparation for your Psychology exam. What type of attention must you use to gain maximum benefit from this session?

- **A** selective attention
- **B** sustained attention
- **C** divided attention
- **D** executive attention

17 Which of the following activities requires selective attention?

- **A** driving on a busy freeway
- **B** reading a book on a train
- **C** solving a mathematics problem in class
- **D** all of the above

18 Jaxon loves to take things apart to see how they work. When he takes the back off a clock, he looks at the inner mechanisms and tries to figure out how they work. Jaxon is using

- **A** divided attention.
- **B** bottom-up processing.
- **C** top-down processing.
- **D** automatic processes.

19 Which visual perception principle is represented in Figure 9.36?

Figure 9.36

- **A** size constancy
- **B** brightness constancy
- **C** shape constancy
- **D** orientation constancy

20 If you were asked whether you saw a young girl or an old woman in Figure 9.37 and your saw a young girl, your perception would be the result of

- **A** bottom-up processing.
- **B** top-down processing.
- **C** sustained attention.
- **D** selective attention.

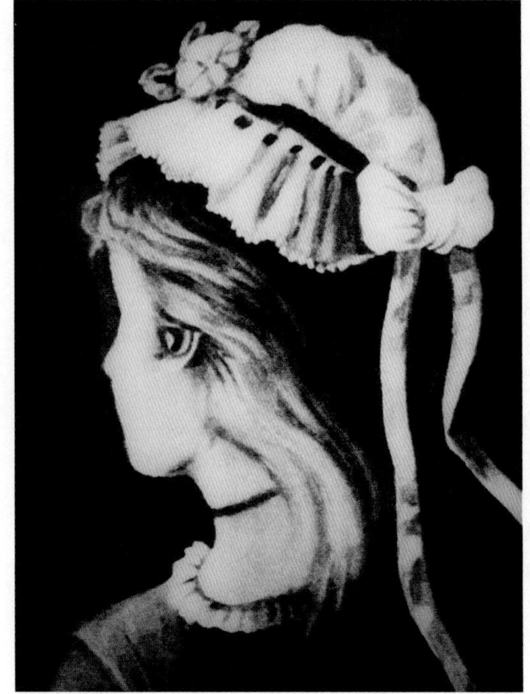

Alamy Stock Photo/Ian Paterson

Figure 9.37

Section B: Short answer

1 When he was chopping wood, Dennis sustained an injury to his right eye, and eventually lost sight in that eye.

a Name a depth perception cue that Dennis would not be able to use after his accident. [1 mark]

b Name a non-pictorial depth cue that Dennis would be able to use after his accident. [1 mark]

[Total = 2 marks]

2 a Name the Gestalt principle operating in the pattern shown in Figure 9.38. [1 mark]

b Explain how this principle operates to produce the perception of a soccer ball. [2 marks]

[Total = 3 marks]

Figure 9.38

3 Explain whether perception is a bottom-up process or a top-down process. [3 marks]

Distortions of perception 10

Key knowledge

- the fallibility of visual perceptual systems, for example, visual illusions and agnosia
- the fallibility of gustatory perception, for example, supertasters, exposure to miraculin and the judgement of flavours
- distortions of perception of taste and vision in healthy individuals, such as synaesthesia and spatial neglect

Key science skills

Develop aims and questions, formulate hypotheses and make predictions

- identify independent, dependent and controlled variables in controlled experiments
- formulate hypotheses to focus investigation

Plan and conduct investigations

- design and conduct investigations; select and use methods appropriate to the investigation, including consideration of sampling technique and size, equipment and procedures, taking into account potential sources of error and uncertainty; determine the type and amount of qualitative and/or quantitative data to be generated or collated

Comply with safety and ethical guidelines

- demonstrate ethical conduct and apply ethical principles when undertaking and reporting investigations

Generate, collate and record data

- record and summarise both qualitative and quantitative data, including use of a logbook as an authentication of generated or collated data
- organise and present data in useful and meaningful ways, including tables, bar charts and line graphs

Analyse and evaluate data and investigation methods

- identify and analyse experimental data qualitatively, applying where appropriate concepts of: accuracy, precision, repeatability, reproducibility and validity of measurements; errors (random and systematic); and certainty in data, including effects of sample size on the quality of data obtained
- evaluate investigation methods and possible sources of error or uncertainty, and suggest improvements to increase validity and to reduce uncertainty

Construct evidence-based arguments and draw conclusions

- evaluate data to determine the degree to which the evidence supports or refutes the initial prediction or hypothesis
- use reasoning to construct scientific arguments, and to draw and justify conclusions consistent with the evidence and relevant to the question under investigation
- identify, describe and explain the limitations of conclusions, including identification of further evidence required
- discuss the implications of research findings and proposals

Analyse, evaluate and communicate scientific ideas

- discuss relevant psychological information, ideas, concepts, theories and models and the connections between them

Source: VCE Psychology Study Design (2023–2027), pp. 30 & 12–13

10 Distortions of perception

In Chapter 9 you learned how our bodies perceive the sensory stimuli in our environment and that a range of factors mean we interpret and perceive the stimuli differently. Now we will look at distortions of perception; that is, situations where our brains can be fooled, causing our perceptions to be inaccurate.

p. 393

10.1 Fallibility of visual perception systems

When assigning meaning to stimuli, the brain applies its rules for constructing perceptions – these are called visual perception principles. Most of the time, these shortcuts that the brain takes to assign meaning quickly work well for us. However, sometimes this can distort our perception of the stimuli, like when we look at a visual illusion and we see the outline of a shape that's not actually there.

iStock.com/francescoch

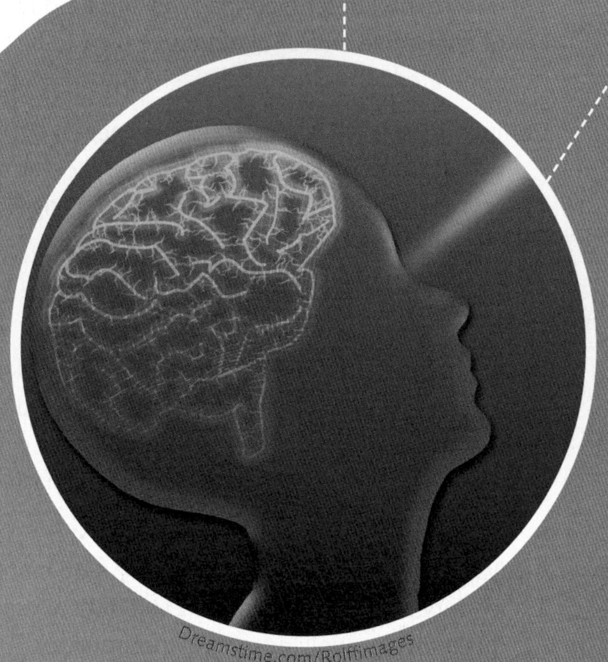

Dreamstime.com/Rolffimages

Shutterstock.com/MrslePew

p. 410

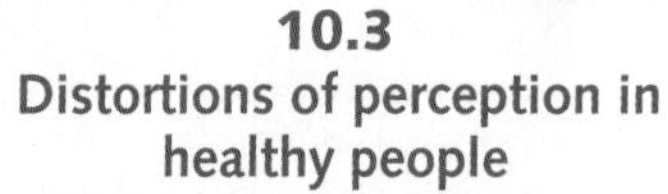

10.3 Distortions of perception in healthy people

There are certain conditions, such as synaesthesia, that give people a unique way of perceiving the world, but the condition is not the result of damage to the brain, nor does it mean the person is unhealthy. Amazingly, synaesthesia is a sensory crossover that can lead someone to associate colours with certain musical sounds or perceive a certain odour when they hear a sound.

p. 406

10.2 Fallibility of gustatory perception

A supertaster might sound like a character out of a superhero film, but there are in fact people who have a stronger perception of taste than others. Most people in most cases perceive food flavours correctly but, like our other senses, sometimes our perception can be inaccurate.

iStock.com/Jobalou

Most of the time, our brains do an excellent job of organising sensory stimuli and assigning meaning to the sensations. However, in both healthy and unhealthy people sometimes the perceptions are inaccurate, and this can lead us to see strange things such as impossible staircases or to taste a fruit and find it sweet when it is actually sour. Don't always believe what you see or experience!

Slideshow
Chapter 10 slideshow

Flashcards
Chapter 10 flashcards

Test
Chapter 10 pre-test

Assessment
- Pre-test
- End-of-chapter exam

Revision
- Chapter map
- Key term flashcards
- Key concept summary
- Slideshow

Investigation
- Investigation: The Müller-Lyer Illusion
- Data calculator
- Logbook template: Controlled experiment

Worksheet
- The effect of exposure to Miraculin on gustatory perception

To access these resources, visit **cengage.com.au/nelsonmindtap**

Nelson MindTap

Know your key terms

Agnosia
Ames room illusion
Gustatory cortex
Miraculin
Müller-Lyer illusion
Non-taster
Perceptual anomaly
Perceptual disorder
Spinning Dancer illusion
Supertaster
Synaesthesia
Taste sensitivity
Visual agnosia
Visual illusion

Recently musician Billie Eilish commented on one of the things that makes her different – her synaesthesia (Figure 10.1). Synaesthesia is a neurological condition in which information meant to stimulate one sense stimulates several senses. While both her brother Finneas and her dad have synaesthesia, it's quite rare. Approximately 4.4 per cent of the general population experience one of over 60 known types of synaesthesia.

In an interview on The Tonight Show Starring Jimmy Fallon (2021), Eilish herself described the condition as 'a thing in your brain where you associate random stuff to everything. So for instance, every day of the week has a colour, a number, a shape. Sometimes things have a smell or a temperature or a texture'. In an earlier interview with *Rolling Stone* (Eells, 2019), Eilish said, 'Every person I know has their own color and shape and number in my head, but it's normal to me.' She described her brother Finneas as an orange triangle, although the name 'Finneas' is to her dark green. She also described her song 'Bad Guy' as 'yellow, but also red, and the number seven. It's not hot, but warm, like an oven. And it smells like cookies' (Eells, 2019).

How can people smell colour or taste a word? Eilish's experience shows how perception is a very unique experience. Sometimes perception can be distorted. In other instances, we simply make a mistake and perceive a stimulus inaccurately. These situations demonstrate that perception is fallible (imperfect and open to error). In this chapter we explore examples of perceptual fallibility as well as distortions of perception of taste and vision.

Shutterstock.com/Tinseltown

Figure 10.1 Musician Billie Eilish feels her synaesthesia is normal.

Adapted from 'Billie Eilish explains her synesthesia and how it influences everything she does' By Maia Kedem, Audacy, August 11, 2021

10.1 Fallibility of visual perception systems

Most people have no trouble sensing and correctly perceiving the constant stream of incoming sensory stimuli they receive from the external environment. Our brain automatically selects the important features of the stimuli and efficiently organises them into a meaningful whole that it interprets correctly. The result of the sensation and perception process is that we form an accurate perception of the world around us.

Before the brain can transform sensory stimuli into meaningful perceptions, it must organise it into something the brain can recognise. Objects must be perceived as distinct from their surroundings and seen as having a meaningful and constant form. How far away they are also needs to be estimated. When assigning meaning to stimuli, the brain applies its rules for constructing perceptions – these are called **visual perception principles**. As we saw in Chapter 9, these are rules our brain automatically uses to help organise and interpret the stream of constantly changing stimuli it receives. The application of perception principles is usually reliable and our perceptions of the outside world are accurate – they match the reality (Figure 10.2).

iStock.com/Pankova

Figure 10.2 If your visual perceptual system is working correctly, you will interpret the incoming features of the visual stimulus from this image as a red rose.

Sometimes, however, we experience perceptual fallibility. That is, our perceptions are not always accurate. Even a normal, healthy, fully functioning brain can make a perceptual error. This is the case with visual illusions.

Visual illusions

A **visual illusion** is a consistent perceptual error in interpreting the features of an external stimulus. Illusions distort real stimuli and lead us to misapply the principles we use to organise stimuli into a stable, consistent or meaningful perception. This causes us to make a false judgement of reality so we misinterpret the visual stimuli. Figure 10.3 is an interesting example of how our brains can be tricked into making a perceptual error.

When we experience a visual illusion, the stimulus provides us with cues that mislead our perception. As a result, a mismatch between our perception and the reality of the actual stimulus occurs. Although perception of common stimuli is often described as unique, this is not the case with a visual illusion. With visual illusions, all individuals tend to be misled by the stimulus in the same way. In a visual illusion, length, position, motion, curvature or direction of the stimulus is consistently misjudged (Gillam, 1980).

Shutterstock.com/Billywolf

Figure 10.3 In this illusion, as the staircases shift orientation, the people start to defy gravity. This is an example of an impossible figure. Pictorial depth cues are used to make the scene appear three-dimensional, yet manipulated so we misapply these cues to make the object appear to have depth even though it is physically impossible. The illusion persists even when your mind knows the explanation.

Weblink
How and why do optical illusions work?

Illusions are a fascinating challenge to our understanding of perception. We know they intentionally involve visual deception by manipulating the cues we are experienced in using to create a perception. Table 10.1 presents some examples of interesting visual illusions.

Table 10.1 Examples of visual illusions

Illusion	Description
The Ponzo illusion Shutterstock.com/NeslihanGorucu	**The illusion:** Both yellow lines are the same length but most people perceive the top line as longer and further away. Explanation: » Your brain takes the context into account. Your brain's experience is that if an object (or a line) is further away, it will cast a smaller retinal image than an object (or a line) of the same length that is closer. » The top horizontal line looks longer because your brain applies the pictorial depth cue of linear perspective. » Since the vertical parallel lines seem to grow closer together as they move further away, your brain applies linear perspective to interpret the top line as being further away. » An object in the distance would need to be longer in order for it to appear the same size as a near object, so the top 'far' line is seen as being longer than the bottom 'near' line, even though they are the same size.
The Kanizsa triangle  Shutterstock.com/Bany's beautiful art	**The illusion:** A white triangle is perceived even though it is not actually there. Explanation: » Because of the Gestalt principle of **closure**, we tend to ignore gaps in a retinal image and mentally fill in the missing contour lines to make the image appear a cohesive whole.
The Zollner illusion 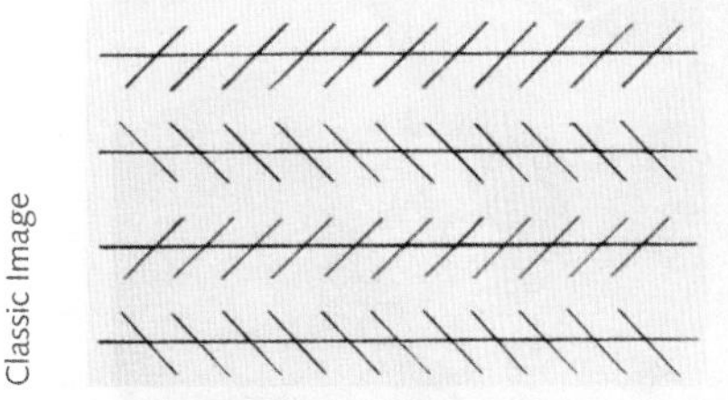  Alamy Stock Photo/ Classic Image	**The illusion:** The horizontal lines are crossed with overlapping short diagonal lines. The horizontal lines are, in reality, parallel, but they look as if they are crooked and will diverge. **Explanation 1:** » The background of an image can distort the appearance of straight lines. » The angle of the short lines compared to the longer lines creates an impression of depth. » One of the lines appears to be nearer to us; the other, further away. **Explanation 2** » The brain attempts to increase the angles between the long and short lines. This results is distortion as the brain tries to bend the lines away and towards each other.
 Shutterstock.com/Pat_Hastings	**The illusion:** A pencil or straw standing in water seems broken at the surface where the air and water meet, and the bottom portion appears closer. **Explanation:** » A ray of light passing from one transparent medium (air) to another (water) is bent as it emerges. The density of the two mediums is different (water being denser than air), which changes the speed of light, and changes the direction of the light ray, thus creating a distortion in the visual system. » The bending light makes the submerged part of the pencil look closer to the surface than it really is. This makes the pencil look like it bends slightly where it enters the water.

The Ames room illusion

In visual perception, expectations based on factors such as your past experience and the context an object appears in can make you ready to perceive visual stimuli in a particular way. Magicians rely on people's expected perceptions when they use sleight of hand to distract observers while performing tricks. Another kind of 'magic' is related to consistency in the environment. For example, based on your past experience with the shape of rooms, it would be safe to assume that rooms you see in the future will be roughly shaped like a box. This need not be true, as illustrated by the Ames room illusion. The **Ames room illusion** is summarised in Table 10.2

Table 10.2 The Ames room illusion

What is an Ames room?	» A room (designed by Adelbert Ames) that appears square when viewed through a peephole located at a certain point in the wall (Figure 10.4). » The true shape of the room is trapezoidal. The walls are slanted and the ceiling and floor are at an incline. » The proportions of the walls, floor, ceiling and windows have been distorted to provide misleading information about depth and distance. » Its trapezoidal shape causes the two back corners to appear an equal distance from the viewer, but the rear-left corner is actually further away from the peephole than the rear-right corner.
What is the Ames room illusion?	» People appear to change in size as they walk diagonally across the room. » Two people, who in reality are of equal size, stand in the back corners – one in the rear-left corner and one in the rear-right corner (Figure 10.4). » When the person standing in the rear-left corner walks across the room to the rear-right corner, they appear to increase in size. They are actually walking diagonally towards the viewer, but the distorted room makes it seem as though they are walking directly from left to right. » Similarly, when the person standing in the rear-right corner crosses from right to left, they appear to decrease in size.
How is it created?	The illusion happens partly because: » The person standing in the rear-left corner is perceived as smaller because they cast a smaller image on the observer's retina than the person in the rear-right corner. In reality they are just further away. » As they walk across the room from left to right, the image they cast on the observer's retina increases and they are perceived as growing taller. In reality, they are just moving closer to the viewer. » This faces the viewer with a conflict: they can either maintain what they know about the shape of the object (known as **shape constancy**) by perceiving the room as square, or they can maintain what they know about the size of the object (known as **size constancy**) by refusing to see the person 'grow'. Most people maintain shape constancy, and so see people 'shrink' and 'grow' before their eyes, but misapply size constancy. And partly because: » The viewer views the room through the peephole, so their brain only receives monocular cues to help judge depth and distance. » This restriction disrupts their understanding of the shape and size of real objects and people to create the impression that people change size as they cross the floor.
Explanation for the Ames room illusion	The apparent-distance hypothesis According to Gregory and Wallace (1963): » When two people of the same size are introduced into the opposite back corners of the room, the person in the left-hand corner comes in further away from the viewer than the person in the right-hand corner. » However, because of the room's distortion, both people appear to be the same distance from the viewer. » These people of the same size and same apparent distance will cast different-sized images on the observer's retinas. The people will, therefore, appear to be of different sizes even though this is at odds with the viewer's past experience of people in general (for example, the person in the rear-left corner may appear to be a giant). » The illusion occurs not because our perceptual processes are faulty but because we try to apply size constancy in the usual way to an extremely unusual situation. » The Ames room causes us to misapply size constancy, but maintain shape constancy.

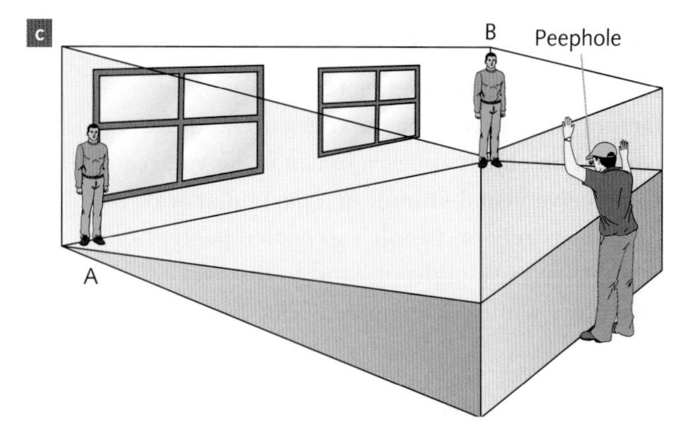

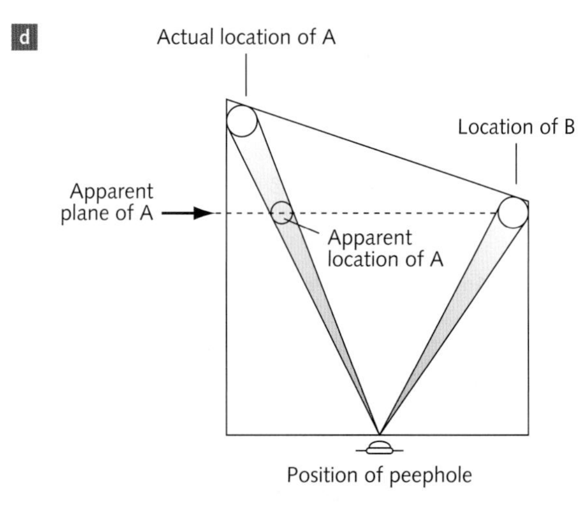

Weblink
What is the Ames room illusion?

Figure 10.4 The Ames room. From the front, as shown in a and b, the room looks normal; the actual right-hand corner is very short and the left-hand corner is very tall. The diagram in c shows the shape of the room and reveals why people appear to 'grow' as they cross towards the nearer, shorter right-hand corner. The floor plan in d shows people's actual positions in the Ames room in relation to the viewer.

The Müller-Lyer illusion

The **Müller-Lyer illusion** is a simple geometrical visual illusion. It was created by Franz Carl Müller-Lyer, a German psychiatrist and sociologist, and first published in an 1889 issue of the German journal *Zeitschrift für Psychologie*.

The Müller-Lyer illusion is shown in Figure 10.5 and summarised in Table 10.3.

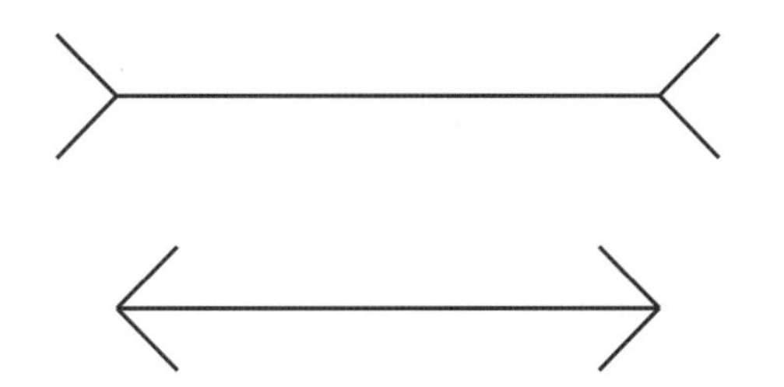

Figure 10.5 The Müller-Lyer illusion. Which of the horizontal lines is the longer?

How can we test whether past experience causes the illusion? If we could test an adult who saw only curves and wavy lines as a child, we would know if experience with a 'square' culture is important. In 1966, Segall et al. tested some Zulu people in South Africa who, at the time, lived in a 'round' culture. In their daily lives, they rarely encountered a straight line. Their houses were shaped like rounded mounds and arranged in a circle, their tools and toys were curved, and there were no straight roads or square buildings (Figure 10.7). When they looked at the Müller-Lyer illusion they did not experience the illusion. At most, they saw the line with inward-pointing arrowheads as only slightly longer than the other (Gregory, 1990). It was suggested that because the Zulus lived in a 'circular culture' they were less susceptible to the illusion than people who lived in a 'carpentered world' of rectangles and parallel lines (Segall et al., 1966).

Table 10.3 The Müller-Lyer illusion

What is the Müller-Lyer illusion?	» Two parallel lines of equal length, one of which ends in inward-pointing arrowheads (> <), the other which ends with outward-pointing arrowheads (< >) (Figure 10.5). » The line with the inward-pointing arrow heads appears to be longer than the line with the outward-pointing arrowheads, but in reality both lines cast the same-sized retinal image.
How is it created?	The illusion of differing lengths is created by the different figures at the ends of each line.
Explanations for the Müller-Lyer illusion	Theory 1: Richard Gregory (1990) » We live in a three-dimensional world where our buildings are basically square or rectangular in shape. This has constantly exposed us to the edges and corners of buildings and rooms. » We see the horizontal line with the inward-pointing arrowheads as though it were the inside corner of a room. » The line with the outward-pointing arrowheads suggests the corner of an outer wall of a building. » Because the lines cast retinal images of the same length, the more 'distant' inside corner line is perceived as being longer. In other words, cues that suggest a three-dimensional space alter our perception of a two-dimensional design (Enns & Coren, 1995). *Note:* This explanation presumes that people have had years of experience with straight lines, sharp edges and corners because they have lived in a built-up environment. Theory 2: Apparent-distance hypothesis (Gregory, 1990) » An apparently more-distant object that has the same-sized retinal image as an apparently nearer object will be perceived as the larger. » If two objects make retinal images of the same size, the more-distant object is perceived as being larger. This is known as size–distance invariance (whereby the size of an object's retinal image is precisely related to the distance from the eyes). » If the line with inward-pointing arrowheads looks further away than the line with outward-pointing arrowheads, then we must compensate by seeing the line with inward-pointing arrowheads as longer (Figure 10.6).

Weblink
Psychology of the Müller-Lyer illusion

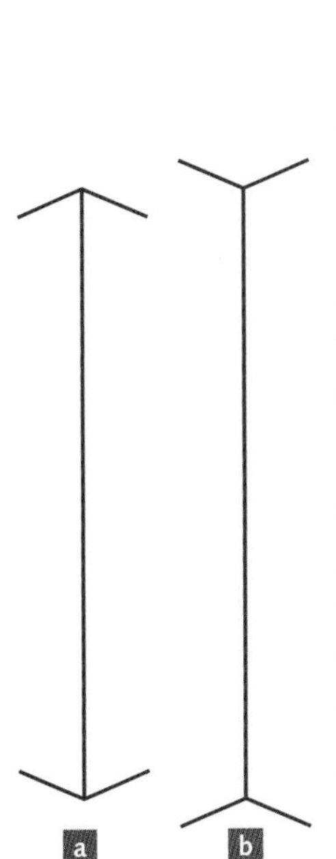

Bill Thomas

Figure 10.6 Why does line **b** in the Müller-Lyer illusion look longer than line a even though the walls of the house are the same height? Probably because it looks more like a distant corner than it does a corner seen close up. As you can see in the photo, additional depth cues accentuate the Müller-Lyer illusion.

Figure 10.7 Zulu villagers who viewed the Müller-Lyer lines in the 1960s did not experience the illusion that one line is longer. This was probably due to their experience of living in an environment where round shapes were most common, and there were no straight roads or square buildings.

The Spinning Dancer illusion

Weblink
The Spinning Dancer illusion

The **Spinning Dancer illusion** (Figure 10.8) is an animated visual illusion created in 2003 by Japanese web designer Nobuyuki Kayahara. The illusion is summarised in Table 10.4.

In an attempt to explain the Spinning Dancer illusion, one popular theory widely circulated on the internet incorrectly suggests that this illusion indicates which hemisphere of the brain is dominant in the viewer. There is no scientific evidence to support this theory.

Figure 10.8 If you stare at the spinning dancer, she appears to spontaneously switch between spinning left and spinning right.

Table 10.4 The Spinning Dancer illusion

What is the Spinning Dancer illusion?	The silhouette of a woman spinning on one foot with her leg extended appears to spontaneously switch direction between left and right.
How is it created?	» If the viewer perceives the foot on the ground to be the left one, the dancer appears to be turning in a clockwise direction. » If they perceive the foot on the ground to be the right one, the dancer appears to be turning in an anti-clockwise direction. » The viewer may also perceive the dancer whirling in both directions. The variation in direction is sometimes influenced by blinking or focusing on the dancer for a long time.
Explanation for the Spinning Dancer illusion	» The spinning dancer is just a two-dimensional black silhouette that provides no clues as to whether the dancer is facing the viewer or looking away from them. » There are no depth cues from the surrounding environment to give our brain cues about the direction the dancer is spinning in. » To make sense of the image, our brains try to create a three-dimensional image by filling in the missing information and assuming the direction the figure is turning.

INVESTIGATION 10.1 THE MÜLLER-LYER ILLUSION

Scientific investigation methodology

Controlled experiment

Aim

To investigate how we misinterpret visual stimuli

Introduction

Data calculator

Logbook template
Controlled experiment

In this investigation you will examine the perceptual fallibility of visual information. You will need to generate appropriate qualitative and/or quantitative data, organise and interpret the data, and research a conclusion in response to your research question.

A visual illusion is a consistent perceptual error in interpreting the features of an external stimulus. Illusions distort stimuli from real images or objects and lead us to misapply the principles we use to organise stimuli into a stable, consistent or meaningful perception. This causes us to make false judgements of reality when we misinterpret the visual stimuli.

The Müller-Lyer illusion is a simple geometrical visual illusion, in which two lines of the same length with different shaped 'inducers' (inwards- or outwards-pointing arrowheads) appear to be of different lengths. This scientific investigation aims to test the influence of the Müller-Lyer illusion on perception.

Pre-activity preparation

You are required to apply the key science skills (listed on page 389) to:

- develop a research question
- state an aim
- formulate a hypothesis
- plan an appropriate methodology and method to answer the question, while complying with safety and ethical guidelines.

Procedure

You will need to design an experiment to test the magnitude of error of the Müller-Lyer illusion. You could use lines with different shaped ends (Figure 10.9) as stimulus material. Alternately, you could create a sliding apparatus to test the illusion (Figure 10.10).

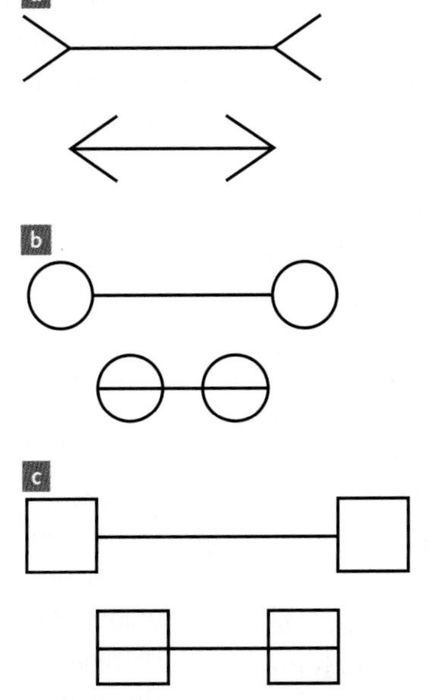

Figure 10.9 The Müller-Lyer illusion with **a** arrowhead ends, **b** circle ends and **c** square ends

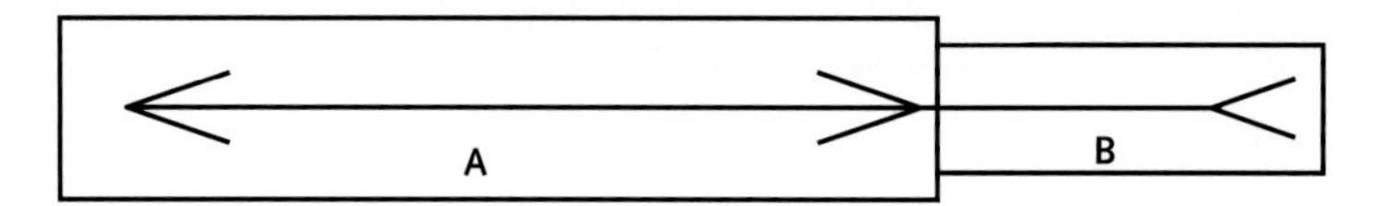

Figure 10.10 An apparatus that can be used for the testing the Müller–Lyer illusion. Participants slide the right half of the apparatus until they perceive that segments A and B are equal in length.

Results

Collate the results for all participants. Process the quantitative data you recorded, using appropriate mathematical calculations and units (for example, descriptive statistics such as a mean). Organise, present and interpret your data using an appropriate table and/or graph.

Discussion

1 State whether your hypothesis was supported or not.
2 Describe the magnitude of error for each of the tests.
3 Discuss any implications of your results.
4 Draw a conclusion in response to your research question.
5 Is it possible to generalise any of your results to the wider population? Why or why not?
6 Discuss any potential extraneous variables and how they could have affected the data.
7 Suggest improvements you could make to the procedure to address the extraneous variables, if the investigation was to be repeated.

Adapted from: VCAA VCE Psychology Unit 4 Advice for Teachers https://www.vcaa.vic.edu.au/

KEY CONCEPTS 10.1a

- Normal, healthy brains usually form correct perceptions of sensory stimuli.
- Visual illusions demonstrate how visual perception can be fallible because it can be intentionally manipulated to form an incorrect perception of a stimulus.
- Visual illusions occur because the stimulus provides misleading cues, and when the brain applies its normal processes during organisation and interpretation it makes an error in perception.
- The Ames room is an intentionally distorted room that creates the illusion that people change size as they cross the room.
- The Müller-Lyer illusion is where two lines of equal length – one tipped with inward-pointing arrowheads, the other with outward-pointing arrowheads, appear to be of different lengths.
- The Spinning Dancer illusion creates the illusion that a silhouette of a woman spinning on one foot spontaneously switches between spinning left and spinning right.

Concept questions 10.1a

Remembering

1 What does fallible mean when applied to perception? **r**
2 What is a visual illusion? **r**

Understanding

3 Explain the apparent-distance hypothesis. **e**
4 How do misinterpreted depth cues cause the Müller-Lyer illusion? **r**
5 Explain how the characteristics of the Ames room contribute to the illusion. **e**

Applying

6 Study Figure 10.11. Use your knowledge of the Müller-Lyer illusion to explain why the woman on the right who is wearing a high-cut bathing suit appears to have longer legs than the woman on the left. **c**

HOT Challenge

7 Last night your sister was looking through the pictures of visual illusions you have collected for a school project. She asked you to explain why she was seeing things she knew could not exist.
What did you tell her? C

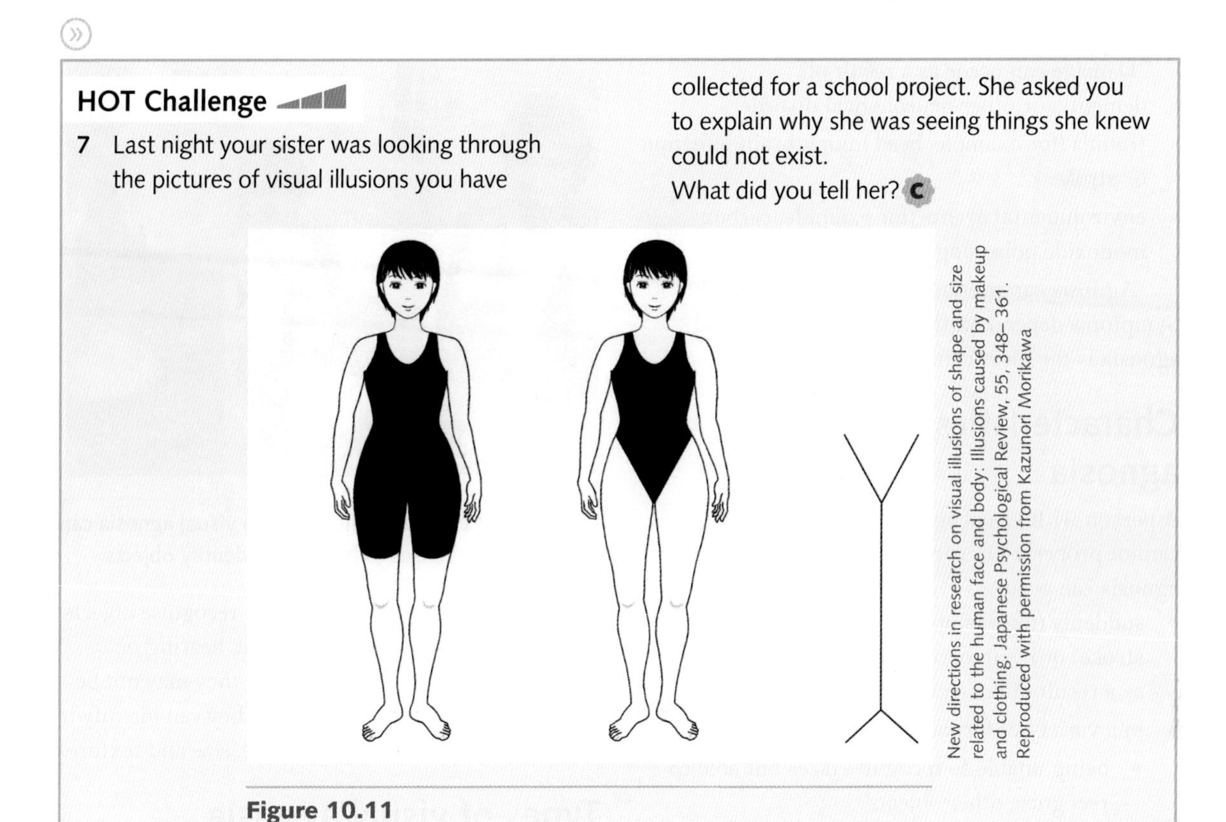

Figure 10.11

New directions in research on visual illusions of shape and size related to the human face and body: Illusions caused by makeup and clothing. Japanese Psychological Review, 55, 348–361. Reproduced with permission from Kazunori Morikawa

Agnosia: an impairment of sensory processing

In his book *The Man Who Mistook His Wife for a Hat*, Dr Oliver Sacks described a patient, Dr P, who, as a result of brain injury, 'mistook his wife for a hat'. Dr P suffered from prosopagnosia (face blindness), a form of visual agnosia. Dr P had great difficulty in identifying objects. His vision was otherwise normal, and he could describe the features of what he saw. When Dr Sacks gave him a rose the patient described it as 'a convoluted red form with a linear green attachment'. Only after Dr P smelt it did he realise it was a rose. On another day when Dr P's wife came to meet him at the doctor's office, he accidentally reached for her head when he wanted to retrieve his hat from the coat rack (Sacks, 1985).

Agnosia is a rare condition that is usually caused by neurological disorder or damage to parts of the brain. It is a **perceptual disorder**, which occurs when a person's perceptual processing is impaired so they have limited ability to make sense of sensory information. People with agnosia fail to *recognise* and *identify* objects, people, smells or sounds despite their senses otherwise functioning normally. Agnosia only affects a single sensory processing system of an affected person; for example, their vision.

The word *agnosia* can be translated from Greek as meaning 'to lack knowledge of', so **visual agnosia** implies a loss of visual knowledge. People with visual agnosia have normal visual field, acuity, colour vision and brightness discrimination. Their language, memory and intelligence remain intact. As damage has occurred in pathways that connect their brain's occipital lobe with other sensory processing areas in their brain's visual centres, particularly those involved in visual memory, they may lose the ability to recognise and identify things that are familiar. These pathways provide access to knowledge and information needed for perception and identification. If they are damaged, the stored information cannot be accessed. Therefore, it cannot be integrated with the sensory information received.

Causes of agnosia

Agnosia is caused by damage along pathways that connect primary sensory processing areas in the brain. These areas typically include the parietal, occipital and temporal lobes. The damage can result in loss of cognitive functions controlled by these areas but leave cognitive abilities controlled by other areas unaffected (Kumar & Wroten, 2021).

10.1.1 VISUAL ILLUSIONS AND VISUAL AGNOSIA

Damage can occur as a result of:

» dementia or other neurological disorders
» trauma (for example, head injury, brain infection or strokes)
» environmental events (for example, carbon monoxide poisoning).

Agnosia can present suddenly or gradually. Symptoms depend on the area involved. Visual agnosia is the most common form of agnosia.

Characteristics of visual agnosia

A person with visual agnosia can see, but they cannot properly interpret what they see. Visual agnosia can occur:

» suddenly (for example, after a head injury or stroke) or it can occur gradually (for example, as a result of dementia or brain cancer)
» in a variety of forms, including:
    » being unable to recognise faces but able to recognise other objects
    » seeing multiple objects but only recognising one object at a time
    » being unable to name or describe familiar objects when looking at them.

Weblink
Visual agnosia

Figure 10.12 People suffering from visual agnosia can use their other senses to help them identify objects.

People with visual agnosia can recognise objects by using other senses such as touch, hearing or smell (Figure 10.12). For example, they may not be able to identify a cup by sight but they can identify it by touching it and feeling its shape, size and texture.

Types of visual agnosia

Agnosia is further divided into two subtypes: apperceptive visual agnosia and associative visual agnosia. These agnosias are summarised in Table 10.5

Table 10.5 Apperceptive and associative visual agnosia

	Apperceptive visual agnosia	Associative visual agnosia
Definition	A form of agnosia in which a person is unable to recognise visual stimuli such as shapes or forms of an object despite having no visual deficits	A form of agnosia in which a person is unable to recall information associated with an object, such as its name or what it is used for
Typical features	» Can see contours and outlines of an object but can't categorise them » Cannot recognise objects or draw or copy a figure although their knowledge of the object is intact (Figure 10.13) » Cannot match two identical stimuli » When a person looks from one object to another they can see the individual elements of the object but they cannot perceive them as forming a recognisable whole » Has difficulty perceiving the difference between features of an object or more than one object at a time » May not be able to recognise pictures of the same object from different angles » Typically results from damage to the rear sections of the right hemisphere's occipital lobes	» Can describe, draw or copy an object but can't name it because they can't access information from brain areas involved in visual memory » Unable to distinguish between objects that are real and those that are not » Will not be able to match different pairs of an item or two photos of the same item pair taken from different angles » Typically associated with damage to both the right and left hemispheres at the border of the occipital and temporal lobes (occipitotemporal border)

	Apperceptive visual agnosia	Associative visual agnosia
Example	If the person was asked to copy a picture of a shoe, they may draw a series of oval scribbles. However, their knowledge of what a shoe is used for remains intact.	» If shown a toothbrush, the person could draw it but could not name it or explain how to use it. If they were given verbal or tactile cues about the toothbrush, they may be able to recognise it and state its use. » If they were shown a drawing of a lion with cow's head, they cannot tell the difference between them or state whether they were real or unreal creatures. » Prosopagnosia (see text)

Prosopagnosia (also known as face blindness) is a form of associative visual agnosia. It is a neurological disorder in which people are unable to recognise familiar faces (sometimes their own) or facial differences and they cannot identify a person by name. However, they can often identify gender and age, and recognise non-facial clues such as hair and clothing. People with prosopagnosia rely on bottom-up processing and what they are sensing in the moment when analysing someone's face. They are not able to use top-down processing to distinguish faces because they cannot mentally store the faces of people they know. They can distinguish different facial features of a person; for example, if the person has a scar or a crooked nose. They are not able to use their memory to put a name to a face. For example, they don't know whether or not they have seen a particular face before, or who it belongs to.

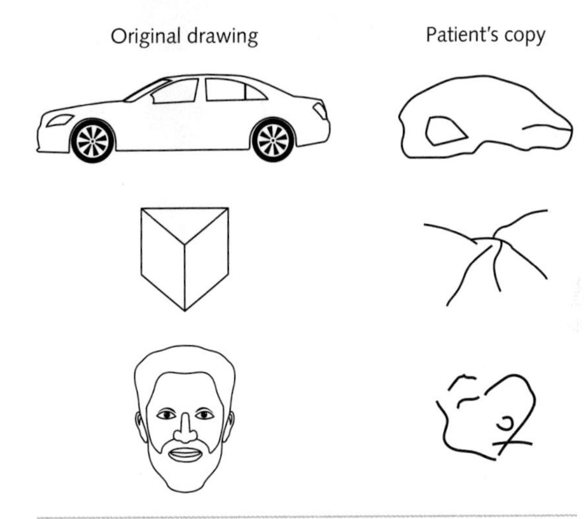

Figure 10.13 People with apperceptive agnosia are not able to copy a drawing of an object. They might produce drawings like these, for example.

WB 10.1.2 WHO AM I?

ACTIVITY 10.1 VISUAL AGNOSIA

Try it

1 Visual agnosia is the inability to recognise familiar objects or faces despite the visual system being intact. There are two main categories of visual agnosia: apperceptive and associative. Visual processing has two broad stages. In the first stage, the brain assembles incoming visual information into an image; in the second stage, the brain associates that image with existing memories to understand the image. Apperceptive agnosia occurs due to impairment at the first stage, and associative agnosia occurs at the second stage.

Two patients with visual agnosia have been asked to copy and name a set of images to determine the type of agnosia they have. Using the results shown in Figure 10.14a and b, determine which patient has apperceptive agnosia and which has associative agnosia.

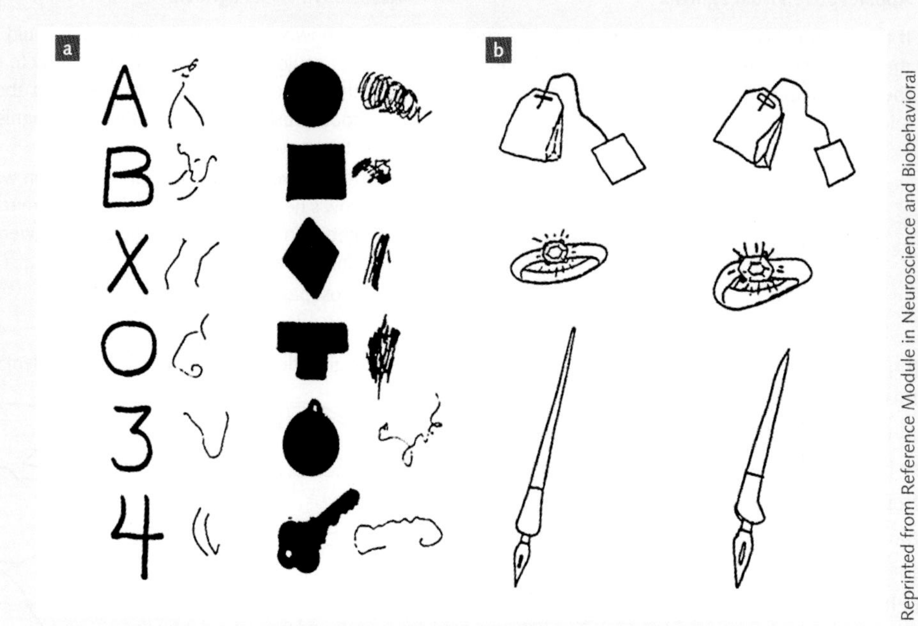

Figure 10.14

Apply it

2 Agnosia is a problem with identification of stimuli, despite the person having properly functioning senses. They do not have a memory problem, just a disconnection between the sensory perception and memory. Patient A could not name an object when shown a picture of the object, but they could immediately recognise the object if granted an opportunity to touch or hear a description of the object's appearance. Patient B was unable to recognise the same object on the basis of touch but could immediately name it when they saw the object.

Identify the type of agnosia demonstrated by Patient A and Patient B.

Exam ready

3 Following brain surgery, a patient is unable to identify a tennis ball by sight but is able to correctly say it is a tennis ball once they reach out to pick the ball up and hold it in their hands. This patient is most likely suffering from

A prosopagnosia.

B tactile agnosia.

C auditory agnosia.

D visual agnosia.

CASE STUDY 10.1

What's it like to be face-blind?

Dr DF, a hospital consultant, has experienced life-long face-blindness: prosopagnosia. He was not formally diagnosed until 2006, at the age of 53. He didn't even realise he had prosopagnosia for 30 years, and he only discovered it was a neuropsychological condition in 2006. 'It now seems remarkable that I lived at least half my life with a socially disabling condition of which not only myself, but also those around me, seemed unaware,' writes Dr DF. Throughout his life Dr DF has developed a complicated strategy for recognising people based on non-facial physical features and context.

Looking back at his childhood, Dr DF can remember details of school buildings, his male friends' clothes, the hairstyles of the girls – but no faces. He remembers one day when he was criticised for not raising his cap to his teacher when he passed by her after school. He was not being rude. He merely failed to recognise her. At secondary school he was successful in making friends. His friendship group was easy to distinguish because they were all physically distinctive.

In his professional life, Dr DF says he has no problems during committee meetings, where everyone keeps the same seat. His job as a hospital doctor also involves working in distinct locations. He uses these environmental contexts to help him judge who he is likely to meet at any given time. Nonetheless, he often walks right past colleagues. This has earned him a reputation as unpredictable and unfriendly.

Before marrying, Dr DF found romantic relationships especially problematic. Although women often dress in more distinctive ways than men, they also vary their appearance more often. 'It seemed to me that girls popped out of the Ether in one place then disappeared perhaps for months or even years before reappearing in another place, often very annoyed with me,' he says.

Dr DF found social situations particularly difficult. Parties were particularly awkward and stressful. 'I once had a long conversation with a man, at a Christmas party,' he recalls. 'We mingled until we met again at the other side of the room. I introduced myself [again]. He looked puzzled until my wife came to my rescue.' Making friends is nigh on impossible. 'Recognising' strangers is a constant embarrassing risk.'

Dr DF has had a successful career in spite of his prosopagnosia, but he feels his success was 'blunted' by the condition. Now aged 60, he has highly developed strategies for coping. For example, he has a better sense of people's age, which is one of the criteria he uses to distinguish people. And he tries to focus on distinct items of jewellery, such as people's rings, that tend to be worn at all times. The increased popularity of tattoos is another help.

Looking to the future, Dr DF is worried that his confusion about people's identity could lead to him being misdiagnosed with dementia. 'During a recent hospital stay I asked the nurses to introduce themselves every time, as I was concerned that I might be misdiagnosed as confused if I muddled them up.'

Experts used to think that the inability to recognise faces was a problem that nearly always arose after brain injury. In recent years, however, it has become apparent that many people are born with prosopagnosia (or develop it early in life), with the prevalence estimated at two per cent of the population.

Adapted with permission from (2012). A life with prosopagnosia. Cognitive Neuropsychology, 1-6 DOI: 10.1080/02643294.2012.736377. Taylor & Francis Ltd, http://www.tandfonline.com Adapted with permission from the author.

Questions

1 What may have caused Dr DF's prosopagnosia?
2 What were the indicators that Dr DF suffered from prosopagnosia?
3 Throughout his life, what difficulties has Dr DF encountered as a result of his prosopagnosia?
4 What coping mechanisms has Dr DF developed to compensate for his prosopagnosia?

HOT Challenge

5 Outline one way a neuropsychologist could test a patient to confirm their suspicions that the patient had prosopagnosia.
6 Look at the person sitting next to you. If you suffered from prosopagnosia, how would you recognise them if you saw them on the weekend wearing different clothes?

Weblink
How being face-blind made it easier to see people

KEY CONCEPTS 10.1b

- Agnosia is a neurological disorder that is usually caused by brain damage.
- People with agnosia are unable to recognise and identify familiar stimuli.
- People with visual agnosia can see clearly but they fail to recognise or correctly interpret what they see.
- There are two types of visual agnosia: apperceptive visual agnosia and associative visual agnosia.

Concept questions 10.1b

Remembering

1 What is agnosia? r

2 What causes agnosia? r

3 What are the two types of visual agnosia? r

Understanding

4 Explain the difference between apperceptive visual agnosia and associative visual agnosia. e

Applying

5 During a series of tests to determine the effects of a traumatic brain injury, Antonio was shown a spoon, a hammer and a pen. He was able to describe these objects and what they were used for but he could not name them. Name and describe the condition Antonio was suffering from. c

6 A neuropsychologist was interviewing a patient suspected of suffering from visual agnosia. Describe two behaviours the neurologist would look for in the patient to confirm a diagnosis. c

HOT Challenge

7 Design tests to determine whether a person is suffering from apperceptive visual agnosia or associative visual agnosia. i

10.2 Fallibility of gustatory perception

Like all forms of perception, the perception of taste – or how we perceive flavour – has a biological basis. As we saw in Chapter 9, taste buds, or gustatory receptor cells, are located on the upper surface of the tongue, soft palate, upper oesophagus, cheek and epiglottis. As we chew, the chemical molecules in food come in contact with taste buds. The sensation of taste is activated and the taste buds send electrochemical signals to the gustatory cortex (taste centre) of the brain about the type of taste being sensed. The **gustatory cortex** (Figure 10.15) is the brain area where taste information is processed. It consists of two small substructures. One substructure, the frontal operculum, is located on the frontal lobe. The other substructure, the anterior insula, is located deep in the cerebral cortex, under the frontal, parietal and temporal lobes.

In most cases, largely because of our prior experience with the food type, our brain correctly perceives the taste. We know whether the food is sweet, salty, sour, bitter or savoury (umami). Sometimes, however, mistakes can be made. In other words, our perception of food flavours can be fallible.

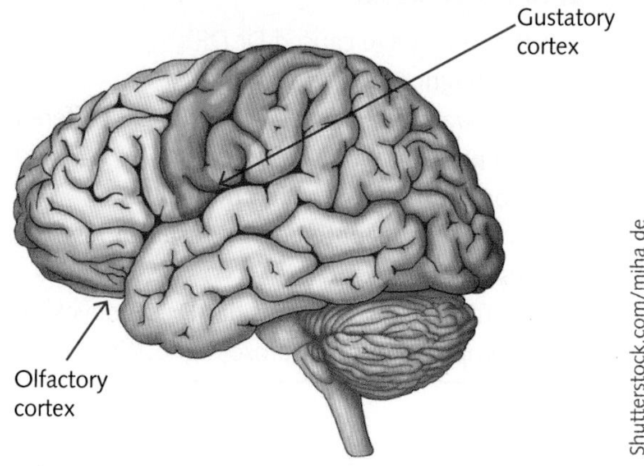

Figure 10.15 The gustatory cortex receives stimuli from taste buds and forms them into a perception of the flavour of a food or drink.

Supertasters

A **supertaster** is a person who is born with more taste buds and taste receptors on their tongue than average, so their perception of taste is stronger (Figure 10.16). Supertasters taste certain flavours and foods more strongly than other people.

Supertasters have about four times as many taste buds per square centimetre as people who are classified as **non-tasters**. This means that supertasters have a higher taste sensitivity. **Taste sensitivity** refers to the intensity with which we perceive the sensation of a taste and flavours. Although both groups may respond similarly to many foods, their higher taste sensitivity means that supertasters find flavours (and smells, too) more intense. Supertasters are also quicker to distinguish individual flavours in a mixture than non-tasters. Non-tasters have a lower taste sensitivity, so they do not distinguish individual flavour as easily as supertasters.

Supertasters tend to avoid powerfully flavoured foods. They tend to have stronger tastes for sweet and bitter, and for irritants such as alcohol and capsaicin (the chemical that makes chillies hot). Generally, they will avoid sugary desserts and they will use more salt than other people because it blocks the bitter taste of certain foods. Women are more often supertasters. Non-tasters tend to prefer sweet and fatty foods, which may be why supertasters tend to be slimmer than non-tasters (Bartoshuk et al., 1994). According to the website Healthline (2019), approximately 25 per cent of the US population are supertasters, 25 per cent are non-tasters, and 50 per cent are average tasters.

10.2.1 ARE YOU A SUPERTASTER?

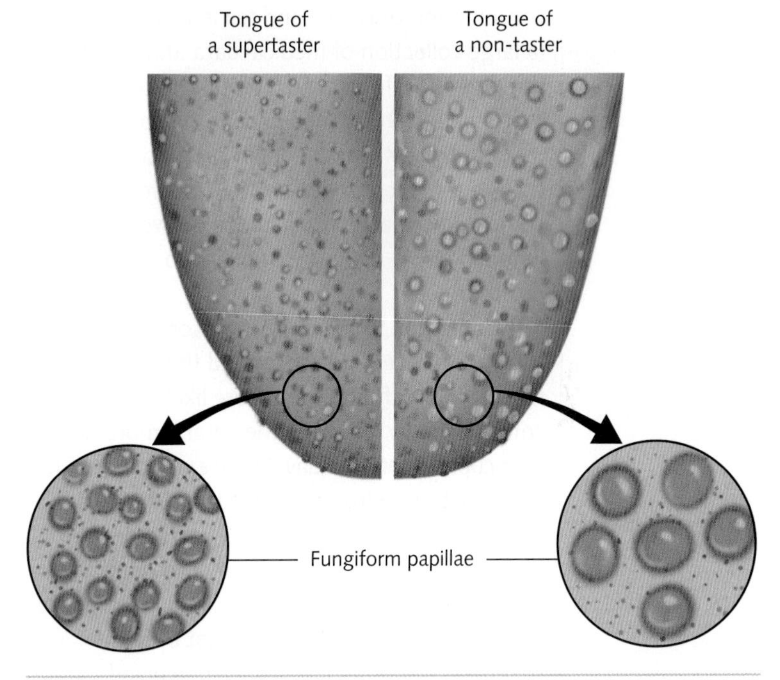

Figure 10.16 Supertasters have more taste buds and taste receptors than non-tasters. This causes them to experience tastes more intensely than non-tasters.

ANALYSING RESEARCH 10.1

Why you like coffee, and I choose tea – it's in the genes

Daniel Liang-Dar Hwang, Postdoctoral Researcher, The University of Queensland

A study by the University of Queensland shows that the likelihood of a person being a coffee drinker or a tea drinker is linked with the presence or absence of key genes that shape how bitter flavours taste.

Tea and coffee generally taste bitter because they contain bitter-tasting substances such as caffeine. Quinine is another substance that contributes to the bitterness of coffee. It is also found in tonic water. The University of Queensland study revealed bitter taste receptor genes that are responsible for the perception of caffeine, quinine and the synthetic bitter substance propylthiouracil (PROP). PROP has the same bitterness as Brussels sprouts.

Previous research indicated that inherited factors play a role in the amount of coffee and tea a person drinks a day, and that the ability to digest caffeine plays an important role in people's consumption of caffeinated beverages.

10.2.2 JUDGING FLAVOURS PART 1

»

However, it was not known whether genes for bitter taste perception were involved in determining consumption of bitter-tasting beverages. Previous studies with small sample sizes reported no or inconsistent relationships.

In the University of Queensland study, the consumption of coffee and tea in a large Biobank (a large collection of medical data and tissue samples, used for research purposes) cohort of more than 400000 men and women aged 37 to 73 in the UK was used along with data about their bitter receptor genes. The researchers compared coffee and tea intake between people who did or did not carry particular bitter taste receptor genes.

Compared to an average person, the study showed that people who carried the bitter taste receptor for caffeine were more likely to be heavy coffee drinkers, meaning they drank more than four cups of coffee a day. Every extra copy of the bitter taste receptor gene lead to a 20 per cent higher chance of being a heavy coffee drinker. These 'super-tasters' of caffeine also drank less tea.

As caffeine contributes to not only the bitterness of coffee but also its perceived strength and texture, people who are better at detecting caffeine may find coffee more enjoyable and flavourful. In contrast, people who carried the bitter taste receptors for quinine or PROP drank less coffee and more tea. Compared to an average person, every extra copy of the quinine or PROP receptor gene was linked with a 9 per cent or 4 per cent higher chance of being a heavy tea drinker (meaning they drank more than 5 cups of tea a day). When there is a need for caffeine, 'super-tasters' of quinine and PROP generally choose tea over coffee because they tend to be more sensitive to overall bitterness.

The study demonstrated that genes for bitter taste perception are linked with the amount of coffee and tea we drink. However, genes may not account for taste and dietary behaviour change over time as people age. A person whose genes predispose them to disliking bitter flavours could still learn to enjoy deliciously bitter-tasting foods and beverages in the future.

Adapted with permission from 'Why you like coffee, and I choose tea – it's in the genes' by Daniel Liang-Dar Hwang, Postdoctoral Researcher, The University of Queensland, The Conversation, November 16, 2018, https://theconversation.com/why-you-like-coffee-and-i-choose-tea-its-in-the-genes-106854

Questions

1 Identify the population of interest for this study.
2 Write a hypothesis for this study.
3 What are the results of this study?
4 What conclusion(s) can be drawn from this study?

Judging flavours: the magic of miraculin

Can you imagine what it would be like to live in a world where you could indulge your love of sweet things without damaging your health? Unfortunately, no magic trick has been invented to make this happen. However, there is a miracle fruit that can trick your taste buds into thinking sour or acidic foods are sweet. Its name is miraculin (Figure 10.17).

Shutterstock.com/claudiah

Figure 10.17 **Miraculin berries are found on the coast of West Africa. When chewed prior to tasting a sour or bitter food, they alter the taste perception of the food.**

Miraculin is a natural sugar substitute that, when consumed with something sour, turns the sour tasting food or beverage into a sweet tasting food or beverage. Miraculin is a soluble protein found in the skin and pulp of the small red miracle berry of *Synsepalum dulcificum*, a shrub native to West Africa. The local population have known about the effects of the berry for decades. For example, they chewed on the pulp of the fruit prior to consuming acidic foods, such as stale and sour maize bread, to make them taste more palatable.

The berry was first documented in the 1700s. The active substance, isolated by Japanese scientist Kenzo Kurihara, was named miraculin after the miracle fruit when Kurihara first extracted it in 1968 (Koizumi et al., 2011). The miraculin itself is not sweet. However, it is a 'taste modifier' that changes the perceived quality and intensity of tastes (Faus & Sisniega, 2003).

How miraculin works

Miraculin alters our sense of taste at a molecular level. The effect is quickly reversible because it is only active at the taste receptor site in the mouth. Miraculin only activates the sweet taste receptor in a sour or acidic environment. The response to miraculin occurs in the following sequence:

- sour and acidic food (low pH) comes into contact with the miraculin-sweet taste receptor complex
- this triggers a cellular response when miraculin binds to and activates the sweet receptors on the taste buds
- neural impulses are transmitted from the tongue to the gustatory cortex of the brain
- this information is integrated with other taste information stored in other brain areas
- a taste perception is formed.

So, ordinarily sour and acidic foods, fruits (such as citrus) and beverages are perceived as sweet without any added sugar or sweetener. The effect lasts as long as the protein is bound to the tongue. This can be up to an hour, until it is destroyed by an enzyme in the saliva.

Miraculin makes most acidic foods taste sweet, but does not improve the taste of bitter things. The berry's pulp can be freeze dried into tablet form that can be consumed with food or beverages. Because of its taste-modifying property, the miracle berries can be used as an alternative to a sweetener or sugar substitute, without any additives, preservatives or added sugar. So, you could eat a slice of lemon without detecting its intense sourness while still detecting its flavour and natural sweetness.

10.2.3 JUDGING FLAVOURS PART 2

Worksheet
The effect of exposure to Miraculin on gustatory perception

ACTIVITY 10.2 MIRACLE BERRY TASTING

Try it

1 Watch the episode of the BBC show *QI: Miracle berry tasting* by clicking on the weblink. Then try the sequence in Figure 10.18.

Figure 10.18

Draw your own miraculin flowchart with a sour food choice of your own (for example, grapefruit, rhubarb, vinegar, raw cranberries, kimchi, sauerkraut).

Apply it

2 The host of *QI*, Steven Fry, said he can make lemons taste like oranges. How would he make that happen?

Exam ready

3 Exposure to miraculin has an impact on

A the judgement of flavour of food.

B the actual flavour of food.

C the judgement of enjoyment of food.

D gustatory sensation.

Weblink
QI: Miracle berry tasting

KEY CONCEPTS 10.2

- Gustation is activated when chemical molecules from food or drink come into contact with taste receptors in the taste buds.
- Our perception of taste is usually correct but we can make mistakes.
- Supertasters perceive the flavour of certain foods more intensely because they have more taste buds than other people.
- Miraculin is a soluble protein that changes the flavour of sour and acidic foods so that we perceive them as tasting sweeter than they really are.

Concept questions 10.2

Remembering

1 Where is taste information processed in the brain? r

2 Identify two differences between a supertaster and a non-taster. r

3 How does miraculin affect taste perception? r

Understanding

4 Explain how consuming miraculin before eating a lemon would alter your perception of the lemon's flavour. e

Applying

5 Create a test to determine whether a person was a supertaster or a non-taster. i

HOT Challenge

6 In the 1964 movie *Mary Poppins*, Mary sings, 'A spoonful of sugar makes the medicine go down'. Use your knowledge of the physical and psychological features of taste perception to explain why Mary was correct. e

10.3 Distortions of perception in healthy people

Perception distortion occurs when a person's response to environmental stimuli is different from what is considered the normal response. This distortion can be the result of physical damage to the brain, psychological impairment or the influence of drugs. However, people with healthy brains can also experience perceptual distortions. This is the case with those people who experience the condition of synaesthesia.

Weblink
What is synaesthesia?

Synaesthesia: a sensory crossover

Do you see colours when you hear music? When you hear specific sounds, do they have a unique odour or taste? If you do, then you are experiencing an unusual perceptual condition known as synaesthesia. The word 'synaesthesia' comes from two Greek words, *syn* (together) and *aisthesis* (perception). Therefore, synaesthesia literally means 'joined perception'. **Synaesthesia** refers to a group of neurological conditions where information taken in by one sense is involuntarily experienced in a way normally associated with another sense. In other words, it is a form of sensory 'crossover'. Sounds, for example, can be experienced as tastes or colour; tastes and smells can be experienced as sounds or associated with visual input (Table 10.6, page 413). Synaesthesia is an involuntary, life-long condition that is usually present from birth. A person who experiences synaesthesia is called a synaesthete.

In 1880, English scientist and explorer Sir Francis Galton reported the first case of synaesthesia. Galton noticed that some people who were otherwise completely normal seemed to have a certain peculiarity: one sensory stimulus caused them to experience a number of different sensory responses. For example, on hearing musical notes they saw distinct colours; F-sharp might be red and C-sharp might be blue. Or they might report that the printed number 5 always 'looks' green, whereas 2 looks red (Ramachandran & Hubbard, 2003).

Generally, however, synaesthesia was largely dismissed as a phantom condition, and it is only relatively recently that it has been accepted as a genuine neurological condition. This acceptance is mostly because of evidence from brain imaging techniques such as PET (positron emission tomography) and fMRI (functional magnetic resonance imaging), which have shown that there is sensory cross-activation in the brains of synaesthetes.

For example, there is activation of brain regions associated with visual perception in blindfolded synaesthetes when they listen to words that evoked visual experiences. These activations were different from those evoked in either non-synaesthetes or the same synaesthetes listening to tones that did not induce visual experiences (Mulvenna & Walsh, 2005).

During normal, everyday conversations, people often use metaphors that are synaesthetic; that is, they describe one sensory experience with vocabulary best suited to a different sense. This includes descriptions such as cool colours, loud wallpaper, bitter cold, sweet silence, a sour facial expression, an icy stare, hot (attractive) people and a rough (bad) day. When we hear these descriptions, they help us visualise what people are talking about, but this does not mean we are synaesthetic.

10.3.1 DISTORTIONS OF PERCEPTION: VISUAL

Synaesthesia can manifest itself in many different forms. It can occur between any two senses, and different forms of synaesthesia can be experienced by different people. However, each individual has a unique experience of synaesthesia.

ANALYSING RESEARCH 10.2

The neuroscience behind the Bouba/Kiki effect

In 2001, Professor Vilayanur Ramachandran, director of the Centre for Brain and Cognition at the University of California, and his associate Edward Hubbard presented American college undergraduates and Tamil Speakers in India with a two-letter alphabet consisting of one round-shaped letter and another letter resembling a pointed star (Figure 10.19). They asked the students 'Which of these shapes is bouba and which is kiki?' The results showed that 95% to 98% of subjects said bouba was Letter A and Kiki was Letter B even though these were completely novel words being matched with drawings that they had never seen before.

Ramachandran concluded that the results were due to the nature of the connections that exist between sensory and motor areas of the brain. He suggested that because the mouth assumes a rounded contour when saying 'booba', it mimics the round contour of the letter. 'Kiki', in contrast, has a sharp verbal edge to it that mimics the sharpness of the pointed star (Ramachandran & Hubbard, 2001). In later work Ramachandran and Hubbard found that damage to an area of the brain important for language called the angular gyrus resulted in a person being much less likely to match the rounded object with the word bouba.

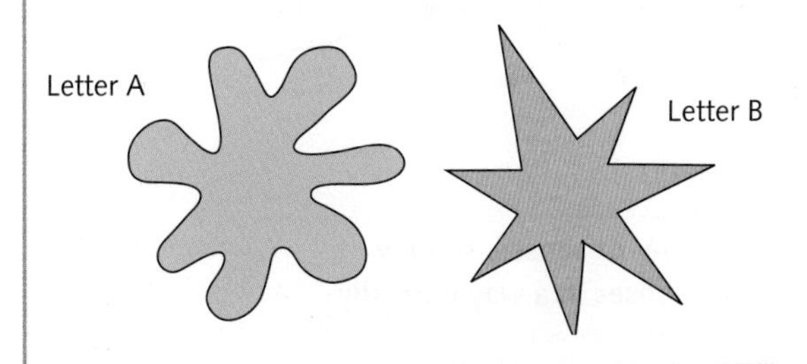

Figure 10.19 A two-letter alphabet

Another group of researchers tested this similar question with toddlers. The finding was that the associations of 'kiki' to jagged shapes and 'bouba' to rounded shapes were consistent even before language development had occurred (Maurer et al., 2006). These results suggested that no matter if test subjects were different native language speakers or very young children, people were always able to make this association. Like Ramachandran, the researchers reasoned that because of the sharp form of the visual shape, subjects tended to map the name 'kiki' onto the right figure, and because of the rounded auditory sound, subjects tended to map the name 'bouba' onto the left figure.

For hundreds of years, we have been investigating how our brain processes sensory information. And this BK effect perhaps now provides us with a unique window to look into how our brain combines all this sensory information to create a coherent picture of how we perceive the world around us.

Source: Adapted from Huang, H. (2019, June 28). What's the neuroscience behind the bouba/kiki effect? NBB in Paris. https://scholarblogs.emory.edu/nbbparis/2019/06/28/whats-the-neuroscience-behind-the-bouba-kiki-effect/

»

Questions

1 What was the aim of Ramachandran's study?
2 Identify the variables of interest in this study, including any that are independent variables, and the dependent variable.
3 What were the results of this study?
4 What conclusion(s) did the researchers draw from these results?

Common characteristics of synaesthesia

Although there are more than 60 known types of synaesthesia, common characteristics of the condition include the following.

WB 10.3.2 EVALUATION OF RESEARCH

- A trigger is required. Triggers can be external (such as a sound or odour) or internal (such as the thought of a number or a food). If the specific stimulus trigger is present, the experience will always occur.
- The synaesthetic perception is stable over time. For example, a synaesthete may always experience seeing yellow when they see the letter 'T', or experience the taste of strawberries when they hear middle C played on a violin.
- Synaesthesia is only experienced in one direction. For example, the taste of a specific food might trigger the perception of a specific colour, but exposure to the colour doesn't trigger the perception of the taste (Ward & Mattingly, 2006).
- The experience is unique to the person (Figure 10.20). Although people may share the same form of synaesthesia, the experience is different for each individual. Some people's synaesthesia is specific to one sense; for example, vision. Others may experience a combination of the senses; for example, vision and taste.
- The experience is vivid for the person and can be remembered.
- Synaesthesia is involuntary or automatic. Synaesthetes report they cannot stop or start their synaesthesia at will.
- Synaesthesia is more common in females than in males, with a ratio of six to one (Baron-Cohen et al., 1996).
- Synaesthesia is more common among artists, poets, novelists and creative people in general. One survey of Australian synaesthetes found that 24 per cent were employed in an artistic profession, whereas in the non-synaesthete population this applies only for 2 per cent (Rich et al., 2005). Synaesthetes are also more likely to be left-handed.

Figure 10.20 **Synaesthetes experience a crossover of senses, so they experience information taken in by one of their senses in a way normally associated with another sense.**

Table 10.6 Common forms of synaesthesia

Type	Description	Example
Grapheme–colour synaesthesia	Specific numbers, letters or words trigger specific colours	ABCDEFGHIJKLMN OPQRSTUVWXYZ
Sound-to-colour (auditory–visual) synaesthesia	A sound automatically triggers the visualisation of coloured, generic shapes	Colour–sound imagery
Number-form synaesthesia	When thinking about numbers, the person automatically visualises a number map	1, 6, 10, 12, 20, 30, 40, 50, 60, 70, 80, 90, 100, 112, 110, 120, etc.
Personification	When presented with an ordered sequence, such as numbers, letters, months and days, they are perceived as having various personalities	When looking at a calendar, Thursdays may be described as being angry or October as nervous.
Lexical–gustatory synaesthesia	When speaking, hearing or writing specific words, a person experiences unique tastes	Hearing the word 'jail' may evoke the taste of pineapple.

What causes synaesthesia?

Synaesthesia is not a perceptual disorder. A perceptual disorder – for example, agnosia – occurs when a person's perceptual processing is impaired so they have limited ability to make sense of sensory information. They have difficulty analysing and interpreting information, and this causes a distorted perception of the information they have sensed. People with synaesthesia do not experience a loss of ability, and the condition rarely interferes with their daily life. Synaesthesia is instead a type of **perceptual anomaly**. Synaesthetes experience irregularities in their perceptual processing that cause them to experience sensations that are not consistent with a stimulus (Figure 10.21).

Despite considerable research, there is still no clear explanation for why synaesthesia occurs. However, there are many popular explanations for this condition. Some of these are outlined in Table 10.7.

Shutterstock.com/s_bukley

Figure 10.21 Singer-songwriter Pharrell Williams says he always visualises specific colours when he hears musical notes.

Table 10.7 Explanations for synaesthesia

Explanation	Description
Synaesthesia is innate.	The experience of synaesthesia is not the same as the experience we have when we learn and remember something. Memory involves learning associations between items of information, such as the word 'beach' being associated with the colour blue, the smell of sunscreen, or the sound of seagulls. Memory is based on experience and learning, but most neuroscientists believe that synaesthesia is not.
Synaesthesia has a genetic basis.	Research suggests that synaesthesia runs in families (Ramachandran & Hubbard, 2003; Hatch, 2002; Baron-Cohen et al., 1996; Cytowic, 1995). For example, Julian Asher of the University of Oxford took genetic samples from 196 individuals belonging to 43 families. From this group, 121 experienced auditory-visual synaesthesia, meaning they 'saw' sounds. The results showed four chromosomal regions where gene variations seemed to be linked to the condition, suggesting that there was a genetic link to synaesthesia (Robson, 2009).
Synaesthesia is linked to learning during childhood.	Children learn to associate numbers or letter with colours, possibly to aid memory. However, this does not explain why siblings with synaesthesia raised in the same or similar environments report different colours for the same trigger, or experience different variants of synaesthesia (Society for Neuroscience, 2012).
Synaesthesia is the result of the brain's failure to remove excess neurons.	During childhood the synaesthete's brain did not efficiently prune excess, weak or unused neuronal connections. Therefore, as adults, they have more neural connections than non-synaesthetes.
Synaesthesia can be the result of external factors.	Hallucinogenic drugs such as LSD (lysergic acid diethylamide) or psilocybin can cause synaesthesia. Brain injuries that leave one part of the brain without its normal innervation (the supply of nerves to a part of the body) can cause neighbouring nerve fibres to invade the damaged area. This rewiring of the brain can result in synaesthesia.

ANALYSING RESEARCH 10.3

Synaesthesia: what colour is that word?

What colour is the letter B? Have you ever smelled the musical note middle C? Around 10000 Australians have the condition synaesthesia, where a sensory experience normally associated with one sensory system is interpreted as belonging to another.

Researchers Anina Rich and Dr Jason Mattingley of the Cognitive Neuroscience Laboratory at the University of Melbourne have collected the world's largest database of people with synaesthesia (Rich et al., 2005). In one experiment, the researchers investigated the relationship between brain activity and the synaesthetic colour experiences caused by viewing letters. This form of synaesthesia is known as grapheme–colour synaesthesia, and means that letters, words and numbers cause vivid experiences of colour.

The participants were seven grapheme–colour synaesthetes (six female; two left-handed; mean age 28 years) and seven non-synaesthetic controls. The two groups were matched on age, sex and handedness. All participants were exposed to three display conditions (Figure 10.22):

1. chromatic letters, where the letters were displayed in a colour that matched the synaesthetic colour
2. achromatic letters, where the letters were displayed without colour
3. neutral rectangles, which were non-coloured rectangles that did not induce synaesthesia.

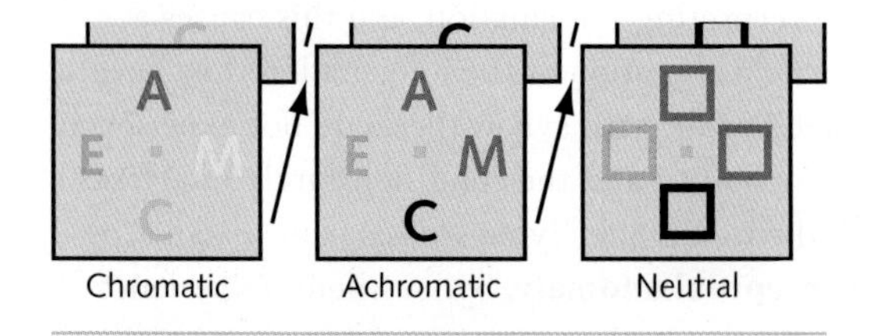

Figure 10.22 Sample stimuli for the conditions of the synaesthetic experience experiment undertaken at the Cognitive Neuroscience Laboratory at the University of Melbourne

An fMRI scanner was used to monitor and record the brain activity of each participant experiencing each condition. Results showed an obvious difference between the the synaesthetes and the matched controls in the type of occipital lobe activity (Figure 10.23). This was particularly evident in the achromatic letter display, where only the synaesthetic participants showed activity in the areas of the brain thought to be involved in synaesthesia experiences.

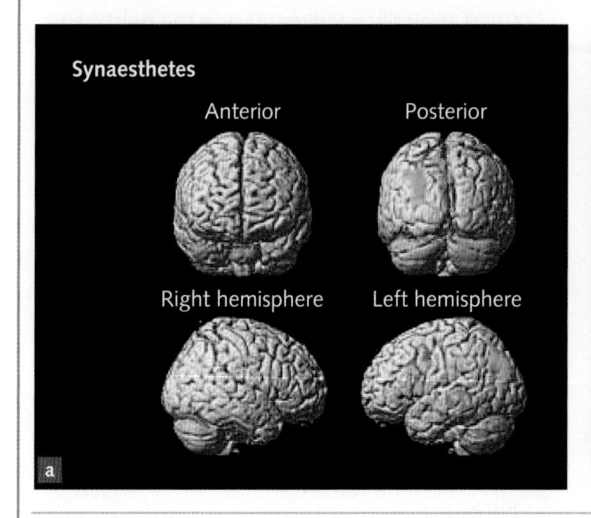

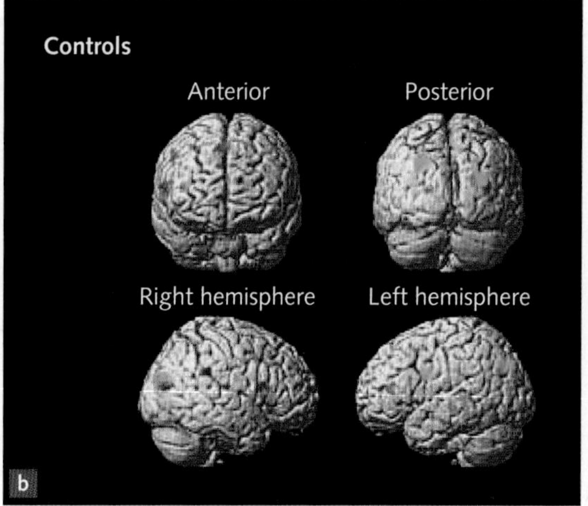

Figure 10.23a Synaesthesia brain activity compared with b normal brain activity: green regions indicate active areas when colour letters were viewed (subtracted from activity when viewing neutral squares); red regions indicate active areas when achromatic letters are viewed (subtracted from activity when viewing neutral squares).

Questions

1 Describe the variables of interest in this study. Identify independent and dependent variables where appropriate.
2 The synaesthetes and control participants were matched on age, sex and handedness. What does this indicate about the type of study design chosen?
3 Synaesthesia is relatively rare. What implications does this have for sampling procedures and for participant selection?
4 What ethical considerations are relevant to this experiment?

Spatial neglect

As we saw in Chapter 5, **spatial neglect** is when a person tends to ignore the left or right side of their body or the left or right side of their visual space. It is not a result of visual impairments. It is a cognitive impairment because it is a disorder of attention. Affected people fail to pay attention to, recognise or respond to stimuli located on one side of their body. Spatial neglect results from damage to one of the cerebral hemispheres (usually the right parietal lobe). It is usually the result of stroke or brain injury that affects the right parietal lobe. However, damage to the left parietal lobe can also result in spatial neglect.

Spatial neglect disrupts a person's everyday life. It can result in people:

» colliding with obstacles on their left (or right)
» eating food only from the left (or right) side of their plate
» ignoring people situated on their left (or right)
» failing to dress or shave their left (or right) side
» possibly behaving as if the left half of their world has ceased to exist, and (in many cases) not even realising that something is wrong.

10.3.4 SPATIAL NEGLECT

ACTIVITY 10.3 SPATIAL NEGLECT

Try it

1 Spatial neglect results from damage to the parietal lobe. Damage to the left parietal lobe would result in the neglect of items on the right side of a person's world. Damage to the right parietal lobe would result in the neglect of the left side of their world. One method used to diagnose spatial neglect is to ask individuals to draw copies of objects as accurately as possible; for example, the clock in Figure 10.24.

Figure 10.24

a What type of image would you expect a person with right spatial neglect to produce? Sketch and annotate your response.

b What type of image would you expect a person with left spatial neglect to produce? Sketch and annotate your response.

Apply it

2 One approach used to treat spatial neglect is constraint-induced movement therapy (CIMT). CIMT typically involves placing the right hand in a sling, while the person with spatial neglect completes a variety of motor tasks with their left hand, such as stacking blocks.

Based on your understanding of the causes and symptoms of spatial neglect, why would the therapist place the patient's right arm in a sling while completing the motor tasks?

Exam ready

3 A patient has experienced damage to his right parietal lobe. To test for possible spatial neglect, the psychologist asks the patient to read a list of words aloud. The word list was:

backyard
high school
basketball

If the patient has spatial neglect, they will read the list as

A yard; school; ball.
B back; high; basket.
C backyard; high school; basketball.
D They would not be able to read the word list.

CASE STUDY 10.2

The man whose brain ignores one half of his world

Thom Hoffman

Alan Burgess doesn't need a rhyme to remember the 5th of November. He'll never forget the day he had his stroke. It left him with a syndrome known as spatial neglect and a strange new perspective. His stroke damaged the parietal lobe on the right side of his brain, the part that deals with the higher processing of attention. The damage causes him to ignore people, sounds, and objects on his left.

'Hemispatial neglect typically occurs after a stroke,' says Dr Paresh Malhotra, senior lecturer in neurology at Imperial College London. 'It is not blindness in one eye, and it's not damage to the primary sensory cortex. It's a process of ignoring one side of space.'

Burgess, 64, originally trained as a tool design draughtsman and, before his stroke, he was working as a driver. His visual neglect makes driving impossible, and he was forced into early retirement. He had never painted before his stroke, but it became an important outlet for him afterwards. Look at his artwork, though, and it is not only the vivid colours that grab your attention. He hands me a sketch of a pig that has half its head missing, and then produces two robins copied from a Christmas card – the robin on the right is full of rich detail, but the one on the left remains unfinished.

'Vision is most strikingly affected because we are visual creatures, but hearing, touch, representation and sense of self are also affected,' says Dr Malhotra. People with hemispatial neglect are often unaware of their condition. Friends or relatives might suggest they look to their neglected side but that instruction misunderstands the problem they have with navigating the space around them. Burgess, and people like him, are not aware that something is missing, so why would they seek it out? Patients might bump into things on their neglected side, shave or apply makeup only on one side of their face, or leave half of the food on their plate.

Alan often has to stop and think to make sense of the world around him. Walking down a street, he hugs the right side of the pavement, brushing up against walls and hedges. He won't notice any potential dangers coming from the left, so he cannot go out on his own.

'I can't describe how the world looks to neglect patients,' says Dr Malhotra. 'Part of the reason it's so difficult is because we don't really appreciate how the world looks to ourselves. We think it's just a nice screen and you can see everything, but that's something that your brain is computing and telling you you're seeing.

'In fact you're attending to specific things at specific times. Your eyes are darting all over the place, but you have a sensation of a static world.'

'It's in the middle of my line, not the middle of your line,' Burgess told Dr Malhotra during one consultation.

Dr Malhotra says that people with visual neglect do realise something is wrong when it is pointed out to them over and over and over again. 'So in a way I think for them it's slightly abstract that they know there's some problem on the left-hand side, and they use strategies to try and overcome the problem, without really being able to appreciate it in its fullness.'

Extract from 'The man whose brain ignores one half of his world' by Thom Hoffman, The Guardian, 23 November 2012

Questions

1 Explain how Alan Burgess' stroke affected him.

2 How does the case of Alan Burgess demonstrate that spatial neglect is not a visual impairment but a cognitive impairment?

3 Identify three ways that spatial neglect has affected Alan Burgess' life.

HOT Challenge

4 If you acquired spatial neglect, what strategies could you design that would allow you to:

a eat all of the meal that was placed in front of you?

b tie the shoelaces on both of your shoes?

KEY CONCEPTS 10.3

- Synaesthesia occurs when information taken in by one sense is involuntarily experienced in a way normally associated with another sense.
- Synaesthesia can manifest itself in many different forms; it can occur between any two senses; and different forms of synaesthesia can be experienced by different people.
- Spatial neglect is a tendency to ignore the left or right side of one's body or the left or right side of one's visual space.
- Spatial neglect is a cognitive impairment, not a visual impairment.

Concept questions 10.3

Remembering

1 What is synaesthesia? r

2 Outline three major characteristics of synaesthesia. r

3 What is spatial neglect? r

Understanding

4 Explain why spatial neglect is a cognitive impairment, but synaesthesia is not. e

Applying

5 If I look at a photo of a child with a dog and I instantly visualise 'Lulu', my much-loved dog from childhood, am I having a synaesthetic experience? Explain. c

HOT Challenge

6 Dr Ahmed suspected that one of her patients, Mary, was synaesthetic. Mary said that when she heard, spoke or wrote specific words she experienced different tastes. For example, every time she heard, saw or wrote the word 'computer' she tasted bacon. Dr Ahmed decided to give Mary a test using an fMRI scan. i

Explain what this test could consist of and how an fMRI scan would assist in diagnosis.

10 Chapter summary

KEY CONCEPTS 10.1a

- Normal, healthy brains usually form correct perceptions of sensory stimuli.
- Visual illusions demonstrate how visual perception can be fallible because it can be intentionally manipulated to form an incorrect perception of a stimulus.
- Visual illusions occur because the stimulus provides misleading cues, and when the brain applies its normal processes during organisation and interpretation it makes an error in perception.
- The Ames room is an intentionally distorted room that creates the illusion that people change size as they cross the room.
- The Müller-Lyer illusion is where two lines of equal length – one tipped with inward-pointing arrowheads, the other with outward-pointing arrowheads, appear to be of different lengths.
- The Spinning Dancer illusion creates the illusion that a silhouette of a woman spinning on one foot spontaneously switches between spinning left and spinning right.

KEY CONCEPTS 10.1b

- Agnosia is a neurological disorder that is usually caused by brain damage.
- People with agnosia are unable to recognise and identify familiar stimuli.
- People with visual agnosia can see clearly but they fail to recognise or correctly interpret what they see.
- There are two types of visual agnosia: apperceptive visual agnosia and associative visual agnosia.

KEY CONCEPTS 10.2

- Gustation is activated when chemical molecules from food or drink come into contact with taste receptors in the taste buds.
- Our perception of taste is usually correct but we can make mistakes.
- Supertasters perceive the flavour of certain foods more intensely because they have more taste buds than other people.
- Miraculin is a soluble protein that changes the flavour of sour and acidic foods so that we perceive them as tasting sweeter than they really are.

KEY CONCEPTS 10.3

- Synaesthesia occurs when information taken in by one sense is involuntarily experienced in a way normally associated with another sense.
- Synaesthesia can manifest itself in many different forms; it can occur between any two senses; and different forms of synaesthesia can be experienced by different people.
- Spatial neglect is a tendency to ignore the left or right side of one's body or the left or right side of one's visual space.
- Spatial neglect is a cognitive impairment, not a visual impairment.

9780170465052

10 End-of-chapter exam

Section A: Multiple-choice

1 Which is the best definition of a perceptual anomaly?
  A a visual illusion
  B a perceptual hallucination caused by a deviation from normal mental processes that give meaning to stimuli
  C a perceptual irregularity caused by a deviation from normal mental processes that give meaning to stimuli
  D a perceptual confusion caused by sensory overload

2 Which of the following statements about a visual illusion is correct?
  A A visual illusion results from a failure in the visual perception process.
  B A visual illusion results from a consistent misinterpretation of correct visual cues.
  C Visual illusions occur in the absence of external stimuli.
  D Visual illusions are the result of a real external stimulus providing misleading visual cues.

3 In the Ames room illusion, the viewer
  A maintains size constancy but misapplies shape constancy.
  B maintains shape constancy but misapplies size constancy.
  C maintains size but misapplies linear perspective.
  D maintains linear perspective but misapplies shape constancy.

4 Molly is planning an experiment on taste perception. She intends to use random sampling to select her participants from the population of VCE students at her school. To select her sample, Molly should
  A ask every student in her Psychology class to participate.
  B ask all her friends at school to volunteer.
  C assign each VCE student a number and select every 10th one.
  D assign each VCE student a number, put all the numbers in a container and draw out the number of participants she needs.

5 Cross-cultural studies indicate that people raised in areas where there are few right angles are _____________ susceptible to the Müller-Lyer illusion than people who are raised in areas where right angles and rectangles are common.
  A much less
  B much more
  C somewhat less
  D equally

6 Apperceptive visual agnosia is typically characterised by
  A an inability to identify familiar objects.
  B an inability to copy drawings or match objects.
  C an inability to recognise pictures of the same object from different angles.
  D all of the above

7 Associative visual agnosia is not typically characterised by an inability to
  A draw or copy a familiar object.
  B recognise familiar objects.
  C recognise familiar faces.
  D distinguish between real and fictional objects.

8 A researcher asked a number of synaesthetes to participate in an experiment on perceptual anomalies. They agreed. When publishing the results, the researcher published the names of the participants. What ethical guidelines have they breached?
  A withdrawal rights
  B confidentiality
  C voluntary participation
  D deception

9 Agnosia can be caused by
  A dementia.
  B physical trauma to the brain.
  C environmental events (for example, carbon monoxide poisoning).
  D all of the above

10 If a person had normal vision but they were unable to recognise and identify familiar objects, they are experiencing
- **A** a visual illusion.
- **B** synaesthesia.
- **C** visual agnosia.
- **D** spatial neglect.

11 Synaesthesia
- **A** can present in many different forms.
- **B** only involves the sense of vision and taste.
- **C** is a perceptual disorder.
- **D** can be experienced in two forms by one individual.

12 Spatial neglect is usually caused by damage to the
- **A** left hemisphere's parietal lobe.
- **B** right hemisphere's parietal lobe.
- **C** left hemisphere's occipital lobe.
- **D** right hemisphere's occipital lobe.

13 The brain area that processes the sensation of the taste of a food or beverage is the
- **A** temporal lobe.
- **B** somatosensory cortex.
- **C** occipital lobe.
- **D** the gustatory cortex.

14 A person with spatial neglect will most likely
- **A** have difficulty coordinating movements as a result of frontal lobe damage.
- **B** have difficulty recognising faces because of temporal lobe damage.
- **C** only pay attention to one side of visual space because of parietal lobe damage.
- **D** not pay attention to stimuli in the environment because of frontal lobe damage.

15 Where in the body are your taste buds located?
- **A** on the surface of your tongue
- **B** on the walls of your mouth
- **C** at the back of your throat
- **D** all of the above

16 Miraculin
- **A** only activates the sweet taste receptor in a sour or acidic environment.
- **B** is a synthetic flavour enhancer.
- **C** changes the taste of sweet foods and makes them taste saltier.
- **D** distorts our perceptual abilities.

17 As we age, we have
- **A** fewer taste buds than at 6 years of age.
- **B** more taste buds than at 6 years of age.
- **C** the same number of taste buds we were born with.
- **D** stronger taste buds than when we were younger.

18 Which of the following statements is not true for someone classed as a supertaster?
- **A** They have more taste buds and gustatory receptor cells than other people.
- **B** They have a higher sensitivity for taste than other people.
- **C** They avoid foods with very intense flavours.
- **D** They are usually male.

19 After suffering brain damage, Jason would only shave one side of his face. He most likely suffers from
- **A** Broca's aphasia.
- **B** Wernicke's aphasia.
- **C** spatial neglect.
- **D** agnosia.

20 Synaesthesia is a type of
- **A** perceptual illusion.
- **B** hallucination.
- **C** perceptual irregularity.
- **D** perceptual error.

Section B: Short answer

1 What is agnosia? [1 mark]

2 Explain the apparent-distance hypothesis. [2 marks]

3 Ever since he could remember, Hamish has experienced grapheme–colour synaesthesia. Whenever he heard certain numbers, letters or words, he would automatically visualise specific colours.

In an experiment investigating the brain function of synaesthetes, Hamish underwent a PET scan. During the scan, Hamish was blindfolded and listened to a list of words, numbers and letters.

What areas of Hamish's brain would be activated during this test? [4 marks]

Unit 2, Area of Study 2 review

Section A: Multiple-choice

Question 1
The ability to focus on an activity or stimulus for a long period of time is known as

A sustained attention.

B divided attention.

C selective attention.

D organisation.

Question 2
Gerard is busy studying for his Psychology test. Because he is so focused, he doesn't realise his little sister has come up behind him to scare him. This is an example of

A sustained attention.

B divided attention.

C selective attention.

D organisation.

The following information relates to questions **3–6**.

Psychologists are interested in studying the effects of divided attention on productivity in the workforce. They use a study of 20 employees of a local fast-food restaurant. One group is exposed to a 2-week training course to develop divided attention skills and the other is not. Their productivity is measured through the amount of money made in a night.

Question 3
What is the hypothesis of this study?

A It is hypothesised that individuals who get divided attention training will make more money than individuals who do not.

B It is hypothesised that individuals who get divided attention training will perform the same as individuals who do not.

C It is hypothesised that individuals who get divided attention training will perform worse than individuals who do not.

D It is too hard to make a prediction.

Question 4
What is the population of this study?

A 20 fast-food restaurant staff

B all fast-food restaurant staff

C all workers

D all humans

Question 5
What is a likely limitation of this study?

A not a large enough sample size

B convenience sample is unrepresentative of population

C lack of minimisation of individual participant differences

D all of the above

Question 6
What are the likely results of this study?

A Individuals who did the training will be unable to focus attention to get tasks done well, so will be less productive.

B Individuals who have the training will be more productive because of their ability to multitask.

C There will be no difference in the two groups.

D The group who did the training will start off with better productivity but this will decline due to fatigue.

Question 7
In the visual perception system, the process of reception involves

A transmitting information from the eye to the visual cortex.

B converting electromagnetic energy to electrochemical energy.

C light entering the eye and being focused on the retina.

D visual stimuli being interpreted.

Question 8
Which of the processes is present in the visual perception pathway but not in the taste pathway?

A transmission

B transduction

C reception

D organisation

Question 9
Which of the following statements is true of sensation?

A it is involving top-down processes only

B it involves bottom up processing only

C it can consist of bottom-up and top-down processing

D it involves neither top-down or bottom-up processing

Question 10
Which of the following options best describes the order in which visual perception takes place?

A reception, organisation, interpretation
B reception, transmission, transduction
C selection, transduction, transmission
D selection, interpretation, organisation

Question 11
Taste perception processes begin when __________ energy is changed into _________ energy.

A chemical; electrochemical
B electromagnetic; electrochemical
C electromagnetic; chemical
D electrochemical; electromagnetic

Question 12
Over time, some people develop a liking for some bitter tastes, such as coffee, because as we age

A the sensitivity of our taste buds decreases.
B we learn that bitter tastes are not harmful.
C the number of taste buds decreases.
D the sensitivity of our taste buds increases.

Question 13
What is a social impact on our taste perception?

A family and friends
B genetics
C age
D culture

Question 14
What is a biological factor that affects taste?

A genetics
B prior experience
C mood
D culture

Question 15
Jenny picks up an olive, thinking it is a grape. She bites into it, then instantly spits it out. Jenny thinks it will be sweet because of

A social factors.
B psychological factors.
C biological factors.
D a combination of all of these factors.

Question 16
Jenny hates coriander and thinks it tastes like soap, whereas Lily thinks it tastes citrusy and loves it in her food. This difference is due to a gene that impacts taste perception of the herb. What type of factor is influencing their taste perception?

A biological
B cultural
C psychological
D social

Question 17
When viewing the Ames room it is difficult to perceive people as getting closer or further away; instead, they appear to be growing and shrinking as they cross the room. Which of the following features of the Ames room creates the illusion?

A the peephole
B the trapezoidal shape of the room
C the distorted window sizes
D all of the above

Question 18
Although the two lines in the Müller-Lyer illusion are of ______________ length, the feather-tailed line is perceived to be ______________ than the arrowhead line.

A equal; longer
B equal; shorter
C unequal; longer
D unequal; shorter

Question 19
A rare condition in which individuals have an inability to recognise objects, people, smells or songs is called

A spatial neglect.
B agnosia.
C super taster.
D cognitive distortion.

Question 20
What are visual illusions?

A a misperception of external visual stimuli
B images interpreted objectively by everyone who views them
C images that appear the same as reality
D all the above

Question 21

Bec is seeing Dr Rossi because she cannot recognise objects shown to her. However, she can identify them when she picks them up and handles them. It is likely Bec will be diagnosed with

A spatial neglect.

B tactile agnosia.

C visual agnosia.

D blindness.

Question 22

Which statement is true of supertasters?

A They enjoy most foods because they have strong taste.

B They have more papillae, which enables them to taste more by having more taste buds.

C They typically enjoy spicy food.

D They tend to not be picky eaters.

Question 23

How does miraculin impact on our taste perception?

A It makes everything taste salty.

B It makes everything taste sweet.

C It makes everything taste bitter.

D It makes everything taste bland.

Question 24

How does prior experience impact on our taste perception?

A We expect food to taste a certain way because it tasted that way in the past.

B If we disliked a food when trying it, we are likely to avoid it in future.

C Our brains struggle to process taste sensations when they are different from our previous experiences.

D All of the above

Question 25

What percentage of the population are supertasters?

A 25 per cent

B 50 per cent

C 75 per cent

D 5 per cent

Use the following information to answer questions **26–29**.

Nick is a psychologist interested in studying the effects of brain trauma in the parietal lobe on a person's spatial awareness. He found five volunteers to participate, which involved having their parietal lobe temporarily probed. One of his subjects accidentally suffered long-term damage from this and now does not notice the left side of his visual field.

Question 26

Is this study ethical?

A yes, because they consented

B no, because they weren't told of the nature of the study

C no, because long-term irreversible harm was caused by the study

D yes, because the benefit to research outweighed the harm caused

Question 27

What is it likely this participant now suffers from?

A agnosia

B spatial neglect

C synaesthesia

D Alzheimer's disease

Question 28

What is a more ethical way Nick could have conducted his experiment?

A Use a larger sample size.

B Get participants to sign a waiver.

C Use patients who already suffer parietal lobe damage and do a case study.

D It is impossible to study this field without the risk of long-term harm.

Question 29

What part of the parietal lobe was damaged in this participant?

A the left parietal lobe, because the left visual field is affected

B the right parietal lobe, because the right visual field is affected

C the left parietal lobe, because the right visual field is affected

D the right parietal lobe, because the left visual field is affected

Question 30

Which of the following is not a type of synaesthesia?

A grapheme–colour

B auditory–visual

C lexical–gustatory

D agnosia

Section B: Short answer

Question 1

a Describe a characteristic of grapheme–colour synaesthesia. [1 mark]

b Explain why synaesthesia can be a perceptual distortion. [2 marks]

[Total = 3 marks]

Question 2

Differentiate between selective attention and divided attention. [2 marks]

Question 3

[3 marks]

Give one example of each of a biological, a social and a psychological factor impacting on visual perception.

Question 4

a What is the Stroop effect? [2 marks]

b Does the Stroop effect use top-down or bottom-up processing? Justify your response. [2 marks]

[Total = 4 marks]

Question 5

Jeremy is a picky eater. He hates foods that are super bitter and usually doesn't eat enough vegetables because he hates the flavour of them.

What is likely to be the cause of Jeremy's dislike of bitter food? Explain what this means. [2 marks]

9780170465052

Scientific investigations 11

Key knowledge

Investigation design
- the role of scientific investigations in reducing uncertainty
- psychological science concepts specific to the selected scientific investigation and their significance, including the definition of key terms
- scientific methodology relevant to the selected scientific investigation, selected from classification and identification; controlled experiment; correlational study; fieldwork; modelling; or simulation
- techniques of primary qualitative and quantitative data generation relevant to the investigation
- accuracy, precision, repeatability, reproducibility and validity of measurements in relation to the investigation
- health, safety and ethical guidelines relevant to the selected scientific investigation

Scientific evidence
- the distinction between an aim, a hypothesis, a model, and a theory
- observations and investigations that are consistent with, or challenge, current scientific models or theories
- the characteristics of primary data
- ways of organising, analysing and evaluating generated primary data to identify patterns and relationships including sources of error and identification of remaining uncertainty
- use of a logbook to authenticate generated primary data
- the limitations of investigation methodologies and methods, and of data generation and/or analysis

Science communication
- the conventions of scientific report writing including scientific terminology and representations, standard abbreviations and units of measurement
- ways of presenting key findings and implications of the selected scientific investigation

Source: VCE Psychology Study Design (2023–2027), p. 30–31
(Binocular Cues Vs Monocular Cues-Definition, Difference and Uses, 2021)

11 Scientific investigations

In Unit 2, Outcome 3 you will be required to conduct your own research to produce your own primary data. This is your opportunity to use your science skills, but enough reading: it is time to apply what you have learned to investigate a topic that relates to perception or behaviour.

p. 431

11.1 Designing and conducting an experiment

Scientific investigations and research can take many different forms. The one thing they have in common is that they all follow the same method: the scientific method.

iStock.com/iambuff

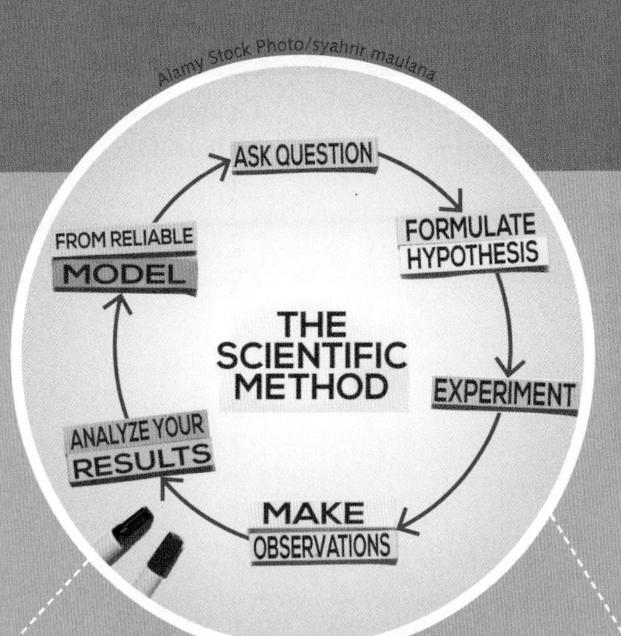

Alamy Stock Photo/syahrir maulana

p. 453

11.2
Science communication

Science is about contributing to our understanding of the world. Therefore, it's important that we can effectively communicate our findings to allow everybody to access the information.

Shutterstock.com/Ruslan Huzau

Being able to follow the scientific method to conduct investigations with primary and secondary data is fundamental to being able to contribute to psychological research. Given how complex our brains are, there will always be more to learn about the brain. What is it about our thoughts, feelings and behaviours that you would now like to know more about?

Slideshow
Chapter 11 slideshow

Flashcards
Chapter 11 flashcards

Test
Chapter 11 pre-test

Know your key terms

Analyse
Evaluating
Limitations
Methodology

Weblink
Selective attention test: Invisible gorilla experiment

What does a person dressed up in a gorilla suit and a game of basketball have to do with psychology? No! It is not the start of a bad joke. The answer is… more than you think! In 1999 psychologists Daniel Simons and Christopher Chabris first conducted their now-famous 'Selective attention: Invisible gorilla experiment' at Harvard University (Figure 11.1). In this experiment, participants were asked to watch a video of people passing a basketball between them. Half of the people in the video wore white shirts and half wore black shirts. The viewer's task was to count how many times members of the 'white' team passed the ball, a relatively simple task for most people. But this was not the whole experiment. In the middle of the 'game' a person in a gorilla suit walked into the playing zone, stood in the centre and beat their chest, and then walked off again. The study found that the majority of the participants watching the video did not notice the gorilla at all. Simons and Chabris concluded that we often overestimate our ability to effectively multitask. More specifically, they thought that when people are asked to attend to one task, they focus so strongly on that element that they may miss other important details (Simons & Chabris, 1999).

You may also be wondering why anyone would have someone dress up in a gorilla suit and join a basketball game in the first place. It all relates back to the goals of psychological science – to understand how and why humans think and behave the way they do. To achieve these goals, scientists conduct research, gather data and analyse them to test theories and models. In this case, the data collected in the Invisible gorilla experiment tested the theory of selective attention and became the groundwork for a lot of other research in the field of attention.

Simons, D. J., & Chabris, C. F. (1999). Gorillas in our midst: Sustained inattentional blindness for dynamic events. Perception, 28, 1059–1074. Figure provided by Daniel Simons, (www.dansimons.com)

Figure 11.1 The Invisible gorilla experiment conducted by Daniel Simons and Christopher Chabris in 1999 demonstrated the limitations of selective attention.

11.1 Designing and conducting an experiment

A scientific investigation is a plan for asking questions and then conducting research to determine possible answers. This plan is called the scientific method. The scientific method is a standardised series of steps that researchers follow to make observations, ask questions, form hypotheses, gather data, test predictions, interpret results and draw evidence-based conclusions.

Each investigation you complete in Psychology will follow the same format. The Unit 1, Outcome 3 task was to complete a research investigation using secondary data. Unit 2, Outcome 3 requires you to generate your own primary data by either adapting an existing study or designing your own scientific investigation. Your investigation will build on the concepts you have already studied in chapters 1 and 6, and requires you to apply this knowledge in a practical setting. This involves:

- selecting a topic of interest
- posing a research question
- researching previous published studies
- formulating a hypothesis
- setting up your investigation
- carrying out your investigation (collecting data)
- organising, analysing and interpreting your data
- drawing relevant generalisations and evidence-based conclusions in response to your research question.

In the next section we will explore what you will need to consider when designing your experimental investigation, and how you will **analyse** and evaluate your results.

There will be a worked example running through this chapter, showing how a student has adapted an existing investigation.

A Psychology student (Alina) has just finished Chapter 9: Perception. For practice, the teacher has asked her to conduct a scientific investigation that examines how depth cues can influence depth perception (Figure 11.2).

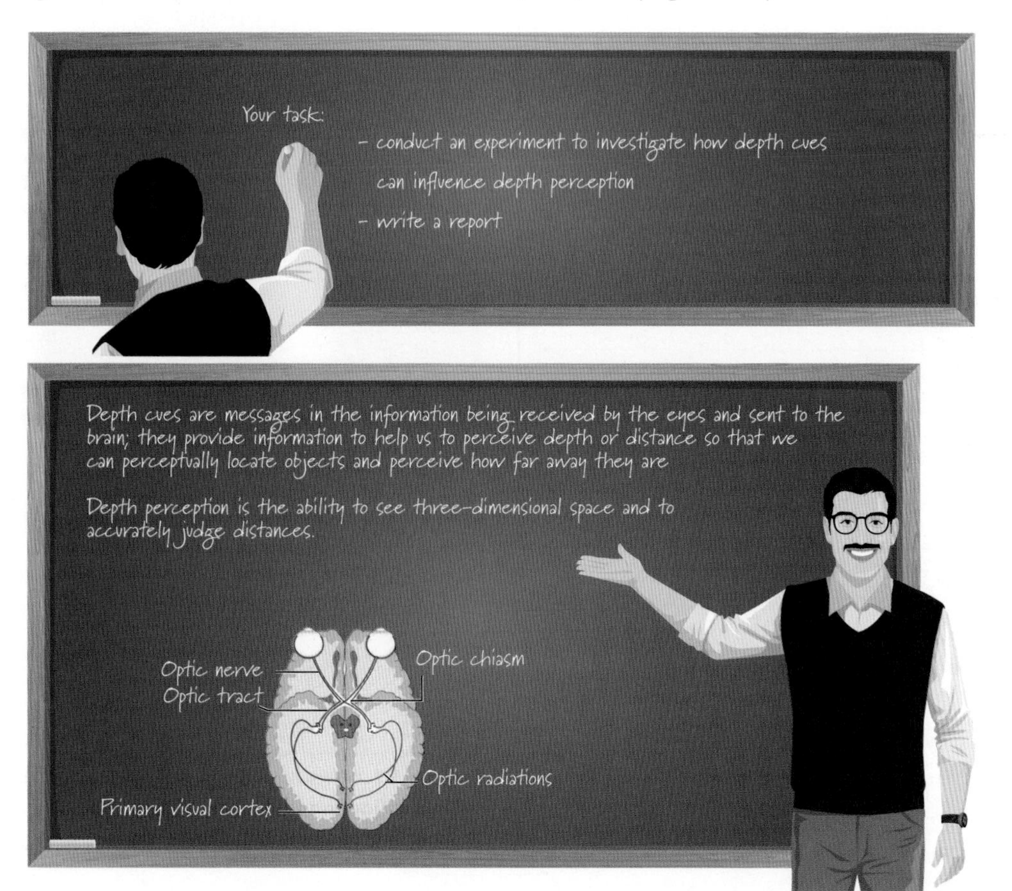

Figure 11.2 Your task for Unit 2 Outcome 3 is to design or adapt an existing investigation and to generate primary data.

Logbook alert!

Make sure you have your logbook handy as you get ready to start your research project (Figure 11.3). It should contain a complete record of your progress, including your initial brainstorming of ideas, your research, procedure, observations, data collection, analysis, discussion ideas and any issues that you encounter along the way. If this is done well, when it comes to writing your report you will only need to transfer the information in your logbook into your report. You will find logbook templates for each type of possible investigation in the online resources, so be on the lookout for the logbook icon alongside the investigations.

INVESTIGATION

Investigation number

Title

Introduction

Aim

Hypothesis

Name: Your signature: Date:
Partner: Partner's signature: Teacher's signature:

Figure 11.3 An example of the logbook templates available in this resource

Setting up your scientific investigation

As mentioned in Chapter 1, a well-planned investigation will help ensure that you can answer your research question (Figure 11.4). There are many possibilities for you to explore, so making sure you have a clear idea of your chosen topic and what you want to investigate is essential. Following the steps of the scientific method will help you focus your research question and work through each step of your experiment successfully. To help you, this chapter will work through a sample of one student's work, in which a student adapted an existing study.

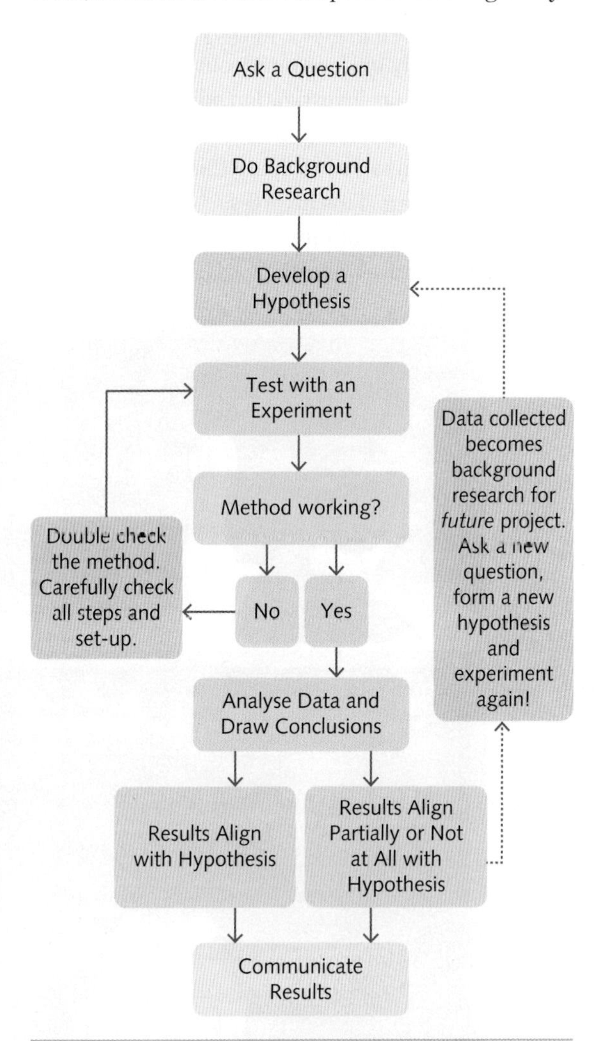

Figure 11.4 A summary of the overall process you will be following. Keep it in mind and refer back to it throughout your investigation.

Step 1: Select a topic of interest

You have now covered the content for Unit 2. For Unit 2, Outcome 3 you need to select one of the topics you have learned about in Unit 2 to focus on for your experimental investigation. You should already have a clear idea about what topics interested you and what it is you want to investigate. Use Table 11.1 to help you.

Table 11.1 Which Area of Study do you want to focus on?

Area of Study	Your choice
AOS 1: How are people influenced to behave in particular ways?	
AOS 2: What influences a person's perception of the world?	

Reduce your options

What part of your selected Area of Study did you find particularly interesting? What exactly do you want to investigate? If you examine the key knowledge below you may notice that some appear easier to investigate. You may find it helpful to discuss with your classmates which options you think would be most suitable. There are also some dedicated websites (see weblink) that have a range of tests that you could potentially use. You may even find a study that combines two concepts you have learned about. Your teacher may also be able to suggest research that previous students have undertaken which you could use as a starting point. Use Table 11.2 to help you.

Step 2: Do your background research

When you have decided on your topic, the next step is to start researching it.

In Chapter 6 you learned how to search for relevant research articles and critically evaluate each one for validity, reliability and integrity. Draw on these research skills once again to form the basis of your investigation. Your background research will inform you about which psychological theories or models relate to your research question. You will then be able to work out a hypothesis that is a testable prediction, to investigate and answer your research question.

Don't forget to check that you have found valid and reliable sources!

Weblink
Psytoolkit

Table 11.2 Which topic within an Area of Study do you want to focus on?

AOS 1: How are people influenced to behave in particular ways?		Your choice
Social cognition	the role of person perception, attributions, attitudes, and stereotypes in interpreting, analysing, remembering, and using information about the social world, including decision-making and interpersonal interactions	
	the avoidance of cognitive dissonance using cognitive biases	
	the positive and negative influences of heuristics as mechanisms for decision-making and problem-solving	
	the influence of prejudice and discrimination within society on a person's and/or groups mental wellbeing and ways to reduce it	
Factors that influence individual and group behaviour	the influence of social groups and culture on individual behaviour	
	the concepts of obedience and conformity and their relative influence on individual behaviour	
	positive and negative influences of different media sources on individual and group behaviour such as changing nature of social connections, social comparison, addictive behaviours and information access	
	the development of independence and anti-conformity to empower individual decision-making when in groups	
AOS 2: What influences a person's perception of the world?		
Perception	the role of attention (sustained, divided, selective) in making sense of the world around us	
	the role of perception in the processing and interpretation of sensory information, as demonstrated through top-down and bottom-up processing	
	the influence of biological, psychological and social factors on visual perception and gustatory perception	
Distortions of perception	the fallibility of visual perceptual systems; for example, visual illusions and agnosia	
	the fallibility of gustatory perception; for example, supertasters, exposure to miraculin and the judgement of flavours	
	distortions of perception of taste and vision in healthy individuals such as synaesthesia and spatial neglect	

As discussed earlier, one of the main aims in this type of research is to build on the work that has already been done. This means you have two options available to you. You can either *adapt* an existing scientific investigation or you can *design* one using previous investigations as a starting point.

Adapting an existing investigation

If you choose to adapt an existing investigation, you will be trying to reproduce the results with a different sample. It is unlikely you will be able to use exactly the same measures used in the original investigation, as you may not have access to the specialised equipment the researchers used. However, you can adapt their approach to create your own achievable measures and still capture key features of the original measures. You would also need to justify these decisions in your report. Part of your evaluation of your findings will be commenting on the extent to which your success or failure to reproduce the original result was because your measurements did not capture what was done in the original study.

Designing your own investigation

Designing your own investigation means defining your own variables and methodology, rather than adapting the methods and measures of a previous study. Designing your own investigation does not mean that you cannot use existing measures though; for example, you might use existing questionnaires. It just means that your aim is not to reproduce a previous result adapted to your context. If you design your own investigation, you will need to carefully consider how you are going to ensure that it has internal validity. This means you will need to make sure your investigation is testing what it is designed to test! You will also need to ensure that your sample size is large enough. In addition, you may need to rerun the investigation to have confidence in the repeatability of your method (Figure 11.5). You will want to determine how likely it is that, having produced a result once, you can try the same investigation with the same setup, and produce the same result again.

Whatever topic you choose for your investigation, it has probably been studied before. You can use this to your advantage and look at previous experiments as a guide to start planning your experiment.

Figure 11.5 The difference between reproducibility and repeatability

This textbook is a great place to start your research. Begin by reading the section that covers your chosen topic. What are the key terms and theories it highlights? Does it mention any studies? Does it mention any experiments you could use? Check the sources in the References section (Chapter 14, pages 481–490). Go online and use your skills from Chapter 6 to research your topic. You will need to find two studies that relate to your topic. You will need to refer to these investigations in your introduction section later, when you write your report.

Comply with safety and ethical guidelines

When choosing and designing your investigation, you must ensure that you are adhering to the ethical concepts and ethical guidelines of psychological research. These are sets of concepts and guidelines that guide research designs and practices. **Ethical concepts** are considerations about how an investigation may affect human or non-human participants that need to be taken into consideration when you are designing your investigation. **Ethical guidelines** are considerations that have to do with your treatment of participants before, during and after your investigation. If you do not meet ethical standards, you will not be able to conduct your investigation (Table 11.3). Scientists and researchers must always follow a code of conduct when collecting data from people or animals. For example, to ensure that your research is ethical you must provide your participants with an informed consent form. Informed consent is discussed in Chapter 1. You will also need to carefully consider whether your investigation has potential to cause participants any physical or unacceptable psychological stress. If it does, you will need to choose another way to investigate your chosen topic. There are a number of other ethical considerations to take into account when conducting your research. You can find these explained in Chapter 1 (pages 38–40). These are all designed to remind you to:

» consider the effects of your investigation on living things and the environment
» apply integrity when recording and reporting the outcomes of your investigations.

If you need your participants to undertake some actions that may cause harm you will need to complete a risk assessment (Figure 11.6). Your school is likely to have one of these documents for you to complete. If you are unsure of either the ethical or the health and safety aspects of your investigation, check with your teacher.

Table 11.3 Ethical concepts and guidelines

	Yes/No
Ethical concepts	
Beneficence	
Integrity	
Justice	
Non-maleficence	
Respect	
Ethical guidelines	
Confidentiality	
Voluntary participation	
Informed consent procedures	
Withdrawal rights	
Use of deception in research	
Debriefing	

Use the checklist in Table 11.3 to ensure that your investigation is ethical.

See Chapter 1 for the description of each ethical concept and guideline.

Knowledge

I/we have read and understood the potential hazards and standard handling procedures of all the equipment, chemicals and living organisms.
I/we have read and understood the (Material) Safety Data Sheets for all chemicals used and produced.
I/we have copies of the (Material) Safety Data Sheets of all the chemicals available in or near the laboratory.

Agreement by student(s)

I/we, Bill Wilkins, Mary Newt, Christina Lee, agree to conduct this experiment safely in accordance with school rules and teacher instructions.

Risk assessment

I/we have considered the risks of:

fire	breakage of equipment	electrical shock	radiation
explosion	cuts from equipment	escape of pathogens	waste disposal
chemicals in eyes	sharp objects	heavy lifting	inappropriate behaviour
inhalation of gas/dust	rotating equipment	slipping, tripping, falling	allergies
chemicals on skin	vibration and noise	falling objects	special needs
runaway reaction	pressure	heat and cold	other risks

Assessment by student(s)

I/we have assessed the risks associated with performing this experiment in the classroom on the basis of likelihood and consequences using the School's risk matrix, according to International Organization for Standardization Standard ISO 31000:2009 and the Risk Management Guidelines, HB 436:2013.

I/we consider the inherent level of risk (risk level without control measures) to be:

Low risk **Medium risk** High risk Extreme risk

Control measures:
Always point test tube away from any person.
Add hydrochloric acid slowly and carefully to avoid vigorous reaction and projection of material from test tube.
Dip matches and tapers in water to ensure extinguished before disposal.
Additional measures: safety glasses, gloves

With the specified control measures in place, I/we have found that all the risks are "low risk". Risks will therefore be managed by routine procedures in the classroom, in combination with the specified control measures.

Certification by teacher

I have assessed the risks associated with performing this experiment in the classroom on the basis of likelihood and consequences using the School's risk matrix, according to International Organization for Standardization Standard ISO 31000:2009 and the Risk Management Guidelines, HB 436:2013. I confirm that the risk level and control measures entered by student(s) above are correct and appropriate.

Name: .... **Signature:** .... **Date:** ....

Certification by Laboratory Technician

I have assessed the risks associated with preparing the equipment, chemicals and living organisms for this experiment and subsequently cleaning up after the experiment and disposing of wastes, on the basis of likelihood and consequences using the School's risk matrix, according to International Organization for Standardization Standard ISO 31000:2009 and the Risk Management Guidelines, HB 436:2013.

I consider the inherent level of risk (risk level without control measures) to be:

Generated with Student RiskAssess (https://www.riskassess.com.au)

Figure 11.6 An example of a risk assessment sheet

Alina is adapting an existing experiment.

Step 1: Select a topic of interest	
Topic	Depth perception
Step 2: Do your background research	
Background research **Note: in this example only one study has been demonstrated.** **Repeat this section for one to two more articles depending on your assessment criteria.**	**Article 1:** The contribution of stereo vision to one-handed catching Reference citation: **Researchers:** Mazyn, L., Lenoir, M., Montagne, G., & Savelsbergh, G. J. P. (2004). https://research.vu.nl/ws/portalfiles/portal/2022889/173426.pdf **Aim:** To examine both monocular and binocular performance in a natural catching task in people with good and weak stereopsis Hypothesis: Since all participants were inexperienced in the monocular test condition, they are expected to perform better binocularly than in the monocular test situation. Approved by ethics committee? Yes **Method:** Methodology: Controlled experiment Participants: Two groups of nine students in physical education (three males and six females in each group) between 18 and 23 years of age with normal or corrected-to-normal vision. All but one of the participants were right-handed catchers. As students in physical education, all participants had experience in ball games such as basketball, volleyball, handball, soccer and tennis. **Materials:** • ball-projection machine • yellow tennis balls **Procedure:** All participants were tested in two sessions with a 1-week interval. One session was executed under a normal binocular viewing condition, while in the other session the balls had to be caught monocularly. The monocular-binocular order was randomised over participants. In the monocular condition, they caught with their preferred eye, while the other eye was covered with an eye patch. In each session, the participants had to catch 90 tennis balls with their preferred hand, in three blocks of 30 for each velocity condition. The sequence of ball speeds was randomised for all sessions to exclude effects of presentation order. Before each block of 30 balls, 10 acclimatisation trials were provided. The first 10 catches in every session were recorded. After a successful catch, the ball was dropped into a basket and the hand returned to the initial position. In case of a failure, the hand was returned into the initial position straight away. The outcome of each trial was immediately registered by one of the researchers as a catch or miss. **Results:** Catchers with good stereopsis showed an increase in catching performance when catching binocularly compared with monocular viewing Whatever experimental condition, participants with weak stereopsis seemed to depend only on monocularly available information to time their catch.

Step 3: Pose your research question and identify your aim

Once you have decided on your investigation topic you will need to construct your research question. Research questions should be written as an open-ended question and usually begin with how, what, when, who, which or why; for example, 'What role do monocular and binocular depth cues play in allowing people to accurately judge distances?'. For further information, refer to Chapter 1.

Now you will establish the aim of your experiment. The aim should state the purpose of the experiment without providing a prediction: for example, 'To investigate whether depth cues affect depth perception'.

Step 4: Formulate your hypothesis and make predictions

The hypothesis is your proposed answer to your research question. It should be formed from a strong rationale that is based on your background research and should also be consistent with it.

In our example on depth perception, you might hypothesise that 'Using monocular depth cues will decrease a person's depth perception ability'. From this you would develop a prediction: an outcome you expect to observe if your hypothesis were correct. A prediction is often written in the form of an 'if, and, then' statement. For example, 'If my hypothesis is true, and I were to do this test, then this is what I will observe.' You could predict that, '**If** the use of both binocular and monocular depth cues allow better accuracy of depth perception, **then** when participants are restricted to the use of just monocular depth cues, they will perform worse on depth perception tasks.'

Writing a hypothesis for a controlled experiment

If you are conducting a controlled experiment, your hypothesis will also generate specific predictions of what you expect for the variables in your experiment. The predictions for your experiment must include:

» the population of research interest
» clearly defined independent and dependent variables
» how these variables will be measured
» the direction or strength of their relationship.

To formulate your hypothesis you will need to:

» *Identify the independent variable (IV) that will be manipulated.* From Chapter 1, we know that the IV is the variable that is manipulated or changed by the researcher. This variable is assumed to have a direct effect on the dependent variable. In the previous example, the variable being manipulated is depth cues.
» *Think about how you will measure the IV.* How are you going to manipulate this? What is going to be the difference between your two groups? In this example, the IV (depth cues) is going to be manipulated in the following way: participants are going to perform a task with either the use of one eye (using an eye patch for monocular cues only) or with two eyes (able to use both monocular and binocular vision cues).
» *Identify the dependent variable (DV).* This is the variable the researcher measures that is assumed to be affected by the independent variable. What are you going to measure? In our example involving depth cues, the DV is depth perception
» *Think about how you will measure the DV.* In the depth cues example, we are going to measure depth perception by how long it takes for participants to successfully throw and catch 10 times.
» *Predict the direction or strength of the relationship between the IV and the DV.* You make a prediction based on evidence provided in previous research. For example, after reading previous research into depth perception and depth cues you would predict that the ability to perceive depth accurately will decrease with the use of only monocular vision.

To make it easier, we generally structure a hypothesis in the following way:

It is hypothesised that population – independent variable (experimental group) – prediction of strength/direction – dependent variable – independent variable (control group).

For example: It was hypothesised that people with the use of monocular depth cues will have less accurate depth perception compared to when they use binocular depth cues.

11.1.1 WRITING A TESTABLE RESEARCH QUESTION

If you are conducting a controlled experiment, you will need to do another step!

Student check:
✓ I have identified, researched and constructed a research question and aim for my investigation.
✓ I have formulated a hypothesis to focus my investigation.
✓ If conducting a controlled experiment, I have identified the independent and dependent variables.
✓ I have predicted a possible outcome for my investigation.

at this stage, it is handy to think about how your variables will be measured

11.1.2 WRITING A RESEARCH HYPOTHESIS

Step 3: Pose your research question and identify your aim	
Research question	What role do monocular and binocular depth cues play in allowing people to accurately judge distance and motion of a moving object?
Aim	To determine whether monocular or binocular depth cues are superior accurately judging distance and motion of a moving object
Step 4: Formulate your hypothesis and make predictions	
Hypothesis	Identify the independent and dependent variables. How you are going to manipulate the independent variable? How you are going to measure the dependent variable? IV: The variable being manipulated is the depth cues Measure IV: Participants will use one eye (monocular) or two eyes (binocular) DV: Depth perception Measure DV: Record how long it takes for participants to throw and catch a tennis ball 10 times against a brick wall that is 3m away. It was hypothesised that people who had the use of binocular depth cues will have better depth perception than when restricted to monocular depth cues.

You do not need to explain how you are going to measure your variables in your hypothesis but it is handy to think about this now.

KEY CONCEPTS 11.1a

- The scientific method is a standardised series of steps that researchers follow to conduct a scientific investigation.
- When you are searching for relevant research to include in your investigation, it is essential that you critically evaluate each source for validity, reliability and integrity.
- When formulating a hypothesis, make sure to include the population of research interest, clearly defined independent and dependent variables, and the direction of their relationship.
- When designing and conducting any investigation you must apply ethical concepts and guidelines.
- When conducting a study, you must apply occupational health and safety guidelines where relevant.

Concept questions 11.1a

Remembering

1 What are the two key objectives that the ethical concepts and guidelines aim to achieve? r

2 What is the purpose of a research question and what is it based on? r

3 What is a hypothesis? r

Understanding

4 Why do scientists use logbooks? List eight items you should record in your logbook. r

Applying

5 A Year 11 student is conducting an experiment to investigate the effect of loud music on people's ability to pay attention when completing a simulated driving test. Their hypothesis is: c I think that the people who perform better on the driving test will be the ones in the 'no music playing' group, and the people who have music playing will be worse. Rewrite this hypothesis using the structure shown below.
It was hypothesised that population – IV–prediction – DV – IV

HOT Challenge

6 More culturally inclusive research is now being conducted that is challenging the idea that one theory can be generalised across all cultures. 'Does language matter? Exploring Chinese–Korean differences in holistic perception' is an example of this research. This example is a cross-cultural study comparing Korean and Mandarin speakers'

relative bias to focus on figure or ground in images, and the effects on memory for images. Access the weblink to read the article. In less than 100 words, use the structure provided below to summarise this article. Include the following: aim, sample, method, findings and one identified limitation. **C**

Aim	
Sample	
Method	
Findings	
Limitations	

Weblink
Does language matter?

Step 5: Plan your investigation

Now it is time to plan your scientific investigation in a way that tests your hypothesis. This involves more than just asking questions or conducting tests and recording information in your logbook.

To begin, you need to decide on the most appropriate **methodology**. You need to decide who your participants are and how they will be selected. Then you need to design your specific methods and procedures. This involves writing all the steps of your planning in detail in your logbook.

Choosing your methodology

There is a variety of methodologies from which you can choose (summarised in Table 11.4).

11.1.3 WHAT DESIGN?

Table 11.4 Some of the methodologies you can choose from

Methodology	Type of research it is used for	Type of research question and hypothesis it could generate
Classification and identification	Classification of mental illness (*DSM-5-TR*™), or classifying culturally based thinking styles, in terms of holistic vs analytic (Eastern vs Western ways of thinking) or collectivist cultures vs individualist cultures Identification could include different attachment behaviours in young children.	Example: When surveying a class of students, can we find evidence of people who can be classified as dominantly analytic or holistic thinkers/perceivers? Research question: Can certain kinds of behaviour or kinds of people be identified with respect to thinking style? Hypothesis: There are two distinct kinds of thinkers/perceivers in the world: analytic types and holistic types. The prediction for the actual study will then be based on this hypothesis.
Controlled experiment	Usually conducted in a controlled environment (e.g. a lab); investigates the relationship between an independent variable and a dependent variable; controls all other variables	Example: A class of students is investigating the impact of colour intensity on the perception of flavour intensity. Research question: Can the intensity of colour affect a person's perception of flavour intensity? Hypothesis: People who are given a drink with a weak colour intensity will perceive the drink as having a weaker flavour than when they are given the same drink with a stronger colour intensity. Note: This is an experimental hypothesis, where the IV and DV have been identified.
Correlational study	Used when researchers want to understand if there is a relationship (association) between two variables. For example, how might one trait or ability or behaviour relate to performance on some measure, or to some other trait or ability? This includes positive and negative correlation, and no correlation or relationship (see Chapter 1, page 15). These studies may also be used to identify which factors may be of greater importance (i.e. more strongly or weakly correlated).	Example: I might be interested in the relationship between people's intelligence and their attitude towards climate change (the extent to which they believe in human-induced climate change). Research question: Are people of higher intelligence more likely to believe in climate change? Hypothesis: There will be a strong positive correlation between intelligence and belief in climate change. The prediction for the actual study will then be based on this hypothesis.
Fieldwork	Usually takes place in a natural or real-world setting, such as a shopping centre. The data collection methods for fieldwork research are varied. Data can be qualitative and/or quantitative. The methods could include a mix of surveys, interviews, questionnaires, case studies, focus groups, yarning circles or participant observations. Involves observing and interacting with a selected individual or group of individuals.	Example: I might be interested to see if encouragement leads to more prosocial behaviour in primary-school-aged children. The data for this could be collected a number of ways; for example, by surveys, interviewing the children or observing them in a variety of settings, such as at home, in a classroom or in a schoolyard. Research question: Do children demonstrate more prosocial behaviours with encouragement? Hypothesis: Primary school children will display more prosocial behaviours when encouraged to do so. The prediction for the actual study will then be based on this hypothesis.

Each type provides a framework for the overall shape of your research. Refer to Chapter 1 (pages 14–20) to read about each methodology in more detail to help you decide which best suits your research. You could also revisit the published research you examined as part of your research for this investigation. What methodology did the researchers use? If they have used a different name for their methodology but recorded its key attributes, revisit Chapter 1 to identify what it is likely to be. Alternatively, you may have chosen a study and want to adapt it to another methodology; for example, the original investigation was based on fieldwork but you choose to adapt it to a controlled experiment.

The best methodology to test the stated hypothesis in our example is a controlled experiment. This is because we want to investigate whether one variable has an effect on another variable. In other words, we want to manipulate the independent variable (depth cues) to see if it has an effect on the dependent variable (depth perception).

11.1.4 SAMPLING PROCEDURE

Once you have identified the best methodology to be used to answer your research question, you need to move on to selecting the participants you are going to use in your experiment.

Identify the population and the sample

Participants are the people you use in your study and it is their responses (data) that you analyse. You must decide what or who the population of interest is. Who do you want to apply your result to? For example, everyone in Australia, your school, or just your class? Or will you focus on people from a specific demographic, region or background? The more precisely you define your population, the easier it will be to gather a representative sample.

Once you have decided on your population you will need to work out *who* will make up the sample. Remember, one of your aims is for your sample to represent your population as closely as possible so you can generalise your results later on. For example, you may decide that you are going to focus on VCE students. It is not possible to test every VCE student in Victoria so you are probably going to take a sample from within your school, or maybe even within your Psychology class. Although it may be difficult to achieve, you should aim to include a diverse range of participants in your study. The more closely your sample represents the overall human population, the more external validity your investigation will have.

You will need to identify both the population and the sample in your logbook.

Sample size

You will not have the resources or the time to test thousands of people in your study. Even with a narrowly defined population, it is rarely possible to collect data from every individual. Instead, you will collect data from a sample. A sample of 20–30 participants for this type of study will be suitable. If you have fewer than 20 participants, you may have problems with reproducibility and random errors.

Sampling procedure

The next step is to think about *how* you are going to sample your participants. This is often spoken about as your sampling procedure. As stated in Chapter 1, the basic sampling procedures are as follows.

» **Random sampling**: If you have access to every member of the population then this would be a good method to use, because it increases the likelihood of the sample being representative of the population. This means that every person in the population has an equal chance of being selected for the study.
» **Stratified sampling**: If you have a variation that you would like to keep consistent from the population to the sample then you could consider this sampling procedure. The population is divided into relevant categories beforehand and then participants are randomly selected based on equivalent proportions of the population. This sampling technique ensures that the sample group contains the same proportions as found in the population.
» **Convenience sampling**: This is the sampling procedure that is typically used in investigations such as the one you will be conducting. This is where participants are selected because they are readily available and known to the researcher. For example, if you are in remote learning and need participants, you are likely to select them because they live with you or are known to you.

What type of data will you use and how will it be generated?

You may have decided to do fieldwork where you are going to conduct a survey. What are your survey questions going to look like? Will responses

Table 11.5 The differences between quantitative and qualitative data collection methods. You can use this information in your discussion section when you look at limitations of your investigation.

Qualitative approach	Quantitative approach
Used to understand subjective experiences, beliefs and concepts	Used to measure variables and describe frequencies, averages and correlations
Semi-structured or unstructured techniques such as open-ended questions	Fixed and more structured techniques such as closed-ended questions
Not instrument or test based	Instrument or test based
Information is collected as text	Information is collected as numbers
Not used for statistical tests	Used for statistical tests

Source: Quantitative and Qualitative Research Approaches, http://kwangaikamed.weebly.com/data-collection-analysis–interpretation.html

be represented as numbers; for example, a rating scale? Or will you use open-ended questions or descriptions of what you observe? You must make sure your data collection method allows you to test your hypothesis. Your choice of method will be based on whether you will be collecting qualitative or quantitative data as part of your research (Table 11.5).

Qualitative data refers to any information that is observed and recorded using words (that is, it is not presented in numerical form). It can be either written or verbal. Open-ended questions in questionnaires or interviews, such as asking someone to describe a behaviour, is an example of qualitative data. If you decide to use a qualitative survey, for example, revisit the investigations you used in your background research. If you look at their appendix section, you should be able to work out how they collected their data (that is, what questions they used).

Quantitative data refers to any information that is observed and recorded in numerical form (numbers and statistics). This could include using a tally of behavioural categories and closed questions in a questionnaire, or timing how long it takes a person to complete a task.

Participant allocation in controlled experiments

If you are conducting a controlled experiment then when you wrote your hypothesis you would have already worked out how you are going to manipulate the independent variable and how you are going to measure the dependent variable. Now you need to decide how participants will be allocated to various groups. To do this you will need to select the experimental research design that is best suited to your investigation. You will recall from Chapter 1 that your options include:

» **between-subjects design** (also known as **independent groups**)
» **within-subjects design** (also known as **repeated measures**)
» **mixed design** (a combination of both within subjects and between subjects design).

11.1.5 MEASURING VARIABLES

Between subjects

Each participant belongs to one group only and is not exposed to the other condition. In other words, they are either in the experimental group (where the independent variable is present) or the control group (where the independent variable is absent). In our previous example on depth cues, the control group would be participants who were not restricted to the use of monocular cues. This is because, generally, people view objects or scenes using two eyes. Participants who were restricted to using only monocular vision (one eye) would be the experimental group. If you choose this design, you will need to randomly allocate your participants to either the control group or the experimental group.

If you are conducting a controlled experiment you will need to do one more step! If not, you can skip this part!

Within subjects

Participants are exposed to both the experimental condition and the control condition. In our example, this would mean that the participants would record a time throwing the ball with the use of both eyes and the use of just one eye. A within-subjects design is used when the researcher believes that individual participant differences must be controlled.

Mixed design

In this design, all your participants would participate in both the control and experimental condition, just as they would in a within-subjects design. However, you may decide that there is an additional factor that might affect your dependent variable which you want to investigate. In our depth cue example, it could be the participants' sporting ability. All your participants would complete the control and experimental condition (that is, binocular and monocular throwing) but you would then include an additional experimental group; for example, participants who play elite cricket and those who do not play any sport.

For example, you could set up three groups:

» control group: binocular vision all participants
» experimental group 1: monocular vision people who play elite cricket
» experimental group 2: monocular vision people who play no sport.

This then allows you to not only compare binocular and monocular cues (comparing the experimental group to the control group), but also to see if there is a difference between those who play elite sport and those who do not (comparing the two experimental groups).

KEY CONCEPTS 11.1b

» To describe your sampling method, include details of the population and the sample, the sample size and the sampling procedure being used.
» Primary data can be either qualitative or quantitative.
» The methodology is the broad framework of a study, and includes classification and identification, controlled experiment, correlational study and fieldwork.
» Once a research population is chosen, the sample size and sampling procedure can influence the validity of the research.
» In controlled experiments, participants may be allocated to groups using between-subjects, within-subjects or mixed study designs.

Concept questions 11.1b

Remembering

1 What is a controlled experiment? r
2 Name three types of sample selection and explain their key features. r
3 List three differences between qualitative and quantitative data. r

Understanding

4 Why is it important that a sample represents a population as much as possible? e
5 Use an example to demonstrate the difference between within-groups and between-groups designs. c

Applying

6 Ava works in marketing and has been asked to conduct research to see people's responses to a new drink line her company had just released. What type of methodology would be most appropriate for this type of research? Explain.

HOT Challenge

7 A student has decided to investigate the impact of smell on flavour perception. They recently had a severe cold and as a result, not only were they unable to smell, but they noticed that there was very little flavour in the food they ate at the time. To investigate, they asked their participants to complete two trials.
In one trial they asked participants to cover their eyes with a blindfold and then eat a skittle and say out loud what flavour they thought it was.
In the second trial they asked participants to cover they eyes with a blindfold, block their nose with one hand, eat a skittle and say out loud what flavour they thought it was.

The results are shown below

Participant	Blindfolded with nose blocked Able to correctly identify the flavour (Y/N)	Blindfolded with no nose blocked Able to correctly identify the flavour (Y/N)
1	N	Y
2	N	Y
3	N	Y
4	Y	Y
5	N	N
6	N	Y
7	N	Y
8	N	N
9	N	Y
10	N	Y

1 Write an aim for this investigation
2 Construct an appropriate hypothesis for this investigation based on the students past experience
3 Using the raw data above, analyse it using the most appropriate descriptive statistic
4 Create a graph to display your descriptive data

Step 5: Plan your investigation

Sampling methods

1 What is your population and sample?
Population: people
Sample: People between the ages of 14 and 48 who are known to me from school
2 How many participants will be involved?
10
3 How will participants be selected from the nominated population?
Convenience sampling

What type of data will be generated?

Quantitative

How will participants be allocated?

Between groups

How will you organise your results (raw data)?

Table 1. Raw scores for time taken to reach 10 successful catches in two eyes used (binocular) and one eye covered (monocular) conditions

			Time taken to reach 10 successful catches (00.00) (sec)	
Participant	Age	Gender	Trial #1 (both eyes used)	Trial #2 (one eye covered)
1				
2				
3				

»

<table>
<tr><td>How you will analyse your results?</td><td>Calculate mean scores of:
– Binocular average time
– Monocular average time</td></tr>
<tr><td>How will you display your analysed data?</td><td>Table 1. Mean scores for time taken to reach 10 successful catches in two eyes used (binocular) and one eye covered (monocular) conditions

<table><tr><td colspan="2">Time taken to reach 10 successful catches (00.00) (sec)</td></tr><tr><td>Trial #1 (two eyes used)</td><td>Trial #2 (one eye covered)</td></tr><tr><td></td><td></td></tr></table>
Time (sec): 0, 5, 10, 15, 20, 25, 30, 35, 40, 45, 50, 55
Binocular | Monocular

Figure 1 Mean scores for time taken to reach 10 successful catches in two eyes used (binocular) and one eye covered (monocular) conditions</td></tr>
<tr><td>Materials</td><td>Flat wall surface
Tennis ball
Measuring tape
Masking tape
Blindfold
Stopwatch
Cones</td></tr>
<tr><td>Procedure</td><td>1 Measure out 3 metres from a flat wall surface and mark with a cone.
2 Get participant to stand at the cone with the ball in their preferred hand.
3 Instruct participants that they need to throw and catch the ball 10 times and that they will be timed.
4 Do step 3 using both eyes, then with one eye covered. Record results.</td></tr>
<tr><td>Ethical and safety check</td><td>Yes teacher signature</td></tr>
<tr><td>Systematic error check</td><td>Yes</td></tr>
</table>

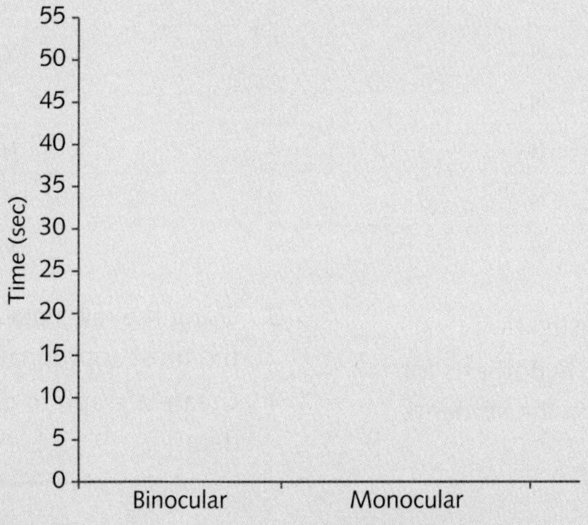

Step 6: Organise and summarise your data

Organising your results

If you are using qualitative data, you may already have a survey or a questionnaire ready to go. But if you are using quantitative data, you need to create a suitable table to record raw data.

Deciding how you will display your data

You will need to display your analysed data in two forms: a table of the descriptive statistics that you have calculated and a graph (typically either bar or line depending on your data).

Setting up your table

Your results table is usually a simpler version of the table you used for your raw data. You will need to include your independent variable but instead of having all your raw data, it will have the descriptive statistic information you have calculated. See Table 11.6 for an example.

Setting up your graph

Your graph(s) will display the descriptive data that you have put into your results table (Figure 11.7). Refer to Chapter 1 to determine the type(s) of graph(s) you will use. Remember to:

» label your graph with the same descriptive title as your table; for example, Mean number of times ball is dropped by control and experimental groups at varying distance of throw. Label it a 'Figure' and put the figure number below the graph; for example, Figure 1
» choose a suitable scale for your axis, so that your graph takes up about half a page
» label your axes clearly, including any units used
» for a controlled experiment, place the independent variable on the horizontal axis (the 'x-axis') and the dependent variable on the vertical axis (the 'y-axis')
» if drawing by hand, use a pencil and a ruler.

Table 11.6 An example of a results table, showing the mean number of times a ball is dropped by control and experimental groups at varying distance of throw

Distance of throw (m)	Mean number of times ball is dropped		
	Control	Group 1	Group 2
1	0	0	2
5	0.6	1.9	2.5
10	1.3	2.3	3.7
15	1.9	2.9	4.8

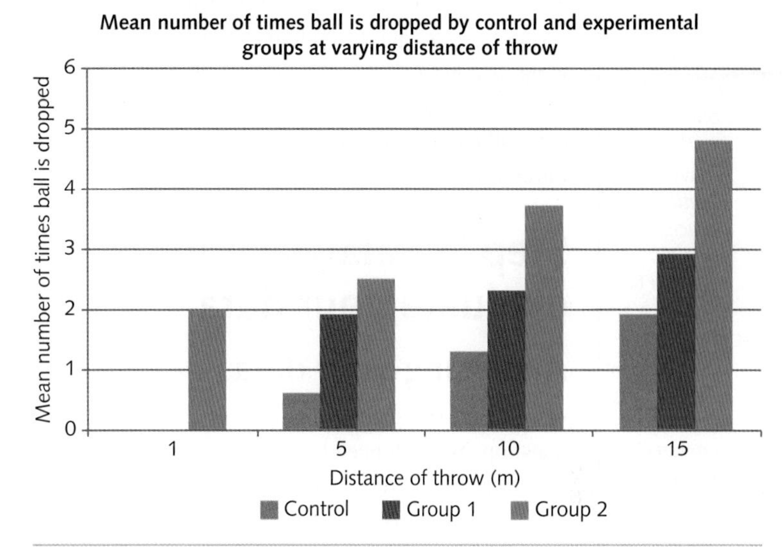

Figure 11.7 A well-designed graph that shows a clear trend

Recording the materials

11.1.6 PRESENTING SUMMARISED DATA

List in your logbook all the materials you used in your investigation, including specific amounts. If possible, use or adapt existing materials designed from one of the studies you are basing your investigation on. Make sure you include any questionnaires or inventories where the reliability and validity has already been established.

Your record needs to include:

» the equipment you used in your chosen investigation, if applicable; for example, three tennis balls, measuring tape, stopwatch
» what you used to record your results; for example, the table or questionnaire you used to collect your data (you will need to keep a copy of this for your appendix when you write your report, so stick one into your logbook for later).

Writing up the procedure

Record in your logbook, in detail, the exact steps you are going to take to run your experiment. You must document specific step-by-step instructions to ensure you carry out the same steps in the same way for each participant. This also means you collect your data in a consistent way and helps to strengthen the repeatability of your data. If you do not structure your procedure carefully, you may create problems with internal validity.

Student check:
- ✓ *I have outlined the selected methodology and data collected to be used.*
- ✓ *I have identified my participants and how they will be used.*
- ✓ *I have outlined how relevant data will be generated.*
- ✓ *I have outlined how relevant data will be analysed.*
- ✓ *I have followed all ethical and safety checks.*

If this happens, no conclusions can be drawn. By carefully recording your procedure in your logbook you provide all the information you need for writing the procedure section of your report. This will also allow you to check to see if there are any systematic errors in your data, if you find the results don't make any sense. If you are conducting a controlled experiment, don't forget to check that you are controlling all your other variables (eg. be mindful to minimise any extraneous variables).

Step 7: Analyse and evaluate your data

Analysing your data

On its own, raw data cannot answer your research question. You will need to summarise and analyse the data to reveal patterns and relationships. To do this you will use descriptive statistics. There is a variety of statistics you can apply to your raw data, so you will need to choose which one is the most appropriate for your investigation. How to apply descriptive statistics is covered in Chapter 1 (pages 31–33); however, as a general rule you will need to apply one or more of the following descriptive statistics:

» calculate percentages or a percentage change
» apply measures of central tendencies (mean, median, mode)
» find out how spread out your data are (standard deviation).

You will also need to consider how you are going to analyse:

» the accuracy, precision, repeatability, reproducibility and validity of your results
» whether there are any obvious errors in the data
» the effects of sample size on the quality of data you have obtained.

Systematic errors affect measurements by the same amount or by the same proportions (provided the reading is taken in the same way each time). Systematic errors affect the accuracy of your results. They are predictable: whatever the error is, it will be present in the same amount across all your data.

Causes of systematic errors

Systematic errors can be the result of:

» observational mistakes; for example, the researcher incorrectly records a catch as 'successful' even if the ball bounces first
» poor calibration of or a faulty instrument; for example, a stopwatch has low battery and readings are 0.5 seconds out each time
» environmental factors; for example, the researcher conducts their investigation with participants looking into the Sun, which makes it difficult for them to see the ball.

How to reduce systematic errors

Because systematic errors are consistent and predictable they can be reduced to minimise their impact. Some ways to do this are:

» be familiar with the **limitations** of your equipment and check that it is working properly
» know how to use your equipment properly
» try to eliminate any undesirable variables at the start of your investigation. These are also known as extraneous variables. You have covered extraneous and confounding variables in Chapter 1 (page 23), but refer to Table 11.7 as a quick guide.

When conducting your investigation, you must be careful to avoid random errors. These are errors that are caused by unpredictable changes *during* an investigation, while you are collecting the data. A random error can cause one measurement to differ slightly from the others. For example, while

Table 11.7 Extraneous variables and how to minimise them

Extraneous variable	How to minimise it
Use of non-standardised instructions and procedures	Use standardised instructions and procedures
Experimenter effect	Double blind
Placebo effect	Single blind
Individual participant differences	Experimental designs: » between subjects » within subjects » mixed designs
Order effect	Counterbalancing
Sampling bias; e.g. convenience sampling	Random or stratified sampling

measuring a person's weight, they may stand on the scales slightly differently each time and this will give a slight difference in the weight recorded on the scales each time. A random error means each measurement will be a variation of the true value (if we could take the person's weight exactly in this case). Clearly, random errors affect the precision of your data collection.

If your results are very inconsistent, you may need to go back, rework your procedure and think about repeating the experiment if necessary to ensure you are being more precise in your data collection.

Remember, these are errors that can happen *while* you are collecting your data.

Causes of random errors

Random errors can be caused by:

» a limitation of the instrument; for example, the stop button on the stopwatch is faulty and sometimes gets a bit stuck

» a limitation of the environment; for example, the experiment is done outside and there are gusts of wind

» a slight variation in procedures; for example, the researcher gives the participants slightly different instructions, so some are very clear on what they have to do and some are not too sure.

How to reduce random errors

Random errors always occur to some extent. You cannot eliminate them entirely from your experiment, but there are ways to reduce them. These include:

» taking multiple data points and then averaging them to gain an idea of the amount of variation and an estimate of the true value

» increasing your sample size so that you can then take an average of the data points.

Outliers

An outlier refers to a result that lies a long way from other results. Outliers may occur by chance, indicating that there are measurement or recording errors, a skewed distribution or an extraneous variable occurring. If you do see any outliers in your results, think about how it may affect the validity of your research. If you have one or more large outlier(s) that will affect the average too much, you may decide to take them out of your data completely. This will mean that your descriptive statistics will be more accurate. If this occurs, you need to find the average (mean) of the remaining data points and substitute this value into where your outlier was.

Missing or incomplete data

If you are doing a survey, did someone miss a question? Did your timer stop and you didn't get an accurate reading? Did the computer fail to register a response? If this is the case, you can remove that participant entirely. If there is a lot of missing data, it would be best to increase your sample size by recruiting more participants. You could also use the mean substitution method described under 'Outliers' above.

When analysing and evaluating your data, you need to think about the following questions.

» What do the results of your descriptive statistics say about your investigation?

» Do they answer the research question and support your hypothesis?

» Does the difference between one measurement and another indicate a real change in what is being measured? For example, are the percentage differences enough to show that there are differences between groups, or do your data show a strong correlation between two variables? Some typical approaches to interpreting the data include:
  » identifying any general patterns or relationships among the data
  » deciding whether there is a significant difference between your groups
  » explaining unexpected results and **evaluating** their significance
  » examining the results to determine whether they met your predictions or supported your hypothesis.

The data you collect are raw data (the actual data you collect from a study before they are sorted or analysed). Remember, for the duration of your data collection, it is really important that you follow your method section closely.

11.1.7 UNDERSTANDING ERRORS

Uncertainty and validity

Before you can draw any conclusions, you need to consider the question, 'Can I be certain the data I have generated are valid when drawing conclusions?'

The validity of a psychological investigation refers to how well the results among the study participants represent true findings among similar individuals outside of a study. There are two types of validity that you will need to consider. If you decide your research investigates what it sets out to do and/or claims to investigate, then you have **internal validity**. To make this decision you need to evaluate the appropriateness of the investigation design, sampling and allocation techniques, as well as the impact of any errors you may have identified. Lack of internal validity implies that the results of the study do not represent what you set out to test. Lack of internal validity means no conclusions can be drawn.

External validity occurs if the results of the research can be applied to similar individuals in a different setting. Lack of external validity implies that the results of the research may not apply to individuals who are different from your investigation population. External validity can be increased by using broad inclusion criteria and sampling techniques that result in your investigation population more closely resembling the overall general population.

When evaluating research, you need to decide what level of uncertainty there is within your results. You need to determine what your results tell you. Likewise, if there were too many errors in how you set up your experiment or how you collected your results, it is unlikely you will be able to generalise your results or draw any conclusions. Even the best research has some limitations, so even if your investigation has some, it has not failed. However, it is important to think about how these limitations have affected your results. These effects could come from a wide range of factors ranging from your methodology choice to any problems you hadn't anticipated. You only need to mention the ones that you think have directly impacted your investigation. Then, you need to evaluate how much impact they had on achieving the aim of your research.

Precision

The precision of your results refers to how close your measured values are to each other. If they are close, then you have high precision. If you think your precision is low, this is likely to mean you have random errors affecting your results. With low precision, your internal validity should still be good but your external validity will likely be low. Some solutions to this problem include:

» aim to reduce discrepancies in measurements
» increase your sample size; a small sample size will often not give enough data for generalisations
» average your results so that any minor errors in your measurements can be combined to create a 'typical score'
» exclude outliers so that they do not drastically skew your data
» substitute any missing or incomplete data with the average score that you have calculated.

Accuracy

The accuracy of results reflects how close the measured values are to the true value. If accuracy is low, then you probably have some systematic errors in your measurements. Unlike precision, accuracy cannot be improved by obtaining more measurements or results. Increasing the sample size will not help, because *how* you are measuring the data is the issue. If accuracy is low, internal validity (and, as a consequence, external validity) will be low. That is, you are probably not testing what you want to test so you cannot make generalisations. If you are conducting an experimental investigation, the way you are measuring the dependent variable may be the problem affecting the accuracy of the data. How much you can reduce the uncertainty about the research question is directly affected by this.

To reduce the problems of accuracy, you should revisit the method section and minimise systematic errors. Most systematic errors can be reduced by making sure you are familiar with the limitations of instruments and experienced with their correct use. This in turn will help ensure you are collecting data accurately. This should also help to reduce uncertainty in your results.

A summary of precision and accuracy can be seen in figure 11.8.

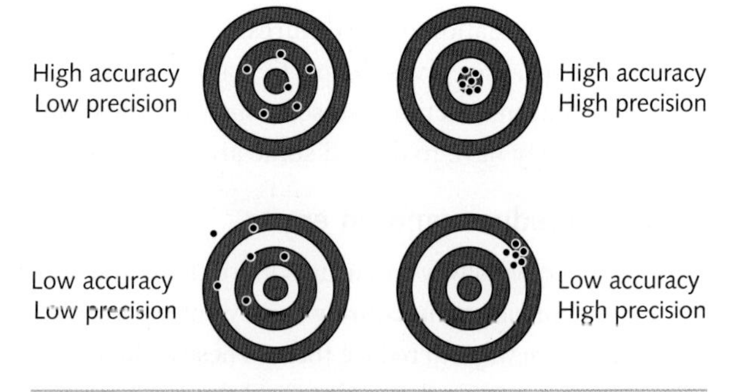

Figure 11.8 **A representation of the precision and accuracy of measurements, where the bullseye of the dartboard represents the true value**

Repeatability

You can also evaluate your investigation in terms of repeatability. **Repeatability** refers to how confident are you that you would get the same results if you carried out multiple trials within your investigation under the same conditions. In the multiple trials, you would use the same measurement procedure, the same measuring instrument(s), the same experimental conditions and the same location. High repeatability is a good indication of high accuracy in your data collection and can indicate a low level of systematic errors.

Reproducibility

When you evaluate the reproducibility of your results, you assess whether your investigation produced similar findings to previous studies under changed conditions. For example, it is most likely that you used a different measuring instrument, and that you conducted your investigation at a different location, at a different time and using a different sample of participants. If the results of your investigation are similar to previous investigations (using the same method) then you would say there is high reproducibility. If you found this, it would also mean that the original study on which you based yours is replicable (it can be replicated with similar findings). If your findings are vastly different, then you would say there is a low level of reproducibility and that the original study is irreplicable and the study's results may lack credibility.

Summary of how to evaluate your data

Some typical approaches to evaluating data include:

» consider the level of precision and likely accuracy of measures
» evaluate the level of internal and external validity
» evaluate the level of repeatability
» evaluate the level of reproducibility (that is, can you compare your results with the previous work that you are replicating or basing your investigation on?)
» evaluate the level of uncertainty: contextualise your findings within previous research and theory
» revisit your sample and decide how much it represents your population. Is your sample big enough? How much can you generalise your results?

By this stage, you evaluated results extensively. Your final step in evaluation is to consider if there is any remaining uncertainty.

An overview of Step 6 and 7 is shown in Figure 11.9.

Student check:
✓ I have presented the generated data/evidence in an appropriate format.
✓ I have provided an explanation of the trends, patterns and/or relationships shown in the data/evidence.
✓ I have provided an interpretation of the analysed primary data.
✓ I have provided an evaluation of the analysed primary data.
✓ I have identified one or more limitations in the data and/or methods.
✓ I have suggested improvements to address the limitation/s identified.

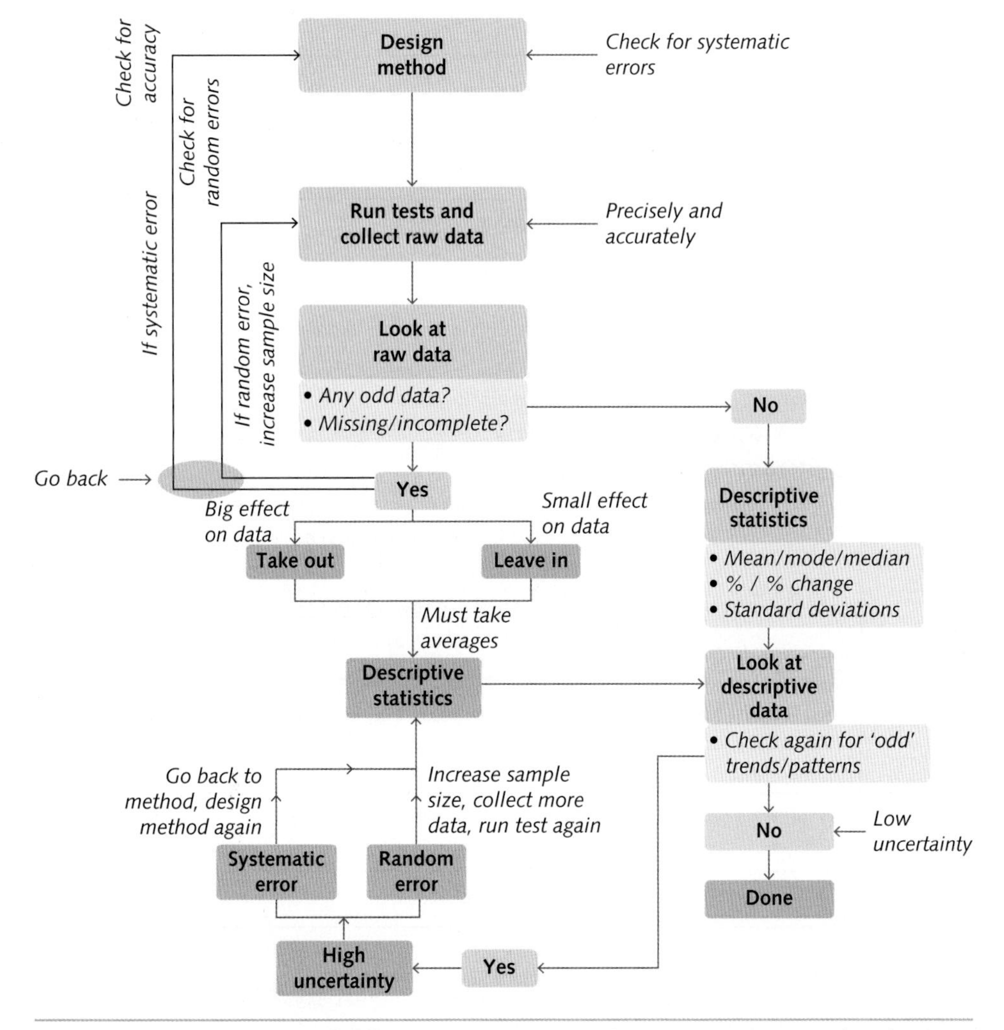

Figure 11.9 Concept map of collecting, organising, analysing, interpreting and evaluating your data

Steps 6 & 7: Organise, summarise, analyse and evaluate your data

Collecting the data

In this example the student has used quantitative data

Table 1. Raw scores for time taken to reach 10 successful catches in both eyes used (binocular) and one eye covered (monocular) conditions

			Time taken to reach 10 successful catches (00.00) (sec)	
Participant	Age	Gender	Trial #1 (both eyes used)	Trial #2 (one eye covered)
1	14	Female	14.41	29.49 *
2	42	Female	14.41	29.49 * oops looks like the same recording has happened twice
3	48	Male	12.35	25.58
4	42	Female	16.41	30.45
5	18	Female	21.23	35.22
6	19	Female	30.05	24.99* error or unusual result?
7	21	Female	11.78	26.44
8	20	Male	13.97	27.86
9	16	Male	10.7	24.89
10	47	Female	24	116* error or unusual result?

Random error check

See annotations in the table above

Analyse the data using descriptive statistics

1 Check for outliers.
Participant 10 looks like an outlier or an error. Will take out and substitute the average score here.
2 Check for missing or incomplete data.
No missing or incomplete data.
3 Generate descriptive statistics.

Mean monocular time: $\frac{\text{add the value of each score}}{\text{Total number of scores}} = 28.10$ sec

Mean binocular time: $\frac{\text{add the value of each score}}{\text{Total number of scores}} = 17.37$ sec

Descriptive data table

Table 1. Mean scores for time taken to reach 10 successful catches in both eyes used (binocular) and one eye covered (monocular) conditions

Time taken to reach 10 successful catches (sec)	
Trial #1 (both eyes used)	Trial #2 (one eye covered)
17.37	28.10

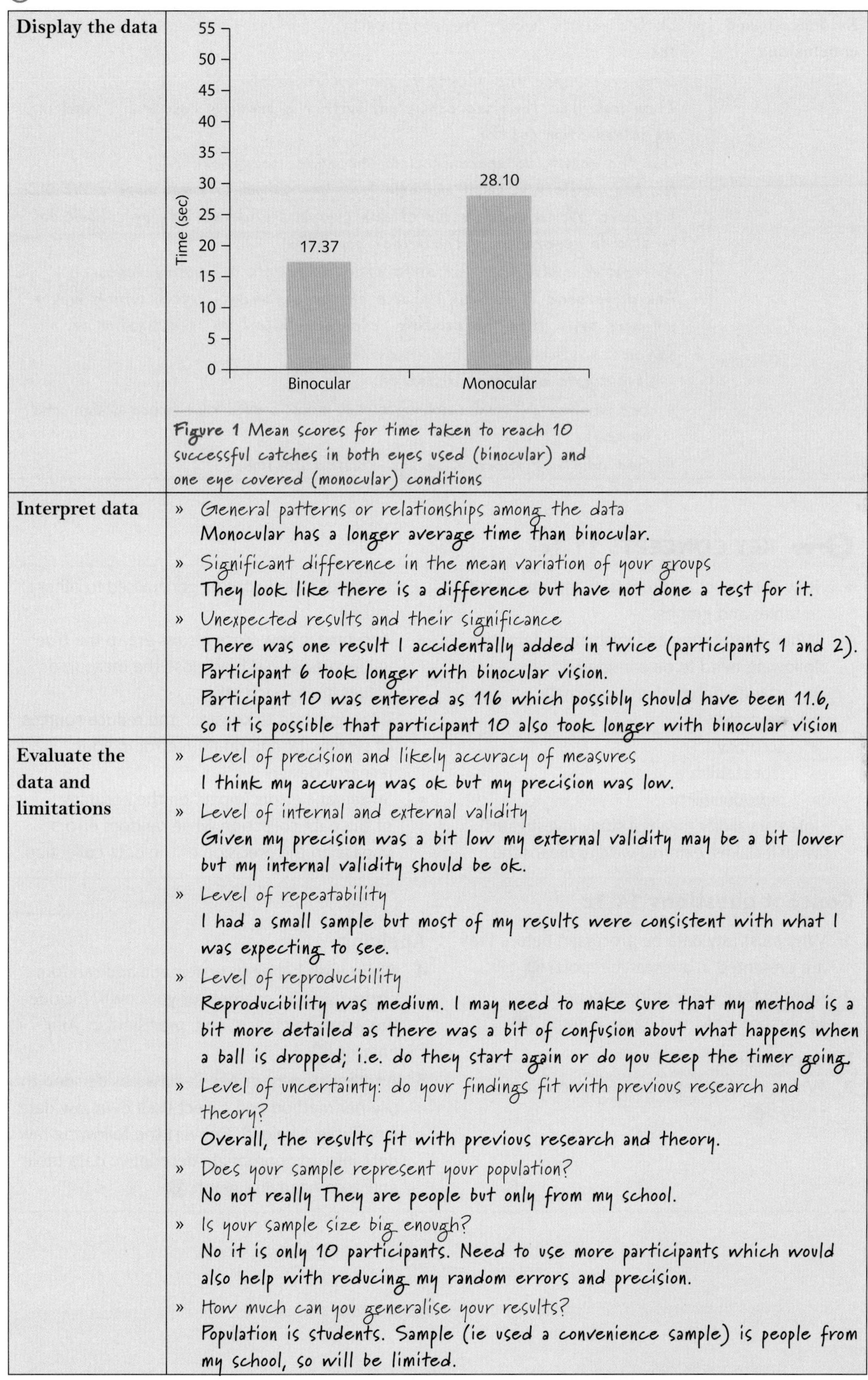

Display the data	Figure 1 Mean scores for time taken to reach 10 successful catches in both eyes used (binocular) and one eye covered (monocular) conditions
Interpret data	» General patterns or relationships among the data Monocular has a longer average time than binocular. » Significant difference in the mean variation of your groups They look like there is a difference but have not done a test for it. » Unexpected results and their significance There was one result I accidentally added in twice (participants 1 and 2). Participant 6 took longer with binocular vision. Participant 10 was entered as 116 which possibly should have been 11.6, so it is possible that participant 10 also took longer with binocular vision
Evaluate the data and limitations	» Level of precision and likely accuracy of measures I think my accuracy was ok but my precision was low. » Level of internal and external validity Given my precision was a bit low my external validity may be a bit lower but my internal validity should be ok. » Level of repeatability I had a small sample but most of my results were consistent with what I was expecting to see. » Level of reproducibility Reproducibility was medium. I may need to make sure that my method is a bit more detailed as there was a bit of confusion about what happens when a ball is dropped; i.e. do they start again or do you keep the timer going. » Level of uncertainty: do your findings fit with previous research and theory? Overall, the results fit with previous research and theory. » Does your sample represent your population? No not really They are people but only from my school. » Is your sample size big enough? No it is only 10 participants. Need to use more participants which would also help with reducing my random errors and precision. » How much can you generalise your results? Population is students. Sample (ie used a convenience sample) is people from my school, so will be limited.

Evidence-based conclusions	» Do the results support the hypothesis? Yes » Cross-reference the results to previous research They look like they are consistent with the previous research I used in my introduction section. » Can the results be generalised to the wider population? No. My sample was small and only from one school so I may have a WEIRD bias here. Would need to use students from a wide variety of schools to be able to generalise to the wider population. » Alternative explanations for differences in results to existing research Any difference is probably because of the equipment I used, which was a lot more basic than the existing research I based my investigation on. » Recommendation for further research » Investigate different distances. » See whether different catching abilities makes a difference (mixed design could be used). » See whether there is an age-related decline.

Student check:
✓ I have cross-referenced the results to relevant psychological concepts.
✓ I have cross-referenced the results to previous research.
✓ I have commented on the extent to which the results can be generalised to the wider population.
✓ I have discussed questions left unanswered and considered any alternative explanations for differences in results to existing research.
✓ I have made a recommendation for further research.

KEY CONCEPTS 11.1c

- Raw data must be processed and presented in tables and graphs.
- When interpreting and evaluating data, the following need to be considered:
  - validity (internal and external)
  - precision
  - accuracy
  - repeatability
  - reproducibility.
- Internal validity means a study investigates what it claims; external validity means the results of a study can be generalised to other settings.
- Accuracy is how close results are to the true value; precision is how close the measured values are to each other.
- It is important to consider and reduce sources of systematic and random error in your research design.
- Systematic errors impact on the accuracy of the data collection while random errors impact on the precision of the data collection.

Concept questions 11.1c

1 Why must raw data be processed before they are presented in a research report? **e**

2 Identify four key features you need to remember when setting up a graph. **r**

Understanding

3 Why is validity important for a research study? **e**

Applying

4 Distinguish between systematic and random errors using an example of your own. Include the words 'accuracy' and 'precision' in your answer. **c**

5 Another student in Alina's class has decided to use her method and collect their own raw data (see Table 1 below). Convert the following raw data into an appropriate descriptive data table and accompanying graph. **d**

Table 1. Raw scores for time taken to reach 10 successful catches in both eyes used (binocular) and one eye covered (monocular) conditions

			Time taken to reach 10 successful catches (seconds)	
Participant	**Age**	**Gender**	**Trial #1 (both eyes used)**	**Trial #2 (one eye covered)**
1	23	Female	15.23	30.46
2	48	Male	14.34	24.49
3	51	Male	12.35	27.58
4	17	Male	13.41	34.22
5	46	Female	15.10	33.34

HOT Challenge

6 While much work has been done to reduce prejudice and discrimination, these attitudes are still present in Australian society. As discussed in Chapter 7 (page 290), the 2021 Social Inclusion Index, conducted by researchers at Monash University, found that many Australians still harbour highly prejudicial views. Sixteen per cent of respondents on average agreed 'moderately' or 'strongly' with statements that indicated prejudice towards young people aged 18–24 years. **C**

A Year 11 Psychology class is given the task of designing a survey on the experiences of prejudice and discrimination in Australian society within this social group. Copy and complete the table below to create three questions for a survey. Collate your questions as a class or in small groups and discuss the level of validity of each question.

Survey question for young people aged 18–24 years	Not true	Somewhat true	True	Very true	Not sure

11.2 Science communication

Congratulations! You have now added to the cycle of psychological research!

If you have heeded the advice and used your logbook effectively, then you have already done the majority of the work. All you will need to do is transfer your work into a report format. As with the scientific method, the report also follows a certain structure. We often refer to it as having an hourglass shape, where we start off broadly with our information, then become more specific as we talk about our method and results, then broaden out again as we get to the discussion section (Figure 11.10).

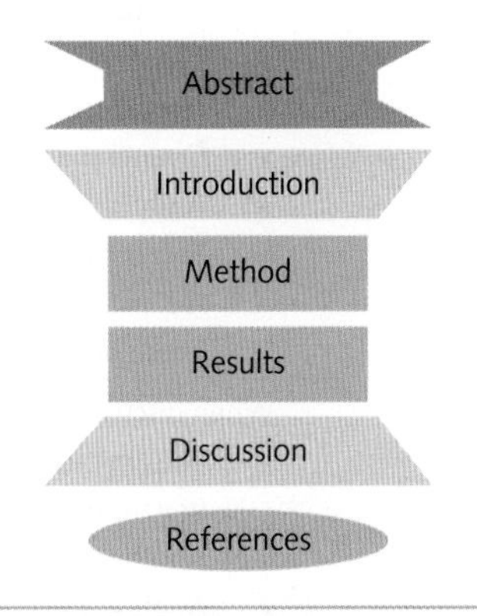

Figure 11.10 The structure of a scientific report

In your discussion and conclusion, your final task is to determine how well the aim of your investigation has been achieved and whether it has answered your research question. For example, has your hypothesis been supported? To do this you will need to consider these points.

11.2.1 CONSTRUCTING A REPORT

Weblink
APA referencing guide

1. How well did your results fit with the theory or model your hypothesis was based on?
2. What contribution has your study made to the existing literature? Does it support the findings in the background research?
3. How much can the results of your investigation be generalised to the wider population? Does your study have a WEIRD bias?
4. If relevant, are there any alternative explanations for your results? Was there a large degree of uncertainty in your results? What other variables may have caused your results to be different from those in the existing research?
5. What recommendations do you have for practical implementation or further research? For example, you could offer concrete suggestions for how future work can build on areas that your research was unable to address (indicate the level of reproducibility of your work).

In psychology, we generally use the APA referencing system to write the references. For more information, refer to the APA referencing guide weblink.

Sometimes the hardest part is to actually start each section; Table 11.8 is a compilation of some handy sentence starters to help you. If possible, see whether your teacher will share the assessment rubric as a guide to determine what needs to be included in each section of the report.

Finally, to assist you and to finish off this chapter there is an example of a student report. This will be annotated to show you how you could structure your report and what to think about at each step. It has been broken up into sections to help you to write your report but in reality it would not be in a table format.

Table 11.8 Helpful sentence starters

Section	Starter
Introduction	» It has been found that ... » Prior research shows ... » This can also be linked to the works of ... » This study aims to investigate/determine ... » The aim of this study is to explore/identify/determine/explore/investigate/establish ...
Discussion	When talking about your results: » The results indicate that ... » The study demonstrates a correlation between ... » This analysis supports the theory that ... » The data suggest that ... » The implications of these findings are ... » The results of the experiment showed clear support for ... » Comparisons between the data sets revealed that ...
	When comparing with prior studies: » These results go beyond previous reports, showing that ... » These results are in line with previous studies ... » Overall these findings are in accordance with findings reported by ... » A similar conclusion was reached by ... » This is consistent with what has been found in previous studies, such as ... » A similar pattern of results was obtained in ... » These basic findings are consistent with research showing that ...
	When discussing limitations of your work: » The limitations of the present studies include ... » Regarding the limitations of ..., it could be argued that ... » Another limitation of this ... » Another limitation in ... involves the issue of ... » The main limitation is the lack of ... » One limitation found in this study is ... » One limitation of these methods, however, is that they ...

9780170465052

Section	Starter
	When making speculations: » This may be due to ... » The results show that ... does not seem to impact ... » Because of this potential limitation, … » Therefore, it remains unclear whether ... » This may explain why ...
	When making deductive arguments: » A difference between these ... can only be attributable to ... » Nonetheless, there is justification to ... » This may raise concerns about ... which can be addressed by ... » As discussed, this is due to the fact that ... » These findings support the notion that ... » This may be the reason that ...
Generalisation	» Due to …, results could be/could not be generalised to the broader population.
Conclusion	» It can be concluded that ... » To truly draw a meaningful conclusion, further research is required to ...

Sample report

Title	The influence of depth cues on depth perception	Title includes IV and DV
Introduction	Visual perception principles are rules our brain automatically applies to visual stimuli to help organise and interpret them in a consistent and meaningful way (Iersel et al., 2022). Once such principle is the theory of depth perception. This refers to the visual ability to perceive the world in three dimensions, coupled with the ability to gauge how far away an object is. Depth perception, size, and distance are ascertained through both monocular and binocular cues (Advanced Topics in Perception, n.d.). Monocular cues require information from one eye only and consist of accommodation and pictorial cues (Iersel et al., 2022). Monocular vision is poor at determining depth. When an image is projected onto a single retina, cues about the relative size of the object compared to other objects are obtained. However, in binocular vision, these relative sizes are compared, since each individual eye is seeing a slightly different image from a different angle (Advanced Topics in Perception, n.d.) Binocular cues require both eyes to work together to provide the brain with information about depth and distance. This allows the brain to use monocular cues in conjunction with binocular cues such as convergence and retinal disparity. Convergence refers to eye movement as an object gets closer. As an object gets closer, the eyes turn inwards, and when the input from both eyes is compared (stereopsis), then an impression of depth is obtained. Depth perception theory suggests that depth perception relies on the convergence of both eyes on an object, the relative differences between the shape and size of the images on each retina, the relative size of objects in relation to each other, and other cues such as texture and constancy (Advanced Topics in Perception, n.d.)	Starts off broad A discussion of relevant terms and theories The use of in-text referencing

Another key factor to consider is motion detection. This is the process of working out the direction and speed of stimuli in the environment based upon the visual information that is being sent to the brain. It can be detected by both monocular and binocular vision. Monocular cues (information received from one eye) can detect nearby motion but the accuracy of depth perception is low. Binocular vision is better and more effective in detecting motion from a distance, mainly due to its increased ability to gauge depth perception.
There is little existing data to support beyond doubt as to how well people are able to discriminate between distances under natural conditions. (Binocular Cues Vs Monocular Cues-Definition, Difference and Uses, 2021)

Previous studies by Gonzalez and Niechwiej-Szwedo (2016) examined the effects of monocular and binocular vision on eye-hand coordination during the placement and grasping of objects. To test their research question, 15 participants were tested on a sequencing task while eye and hand movements were recorded binocularly using a video-based eye tracker and a motion capture system. They found that monocular viewing disrupted the temporal (timing) coordination between the eyes and the hand during the place-to-reach transition phase. Gonzalez and Niechwiej-Szwedo (2016) concluded that binocular vision is ultimately more effective at perceiving depth in everyday life.

Two pieces of research discussed; where possible, include a brief explanation of aim and hypothesis, research method used, participants, materials, procedure and results

The investigation into the contribution of stereo vision with one-handed catching undertaken by Mazyn et al. (2004) also supports this theory. To test this they examine both monocular and binocular performance in a natural catching task in people with good and weak stereopsis. They hypothesised that since all participants were inexperienced in the monocular test condition, they were expected to perform better binocularly than in the monocular test situation. Results indicated that monocular and binocular vision did not affect catching performance for participants with weak stereoscopic vision; however, participants with normal stereopsis caught more balls with binocular vision than monocular vision. The study concluded that binocular vision was more effective when determining depth perception for individuals with normal vision.

The original experimental methodology devised by Mazyn and associates (2004) was modified for the current experiment due to the inaccessibility of materials used in the original study. The modified experiment aims to determine whether depth perception is impacted by the use of monocular and binocular cues using a different method and materials. These changes should still result in similar trends and relationships between depth perception and depth cues.

The rationale explicitly communicates the reasons for modifying the original experiment.

Research question: What role do monocular and binocular depth cues play in allowing people to accurately judge distances?
The aim of this study is to investigate whether monocular or binocular depth cues are superior when accurately judging distance and motion of a moving object. It was hypothesised that participants who had the use of binocular depth cues would be more accurate in their judgement of depth perception than when they were restricted to monocular depth cues.

The research question is connected to the rationale and enables effective investigation of the influence of depth cues on depth perception.

Clear aim, a hypothesis and/or prediction

Method	Modifications to methodology As with the original methodology, a controlled experiment was selected. However, the original experiment was modified to be able to be conducted using everyday materials. In this case, the independent variable was the use of depth cues: monocular cues (as the experimental condition) or binocular cues (as the control condition). Depth cues were operationalised by the use of an eye patch over one eye for the monocular condition and no eye patch for the binocular condition. The dependent variable was depth perception. This was measured by recording how long it took for participants to throw and catch a tennis ball 10 times against a brick wall that was 3m away. In order to minimise systematic errors new batteries were installed in the stopwatch. Safety and ethical considerations During the planning of the methodology for this experiment, ethical issues (e.g. the need for informed consent) were identified and managed. Participants The participants were 10 individuals who were known to the researcher. In order to recruit participants, a convenience sample method was used. The sample consisted of 10 (male= 3, females= 7) people, A between-groups design was used. Materials » tablet stopwatch/timer » 1 tennis ball » 2 eye patches » flat wall surface » masking tape Method 1 A cone was set up 2m from a brick wall, where participants were to stand facing the wall. 2 Participants were instructed to throw the ball against a the wall and successfully catch it 10 times, then the timer would be stopped. 3 The timer was started by the experimenter when the participant began the first throw with the experimenter counting down the number of successful throws left from 10. 4 If the ball is was dropped, the timer was continued and the counting of the amount of successful throws was restarted. 5 The experimenter stopped the timer once the participant reached 10 successful throws and catches in succession. 6 The experiment was then repeated; however, the participant then had a eye patch over their left eye	Methodology modification is justified A methodology has been chosen that enables the collection of sufficient, relevant data. The methodology shows careful and deliberate thought. It enables collection of adequate data so an informed conclusion to the research question can be drawn. Management of risks and ethical or environmental issues Ethical issues have been managed. However, the response does not show careful or deliberate identification and planning.

<table>
<tr>
<td>Results</td>
<td>
An analysis of the data was conducted to determine the mean of each condition. As the data was in intervals there was one obvious outlier identified from the raw data, which was removed and the average was substituted.
Table 1. Mean scores for time taken to reach 10 successful catches in both eyes used (binocular) and one eye covered (monocular) conditions
<table>
<tr><td colspan="2">Time taken to reach 10 successful catches (sec)</td></tr>
<tr><td>Trial #1 (both eyes open)</td><td>Trial #2 (one eye covered)</td></tr>
<tr><td>17.37</td><td>28.10</td></tr>
</table>

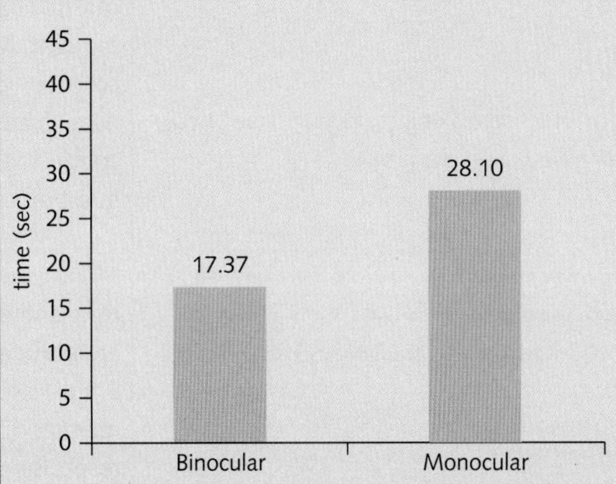

Figure 1 Mean scores for time taken to reach 10 successful catches in both eyes used (binocular) and one eye covered (monocular) conditions
Results summary
The results show that the mean score for the binocular group (17.37 seconds) was lower than the mean score for the monocular group (28.10 seconds).
</td>
<td>
Presentation of generated data/evidence in appropriate format to illustrate trends, patterns and/or relationships

Raw data are manipulated accurately to provide evidence that is applicable to the research question.

Appropriate use of measurement conventions

The response follows scientific conventions of the construction of tables, and graphs.
</td>
</tr>
<tr>
<td>Discussion</td>
<td>
The aim of this study was to determine whether monocular or binocular depth cues are superior in accurately judging distance and motion of a moving object. The hypothesis that participants who had the use of binocular depth cues would be more accurate in their judgement of depth perception (as tested by recording the time it took them to throw and catch a ball 10 times), than when they were restricted to monocular depth cues was supported. Results showed that, on average, participants were quicker to successfully catch a ball 10 times using binocular vision when compared to when they used monocular vision (17.37 seconds and 28.10 seconds respectively). This is in line with depth perception theory, which states that binocular vision allows for better judgement of depth perception via a number of depth cues including convergence.

These results are consistent with Mazyn et al. (2004) as, in both studies, participants with monocular vision dropped the ball more times and did not catch as many balls in general and in succession compared to when the participants had the use of both eyes. These findings support the theory that binocular vision is superior at determining depth perception because when the information from both eyes is compared, it creates the perception of depth that is significantly harder to accomplish with one eye. As the ball came closer to the participants, the participants' eyes turned inwards: convergence. This made catching the ball even harder.
As suggested by Gonzales et al. (2016), the coordination between the hands and the eyes is delayed during monocular vision due to the uncertainty of depth perception.
</td>
<td>
Aim restated

Hypothesis restated supported/rejected

Linking of results to investigation question and to the aim to explain whether or not the investigation data and findings support the hypothesis

Cross-referencing of results to relevant psychological concepts
</td>
</tr>
</table>

	By analysing the evidence obtained in the experiment, it was clear that the experimental processes lacked precision, accuracy and validity. Although experimenters sought to reduce errors in the planning and data collection phases, it is likely that the uncertainty observed in the data was due to a lack of reducing them adequately. Systematic errors such as slight variations in participant instructions lead to confusion for some participants, which impacted on their first attempt at the task.	Identification of limitations in data and methods, and suggested improvements
	Using better technology to accurately time how long it took for the participants to complete the task is also recommended for future studies to reduce random errors in the collection of the data. A number of outliers were also recorded, which may lead to misrepresentation of the results. This experiment is limited in its ability to be generalised due to the small sample size (n = 10), and the sampling procedure which is not representative of the overall population. An additional measure would be to increase the sample size and use random or stratified-random sampling techniques, as they increase the representativeness of the sample to the population. This measure would also help to improve the validity of the current study.	Extent to which the results can be generalised to the wider population identified
	A further consideration should be the effects of the adaptions made to the original experimenter's methodology. Although the investigation examined depth perception while catching, participants were not tested for normal or weak stereoscopic vision. In addition, the current study did not use the same distances or high tech equipment as the original study, which is likely to have affected the reliability of the experimental methodology.	Identification of the extent to which the analysis has answered the investigation question, with no new information being introduced
	An extension to the experiment that would increase the validity would be to increase the sample size and test on a more diverse population group. Lastly, a suggestion for future research would be to use a mixed design to group participants into elite ball sport and non-sport people. This would allow experimenters to discover whether the effect of practice (i.e. the elite sports people with good hand and eye coordination) could also impact on people's depth perception.	Implications of the investigation and/or suggestions as to further investigations that may be undertaken
	It can be concluded that binocular vision is more effective than monocular vision when determining depth perception. The vast majority of participants were faster at catching a ball with binocular vision, suggesting that it is more adept for the perception of depth. However, the experimental result should be considered with caution due to the uncertainty and limitations identified in the data, most likely caused by a lack of reliability and validity in the experimental process.	Conclusion that provides a response to the investigation question

References		
References	Advanced Topics in Perception. (n.d.). Course Hero. Retrieved Nov 1, 2021, from https://www.coursehero.com/study-guides/boundless-psychology/advanced-topics-in-perception/ Binocular Cues vs Monocular Cues-Definition, Difference and Uses. (2021). Binoculars Guru. Retrieved Nov 1, 2021, from https://www.binocularsguru.com/binocular-cues-vs-monocular-cues-difference-and-uses/ Gonzalez, D. A., & Niechwiej-Szwedo, E. (2016). The effects of monocular viewing on hand-eye coordination during sequential grasping and placing movements. Vision Research, 128, 30–38. Iersel, H.V., Blunden, A., Diamond, M., Hutchison, K., Park, L., Scanlon, A., Wallace, C., Wolters, A., Young, N. (2022). VICscience Psychology VCE Units 1 & 2 4th. Cengage Learning Australia. Mazyn, L., Lenoir, M., Montagne, G., & Savelsbergh, G. J. P. (2004). The contribution of stereo vision to one-handed catching. Experimental Brain Research, 157, 383–390. https://doi.org/10.1007/s00221-004-1926-x	Referencing and acknowledgement of all quotations and sourced content relevant to the investigation

KEY CONCEPTS 11.2

- A scientific report is structured using sections, in this order:
  - abstract
  - introduction
  - method
  - results
  - discussion
  - references.
- Each section of a scientific report contains different types of information.
- The aim of a scientific report is to share research findings and suggest recommendations for future research.

Concept questions 11.2

Remembering

1 In which section of a scientific report would you see references to validity?

2 Identify the purpose of the abstract.

Understanding

3 In the discussion section, why is it important to compare the results of the investigation to prior studies?

Applying

4 It is important to be able to communicate scientific information to different audiences. Consider the following section from the sample report:

'One such principle is the theory of depth perception. This refers to the visual ability to perceive the world in three dimensions, coupled with the ability to gauge how far away an object is.'

Reword this as if you were trying to communicate it to a primary school student.

HOT Challenge

5 In Unit 4 Psychology you will be required to display your scientific investigation in a poster format. As practice for this, your challenge is to use the poster format in Figure 11.11 and convert your report into the structure outlined. C

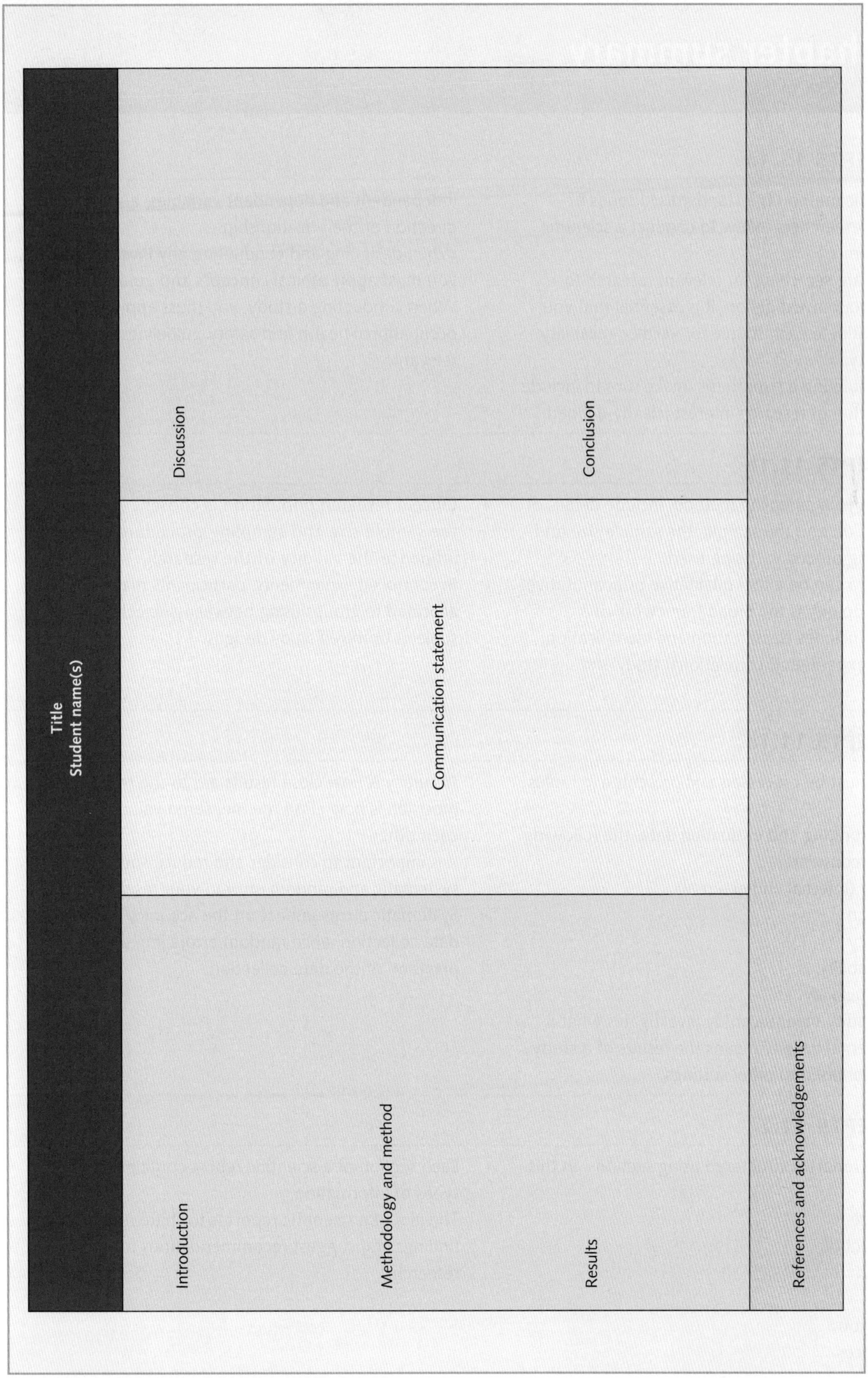

Title
Student name(s)

Introduction

Methodology and method

Results

Communication statement

Discussion

Conclusion

References and acknowledgements

Figure 11.11 Sample poster format

11 Chapter summary

KEY CONCEPTS 11.1a

- The scientific method is a standardised series of steps that researchers follow to conduct a scientific investigation.
- When you are searching for relevant research to include in your investigation, it is essential that you critically evaluate each source for validity, reliability and integrity.
- When formulating a hypothesis, make sure to include the population of research interest, clearly defined independent and dependent variables, and the direction of their relationship.
- When designing and conducting any investigation you must apply ethical concepts and guidelines.
- When conducting a study, you must apply occupational health and safety guidelines where relevant.

KEY CONCEPTS 11.1b

- To describe your sampling method, include details of the population and the sample, the sample size and the sampling procedure being used.
- Primary data can be either qualitative or quantitative.
- The methodology is the broad framework of a study, and includes classification and identification, controlled experiment, correlational study and fieldwork.
- Once a research population is chosen, the sample size and sampling procedure can influence the validity of the research.
- In controlled experiments, participants may be allocated to groups using between-subjects, within-subjects or mixed study designs.

KEY CONCEPTS 11.1c

- Raw data must be processed and presented in tables and graphs.
- When interpreting and evaluating data, the following need to be considered:
  - validity (internal and external)
  - precision
  - accuracy
  - repeatability
  - reproducibility.
- Internal validity means a study investigates what it claims; external validity means the results of a study can be generalised to other settings.
- Accuracy is how close results are to the true value; precision is how close the measured values are to each other.
- It is important to consider and reduce sources of systematic and random error in your research design.
- Systematic errors impact on the accuracy of the data collection while random errors impact on the precision of the data collection.

KEY CONCEPTS 11.2

- A scientific report is structured using sections, in this order:
  - abstract
  - introduction
  - method
  - results
  - discussion
  - references.
- Each section of a scientific report contains different types of information.
- The aim of a scientific report is to share research findings and suggest recommendations for future research.

Glossary

Abstract thinking Thinking in terms of general concepts rather than specific objects, experiences or events; characteristic of thinking in older children and adults

Accommodation A monocular cue of visual depth perception involving voluntary or involuntary adjustments by the muscles attached to the lens in each eye that alter the shape of the lens to keep an object in focus as its distance from the viewer changes

Accommodation In Piaget's theory, the modification of existing psychological concepts or processes (e.g. mental schemas) to adapt understanding and behaviour in response to new information or experiences; for example, occurs when a child alters their existing understanding and behaviour to be able to interact with a new toy

Accuracy How close a measurement is to the true value of the quantity being measured

Acquired brain injury (ABI) Brain damage caused by events after birth that affects the functional ability of the brain's nerve cells

Actor–observer bias The bias that occurs when a person attributes their behaviours to external (situational/ environmental) factors while attributing other peoples' behaviours on internal (dispositional) factors

Adaptive behaviour Behaviour that allows a person to meet and respond effectively to the demands of everyday living

Adaptive plasticity The ability of brain neurons to grow new connections between the synapses to suit the environmental conditions, such as when learning something new or when relearning something after brain injury

Affect heuristic An approach that is used when a person's decision-making is influenced by their current emotional state or mood

Agnosia Loss or impairment of the ability to recognise or appreciate certain sensory experiences due to brain damage or disorder, specific to a particular sense

Aim A broad statement about what an investigation intends to find out

Ames room illusion A visual illusion affecting the perceived size or shape of people or objects within a room when they are viewed with one eye through a peephole; it is caused by the trapezoidal shape of the floor, ceiling and side walls of the room, which disrupts perceptual constancies, misleading the viewer to perceive people as either shrinking or growing as they cross the floor diagonally from one corner to another

Amygdala An almond-shaped brain structure, located deep within each temporal lobe, that is responsible for the physiological experience of emotions, especially the emotional response of fear, and vital for emotional learning and memory

Anchoring bias The tendency to rely on the first piece of information offered (the 'anchor') when making decisions

Anecdote A factual claim based only on personal observation, collected in a casual or non-systematic way

Animism The tendency of children (during the preoperational stage) to believe that any inanimate object possess lifelike characteristics, such as feelings and emotions

Anti-conformity A deliberate refusal to comply with social norms, demonstrated by a person expressing ideas, beliefs or judgments that challenge these social norms; also known as counter-conformity

Aphasia An acquired language impairment that results from brain damage

Assimilation In Piaget's theory, a process through which new experiences are approached by applying existing mental structures or processes; for example, occurs when a child interacts with a new toy by treating it like another toy that is already familiar

Attachment In development, the close emotional bond shared between an infant and their primary caregiver; also, the tendency to seek emotionally supportive relationships in adulthood

Attention A voluntary or involuntary orientation of awareness towards a particular stimulus, ignoring other stimuli

Attentional bias The tendency to focus on particular stimuli whilst overlooking or ignoring other relevant pieces of information

Attitudes The learned ideas a person holds about themselves, others, objects and experiences

Attributions The process of attaching meaning to behaviour by looking for a cause or causes to explain the behaviour

Atypical behaviour Behaviour that differs from the norm in being unusual or unrepresentative

Automatic processes Actions that require little awareness or mental effort, and do not interfere with performance of other activities

Availability heuristic A mental shortcut in which a person uses the information that they first think of to make a judgement

Basal ganglia A group of brain structures located at the base of the forebrain and in the midbrain that play important roles in controlling voluntary movement and processing rewarding stimuli

Beneficence An ethical concept that means having a commitment to do good (and minimise risks and harms)

Between-subjects design A type of experimental design in a controlled experiment in which scores are compared between different participants

Binocular cues A group of depth cues that require both eyes to work together to provide the brain with information about depth and distance

Biopsychosocial model An approach that proposes that health and illness outcomes are determined by the interaction and contribution of biological, psychological and social factors

Bottom-up processing Information processing in which incoming data from a stimulus initiate and determine the higher-level processes involved in organising and interpreting the data

Broca's aphasia A disorder of language characterised by non-fluent conversational speech that is slow and halting; writing is also usually impaired; caused by damage to Broca's area

Broca's area A region within the frontal lobe of the brain, in the left hemisphere for right-handed people and most left-handed people, that controls speech production

Case study An analysis of one particular example in an area of interest that is carried out to develop our understanding of a whole process

Centration The tendency of children during the preoperational stage to narrowly focus on one aspect of a problem, object, or situation at a time, and exclude all others

Cerebellum A hindbrain structure located on top of (dorsal to) rest of the brainstem, essential for coordination of voluntary movement, balance and some forms of conditioning

Cerebral hemispheres The two halves of the cerebrum that cover the upper part of the brain

Cerebrum The largest part of the forebrain, located in front of and above the cerebellum, consisting of two cerebral hemispheres bridged by the corpus callosum

Chronic traumatic encephalopathy (CTE) A progressive, degenerative and fatal brain disease linked to repeated blows to the head; characterised by behavioural problems, personality changes and deficits in thinking

Classification The ability (developed during childhood) to be able to group together or categorise objects based on similar properties

Closure A Gestalt principle involving the viewer's tendency to perceptually complete an incomplete figure by filling in an imaginary contour line so that the figure has a consistent overall form

Cognition A general term that refers to all mental (or thought) processes engaged in by an organism to interpret experience and respond adaptively; including perceiving, conceptualising, remembering, reasoning, imagining, judging and problem-solving

Cognitive bias The systematic errors (consistent mistakes) that occur in a person's decision-making

Cognitive dissonance The psychological discomfort that a person experiences when there is inconsistency between their attitudes and their behaviours

Collectivist culture A culture that values group needs or interests over the interests of individuals

Conclusions The descriptions and explanations drawn from the results of a study to discuss how they relate to the aim and hypothesis of the study

Concrete thinking Thinking that relates to direct experiences or specific objects or events; the opposite of abstract thinking; characteristic of young children's thinking

Confidentiality An ethical guideline that means ensuring the privacy of participants' personal information

Confirmation bias The bias that occurs when a person focuses on and favours information that supports their perspectives whilst ignoring contradictory information or views

Conformity The act of an individual changing their behaviour as the result of real or implied pressure from others

Confounding variable A variable other than the independent variable that has systematically affected the dependent variable because its influence is not evenly distributed across the levels of the independent variable

Conservation The knowledge that the physical quantities of weight, mas, or volume of matter remain unchanged (are conserved) even when their shape or appearance changes; for example, when a quantity of water is poured from a short, wide container into a tall, thin container; achieved in the concrete operational stage of Piaget's theory of cognitive development

Context Information (conditions or circumstance) that surrounds a stimulus and that influences the perception of the stimulus

Control group A group in a controlled experiment that does not receive or experience the experimental condition

Controlled experiment An experimental methodology used to test a hypothesis in which the researcher systematically manipulates (changes) one or more variables to investigate the effect of these manipulations on another variable

Controlled processes Conscious, voluntary actions or cognitive processes that require a high level of attention and monitoring

Controlled variable A variable that is held constant in a controlled experiment to prevent it from affecting the dependent variable; this is not the same as a controlled experiment

Convenience sampling A method of sampling in which the researcher recruits a sample of participants that is convenient to recruit

Convergence A binocular depth cue that involves both eyes simultaneously turning inwards so that the image falls on corresponding points of the fovea of each eye, allowing the two slightly different views of an object seen by each eye to form a single image

Corpus callosum The largest tract of nerve fibres (white matter) between the left and right cerebral hemispheres, providing the main connection between them

Correlational study A non-experimental study where the researcher investigates relationships between variables in a sample of participants by measuring rather than manipulating any variables

Counterbalancing A technique used in within-subjects design where the order in which participants experiece the conditions is split, so not everybody completes the same conditions in the same order

Critical period A stage, usually early in the lifespan, when an organism is most open to acquiring a specific cognitive or motor skill, or socio-emotional competence as a normal part of development and which cannot normally be acquired at a later stage of development

Cultural norms behaving in a way that is acceptable to a society

Culture The distinctive beliefs, values, customs, knowledge, art and language that form the foundation of the everyday behaviours and practices of a society, and that are transmitted from one generation to the next

Data Any information collected in scientific investigations

Debriefing An ethical guideline in psychology research that requires the researcher, after the experiment, to disclose the aim, results and conclusions, answer participants' questions and provide support

Deception In psychological research, concealing aspects of a study; its use must comply with ethical guidelines by only occuring when absolutely necessary, and being accompanied by debriefing

Decision-making An executive function that results in the selection of a course of action from a number of possible alternatives

Deindividuation The loss of social identity and inhibition a person may experience when in a group, which results in them behaving with less care for potential consequences

Dependent variable The outcome variable that the researcher measures to determine whether manipulating the independent variable had an effect; sometimes called the outcome variable

Depth cue Any of a variety of internal and external mechanisms, stimuli or processes that tell a person's visual system about the depth of an object (its three-dimensionality) or its distance away; includes monocular and binocular cues

Depth perception The ability to see three-dimensional space and to accurately judge distances

Descriptive statistics Statistics that *describe* the data set by condensing a set of values down to a single numerical value

Discrimination The action of being prejudiced or treating others in an unfair manner based on the negative attitude held about them or the group they belong to

Distress A negative psychological response to a stressor that results from being overwhelmed by the perceived demands of a situation, or loss, or threat

Divided attention The type of attention used when we rapidly switch the focus of our awareness between two (or more) sources of information so that we can perform two (or more) tasks (or components of the one task) at the same time

Dunning–Kruger effect An effect that occurs when people with low ability at a task overestimate their own skill-set, and people with high ability at a task underestimate their own skill-set

Egocentrism The tendency of children, during the preoperational stage, to view situations and events only from their own perspective, with the belief that others will see things from the same point of view as them

Emotional regulation The ability to apply procedures to control an emotion or set of emotions

Emotions The set of complex and intense physiological, psychological and behavioural reactions to a significant event, which cause specific kinds of experience, such as happiness or fear

Endocrine system The network of glands that produce and secrete hormones, which are carried via the bloodstream to different parts of the body; crucial in regulating moods, metabolism, growth and development

Ethical concepts Considerations about how an investigation may affect human or non-human participants that must be taken into account before an investigation is carried out

Ethical guidelines Considerations about how an investigation may impact a human participant that must be taken into account before, during and after an investigation is carried out

Evaluating Determining the significance of something by careful review

Executive functions Higher level cognitive processes that require flexible thinking and self control, such as planning, analysing and problem-solving

Experimental Group The group in a controlled experiment that experiences or receives the experimental condition

External attributions The environmental factors that are external to an individual, such as their location or the people around them; also known as situational attributions

External validity A measure that indicates how well the results of a study can be applied meaningfully to real-world contexts, situations and behaviours

Extraneous variable Any variable other than the independent variable that *may* affect the dependent variable

False-consensus bias The tendency for people to assume that their attitudes, beliefs and behaviours are relatively common and more widely shared by others than is actually the case

Fieldwork A research methodology that involves observing and interacting with a selected environment beyond the classroom or laboratory

Figure–ground A Gestalt principle involving the viewer perceptually grouping and separating some features of a stimulus so that part of a stimulus appears to stand out as an object (the figure) against less distinct background (the ground)

Forebrain The largest region of the brain, comprising the entire cerebrum, the thalamus and hypothalamus, much of the basal ganglia, and the pineal gland

Frontal lobe The front-most lobe of the brain; associated with complex mental abilities, such as decision-making and attentional control, and with the control of voluntary movement

Functional fixedness A cognitive bias whereby a person believes an object or item can only be used in a particular way (its intended use)

Generalisability The extent to which research findings (found using a sample) can be applied to the population of interest

Generativity In Erikson's theory of psychosocial development, the positive goal of middle adulthood in which a person's focus shifts from narrow self-interest towards fulfilling family and wider social responsibilities to nurture the next generation

Geschwind's territory A region of the brain that helps Wernicke's area to allow comprehension of spoken or written language by integrating the different properties of a word

Gestalt principles The set of laws (principles) of human perception described by the Gestalt psychologists to explain how the perception of an experience is organised in terms of whole forms

Glossary A list of words with definitions

Goal-directed behaviour In Piaget's theory, the behaviour of infants when they learn the relationship between their actions and obtaining what they desire in the external world

Group A collective of two or more people that interact, influence each other and share a common objective

Group norms Standards that are set by the groups to which we belong

Group shift The tendency for group members discussing an issue or dilemma to adopt a more extreme position than their individual position before the discussion; also known as group polarisation

Groupthink An effect that occurs when the members of a group prioritise the strong bonds of the group over clear decision-making

Gustation The sensory experience of a food or drink that is perceived as flavour

Gustatory cortex The area of the brain where taste information is processed; consists of the frontal operculum and the anterior insula

Gustatory receptors Proteins found in the cell membranes of taste bud cells of the tongue and in taste receptor cells within the mouth and throat that detect taste sensations and begin the neural processes that send information to the taste processing areas of the brain

Halo effect A cognitive bias whereby the positive evaluation we hold about one quality of a person influences our beliefs and expectations regarding other qualities of that person

Height in the visual field A monocular pictorial cue whereby the height of objects in the visual field (either above or below the horizon) acts as a depth cue, so that objects close to the horizon appear further away than objects located below the horizon; also called elevation

Hemispheric specialisation The strong tendency for each cerebral hemisphere to be specialised for processing particular cognitive or motor functions; for example, the left hemisphere is dominant (specialised) for language processing

Heuristics The mental shortcuts we use to make quicker, more efficient decisions

Hindbrain The lower back portion of the brain, including the brainstem (pons and medulla) and cerebellum; connects the brain to the spinal cord and controls vital functions such as heartbeat and breathing, as well as coordination of voluntary movements, balance and posture

Hindsight bias A cognitive bias that occurs when a person suggests that an outcome was more predictable than before it occurred

Hippocampus The seahorse-shaped structure located deep within each temporal lobe that is vital for building memories, especially fact- and event-related memories

Hypothalamus A small structure located deep in the forebrain responsible for a variety of involuntary functions, including regulation of body temperature, sex drive, appetite and the sleep–wake cycle

Hypothesis a statement that expresses a possible (that is, hypothetical) answer to a research question

Hypothetico-deductive reasoning The capacity to test a logical hypothesis through abstract thought

Independence The behaviour that occurs when an individual is aware of how a group expects them to act or respond, but their decision-making is not affected

Independent variable the variable that the researcher manipulates

Individualist culture A culture that values individual interests over the interests of groups

Informational influence The social environment or situational cues used by people to help them monitor or adapt their behaviour to fit with the behaviour of those around them

Informed Consent An ethical principle that means Ensuring participants understand the nature, purpose and risks of the study before agreeing to participate

Integrity An ethical concept that means acting with honesty and transparency

Internal attribution The factors within a person that shape their behaviour, including personality characteristics, motivation, ability and effort; also known as dispositional **attributions**

Internal validity A measure that indicates how well the design of a scientific investigation and the measurements it uses provide meaningful and accurate information about the psychological constructs and the population being studied

Interposition A monocular pictorial cue that occurs when one object partially blocks another object and is perceived as being in front of, and therefore closer than, the object it covers

Justice An ethical concept that means ensuring fair distribution of benefits, risks, costs and resources

Legitimacy of authority The status attributed to an individual who is seen as an authentic figure whose orders should be obeyed

Limbic system A network of brain structures, including parts of the cortex, the thalamus, hippocampus and amygdala, that is involved in emotion processing and in learning and remembering associations between emotional responses and objects, people, places and events

Limitations The factors that control or reduce something

Linear perspective A monocular pictorial depth cue where the apparent convergence of parallel lines creates the illusion of increasing distance

Literature review A method of scientific inquiry that involves collating, analysing and synthesising the existing scientific literature on a specific topic, with the goal of determining the 'state of the art' in the field, to provide the background to a new investigation, and/or to determine the extent of current consensus of views on the topic, articulate the other viewpoints, and recommend directions for future research in the field

Lobotomy A drastic surgical procedure once used to treat mental illness that involves severing neural connections in the prefrontal cortex of the brain; now viewed as an inhumane and unethical treatment

Logbook A complete, permanent record of how an experiment or research project was conducted; it shows what was done at every step along the way

maladaptive behaviour Behaviour that is potentially harmful and prevents a person from meeting and adapting to the demands of everyday living

Mean A measure of central tendency that gives the numerical average of a set of scores, calculated by adding all the scores in a data set and then dividing the total by the number of scores in the set

Measure of central tendency Measures that provide a number that describes a 'typical' score around which other scores lie; they include mean, median, and mode

Median A measure of the middle score in a data set; it is calculated by arranging scores in a data set from the highest to the lowest and selecting the middle score

Medulla A hindbrain structure responsible for regulating internal bodily systems necessary for survival, such as heart rate and breathing

mental wellbeing (mental health) A state of well-being in which the individual realises his or her own abilities, can cope with the normal stresses of life, can work productively and fruitfully, and is able to make a contribution to his or her community (World Health Organization); a state of mind characterized by emotional well-being, good behavioural adjustment, relative freedom from anxiety and disabling symptoms, and a capacity to establish constructive relationships and cope with the ordinary demands and stresses of life (American Psychological Association)

Mental wellbeing A state of well-being in which the individual realises their own abilities, can cope with the normal stresses of life, can work productively and fruitfully, and is able to make a contribution to his or her community (World Health Organization); a state of mind characterised by emotional well-being, good behavioural adjustment, relative freedom from anxiety and disabling symptoms, and a capacity to establish constructive relationships and cope with the ordinary demands and stresses of life (American Psychological Association)

Methodology The broad type of design of a research investigation

Midbrain A small region at the top of the brainstem that connects the hindbrain and the forebrain and contains important bundles of white and gray matter that have a crucial role in processing information related to hearing, vision, movement, pain, sleep and arousal; its systems help to keep us alert, awake and attentive

Miraculin A natural sugar substitute that, when consumed with something sour, turns the sour taste of the food or beverage into a sweet taste

Misinformation effect An effect that occurs when a person demonstrates poor recall of events following exposure to additional information after the event took place

Mixed design A type of experimental design in a controlled experiment which uses a mix of between-subjects and within-subjects comparisons

Mode A measure of central tendency found by selecting the most frequently occurring score in a set of scores

Monocular cues A group of depth perception cues that require information from one eye only

Motivation An internal psychological state that activates, directs and sustains behaviour in relation to achieving a specific goal that may be conscious or unconscious

Müller-Lyer illusion A visual illusion in which two lines of equal length – one capped with inward-pointing arrowheads, the other capped with outward-pointing arrowheads – are perceived as being of different lengths

Myelin The fatty white substance that covers some axons, providing protection and insulation, enhancing the speed of neural transmission and giving the whitish appearance of white matter

Nature In the nature-versus-nurture debate, the inborn, inherited factors, gained genetically from biological parents, that influence the development of our physical, psychological, and behavioural characteristics and development

Nature versus nurture The debate regarding the extent to which hereditary factors (nature) and environmental factors (nurture) each influence development

Neural maturation A continuous developmental process whereby dendrites grow and extend to axons of other neurons to form pathways between neurons

Neural migration The movement of newly formed neurons to their final destination in the nervous system

neurodivergent having atypical patterns of thought and behaviour in individual development that result from a spectrum of normal variations in brain functioning, with variations producing certain strengths in neurodiverse individuals; examples of neurodiverse people include those on the autism spectrum, people with attention deficit hyperactivity disorder (ADHD), and people with specific learning difficulties such as dyslexia

Neurogenesis The production of new neurons during early development of the nervous system and throughout the lifespan

Neuroimaging Various technologies used to generate images that enable noninvasive study of the structures and functions of the brain

Neurological disorders Any of the disorders of the nervous system that affects cognitive, affective and/or motor functioning

Neuroplasticity The capacity of the nervous system to modify its structure and function as a result of experience and in response to injury

neurotypical describes individuals who display typical neurological development in their patterns of thought and behaviour; without the characteristics of individuals on the autism spectrum, or other developmental differences

Non-maleficence An ethical concept that means avoiding harm or ensuring potential harm is **outweighed** by benefits

Non-scientific ideas Knowledge that has not be obtained through the use of the scientific method

Non-taster A person who has fewer taste buds than normal and find tastes less intense than others

Non-traumatic brain injury (NTBI) A form of acquired brain injury that occurs slowly over time as a result of internal factors

Norm A socially defined rule, standard or value that governs expected behaviours within groups

Normality patterns of behaviour or personality traits that are typical, or that conform to some standard of acceptable ways of behaving

Normative influence The impact of the established behaviour of a group (the **group norms**) on the likelihood of a person conforming

Nurture In the nature-versus-nurture debate, the effect of external biological and social environmental factors on the development of our physical, psychological and behavioural characteristics; includes the environment within the womb, child-rearing practices, exposure to environmental toxins, education, peer-group, etc.

Obedience The behaviour that occurs in situations in which people change their actions in response to direct commands from others

Object permanence In Piaget's theory, the knowledge that an object continues to exist even when it is hidden from view

Occipital lobe The cerebral lobe located at the back of each cerebral hemisphere; contains the primary visual cortex and is responsible for decoding visual signals received from the retina, via the thalamus

Opinion A statement describing a personal belief or thought that cannot be tested (or has not been tested) and is unsupported by evidence unless it is an opinion that is provided by an expert who uses evidence to support their opinion

Optimism bias The tendency to overestimate the likelihood of experiencing positive events and underestimate the likelihood of experiencing negative events

Order effect An effect that can happen in experiments using within-subjects design, in which the order a participant participates in the conditions can affect their scores

Outlier Data points that differ substantially from the rest of the collected data

Outliers data points that differ substantially from the rest of the collected data

Paraphrase Writing something another person has said or written in a different way

Parietal lobe The lobe of each cerebral hemisphere that is located between the frontal and occipital lobes and above the temporal lobe; it contains the somatosensory cortex, and plays important roles in attention, spatial cognition, language, and in associating information from multiple senses and brain regions

Participants People who participate in a study or experiment

Past experience Our prior exposure to stimuli and previous life experiences

Peer review A review of a scientific report by independent experts in the field, who may comment on its quality and suitability for publication

Perception The psychological activity of becoming aware of objects and events that have been registered by the senses, which enables further cognitive processing to interpret the meaning of perceived objects and events and respond to them adaptively

Perceptual anomaly A perceptual irregularity caused by a deviation from the normal mental processes used to give meaning to stimuli

Perceptual disorder An impairment of perceptual processing that affects a person's ability to recognise and interpret sensory information, such as an inability to filter out irrelevant sounds or sights, or reduced awareness of objects in one side of space (see **spatial neglect**)

Perceptual hypothesis A hypothesis formed when your brain makes an 'educated guess' about how to interpret a particular pattern of sensory stimulation

Perceptual organisation The assembling of features of sensory stimuli to form a coherent whole or pattern that can be given meaning

Person perception The processes by which people think about, appraise and evaluate other people

Philosophy An intellectual approach to understanding the nature of things and behaviours that is based on reasoning and argument, rather than on scientific investigation

Physical cues The prompts such as physical appearance, facial expressions and overall manner that serve as signals that allow us to draw conclusions about a person

Pictorial cues A group of monocular depth cues present when looking at two-dimensional (pictorial) images that allow the brain to perceive apparent three-dimensional depth

Plagiarism Presenting someone else's work or ideas as your own, with or without their consent, by incorporating it into your work without full acknowledgement

Pons The largest part of the brain stem, it is a horseshoe-shaped mass of nerve fibres that connects the medulla with the cerebellum

Population A large group of people from which a sample is selected for study

Precision how close a set of measurement values are to one another

Prejudice A negative preconceived notion that we hold towards individuals because of their membership of a particular group

Primary auditory cortex The area within each temporal lobe that registers and processes auditory (sound) information

Primary data A type of data that is collected by researchers directly from main sources through interviews, surveys, experiments

Primary motor cortex The area at the rearmost portion of each frontal lobe that directs the body's skeletal muscles and controls voluntary movement

Primary somatosensory cortex The area at the frontmost portion of each parietal lobe, adjacent to the primary motor cortex, which registers and processes touch sensations from receptors in the body

Primary visual cortex The area at the base of each occipital lobe that registers, processes and interprets visual information sent from each eye

Proximity A Gestalt principle whereby stimuli that are close together in space are perceived as belonging together and forming a meaningful single unit or group

Psychological construct A concept used in psychology to describe a mental process, psychological state or trait; they are used to describe something that is believed to exist, because we can measure its effects, but we cannot directly observe or measure it

Psychological development The process of growth and change in humans' cognitive, emotional and social capabilities and functioning over the life span, from conception to old age, influenced by the interaction of biological, genetic, social, cultural and environment factors; studied in the discipline of developmental psychology

Psychological model Constructs built from current theoretical understandings to make theory more concrete and testable; models can have limitations, including incorrect assumptions or oversimplifications

Psychological theory An organised set of interrelated psychological constructs, mechanisms and processes that describes and/or explains a psychological system, process or experience

Psychosocial dilemma A conflict between personal impulses and the social world

Qualitative data Non-numerical data; captured using verbal descriptions

Quantitative data Numerical data; can be either counted or measured

Random allocation The allocation of participants to groups within an experiment by chance; it minimises the likelihood of extraneous participant variables (for example, age or level of stress) becoming confounding variables

Random error Unpredictable variations that can happen during measurement; they can be caused by limitations of instruments, environmental factors (such as sudden noises or interruptions) and slight variations in procedures

Random Sampling A sampling technique that uses a chance process to ensure that every member of the population of interest has an equal chance of being selected for the sample

Reception The process of sensory receptors detecting the presence of a stimulus or changes to a stimulus

Relative size A monocular pictorial depth cue where the smaller retinal image of two objects is perceived as being further away, and the larger retinal image of two objects is perceived as being closer

Repeatability The closeness of the agreement between the results of successive measurements of the same quantity, being measured under the same conditions of measurement

Representative heuristic A mental shortcut in decision-making where we estimate the likelihood of something occurring or being true based on its similarity to our existing understanding and expectations

Representative sample A sample that is representative of the population of interest and therefore enables generalisation of study results from the sample to that population

Reproducibility The closeness of the agreement between the results of measurements of the same quantity being measured under changed conditions of measurement

Rerouting The process by which an undamaged neuron that has lost connection with a damaged neuron connects with another undamaged neuron, so that a lost function moves from a damaged area to an undamaged area; creates an alternative pathway between active neurons that compensates for the loss of function

Research question A research topic expressed as a question that requires research and analysis to be answered; usually begins with 'how', 'what' or 'why'

Respect An ethical concept that means giving due regard to individual difference and ensuring the right to autonomy and choice

Reticular formation A dense and complex network of neurons within the brainstem that has connections with the spinal cord, thalamus, and cortex; it includes the reticular activating system, which plays a key role in regulating the sleep–wake cycle, arousal and alertness

Retinal disparity A binocular depth cue created by the slight difference between the retinal images of the left and right eye; also called binocular disparity

Reversibility of thought The recognition that relationships involving equality or identity can be reversed

Risk assessment A process in which the health and safety risks of an experiment are evaluated and steps are taken to mitigate them; may involve the use of safety data sheets for chemicals

Saliency detection The tendency to notice physical features that are unique, novel or stand out from the norm

Sample a group of people who are recruited from a larger population of interest to participate in an experiment

Sample size The size of the selected group (the sample) chosen from the population of interest to participate in an experiment

Sampling the process of selecting participants from a population of interest to participate in a research investigation

Schema A memory structure (mental representation) that represents a person's general knowledge about kinds of objects, entities and events that are developed through experiences with specific instances; used to support perception, recognition, thinking, and problem-solving

Secondary data The data that has already been collected through primary sources and made readily available for researchers to use for their own research

Selection A mechanism in the sensory perceptual system through which specialised neurons act as feature detectors that respond only to specific features within a stimulus pattern, ignoring others

Selective attention The ability to focus our awareness towards specific stimuli in the environment, or in mind, and to suppress awareness of other stimuli; can be consciously directed to serve a specific goal or captured involuntarily by a sudden event in the environment

Self-serving bias The bias that leads a person to attribute a positive outcome to their internal (dispositional) factors, yet attribute negative outcomes to external (situational) factors

Sensation The process through which sensory receptors detect and respond to sensory stimuli and transmit signals to specific sensory areas of the brain to generate sensory experiences

Sensitive period A stage during biological maturation when an organism is most able to acquire a particular skill or characteristic; a period of maximal plasticity, after which development within the area will take more effort and be slower and incomplete; used less restrictively than the term **critical period**, but they are similar concepts

Sensory receptors Specialised neurons located in sense organs that detect and respond to information (physical energy) from the environment and transmit it to the central nervous system

Similarity A Gestalt principle involving the viewer's tendency to perceive stimuli that have similar visual feature(s) as belonging to a group

Social categorisation The process by which we group individuals based on the perceived social category they belong to

Social cognition The way that we behave in social settings and also how we interpret the behaviours of others

Social comparison theory A theory that suggests that to gain an accurate understanding of who we are, we have a drive to make decisions, evaluations and judgments about ourselves in relation to those around us; we make upward, downward, and lateral comparisons to others

Social connection The belief that we belong to a group and generally feel close to other people; also known as social connectedness

social nonconformity Performing behaviours, expressing opinions, or making judgments that are not consistent with the normal standards of a social group or situation; can reflect ignorance or inability to conform, independence (retaining a preferred option), or anticonformity (deliberate, antisocial behaviour)

Social power The amount of influence that an individual can exert over another person

Spatial neglect A tendency to ignore the left or right side of one's body or the left or right side of visual space as a result of damage to one of the cerebral hemispheres (usually the right parietal lobe)

Spatial neglect A tendency to ignore the left or right side of one's body or the left or right side of visual space; it results from damage to one of the cerebral hemispheres (usually the right parietal lobe)

Spinning Dancer illusion A visual illusion in which the silhouette of a woman spinning on one foot with her leg extended appears to spontaneously change direction between turning left and turning right

Split-brain operation A surgical procedure to treat extreme epilepsy where the corpus callosum is severed to either partially or completely disconnect the two hemispheres of the brain

Sprouting The process by which a neuron grows new, bushier dendritic spines with more branches that enable it to connect to other active neurons

Standard deviation (SD) A measure of variability that describes a set of scores' average deviation (or distance) from the mean (that is, the average distance that a data point is from the mean)

Statistical rarity someone whose behaviour deviates from what is considered average

Status A person's position in the hierarchy of a group

Stereotype Generalised views about the personal attributes or characteristics of a group of people

Strange situation test An experiment designed to measure the quality of an infant's attachment to the primary caregiver

Stratified sampling A sampling technique used to ensure that a sample contains the same proportions of participants from each social group (that is, strata or subgroup) present in the population of interest

stress The physiological and psychological responses that a person experiences when confronted with a situation that is threatening or challenging

stressor The object, entity, or event that causes a feeling of stress

Supertaster A person who has more taste buds than normal and find tastes more intense than others

Sustained attention The type of attention used when we maintain the focus of our awareness on a task for an extended period of time; also called attentional vigilance

Symbolic thinking A cognitive skill that enables us to use symbols, gestures or images to mentally represent people, objects and events that may not be present

Synaesthesia A perceptual anomaly where information taken in by one sense is experienced in a way normally associated with another sense

Synaptic pruning The process of removing extra, weak or unused synaptic connections to increase the efficiency of neural transmission

Synaptogenesis The process by which new synapses are formed between neurons

Systematic errors Errors that affect the accuracy of a measurement by causing all of the readings to differ from the true value by a consistent amount; may be caused by measuring instruments not being correctly calibrated or by environmental interference

Taste buds Tiny structures located on the papillae on the tongue's surface

Taste sensitivity The intensity with which we perceive tastes and flavours

Taste The sensory experience of a food or drink that is put in the mouth and perceived as a flavour

Temporal lobe The lobe of the brain located on the lower side of each cerebral hemisphere; it is specialised for processing auditory information and comprehending language, and houses key structures involved in building memories (the hippocampus), emotional processing (amygdala), and recognition of objects, places and faces

Texture gradient A monocular pictorial depth cue whereby the surface features of an object become smaller and less detailed the more distant an object becomes

Thalamus A forebrain structure that sits on top of the brainstem and through which all sensory information (except smell) passes; it redirects this information to the appropriate sensory area of the cerebral cortex for processing

Top-down processing Information processing in which an overall hypothesis about a stimulus is applied to and influences the analysis of incoming information from a stimulus into a meaningful perception

Transduction The process of sensory receptors converting stimulus energy into impulses of electrochemical energy

Transmission The process whereby neural impulses leave sensory receptor sites and travel to the areas in the brain that are specialised to receive them

Traumatic brain injury (TBI) A form of acquired brain injury that results from injury caused by an external force

True value the value or range of values that would be found if a quantity could be measured perfectly

typical behaviour Behaviour that represents what most people do

Unanimity Agreement amongst all members of a group

Uncertainty A measurement of data or an evaluation that states how well something is known

Validity how well the design of a scientific investigation and its measurements (for example, questionnaire results or response times) provide meaningful and generalisable information about the psychological constructs of interest

Variability how spread out or clustered together a series of experimental scores or measurements are

Variable any factor in a study that can vary in its score, amount or type and that can be measured, recorded or manipulated

Visual agnosia A neurological disorder in which the sufferer has total or partial loss of the ability to recognise and identify familiar objects

Visual constancies The group of perception principles that allow us to view objects as unchanging in terms of their actual size, shape, brightness and orientation, even when there are changes to the image that the object casts on the retina

Visual illusion A consistent perceptual error in interpreting a real external stimulus

Visual perception principles Rules that our brains apply automatically to organise and interpret visual stimuli in a consistent and meaningful way

Voluntary participation An ethical guideline that means Ensuring there is no coercion or pressure to participate

Wernicke's aphasia An impairment in the ability to understand or repeat spoken language and to name objects; characterised by fluent but disorganised and inappropriate speech; caused by damage to Wernicke's area

Wernicke's area An area toward the back of the temporal lobe associated with the interpretation of speech and sounds

Withdrawal rights An ethical guideline that means Allowing participants to discontinue involvement in an experiment, without penalty

Within-subjects design A type of experimental design in a controlled experiment in which each participant is exposed to both the experimental condition and the control condition also known as a repeated-measures design

Answers

Chapter 2

ACTIVITY 2.1

Apply it

3 Heredity refers to a person's genetic code or DNA passed from one generation to the next. Hal and Mal are identical twins so they will have identical DNA.
Environment refers to the way in which your behaviour and your environment have an impact on your genetic expression. If Hal makes lifestyle choices such as smoking marijuana, or has work environments that are highly stressful, or experiences greater exposure to dysfunctional parents, even though Mal has the same genetic composition and predisposition to schizophrenia, those different lifestyle or environmental conditions could mean that this disease never develops in Mal. Mal will still have the gene(s) connected with the development of schizophrenia, but these gene(s) will remain 'turned off'.

Exam ready

4 C. It refers to the impact of a person's experiences and environment on genetic expression. Two people with identical genetics can have different life experiences that trigger the activation or suppression of different parts of their genetic code.

ACTIVITY 2.2

Apply it

2 Biological: genetic inheritance from her mother
Psychological: anxiety, stress of starting a new job and long travel times, poor coping strategies (worried that she will not cope)
Social: new workplace, support network of family and friends

Exam ready

3 D

ACTIVITY 2.3

Apply it

2 The concrete operational stage (7–11 years old) is typically when children develop the cognitive understanding of conservation of volume. At 9 years old, Declan is within the age range that should allow him to answer that the glasses hold the same amount of water (despite apparent differences in the height of the water levels in the new containers). Iris is still in the preoperational stage (2–7 years old), so she is likely to say that the tall, thin glass has more water in it because its water level is higher than the level in for the short, wide glass.

Exam ready

3 D. Object permanence is the understanding that something still exists even if it can't be seen. Dylan has not yet achieved this milestone, which should be reached during the sensorimotor stage (birth – 2 years).

ACTIVITY 2.4

Apply it

2 Imprinting is a type of learning that occurs when a very young animal forms an attachment to an object, person or animal with which it has its first sensory experience. Imprinting in ducklings occurs during a critical period (30 hrs). If this imprinting doesn't occur during a specific window, it is very difficult for this behaviour to be learned.

Exam ready

3 B

2 End-of-chapter exam

Section A: Multiple-choice

1 B	2 C	3 C	4 C	5 B
6 D	7 B	8 A	9 D	10 D
11 D	12 C	13 C	14 C	15 A
16 A	17 B	18 B	19 D	20 A

Section B: Short answer

1 There is an important interaction between hereditary factors and environmental factors in the development of an individual. It is not the case that one is more important than the other in determining an individual's psychological development. We know from research that our genes and hormones as well as our environmental experiences and stimuli are interrelated factors contributing to our development.

2 Insecure avoidant attachment: is not affected by mother's presence or absence [1 mark]
Secure attachment: distressed when mother leaves the room but happy and comforted by her return [1 mark]
Insecure resistant attachment: very distressed when mother leaves the room and is not comforted by her when she returns [1 mark]

3 a Concrete thinking involves thinking in terms of direct experiences or specific objects or events; characteristic of young children's thinking (sensorimotor and early stages of preoperational). For example, if an adult says

'It's raining cats and dogs', the young child who only has concrete thinking capacities may look up into the sky for the cats and dogs. [2 marks]

b Symbolic thinking is a cognitive skill that enables us to use symbols, gestures or images to mentally represent people, objects and events that may not be present; characteristic thinking in the preoperational stage of development. For example, children use household objects for pretend play. [2 marks]

c Abstract thinking is thinking in terms of general concepts rather than specific objects, experiences or events; characteristic thinking in the formal operational stage of development. For example, adolescents about to conduct a science experiment hypothesise the results of the study. [2 marks]

Chapter 3

ACTIVITY 3.1

Apply it

2 Statistical rarity:

- Anthony is of high intelligence.

Personal distress:

- Anthony has taken a number of days off and has reported that he struggles to get out of bed most days.
- He is experiencing work-related stress.

Maladaptive behaviours:

- Anthony has reported that he tends to lash out at loved ones when they try to help him.
- Anthony has avoided seeing friends and has reported that he drinks alcohol almost every night.

Exam ready

3 C

ACTIVITY 3.2

Apply it

2 The student would likely be considered neurotypical.

Emotional responses:

- 'When stressed, the student takes breaks as needed and asks for support.'
- 'They support their peers'

Behavioural responses:

- 'The student arrives on time, with all required materials.'

Cognitive responses:

- 'The student's academic results are consistently near the top of the class.'

Exam ready

3 B

ACTIVITY 3.3

Apply it

2 The neurodiversity view is that there is a natural difference that occurs between people's brains; some brains are just wired differently during development. Even though ADHD is a functional problem, it is linked to structural differences in the brain.

Exam ready

3 D. This is a false statement about neurodiversity. The opposite is true. Neurdiversity encourages people to focus on the strengths rather than the challenges that come from difference.

ACTIVITY 3.4

Apply it

2 The GP or family doctor is the first place PJ should go. They can suggest a psychologist or refer you to a psychiatrist. You do not need a referral to see a psychologist but your GP can advise you on the best person for your situation.

Exam ready

3 D. While both psychologists and psychiatrists understand how the brain works and how thoughts, feelings and behaviours interact, and both can treat mental illness with psychotherapies, only a psychiatrist can offer drug treatments so their clients tend to be people whose situation would most benefit from medical interventions.

3 End-of-chapter exam

Section A: Multiple-choice

1 B	**2** D	**3** C	**4** D	**5** C
6 B	**7** C	**8** D	**9** C	**10** C
11 A	**12** B	**13** B	**14** B	**15** B
16 D	**17** C	**18** D	**19** D	**20** A

Section B: Short answer

1 A social norm is a socially defined rule, standard or value that governs expected behaviours within groups. Behaviour can be considered typical according to whether the behaviour fits in with the norms of that society [1 mark]. For example, looking someone in the eye when having a discussion is considered normal in Australia. It indicates to the speaker that you are listening. However, for some Asian cultures eye contact is used to intimidate someone. [1 mark]

2 Adaptive behaviour is when an individual can interact with others and adjust their behaviour to meet the changing demands of everyday living [1 mark]. Maladaptive behaviour is behaviour that interferes with an individual's ability to complete daily tasks and it can lead to emotional, social and health problems. [1 mark]

3 Student answers may vary, but should be similar to the following. [5 × 1 mark]
- Contact parents to gain consent for further investigation.
- Contact GP and make an appointment.
- GP recommends a psychologist.
- Psychologist observes behaviour and interviews student.
- Psychologist consults *DSM-5-TR*™ to make a diagnosis.
- Psychologist creates a behaviour management plan for the student.

Unit 1 Area of Study 1 review

Section A: Multiple-choice

1 B	**2** C	**3** C	**4** B	**5** B
6 D	**7** C	**8** A	**9** C	**10** C
11 B	**12** A	**13** B	**14** B	**15** D
16 C	**17** C	**18** D	**19** B	**20** A
21 A	**22** C	**23** B	**24** C	**25** C
26 A	**27** A	**28** D	**29** B	**30** C

Section B: Short answer

1 a Maladaptive [1 mark], because staying in bed all day is affecting her attendance at school and shows a lack of coping in everyday life. [1 mark]

b Student answers may vary, but could include providing strong social supports to encourage her to go to school and to help her with her lack of coping. [2 × 1 mark]

c Aya's behaviour is considered abnormal or atypical [1 mark] because she is showing a lack of being able to do tasks that are considered socially normal for her age, such as attending school daily. [1 mark]

2 a The nature-versus-nurture debate aimed to determine which has more effect on our development. On one side of the debate (nature), it was believed that heredity had the most impact on our development [1 mark]; on the other side (nurture), it was believed development was all up to the environment in which the individual grew up. [1 mark]

b The conclusion drawn was that both nature and nurture have a role in our development. [2 marks]

3 Social development in children is primarily focused on interactions with family and close friends [1 mark]. As children develop further and become adolescents, this shifts to where there is more social reliance on friends and peers than family. [1 mark]

4 Student answers may vary. One strength of using psychological criteria is it makes it clear to psychologists which signs and symptoms must be present for individuals [1 mark] to have particular disorders [1 mark]; however, a limitation is they are objective measures [1 mark] and don't account for individual differences between people. [1 mark]

5 A person who is neurotypical has generally what people would consider as normal brain functioning and behaviours [1 mark], whereas someone who is neurodivergent differs in their brain functioning and behaviours from most other people. [1 mark]

Chapter 4

ACTIVITY 4.1

Apply it

2 Student responses could include:
- ablation would be the preferred method
- lobotomies are outdated approaches that are not as targeted as ablation
- ablation can remove tumours via laser, surgery or vaporisation.

Exam ready

3 B

ACTIVITY 4.2

Apply it

2 a Exercise 1
b Exercise 4
c Exercise 3
d Exercise 2

Exam ready

3 C

4 End-of-chapter exam

Section A: Multiple-choice

1 C	**2** B	**3** B	**4** B	**5** C
6 A	**7** A	**8** B	**9** C	**10** C
11 B	**12** A	**13** D	**14** B	**15** A
16 C	**17** D	**18** B	**19** B	**20** A

Section B: Short answer

1 Hypothalamus, thalamus, cerebrum [3 × 1 mark]

2 Psychological (any one of these): mood swings, decreased decision making, planning and organisational skills; poor impulse control; inability to concentrate; poor problem solving skills; [1 mark]

Physiological (any one of these): problems with voluntary movement, balance or posture; inability to coordinate movement, loss of motor skills. [1 mark]

3 Severing a section of the corpus callosum [1 mark] would interrupt the transfer of sudden random bursts of electrical impulses generated in one hemisphere [1 mark] from being transferred to the other hemisphere [1 mark] and causing seizures. [1 mark]

Chapter 5

ACTIVITY 5.1

Try it

1 In response to neuron i losing its connection with the purple neuron shown below it, some of the existing axon terminals of neuron ii have connected with the dendrites of this purple neuron. Also, neuron ii has grown a new axon branch from the side of its axon across to this neuron.

Apply it

2 a This is adaptive plasticity, because it occurs in response to learning something new, rather than being a change that occurs through normal development and maturation.

b Sprouting would occur. Sprouting refers to the creation of new connections between neurons. Sprouting results in an increase in the number of synapses between neurons.

As Mitchell uses the new bowling technique, neurons would form new connections with other neurons. As he continues to practise, new axon extensions would grow from the presynaptic neuron and more dendritic spines would allow for stronger connections along this neural pathway

Exam ready

3 D. The frontal lobe, in particular the prefrontal cortex, is slower to develop than other areas of the cerebral cortex.

ACTIVITY 5.2

Apply it

2

- Gage changed from friendly and quietly spoken to impatient and aggressive. (P)
- Gage had trouble maintaining friendships. (S)
- Gage changed from considerate to irresponsible and impulsive. (P)
- Gage developed difficulties with problem-solving. (P)
- Gage could no longer manage his job as a railway worker. (S)
- Gage had impaired ability to sustain his attention. (P)
- Gage had difficulty moving some facial muscles. (B)

Exam ready

3 D

ACTIVITY 5.3

Try it

1 Patient 1 has Wernicke's aphasia. Patient 2 has Broca's aphasia.

Apply it

2 Wernicke's aphasia is a language deficit linked to the *comprehension* of speech. It is also called 'fluent aphasia', because the patient has difficulty understanding the written and spoken word. The response from Patient 1 is made up of long but nonsensical sentences.

Broca's aphasia is a language deficit linked to the *production* of speech. The patient will have difficulty moving the muscles required for producing fluent speech and will often speak in short, disjointed sentence fragments that lack grammatical structure. The response from Patient 2 is made up of short sentences with a large number of pauses, indicating difficulty with speech production.

Exam ready

3 C. Wernicke's area is not in the occipital lobe.

ACTIVITY 5.4

Try it

1 a stage 3

b stage 1

c stage 4

d stage 2

Apply it

2 CTE develops as a result of consistent, forceful blows to the head. By limiting the number of times a young player heads the ball, there will be fewer potential forceful hits to their head. By reducing the number of potential hits to the head, these rule changes could help protect the developing brains of young soccer players.

Exam ready

3 C. It is false to say that CTE can be accurately diagnosed whilst the person is alive; CTE can only be accurately diagnosed after an autopsy has been completed.

5 End-of-chapter exam

Section A: multiple-choice questions

1 A	**2** C	**3** D	**4** A	**5** B
6 A	**7** D	**8** D	**9** D	**10** D
11 B	**12** C	**13** A	**14** C	**15** A
16 B	**17** D	**18** C	**19** C	**20** C

Section B: short-answer questions

1 Any two of the following principles identified: confidentiality, debriefing, informed consent procedures, voluntary participation, withdrawal rights. [2 marks]
Answers must state what this neurosurgeon should do to implement the stated guideline; for example, for voluntary participation the surgeon should ensure that the patient consented to the surgery free of coercion or pressure so that they freely choose to be involved. [2 marks]

2 Myelin protects and insulates a neuron's axons. This increases the speed at which the electrical impulse travels within the neuron [1 mark]. If there is a decline in myelination, the impulse travels more slowly. This results in neural pathways transmitting information less efficiently and more slowly than they previously did. [1 mark]

3 Identify by means of either an MRI or CT scan. [1 mark]
Identify what this scan would be used for; that is, to determine whether Broca's area (left frontal lobe) or Wernicke's area (left temporal lobe) was damaged, as well as the extent of the damage. [1 mark]
You would also test their ability to understand and produce language by getting them to answer questions, name objects or have a conversation with them. [2 marks]

Unit 1 Area of Study 2 review

Section A: Multiple-choice

1 C	**2** A	**3** C	**4** D	**5** A
6 B	**7** C	**8** A	**9** B	**10** B
11 C	**12** C	**13** A	**14** B	**15** D
16 A	**17** D	**18** B	**19** B	**20** D
21 D	**22** A	**23** D	**24** D	**25** A
26 C	**27** C	**28** D	**29** C	**30** D

Section B: Short answer

1 [1 mark for image and 1 mark for each brain area correctly identified]

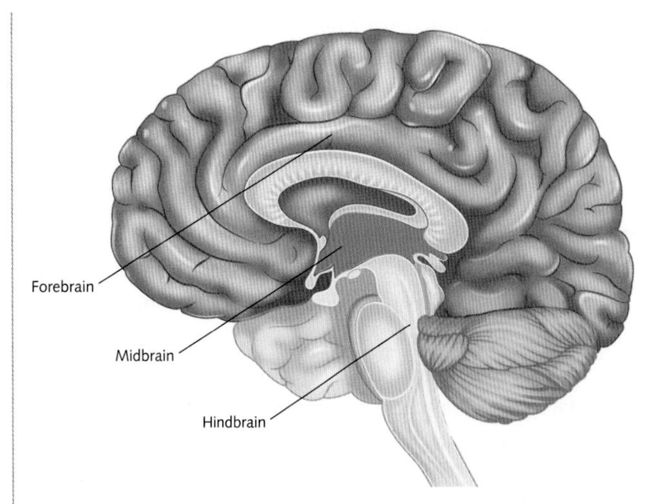

2 a Jake's injury affected his biological functioning in terms of motor (movement) control, as Jake was unable to move his right arm for 2 months after the incident. [2 marks]

b Plasticity was responsible for Jake regaining function. The neurons damaged in his left motor cortex could no longer provide the function of moving his arm, so surrounding neurons needed to reroute and sprout to make new connections and take on the role of the lost function. [3 marks]

c Chronic traumatic encephalopathy (CTE) is the term used to describe brain degeneration caused by repeated head traumas. [2 marks]

3 Students must identify one of the following [1 mark for region and 1 mark for role]

- The frontal lobe plays a role in motor functioning, controlling voluntary bodily movements.
- The occipital lobe processes visual information received from the eyes.
- The parietal lobe processes sensory information felt by the body.
- The temporal lobe processes sound received via the ears.

4 a The hindbrain was most likely to have been affected, because this is the region of the brain located at the posterior of the head. [2 marks]

b Any three of the following: [1 mark each]

- lack of balance
- struggles with voluntary movement
- respiratory issues
- irregular sleep patterns
- issues with reflexes.

5 Student answers may vary, but could include: [1 mark for each example and 1 mark for accurate description]
- Biological: sleep issues, because regulation of sleep is controlled in the brain; if this is damaged we may struggle to fall asleep, stay asleep or may struggle to wake up
- Psychological: memory loss, because we will experience memory loss of events before and after trauma as well as possibly struggling to commit new things to memory
- Social: maintaining friendships, because of a range of factors such as changes in personality and also stigma associated with the brain damage

Chapter 7

ACTIVITY 7.1

Apply it

2 Person perception is the process by which people think about, appraise and evaluate other people. Physical cues, saliency detection and social categorisation all combine to allow us to make judgements about others. Physical cues refer to the physical appearance, demeanour, facial expressions etc. of people. Further to this, physical features that are salient – that is, stand out – capture our attention. Finally, we tend to make quick judgements about the categories that people may belong to (i.e. race, age, sex etc.), and this is known as social categorisation.

Exam ready

3 A. Person perception allows us to make quick judgements about others without much mental effort; it can lead to inaccurate judgements about stereotyping.

ACTIVITY 7.2

Try it

1 Multiple answers to the nine-dot problem can be found online.

Apply it

2 a Functional fixedness is a cognitive bias that suggests that an object or item can only be used in a particular way (its most common use).

b, c Student answers will vary.

Exam ready

3 D. Dan initially demonstrated functional fixedness: he thought only a screw-driver would be able to turn the screws. Then, when he saw Jen successfully turn the screws with a knife, he said that he thought that would work (hindsight bias).

ACTIVITY 7.3

Apply It

2 In this example, the availability heuristic has generated the correct answer. In solving the problem, most people would have spent a few seconds thinking of words that start with 's' and words that start with 'x' to guide their answer. In both situations, we quickly considered our responses based on the information that was easily recalled, or most readily available to us, to arrive at our answer.

Exam ready

3 C. Heuristics are useful in arriving at a quick decision; however, our responses are not always correct.

ACTIVITY 7.4

Try it

1 a First Nations peoples

b Older people

c First Nations peoples; young people; religious minorities; people with a disability; LGBTQI+ people; racial minorities; people with low income; and women.

d Student answers will vary.

Apply it

2 Stereotyping places people in categories based on their shared characteristics due to their membership of a particular group. Stereotypes are generalised views that we hold about others, which don't take into consideration the uniqueness of each person.

Prejudice occurs because of stereotyping. Prejudice is a negative preconceived notion that we hold towards individuals due to the membership of a particular group. Prejudice is marked by suspicion, fear or hatred. Discrimination occurs when these prejudiced views are demonstrated as behaviours (either direct or indirect).

Exam ready

3 C. Indirect discrimination is characterised by the application of a blanket rule that disadvantages one group more than another; while direct discrimination is the application of a policy or approach that is designed to disadvantage particular groups.

7 End-of-chapter exam

Section A: Multiple-choice

1 C	2 C	3 C	4 B	5 C
6 B	7 C	8 D	9 C	10 A
11 D	12 C	13 B	14 C	15 C
16 B	17 C	18 B	19 D	20 C

Section B: Short answer

1 Clear definitions/understanding of prejudice and discrimination. [1 mark]
The difference between the two is that prejudice is a negative view held towards a group, whereas discrimination is an action/behaviour (direct or indirect) that negatively affects a specific group. [1 mark]

2 Affective component: Rashad is opposed to the detention of refugees and asylum seekers. [1 mark]
Behavioural component: Rashad donates money to charities that support refugees and asylum seekers. [1 mark]
Cognitive component: Rashad believes that the government has the resources available to support refugees and asylum seekers. [1 mark]

3 Clear definitions/understanding of cognitive dissonance and cognitive biases: cognitive dissonance is a state of psychological discomfort that we experience when our behaviours don't align with our beliefs or attitudes. Cognitive biases are systematic errors that occur in our decision-making. [1 mark]
Explanation of how cognitive biases reduce cognitive dissonance: Cognitive biases reduce our experience of cognitive dissonance as they can serve to justify our behaviours or reduce the importance of conflicting information. [1 mark]
Example of the application of a cognitive bias: The optimism bias is used by individuals to downplay the potential negative effects of a particular behaviour while also emphasising the positive elements of that behaviour. For example, knowing the potential dangers of smoking, a smoker might highlight that smoking reduces their levels of stress and that their grandfather smoked every day and lived well into his 90s. [1 mark]

Chapter 8

ACTIVITY 8.1

Apply it

2 Broadly speaking, members of individualist cultures are encouraged to stand out from the crowd, whereas this is less encouraged in collectivist cultures. Members of individualist cultures were more likely to select the unique pen colour, while members of collectivist cultures were less likely to select the unique pen colour and, in turn, 'stand out'.

Exam ready

3 B. Both vertical-individualist cultures and vertical-collectivist cultures accept that hierarchies will exist because of differences in levels of status and power.

ACTIVITY 8.2

Apply it

2 Rationale: Members of the group discount warnings and downplay negative feedback that could challenge the assumptions of the group. Even though warnings about the safety of the mission had been put forward, NASA decided to continue with the mission.
Pressure: Members of the group apply direct pressure to any member who entertains doubt, dissenting views or opinions. NASA asked members of Morton Thiokol to change their evaluation of the safe temperatures of the rubber seals.

Exam ready

C. During the debriefing session, participants should be made aware of any deception that was used in the study.

ACTIVITY 8.3

Apply it

2 Benefits (any two of the following):
- can serve as a source of inspiration
- can show us how to regulate our emotions or behave in certain situations
- helps in decision-making
- any other applicable response.

Downsides (any two of the following):
- can encourage pursuit of unattainable beauty standards
- can lead us to engage in unhealthy or dangerous behaviours to be more like the person who we are comparing ourselves to

- can generate feelings of inferiority
- can lead to development of mental health conditions, including depression and anxiety
- any other applicable response.

Exam ready

3 B. The question was asking which statement is not true about social comparison theory. Lateral social comparison occurs when we compare ourselves to someone who we perceive to be relatively equal to us.

ACTIVITY 8.4

Apply it

2 Sienna demonstrates anti-conformity by getting her nose pierced despite her parents telling her not to. On the other hand, Sienna's behaviour can also be described as conformist: she stated that she feels like the odd one out in her friendship group because she doesn't have her nose pierced. As such, she got her nose pierced while her friends were getting more piercings to fit in with the group.

Exam ready

3 B

8 End-of-chapter exam

Section A: Multiple-choice

1 D	2 A	3 D	4 A	5 C
6 D	7 A	8 D	9 B	10 C
11 D	12 D	13 A	14 B	15 C
16 B	17 C	18 D	19 D	20 C

Section B: Short answer

1 Any three of the following [1 mark each]:
- The prisoners were 'arrested' at their homes and taken to the 'prison' via police cars.
- The prisoners were 'processed' once they arrived at the prison (ordered to undress and were sprayed for lice).
- The prisoners were assigned numbers and had to respond to these numbers throughout the experiment.
- Different uniforms were allocated to the prisoners and the guards.
- The guards wore reflective sunglasses and were given batons.
- Any other acceptable response.

2 Two differences are [1 mark each]:
- When acting independently, the individual attempts to resist the influence of the group; in contrast, the response of an anti-conformist is heavily influenced by the group.
- People who use anti-conformity might agree with the group privately, but will disagree with them publicly. People who act independently are consistent in their public and private beliefs.

3 Any one of the following advantages [1 mark]:
- enables people to maintain social connections and communicate across geographic barriers
- allows greater social inclusion for those who may be otherwise excluded in their day-to-day lives; for example, sexually diverse individuals
- may serve as a source of inspiration or creativity
- any other appropriate response.

Any one of the following disadvantages [1 mark]:
- increased risk of developing depression and anxiety due to upward social comparison
- cyberbullying
- may become disconnected or withdrawn from others
- any other appropriate response.

Unit 2 Area of Study 1 review

Section A: Multiple-choice

1 D	2 A	3 C	4 D	5 A
6 A	7 A	8 B	9 C	10 C
11 A	12 A	13 C	14 D	15 D
16 B	17 C	18 B	19 C	20 B
21 C	22 D	23 A	24 C	25 D
26 D	27 A	28 D	29 A	30 C

Section B: Short answer

1 a A: he doesn't like his boss [1 mark], B: he continues to work in his job [1 mark], C: he believes that his boss plays favourites. [1 mark]

b Cognitive dissonance [1 mark], because his behaviour is not matching his attitudes: he does not like his boss yet continues to be nice to him and work in the same job. [1 mark]

2 a Asch aimed to investigate whether people would conform [1 mark] when there is clearly a right and wrong answer. [1 mark]

b People will conform with the majority [1 mark] even when they are aware that the majority are wrong. [1 mark]

3 Prejudice is a negative attitude or belief held against a particular group [1 mark], whereas discrimination is unjust treatment of a particular person because of their membership of a particular group. [1 mark]

4 Individuals can use cognitive bias by creating their own subjective reality [1 mark], which enables them to have congruent beliefs and attitudes. [1 mark]

5 Obedience can be useful in society because it can keep us safe. For example, it allows us to follow directions from doctors and police that help ensure our safety [1 mark]; defying these instructions may have negative consequences, so it is in our best interests in these cases to be obedient. [1 mark]

Chapter 9

ACTIVITY 9.1

Apply it

2 The pain signals being sent from your tongue to your brain are an example of bottom-up processing.
Because you remember how painful the first sip was, and have no desire to repeat the experience, blowing on your coffee before the second sip is an example of top-down processing.

Exam ready

3 A. The other options are describing features of top-down processing.

ACTIVITY 9.2

Apply it

2 Divided attention is when we rapidly switch the focus of our attention between two or more stimuli so we can perform two or more tasks at the same time. When you are in class you are using divided attention because you are quickly switching between listening, reading and writing so you can achieve all of these tasks at the same time.
Sustained attention is the ability to focus on a stimulus for an extended period of time. When you are studying at home for hours you are using sustained attention.

Exam ready

3 B. As you are rapidly switching the focus of your attention between two or more stimuli – finishing the customer's order and thinking about the answer to their question about the store's opening hours – this is an example of divided attention; you are aiming to perform two or more tasks at the same time.

ACTIVITY 9.3

Apply it

3 The tendency to perceive stimuli in a particular way is referred to as perceptual set. This predisposition is a psychological factor that can influence each participant's interpretation of the misspelled word. Sails are linked to people's knowledge of boats and seals are linked to people's prior knowledge of animals.

Exam ready

4 A. Changes in vision as a result of ageing is a biological influence on visual perception. The others are all psychological factors that can influence visual perception.

9 End-of-chapter exam

Section A: Multiple-choice

1 C	2 C	3 B	4 B	5 C
6 B	7 D	8 D	9 B	10 D
11 D	12 D	13 C	14 C	15 B
16 B	17 D	18 C	19 C	20 B

Section B: Short answer

1 a Retinal disparity [1 mark]
b Accommodation [1 mark]

2 a Closure [1 mark]
b The brain mentally fills in the blanks to complete an incomplete shape. [2 marks]

3 Perception is both a bottom-up process and a top-down process. [1 mark]
When we process new or complex stimuli, we use bottom-up processing because we process the basic features separately. Then we integrate them to form a perception. [1 mark]
When processing simple stimuli or stimuli that we have previously encountered, we use top-down processing. We apply pre-existing schemas to the stimuli, based on prior knowledge and experience and expectations, to form a perception. [1 mark]

Chapter 10

ACTIVITY 10.1

Apply it

2 Patient A has visual agnosia, Patient B has tactile agnosia.

Exam ready

3 D. This patient cannot name the tennis ball by sight despite his visual system being intact and his knowledge of what a tennis ball is (demonstrated by his ability to name it by touch).

ACTIVITY 10.2

Apply it

2 Steven Fry would need to give his guests miraculin. Miraculin (from the miracle fruit or berries) works by binding to the sweet taste receptors on the tongue. This changes the shape of the receptors so they are activated by acids that normally result in a sour taste but now result in a sweet taste. It activates receptors that send messages to the brain via neural signals, to perceive the current flavour as sweet.

Exam ready

3 A

ACTIVITY 10.3

Apply it

2 Spatial neglect most commonly occurs as a result of damage to the right parietal lobe. As a result, individuals neglect the left side of their world. As spatial neglect is an attentional disorder, the therapist would place the patient's right arm in a sling while completing the motor tasks to encourage them to pay greater attention to the left-hand side of their body.

Exam ready

3 A

10 End-of-chapter exam

Section A: Multiple-choice

1 C	2 D	3 B	4 D	5 A
6 D	7 A	8 B	9 D	10 C
11 A	12 B	13 D	14 C	15 D
16 A	17 A	18 D	19 C	20 C

Section B: Short answer

1 Agnosia is a rare neurological disorder that causes an inability to recognise and identify sensory stimuli (e.g. objects, people, smells or sounds) despite otherwise normally functioning senses. [1 mark]

2 The apparent-distance hypothesis suggests that an apparently more-distant object that has the same-sized retinal image as an apparently nearer object will be perceived as the larger. [2 marks]

3 Hamish's primary auditory cortex in his temporal lobe would be activated (show red on the PET scan) by the sound of all the numbers, letters or words. [2 marks]
When specific trigger numbers, letters or words were read out, his primary visual cortex in his occipital lobe would be activated (show red on the PET scan). [2 marks]

Unit 2 Area of Study 2 review

Section A: Multiple-choice

1 A	2 C	3 A	4 B	5 D
6 B	7 C	8 A	9 B	10 A
11 A	12 C	13 D	14 A	15 B
16 A	17 D	18 A	19 B	20 A
21 C	22 B	23 B	24 D	25 A
26 C	27 B	28 C	29 D	30 D

Section B: Short answer

1 a Any one of:
Looking at a letter or number produces the experience of colours
May not see a colour but may just 'know' that a particular letter or number is a particular colour

b A perceptual distortion involves a mismatch between a perceptual experience and physical reality [1 mark]. Synaesthesia can be a perceptual distortion because stimulation of one sense produces additional unusual experiences in another sense, hence there is a mismatch. [1 mark]

2 During selective attention, we focus on one thing and suppress other stimuli [1 mark], whereas divided attention allows us to attend to multiple stimuli at once. [1 mark]

3 Biological: ageing – visual decline as we age [1 mark]
Psychological: experience – we expect to see something like what we have seen before [1 mark]
Social: group pressure – can influence us to perceive in a particular way [1 mark]

4 a People are shown a list of words printed in different colours. They're then asked to name the ink colour, rather than the word itself. [1 mark]
People are much slower and make more mistakes when the meaning of the word and the ink colour don't match. [1 mark]

b Top-down processing is when we use prior expectation to interpret incoming stimuli. This is top-down processing [1 mark] because people automatically read the words before thinking about the features of the words, such as colour, which is what makes this task so difficult. [1 mark]

5 It is likely that he is a supertaster. [1 mark]
This means that he has more papillae and ultimately more taste buds, so that he tastes things more intensely than the average person. As such, he will tend to have aversions to bitter tasting foods because they taste even more bitter than they do for the average person. [1 mark]

References

Aarseth, E., Bean, A. M., Boonen, H., Colder Carras, M., Coulson, M., Das, D., Deleuze, J., Dunkels, E., Edman, J., Ferguson, C. J., et al. (2017). Scholars' open debate paper on the World Health Organization ICD-11 Gaming Disorder proposal. *Journal of behavioural addiction, 6*, 267–270.

Abello-Contesse, C. (2009). Age and the critical period hypothesis. *ELT journal, 63*(2), 170–172.

Ainsworth, M. D. S. (1972). Attachment and dependency: a comparison. In J. L. Gerwirtz (ed.), *Attachment and dependency*, 97–137. Washington, DC: V. H. Winston & Sons.

Agarwal, P. K. (2019). Retrieval practice & Bloom's taxonomy: Do students need fact knowledge before higher order learning? *Journal of Educational Psychology, 111*(2), 189–209. https://doi.org/10.1037/edu0000282

Akatemia Suomen (Academy of Finland). (2016, February 8). Sustained aerobic exercise increases adult neurogenesis in brain. *ScienceDaily*. www.sciencedaily.com/releases/2016/02/160208083606.htm

Akhmad, M., Chang, S., & Deguchi, H. (2021). Closed-mindedness and insulation in groupthink: their effects and the devil's advocacy as a preventive measure. *Journal of computational social science, 4*, 455–478.

Allport, G. (1954). *The nature of prejudice*. Reading, MA: Addison-Wesley.

Almond, S. (2008). The influence of the media on eating disorders. *Journal of human nutrition and dietetics, 13*(5), 363–371.

Arioli, M., Crespi, C., & Canessa, N. (2018). Social cognition through the lens of cognitive and clinical neuroscience. *BioMed research international, 2018*. 4283427. doi: 10.1155/2018/4283427

Armstrong, T. (2015). The myth of the normal brain: embracing neurodiversity. *AMA Journal of ethics*, 17(4), 348–352. doi: 10.1001/journalofethics.2015.17.4.msoc1-1504.

Asch, S. (1952). Effects of group pressure upon the modification and distortion of judgements. In H. Guertzkow (Ed.), *Groups, leadership and men*, Pittsburgh: Carnegie Press.

Asch, S. E. (1955). Opinions and social pressure. *Scientific American, 193*, 31–35.

Atkinson, R. C., & Shiffrin, R. M. (1968). Human memory: A proposed system and its control processes. In K. W. Spence & J. T. Spence (Eds), *The psychology of learning and motivation: II*. pp. 89–195). Academic Press. doi: 10.1016/S0079-7421(08)60422-3

Australian Associated Press. (2021, October 9). Late Collingwood great Murray Weideman Diagnosed with CTE. *The Guardian*. https://www.theguardian.com/sport/2021/oct/09/late-collingwood-great-murray-weideman-diagnosed-with-cte

Australian Bureau of Statistics (ABS). (2021). First insights from the National Study of Mental Health and Wellbeing, 2020–21. https://www.abs.gov.au/articles/first-insights-national-study-mental-health-and-wellbeing-2020-21

Australian Bureau of Statistics. (2012). 4338.0 – Profiles of Health, Australia, 2011–13. https://www.abs.gov.au/ausstats/abs@.nsf/lookup/4338.0main+features212011-13

Australian Human Rights Commission. (2021). https://humanrights.gov.au/quick-guide/12049

Bachman, J. G., & Johnson, L. D. (1979). The freshmen. *Psychology today, 13*, 78–87.

Bailes, J. E., Petraglia, A. L., Omalu, B. I., Nauman, E., & Talavage, T. (2013). Role of subconcussion in repetitive mild traumatic brain injury. *Journal of neurosurgery, 119*(5), 1235–1245.

Barkley, R. A. (2006). A theory of ADHD. In R. A. Barkley, *Attention-deficit hyperactivity disorder: A handbook for diagnosis and treatment*, 297–334.

Baron-Cohen, S. (2006). Autism, hypersystemizing, and truth. *Quarterly journal of experimental psychology, 61*, 64–75. doi 10.1080/17470210701508749

Baron-Cohen, S., Ashwin, E., Ashwin, C., Tavassoli, T., & Chakrabarti, B. (2009). Talent in autism: hyper-systemizing, hyper-attention to detail and sensory hypersensitivity. *Philosophical transactions of the Royal Society B: Biological sciences, 364*(1522), 1377–1383.

Baron-Cohen, S., Burt, L., Smith-Laittan, F., Harrison, J., & Bolton, P. (1996). Synaesthesia: Prevalence and familiarity. *Perception, 25*(9), 1073–1080.

Bartels, J. M., & Griggs, R. A. (2019). Using new revelations about the Stanford prison experiment to address APA undergraduate psychology major learning outcomes. *Scholarship of teaching and learning in psychology, 5*(4), 298–304. https://psycnet.apa.org/record/2019-50840-001

Bartoshuk, L. M., Duffy, V. B., & Miller, I. J. (1994). PTC/PROP taste: Anatomy, psychophysics, and sex effects. *Physiology & behaviour, 56*(6), 1165–1171.

Bateman, I., Dent, S., Peters, E., Slovic, P., & Starmer, C. (2007). The affect heuristic and the attractiveness of simple gambles. *Journal of behavioral decision making, 20*(4), 365–380.

Baumeister, R. F., & Leary, M. R. (1995). The need to belong: Desire for interpersonal attachments as a fundamental human motivation. *Psychological bulletin, 117*(3), 497–529.

Beaulieu, C. (2002). The basis of anisotropic water diffusion in the nervous system – a technical review. *NMR in biomedicine*. 15, 435–455. doi: 10.1002/nbm.782

Beeman, M. J., & Chiarello, C. (1998). Complementary right-and left-hemisphere language comprehension. *Current directions in psychological science, 7*(1), 2–8.

Bergman, E. (2022, May 2). Study: Autistic and non-autistic people share more in common than we think. *Autism parenting magazine*. https://www.autismparentingmagazine.com/similarities-between-autistic-individuals/

Berk, L. E. (2015). *Child development*. Australia: Pearson Higher Education.

Berne, R. M., & Levy, M. N. (1996). *Principles of Physiology.* St Louis, MO: Mosby.

Berns, G. S., Blaine, K., Prietula, M. J., & Pye, B. E. (2013). Short- and long-term effects of a novel on connectivity in the brain. *Brain connectivity, 3*(6), 590–600. doi:10.1089/brain.2013.0166

Beyond Blue. (2022). Problem gambling during the coronavirus pandemic and how to seek support. https://coronavirus.beyondblue.org.au/Managing-my-daily-life/Coping-with-isolation-and-being-at-home/problem-gambling-during-the-coronavirus, accessed 27 September 2022.

Birchnell, J. (1993). *How humans relate: A new interpersonal theory (human evolution, behaviour and intelligence).* East Sussex: Psychology Press.

Black, J. M., & Hoeft, F. (2015). Utilizing biopsychosocial and strengths-based approaches within the field of child health: what we know and where we can grow. *New directions for child and adolescent development, 2015*(147), 13–20. doi: 10.1002/cad.20089

Blanchette Sarrasin, J., Nenciovici, L., Brault Foisy, L.-M., Allaire-Duquette, G., Riopel, M., & Masson, S. (2018). Effects of teaching the concept of neuroplasticity to induce a growth mindset on motivation, achievement, and brain activity: A meta-analysis. *Trends in neuroscience and education, 12,* 22–31. doi: 10.1016/j.tine.2018.07.003

Boomsma, D. (2008). Identical strangers: a memoir of twins separated and reunited. *Twin research and human genetics, 11*(4), 478–479.

Bornstein, M. H. (1989). Sensitive periods in development: Structural characteristics and causal interpretations. *Psychological bulletin, 105*(2), 179–197.

Bornstein, M. H. (1995). Parenting infants. In M. H. Bornstein (ed.), *Handbook of parenting.* Mahwah, NJ: Erlbaum.

Borod, J. C., Cicero, B. A., Obler, L. K., Welkowitz, J., Erhan, H., Santschi, C., Grunwald, I., Agosti, R., & Whalen, J. (1998). Right hemisphere emotional perception: Evidence across multiple channels. *Neuropsychology, 12*(3), 446–458.

Bouchard, T. J. Jr (1983). Twins – nature's twice told tale. In *Yearbook of science and the future,* 66–81. Chicago: Encyclopaedia Britannica.

Bransford, J., Sherwood, R., Vye, N., & Reiser, J. (1986). Teaching thinking and problem solving. *American Psychologist, 41*(10), 1078–1089.

BPS Research Digest. (2012, December 20). What's it like to be face-blind? Research Digest, The British Psychological Society. https://digest.bps.org.uk/2012/12/20/whats-it-like-to-be-faceblind/

Broche-Pérez Y., Herrera Jiménez, L. F., & Omar-Martínez, E. (2016). Neural substrates of decision-making, *Neurología* (English edition), *31*(5), 319–325. doi: 10.1016/j.nrleng.2015.03.009.

Brown, R. (1965). *Social psychology.* New York: Free Press.

Brown, T. D., Dane, F. C., & Durham, M. D. (1998). Perception of race and ethnicity. *Journal of social behavior and personality, 13*(2), 295–306.

Brüne, M., Belsky, J., Fabrega, H., et al. (2012). The crisis of psychiatry – insights and prospects from evolutionary theory. *World psychiatry, 11*(1), 55–57.

Bruner, J. S., & Minturn, A. L. (1955). Perceptual identification and perceptual organisation. *Journal of general psychology, 53,* 211–228.

Buckland, M. E., Sy, J., Szentmariay, I. et al. (2019). Chronic traumatic encephalopathy in two former Australian National Rugby League players. *Acta neuropathologica communications,* 7(97). doi: 10.1186/s40478-019-0751-1

Bugelski, B. R., & Alampay, D. A. (1961). The role of frequency in developing perceptual sets. *Canadian Journal of Psychology, 15*(4), 205–211.

Burnstein, E., & Vinokur, A. (1977). Persuasive argumentation and social comparison as determinants of attitude polarization. *Journal of experimental social psychology, 13,* 315–332.

Cain, M. S., Leonard, J. A., Gabrieli, J. D., & Finn, A. S. (2016). Media multitasking in adolescence. *Psychonomic bulletin & review, 23*(6), 1932–1941.

Callan, D. E., & Schweighofer, N. (2010). Neural correlates of the spacing effect in explicit verbal semantic encoding support the deficient-processing theory. *Human brain mapping, 31,* 645–659. doi: 10.1002/hbm.20894

Carlsmith, J. M., & Aronson, E. (1963). Some hedonic consequences of the confirmation and disconfirmation of expectancies. *The journal of abnormal and social psychology, 66(2),* 151–156. doi: 10.1037/h0042692

Carter, R., Aldridge, S., Page, M., & Parker, S. (2009). *The human brain book.* New York, NY: DK Publishing.

Carter, R. (2014). *The brain book.* London: Dorling Kindersley.

Caruso, J. P., & Sheehan, J. P. (2017). Psychosurgery, ethics, and media: A history of Walter Freeman and the lobotomy. *Neurosurgery focus, 43*(3), E6. doi: 10.3171/2017.6.FOCUS17257

Charlton, B. (2012, May 23). Defining my dyslexia. *New York Times.* http://www.nytimes.com/2013/05/23/opinion/defining-my-own-dyslexia.html

Cheng, C., Lau, Y. C., Chan, L., & Luk, J. W. (2021). Prevalence of social media addiction across 32 nations: Meta-analysis with subgroup analysis of classification schemes and cultural values. *Addictive behaviors, 117,* 106845.

Christianson, S., Saisa, J., & Silfvenius, H. (1995). The right hemisphere recognises the bad guys. *Cognition & emotion, 9*(4), 309–324.

Clark, C. (2013). A novel look at how stories may change the brain. eScienceCommons. Emory University. http://esciencecommons.blogspot.com.au/2013/12/a-novel-look-at-how-stories-may-change.html

Collins, P. (2004). *Not even wrong: Adventures in autism.* Bloomsbury.

Comstock, R. D., Currie, D. W., Pierpoint, L. A., Grubenhoff, J. A., Fields, S. K. (2015). An evidence-based discussion of heading the ball and concussions in high school soccer. *JAMA Paediatrics, 169*(9), 830–837. doi:10.1001/jamapediatrics.2015.1062

Concussion Foundation (2020). https://concussionfoundation.org/CTE-resources/what-is-CTE

Cotman C. W., Berchtold, N. C., & Christie, L. A. (2007). Exercise builds brain health: Key roles of growth factor cascades and inflammation. *Trends in neurosciences, 30*(9), 464–472. doi: 10.1016/j.tins.2007.06.011.

Criscuolo, A., Pando-Naude, V., Bonetti, L., Vuust, P., & Brattico, E. (2022). An ALE meta-analytic review of musical expertise. *Scientific reports, 12*(1), 1–17.

Cuddy, A. J. C., Fiske, S. T., & Glick, P. (2008). Warmth and competence as universal dimensions of social perception: The stereotype content model and the BIAS map. In M. P. Zanna (Ed.), *Advances in experimental social psychology*, vol. 40, 61–149. Elsevier Academic Press.

Cytowic, R. E. (1995). Synesthesia: Phenomenology and neuropsychology – a review of current knowledge. *Psyche: An interdisciplinary journal of research on consciousness, 2*(10).

Darling, N. (2017, October 28). Attracting WEIRD samples. *Psychology Today.* www.psychologytoday.com. https://www.psychologytoday.com/us/blog/thinking-about-kids/201710/attracting-weird-samples

Darwish, A. F., & Huber, G. (2003). Individualism vs. collectivism in different cultures: A cross-cultural study. *Journal of intercultural education, 14*, 47–56

David, M. E., & Roberts, J. A. (2021). Smartphone use during the COVID-19 pandemic: Social versus physical distancing. *International journal of environmental research and public health, 18*(3), 1034–1041.

DeKosky, S. T., Blennow, K., Ikonomovic, M. D., & Gandy, S. (2013). Acute and chronic traumatic encephalopathies: Pathogenesis and biomarkers. *Nature reviews neurology, 9*(4), 192–200.

Deutsch, M. (1993). Educating for a peaceful world. *American Psychologist, 48*(5), 510–517.

Deverett, B., Koay, S. A., Oostland, M., & Wang, S. S. (2018, August 13). Cerebellar involvement in an evidence-accumulation decision-making task. *eLife, 7*, e36781. doi: 10.7554/eLife.36781

Diener, E., Fraser, S. C., Beaman, A. L., & Kelem, R. T. (1976). Effects of deindividuation variables on stealing among Halloween trick-or-treaters. *Journal of personality and social psychology, 33*(2), 178–183.

Dobia, B., & Roffey, S. (2017). Respect for culture : social and emotional learning with Aboriginal and Torres Strait Islander youth. In E. Frydenberg, A. J. Martin, & R. J. Collie (Eds.), *Social and emotional learning in Australia and the Asia-Pacific: Perspectives, programs and approaches* (pp. 313–334). https://doi.org/10.1007/978-981-10-3394-0_17

Doidge, N. (2007). *The brain that changes itself: Stories of personal triumph from the frontiers of brain science.* USA: Penguin Group.

Douvan, E. (1997). Erik Erikson: critical times, critical theory. *Child psychiatry & human Development, 28*(1), 15–21.

Down Syndrome Australia. (2022). *What is Down syndrome?* http://www.downsyndrome.org.au/what_is_down_syndrome.html

Dudgeon, P. & Walker, R. (2015). Decolonising Australian psychology: Discourses, strategies, and practice. *Journal of social and political psychology, 3*(1), 276–297. doi: 10.5964/jspp.v3i1.126

Eells, J. (2019, July 31). Billie Eilish and the Triumph of the Weird. Rolling Stone. https://www.rollingstone.com/music/music-features/billie-eilish-cover-story-triumph-weird-863603/

Ehardt, K. (2009). Dyslexia, not disorder. *Dyslexia, 15*(4), 363–366.

Eisele, A., Hill-Strathy, M., Michels, L., & Rauen, K. (2020). Magnetic resonance spectroscopy following mild traumatic brain injury: A systematic review and meta-analysis on the potential to detect posttraumatic neurodegeneration. *Neurodegenerative diseases, 20*(1), 2–11. doi: 10.1159/000508098

Elliot, T. (2012, April 26). La Trobe 'torture' study anguish. *The Age.* https://www.theage.com.au/national/victoria/la-trobe-torture-study-anguish-20120425-1xlmf.html

Engemann, K. M., & Owyang, M. T. (2005). So much for that merit raise: The link between wages and appearance. *The regional economist, 13*(2), 10–11.

Enns, J. T., & Coren, S. (1995). The box alignment illusion. *Perception & psychophysics, 57*(8), 1163–1174.

Enzmann, D. (2012). Social responses to offending. In J. Junger-Tas, I. H. Marshall, D. Enzmann, M. Killias, M. Steketee, & B. Gruszczyńska (Eds.), *The many faces of youth crime: Contrasting theoretical perspectives on juvenile delinquency across countries and cultures* (pp. 143–182). New York: Springer.

Falster, K., Hanly, M., Edwards, B., Banks, E., Lynch, J. W., Eades, S., Nickel, N., Goldfield, S., & Biddle, N. (2021). Preschool attendance and developmental outcomes at age five in Indigenous and non-Indigenous children: a population-based cohort study of 100 357 Australian children. *Journal of epidemiology & community health, 75*(4), 371–379.

Fang, X., van Kleef, G. A., & Sauter, D. A. (2018). Person perception from changing emotional expressions: Primacy, recency, or averaging effect? *Cognition and emotion, 32*(8), 1597–1610.

Faulkner, N., Borg, K., Zhao, K., & Smith, L. (2021). *The Inclusive Australia Social Inclusion Index: 2021 Report*, Monash University, p. 17. https://inclusive-australia.s3.amazonaws.com/files/Inclusive-Australia-2020-21-Social-Inclusion-Index-min.pdf

Faus, I., & Sisniega, H. (2003). Sweet-tasting proteins. In S. R. Fahnestock, A. Steinbuechel (Eds.), *Biopolymers vol. 8: Polyamides and complex proteinaceous materials II* (pp. 204–221). Weinheim, Germany: Wiley-VCH Verlag GmbH & Co. KGaA.

Federation of American Societies for Experimental Biology (FASEB) (2014, April 29). Don't like the food? Try paying more. *ScienceDaily.*

Ferdinand, A., Paradies, Y., & Kelaher, M. (2015). Mental health impacts of racial discrimination in Australian culturally and linguistically diverse communities: a cross-sectional survey. *BMC public health, 15*, 1–14.

Festinger, L. (1954). A theory of social comparison processes. *Journal of human relations, 7*, 117–140.

Festinger, L. (1957). *A theory of cognitive dissonance.* Stanford, CA: Stanford University Press.

Finucane, M. L., Alhakami, A., Slovic, P., & Johnson, S. M. (2000). The affect heuristic in judgments of risks and benefits. *Journal of behavioral decision making, 13*, 1–17.

Fiske, S. T. (1993). Controlling other people. *American psychologist, 48*(6), 621–628.

Fitzsimons, B. (2021, January 21). 'I'm grateful that I was diagnosed quite late.' Celeste Barber on her ADHD. *Mamamia.* https://www.mamamia.com.au/celeste-barber-adhd

Flannery, K. A., & Wisner-Carlson, R. (2020). Autism and education. *Child and adolescent psychiatric clinics, 29*(2), 319–343.

Flavell, J. H. (1977). *Cognitive development.* Englewood Cliffs, NJ: Prentice Hall.

Flavell, J. H. (1992). Cognitive development: Past, present, and future. *Developmental psychology, 28*(6), 998–1005.

Forestell, C. A. (2017). Flavor perception and preference development in human infants. *Annals of Nutrition and metabolism, 70*(3), 17–25. doi: 10.1159/000478759.

Fortenbaugh, F. C., DeGutis, J., Germine, L., et al. (2015). Sustained attention across the life span in a sample of 10,000: Dissociating ability and strategy. *Psychological science, 26*(9), 1497–1510. doi:10.1177/0956797615594896

Frager, R. (1970). Experimental social psychology in Japan: Studies in social conformity. *Rice Institute Pamphlet Rice University Studies, 56.*

Fragkaki, I., Maciejewski, D. F., Weijman, E. L., Feltes, J., & Cima, M. (2021). Human responses to Covid-19: The role of optimism bias, perceived severity, and anxiety. *Personality and individual differences, 176,* 110781.

Franzini A., Moosa S., Servello, D., Small, I., DiMeco, F., Xu, Z., Elias, W. J., Franzini, A., & Prada F. (2019). Ablative brain surgery: an overview. *International journal of hyperthermia, 36*(2), 64–80. doi: 10.1080/02656736.2019.1616833.

Fuster, J. (2015). *The prefrontal cortex* (5th ed.). Elsevier Science. Retrieved from https://www.perlego.com/book/1834696/the-prefrontal-cortex-pdf

Gaertner, S. L., Mann, J. A., Dovidio, J. F., Murrell, A. J., & Pomare, M. (1990). How does cooperation reduce intergroup bias? *Journal of personality and social psychology, 59*(4), 692–704.

Gainsbury, S. M. (2015). Online gambling addiction: the relationship between internet gambling and disordered gambling. *Current addiction reports, 2,* 185–193.

Gambrel, P. A., & Cianci, R. (2003). Maslow's Hierarchy of Needs: Does it apply in a collectivist culture. *The journal of applied management and entrepreneurship, 8,* 143.

Geiger, G., Cattaneo, C., Galli, R., et al. (2008). Wide and diffuse perceptual modes characterize dyslexics in vision and audition. *Perception, 37*(11), 1745–1764.

German, T. P., & Defeyter, M. A. (2000). Immunity to functional fixedness in young children. *Psychonomic bulletin & review, 7*(4), 707–712.

Gibson, J. J. (1966). *The senses considered as perceptual systems.* Boston: Houghton Mifflin.

Gillam, B. (1980). Geometrical illusions. *American psychologist, 242,* 102–111.

Gogtay, N., Giedd, J. N., Lusk, L., Hayashi, K. M., Greenstein, D., Vaituzis, A. C., Nugent, T. F., Herman, D. H. et al. (2004, May 17). Dynamic mapping of human cortical development during childhood through early adulthood, *Proceedings of the National Academy of Sciences, 101*(21), 8174–8179. doi: 10.1073/pnas.0402680101

González-García, C., Flounders, M. W., Chang, R., Baria, A. T., & He, B. J. (2018). Content-specific activity in frontoparietal and default-mode networks during prior-guided visual perception. *eLife, 7,* e36068. doi: 10.7554/eLife.36068.001

Gorgoraptis, N., Li, L. M., Whittington, A., Zimmerman, K. A., Maclean, L. M., McLeod, C., Ross, E. Heslegrave, A., Zetterberg, H., Passchier, J., Matthews, P. M., Gunn, R. N., McMillan, T. M., & Sharp D. J. (2019). In vivo detection of cerebral tau pathology in long-term survivors of traumatic brain injury, *Science translational medicine, 11*(508), eaaw1993.

Grandin, T. (2006). *Thinking in pictures: And other reports from my life with autism.* Vintage.

Gray, P. (1994). *Psychology* (2nd ed.). Boston, MA: Worth Publishers.

Gray, R. (2016, March 18). Could we soon have superhero night vision? Brain implants could give us a 'sixth sense' by making us see infrared. *Daily Mail online.* https://www.dailymail.co.uk/sciencetech/article-3496895/Could-soon-superhero-NIGHT-VISION-Brain-implants-rats-sixth-sense-making-infrared.html

Gregory, R. (1970). *The intelligent eye.* London: Weidenfeld and Nicolson.

Gregory, R. L. (1990). *Eye and brain: The psychology of seeing.* Princeton, NJ: Princeton University Press.

Gregory, R. L., & Wallace, J. G. (1963). Recovery from early blindness. *Experimental psychology society monograph,* no. 2.

Grundy, J. G., Anderson, J. A. E., & Bialystok, E. (2017). Bilinguals have more complex EEG brain signals in occipital regions than monolinguals. *Neuroimage, 159,* 280–288.

Gupta R., Koscik, T. R., Bechara, A., & Tranel, D. (2011). The amygdala and decision-making. *Neuropsychologia, 49*(4), 760–766. doi:10.1016/j.neuropsychologia.2010.09.029

Hall, J. (2010). *Guyton and Hall textbook of medical physiology* (12th ed.), USA: Saunders (Elsevier).

Hamilton, V., & Sanders, J. (1995). Crimes of obedience and conformity in the workplace: Surveys of Americans, Russians, and Japanese. *Journal of social issues, 51.* doi 10.1111/j.1540-4560.1995.tb01335.x

Haregu, T., Jorm, A. F., Paradies, Y., Leckning, B., Young, J. T., & Armstrong, G. (2021). Discrimination experienced by Aboriginal and Torres Strait Islander males in Australia: Associations with suicidal thoughts and depressive symptoms. *The Australian and New Zealand journal of psychiatry, 56*(6), 657–666.

Harlow, H. F. (1958). The nature of love. *American psychologist, 13,* 673–85.

Harlow, H. F., Dodsworth, R. O., & Harlow, M. K. (1965). Total social isolation in monkeys. *Proceedings of the National Academy of Sciences, 54*(1), 90–97.

Harlow, J. M. (1848). Passage of an iron rod through the head. *Boston medical and surgical journal, 39,* 389–393. Viewed at https://neurophilosophy.files.wordpress.com/2006/12/harlowbmsj1860.pdf

Hartmann, M., Graff, M., Beseeyan, C., & Yan, S. (2016, August 24). *Science advances, 2*(8). doi: 10.1126/sciadv.1600716

Hatch, S. (2002, November 14). Some sound red while others just taste green. *Canberra Times*.

Healthline (2019). Are you a supertaster? https://www.healthline.com/health/food-nutrition/supertaster#who-is-a-supertaster

Heider, F. (1958). *The psychology of interpersonal relations*. New York: Wiley.

Heinze, H. J., Hinrichs, H., Scholz, M., Burchert, W., & Mangun, G. R. (1998). Neural mechanisms of global and local processing. *Journal of cognitive neuroscience, 10*(4), 485–498.

Hellige, J. B. (1993). *Hemispheric asymmetry*. Cambridge, MA: Harvard University Press.

Henderson, N. D. (1982). Human behavior genetics. *Annual review of psychology, 33*, 403–440.

Henrich, J., Heine, S. J., & Norenzayan, A. (2010). The weirdest people in the world?. *The behavioral and brain sciences, 33*(2–3), 61–135. doi: 10.1017/S0140525X0999152X

Hess, E. H. (1959). Imprinting. *Science, 130*, 133–141.

Hoffman, L. & Oppenheim, L. (2019). Three identical strangers and the twinning reaction –clarifying history and lessons for today from Peter Neubauer's Twins Study. *The journal of the American Medical Association, 322*(1), 10. doi:10.1001/jama.2019.8152

Hoffman, T. (2012, November 24). The man whose brain ignores one half of his world. *The Guardian*. https://www.theguardian.com/science/blog/2012/nov/23/man-brain-ignores-half-world

Hoogman, M., Bralten, J., Hibar, D. P., Mennes, M., Zwiers, M. P., Schweren, L., van Hulzen, K., Medland, S. E., Shumskaya, E., Jahanshad, N., Zeeuw, P., Szekely, E., Sudre, G., Wolfers, T., Onnink, A., Dammers, J. T., Mostert, J. C., Vives-Gilabert, Y., Kohls, G., Oberwelland, E., … Franke, B. (2017). Subcortical brain volume differences in participants with attention deficit hyperactivity disorder in children and adults: a cross-sectional mega-analysis. *Lancet psychiatry, 4*(4), 310–319. doi: 10.1016/S2215-0366(17)30049-4. Erratum in: *Lancet psychiatry*. (2017, June). *4*(6), 436.

Horn, J. M., Loehlin, J. C., & Willerman, L. (1979). Intellectual resemblance among biological and adoptive relatives: the Texas adoption project. *Behavior genetics, 9*, 177–207.

Huang, H. (2019, June 28). What's the neuroscience behind the bouba/kiki effect? *NBB in Paris*. https://scholarblogs.emory.edu/nbbparis/2019/06/28/whats-the-neurosciencebehind-the-bouba-kiki-effect/

Hubel, D. H., & Wiesel, T. N. (1962). Receptive fields, binocular interaction and functional architecture in the cat's visual cortex. *Journal of physiology, 160*(1), 106–54.

Hübner, R. (1998). Hemispheric difference in global/local processing revealed by same-different judgements. *Visual cognition, 5*(4), 457–478.

Hudson, W. (1960). Pictorial depth perception in sub-cultural groups in Africa. *The journal of social psychology, 52*, 183–208. doi: 10.1080/00224545.1960.9922077

Hwang, D. L.-D. (2018, November 16). Why you like coffee, and I choose tea – it's in the genes. *The Conversation*. https://theconversation.com/why-you-like-coffee-and-ichoose-tea-its-in-the-genes-106854

Indigenous Allied Health Australia. (2015). Cultural responsiveness in action: an IAHA framework. https://www.accesseap.com.au/images/2015-IAHA-Cultural-Responsiveness-Framework-WEB.pdf

Innocence project. (2022). https://innocenceproject.org/eyewitness-identification-reform/ (accessed 24 November 2021.

Iverson, G. L., Keene, C. D, Perry, G., & Castellani, R. J. (2018). The need to separate chronic traumatic encephalopathy neuropathology from clinical features. *Journal of Alzheimer's Disease, 61*(1), 17–28. doi: 10.3233/JAD-170654

Janis, I. (1971, November). Groupthink, *Psychology today*, pp. 43–44, 46, 74–76.

Jensen, P. S., Mrazek, D., Knapp, P. K., Steinberg, L., Pfeffer, C., Schowalter, J., & Shapiro, T. (1997). Evolution and revolution in child psychiatry: ADHD as a disorder of adaptation. *Journal of the American Academy of Child and Adolescent Psychiatry, 36*(12), 1672–1679.

Jo, Y. S., Bhang, S. Y., Choi, J. S., Lee, H. K., Lee, S. Y., & Kweon, Y.-S. (2019). Clinical characteristics of diagnosis for internet gaming disorder: Comparison of DSM-5 IGD and ICD-11 GD diagnosis. *Journal of clinical medicine, 8*(7), 945. doi: 10.3390/jcm8070945

Jordan, B. D. (2013). The clinical spectrum of sport-related traumatic brain injury. *Nature reviews neurology, 9*(4), 222–230.

Kahneman, D., & Tversky, A. (1973). On the psychology of prediction. *Psychological review, 80*(4), 237–251.

Kamin, L. J. (1981). *The intelligence controversy*. New York: Wiley.

Kania, B. F., Wronska, D., & Zieba, D. (2017). Introduction to neural plasticity mechanism. *Journal of Behavioral and Brain Science, 7*, 41–8. doi: 10.4236/jbbs.2017.72005

Karpicke, J. D., Blunt, J. R., & Smith, M. A. (2016) Retrieval-based learning: positive effects of retrieval practice in elementary school children. *Frontiers in psychology*, 7:350. doi: 10.3389/fpsyg.2016.00350

Kavanagh-Hall, E. (2020, April 26). A different headspace: Six people on being neurodivergent during lockdown. *The Spinoff*. https://thespinoff.co.nz/society/26-04-2020/a-different-headspace-being-neurodivergent-during-covid-19

Kelman, H. C., & Hamilton, V. L. (1989). *Crimes of obedience*. New Haven: Yale University Press.

Kim, H., & Markus, H. R. (1999). Deviance or uniqueness, harmony or conformity? A cultural analysis. *Journal of personality and social psychology, 77*(4), 785–800.

Kinnunen, K. M., Greenwood R., Powell J. H., Leech R., Hawkins P. C., Bonnelle V., et al. (2011). White matter damage and cognitive impairment after traumatic brain injury. *Brain, 134*, 449–463. doi: 10.1093/brain/awq347

Kleinman, A., Eisenberg, L., & Good, B. (1978). Culture, illness and care: Clinical lessons from anthropological and cross-cultural research. *Annals of Internal Medicine, 88*(2), 251–258.

Koizumi, A., Tsuchiya, A., Nakajima, K., et al. (2011). Human sweet taste receptor mediates acid-induced sweetness of miraculin. *Proceedings of the National Academy of Sciences, 108*(40), 16819–16824. doi: 10.1073/pnas.1016644108

Kolb, B., & Gibb, R. (2011). Brain plasticity and behaviour in the developing brain. *Journal of the Canadian Academy of Child and Adolescent Psychiatry, 20*(4), 265–76.

Kolb, B., & Whishaw, I. Q. (2014). *An introduction to brain and behavior.* New York, NY: Worth Publishers.

Konarski, J. Z., McIntyre, R. S., Grupp, L. A., & Kennedy, S. H. (2005). Is the cerebellum relevant in the circuitry of neuropsychiatric disorders? *Journal of psychiatry & neuroscience, 30*(3), 178–186.

Kotlaja, M. M. (2020). Cultural contexts of individualism vs. collectivism: Exploring the relationships between family bonding, supervision and deviance. *European journal of criminology, 17*(3), 288–305.

Krueger, T., Szwabiński, J., & Weron, T. (2017). Conformity, anticonformity and polarization of opinions: Insights from a mathematical model of opinion dynamics. *Entropy, 19*(7), 371–392.

Kruger, J., & Dunning, D. (1999). Unskilled and unaware of it: How difficulties in recognizing one's own incompetence lead to inflated self-assessments. *Journal of personality and social psychology, 77*(6), 1121–1134.

Kumar, A., & Wroten, M. (2021, updated July 26). Agnosia. In: *StatPearls.* Treasure Island (FL): StatPearls Publishing. https://www.ncbi.nlm.nih.gov/books/NBK493156/

Kuwabara, M., & Smith, L. B. (2016). Cultural differences in visual object recognition in 3-year-old children. *Journal of experimental child psychology, 147,* 22–38. doi: 10.1016/j.jecp.2016.02.006

Kyoto University (2017, May 23). Visual perception may depend on birthplace and environment. *ScienceDaily.* www.sciencedaily.com/releases/2017/05/170523095013.htm

LaPiere, R. (1934). Attitudes vs. actions. *Social Forces, 13*(2), 230–237.

Le Texier, T. (2019). Debunking the Stanford Prison Experiment. *American psychologist, 74*(7), 823–839.

Lee, R. M., Draper, M., & Lee, S. (2001). Social connectedness, dysfunctional interpersonal behaviors, and psychological distress: Testing a mediator model. *Journal of counseling psychology, 48*(3), 310–318.

Lefkowitz, M., Blake, R. R., & Mouton, J. S. (1955). Status factors in pedestrian violation of traffic signals. *Journal of abnormal and social psychology, 51,* 704–706.

Li, K., & Malhotra, P. A. (2015). Spatial neglect, *Practical neurology, 15,* 333–339.

Liberman, Z., Woodward, A. L., & Kinzler, K. D. (2017). The origins of social categorization. *Trends in cognitive sciences, 21*(7), 556–568.

Lorenz, K. (1937). Imprinting. *Auk, 54*(3), 245–273.

Love, T. & Brumm, K. (2012). *Cognition and acquired language disorders* (p. 210). doi: http://dx.doi.org/10.1016/B978-0-323-07201-4.00019-2

Lowengart, O. (2012). The effect of branding on consumer choice through blind and non-blind taste tests. *Innovative marketing, 8*(4), 7–18.

MacDonald, C. L., Dikranian, K., Song, S. K., Bayly, P. V., Holtzman, D. M., & Brody, D. L. (2007). Detection of traumatic axonal injury with diffusion tensor imaging in a mouse model of traumatic brain injury. *Experimental Neurology, 205,* 116–131.

Maguire, E. A., Gadian, D. G., Johnsrude, I. S., Good, C. D., Ashburner, J. Frackowiak, R. S. J., & Frith, C. D. (2000). Navigation-related structural change in the hippocampi of taxi drivers. *Proceedings of the National Academy of Sciences, 97*(8), 4398–4403. doi: 10.1073/pnas.070039597

Maio, G. R., & Augoustinos, M. (2005). Attitudes, attributions and social cognition. In M. Hewstone, F. D. Fincham, & J. Foster (Eds.), *Psychology* (pp. 360–382), Malden, MA: Blackwell.

Martin, R., & Hewstone, M. (2001). Determinants and consequences of cognitive processes in majority and minority influence. In J. P. Forgas & K. D. Williams (Eds.), *Social influence: Direct and indirect processes* (pp. 315–330). Psychology Press.

Maurer, D., Pathman, T., & Mondloch, C. J. (2006). The shape of boubas: sound–shape correspondences in toddlers and adults. *Developmental Science, 9*(3), 316–22. doi: 10.1111/j.1467-7687.2006.00495.x.

McArthur, L. Z., & Ginsberg, E. (1981). Causal attribution to salient stimuli: An investigation of visual fixation mediators. *Personality and social psychology bulletin, 7*(4), 547–553.

McCormack, A., & Griffiths, M. D. (2012). Motivating and inhibiting factors in online gambling behaviour: a grounded theory study. *International journal of mental health addiction, 10*(1), 39–53.

McKee, A. C., Stein, T. D., Nowinski, C. J., Stern, R. A., Daneshvar, D. H. et al. (2013). The spectrum of disease in chronic traumatic encephalopathy. *Brain, 136*(1), 43–64. https://doi.org/10.1093/brain/aws307

McKee, A. C., Stein T. D., Kiernan P. T., & Alvarez V. E. (2015). The neuropathology of chronic traumatic encephalopathy. *Brain pathology, 25*(3), 350–364. doi: 10.1111/bpa.12248

McLeod, S. A. (2018). Preoperational stage. *Simply Psychology.* www.simplypsychology.org/preoperational.html

McLeod, S. A. (2020, December 29). Maslow's Hierarchy of Needs. *Simply Psychology.* https://www.simplypsychology.org/maslow.html

McLeod, S. A. (2021). *Concrete operational stage. Simply Psychology.* www.simplypsychology.org/concrete-operational.html

Mental Health First-Aid USA (MHFA) (2019). Four ways culture impacts mental health. https://www.mentalhealthfirstaid.org/2019/07/four-ways-culture-impacts-mental-health/

Meppelink, C. S., Smit, E. G., Fransen, M. L., & Diviani, N. (2019). 'I was right about vaccination': Confirmation bias and health literacy in online health information seeking. *Journal of health communication, 24,* 129–140.

Mesulam, M.-M. (Ed.). (2000). *Principles of behavioral and cognitive neurology* (2nd ed.). Oxford University Press.

Mez, J., Daneshvar, D. H., Kiernan, P. T., et al. (2017). Clinicopathological evaluation of chronic traumatic encephalopathy in players of American football. *Journal of the American Medical Association, 318*(4), 360–370. doi:10.1001/jama.2017.8334

Milgram, S. (1963). Behavioral study of obedience. *The journal of abnormal and social psychology*, 67(4), 371–378.

Miyake, K., Chen, S., & Campos, J. (1985). Infant temperament, mother's mode of interaction and attachment in Japan. *Monographs of the Society for Research in Child Development, 50*, 276–97.

Monell Chemical Senses Center (2015, July 17). Some like it sweet, others not so much: It's partly in the genes: Twin study suggests a common genetic pathway underlies sweet taste perception of natural, non-caloric sweeteners. *ScienceDaily.* https://www.sciencedaily.com/releases/2015/07/150717091937.htm

Monin, B., & Norton, M. I. (2003). Perceptions of a fluid consensus: Uniqueness bias, false consensus, false polarization, and pluralistic ignorance in a water conservation crisis. *Personality and social psychology bulletin, 29*(5), 559–567.

Montenigro, P. H., Corp, D. T., Stein, T. D., Cantu, R. C., & Stern, R. A. (2015). Chronic traumatic encephalopathy: historical origins and current perspective. *Annual Review of Clinical Psychology, 11*, 309–30. doi: 10.1146/annurev-clinpsy-032814-112814

Morrison, J. H., & Baxter, M. G. (2012). The ageing cortical synapse: hallmarks and implications for cognitive decline. *Nature reviews neuroscience, 13*(4), 240–250. doi: 10.1038/nrn3200

Moskowitz, H. R., Kumriach, V., & Sharma, S. D. (1975). Cross-cultural differences in simple taste preference. *Science, 190*, 1217–1218.

Moutelikova, I. (2021, November 30). Through my eyes: Epilepsy diagnosis in adulthood, *Medical News Today.* https://www.medicalnewstoday.com/articles/through-my-eyes-epilepsy

Mueller, J. (2000–2022). Resources for the teaching of social psychology. http://jfmueller.faculty.noctrl.edu/crow/group.htm

Mueller, R., Behen, M. E., Rothermel, R. D., Muzik, O., Chakraborty, P. K., & Chugani, H. T. (1999). Brain organisation for language in children, adolescents, and adults with left hemisphere lesion: a PET study. *Progress in neuro-psychopharmacology and biological psychiatry, 23*(4), 657–668.

Mulvenna, C., & Walsh V. (2005). Synaesthesia. *Current biology, 15*(11), R399–R400.

Nail, P. R., Di Domenico, S. I., & MacDonald, G. (2013). Proposal of a double diamond model of social response. *Review of general psychology, 17*(1), 1–19.

Neville, H, J., Stevens, C., Pakulak, E., Bell, T. A., Fanning, J., Klein, S., & Isbell, E. (2013). Family-based training program improves brain function, cognition, and behavior in lower socioeconomic status pre-schoolers. *Proceedings of the National Academy of Sciences, 110*(29), 12138–12143.

Nisbett, R. E., & Miyamoto, Y. (2005). The influence of culture: Holistic versus analytic perception. *Trends in cognitive sciences, 9*(10), 467–473. doi: 10.1016/j.tics.2005.08.004

NYU Langone Health / NYU School of Medicine. (2018, July 31). Past experiences shape what we see more than what we are looking at now. *ScienceDaily.* www.sciencedaily.com/releases/2018/07/180731104224.htm

O'Day, G. M., & Karpicke, J. D. (2021). Comparing and combining retrieval practice and concept mapping. *Journal of educational psychology, 113*(5), 986–997. doi: 10.1037/edu0000486

Olson, J. M., & Maio, G. R. (2003). Persuasion and attitude change. In T. Millon & M. J. Lerner (Eds.), *Comprehensive handbook of psychology*, vol. 5, Personality and social psychology. New York: Wiley.

Oxley, T. J., Yoo, P. E., Rind, G. S., et al. (2021). Motor neuroprosthesis implanted with neurointerventional surgery improves capacity for activities of daily living tasks in severe paralysis: first in-human experience. *Journal of neurointerventional surgery, 13*, 102–108. doi: 10.1136/neurintsurg-2020-016862

Padalia, D. (2014). Conformity bias: A fact or an experimental artifact? *Psychological studies, 59*(3), 223–230.

Pallister, T., Sharafi, M., Lachance, G., Pirastu, N., Mohney, R., MacGregor, A., & Menni, C. (2015). Food preference patterns in a UK twin cohort. *Twin research and human genetics, 18*(6), 793–805. doi:10.1017/thg.2015.69

Palombo, D. J., Keane, M. M., & Verfaellie, M. (2015). How does the hippocampus shape decisions? *Neurobiology of learning and memory, 125*, 93–7. doi: 10.1016/j.nlm.2015.08.005.

Paul, A. M. (2012, February 5). The upside of dyslexia. *New York Times.* http://www.nytimes.com/2012/02/05/opinion/sunday/the-upside-of-dyslexia.html?_r=0. 21 September 2022.

Pearce, M., & Wong, L. (2016, November 1). Living with half a brain: seeing life anew through the camera lens. *ABC News.* https://www.abc.net.au/news/2016-11-01/living-with-half-a-brain-seeing-life-through-a-new-lens/7980762

Perry, G. (2012). *Behind the shock machine.* Melbourne: Scribe Publications.

Peters, W. A. (1971). *A class divided.* Garden City, NY: Doubleday.

Peterson, B. E., & Klohnen, E. C. (1995). Realization of generativity in two samples of women at midlife. *Psychology & aging, 10*(1), 20–29.

Peterson, C. (1989). *Looking forward through the life span: developmental psychology.* Sydney: Prentice Hall.

Petrides, M., Tomaiuolo, F., Yeterian, E. H, & Pandya, D. N. (2012). The prefrontal cortex: comparative architectonic organization in the human and the macaque monkey brains. *Cortex, 48*(1), 46–57. doi: 10.1016/j.cortex.2011.07.002.

Petty, R. E., Wegener, D. T., & Fabrigar, L. R. (1997). Attitudes and attitude change. *Annual Review of Psychology, 48*, 609–647.

Piaget, J. (1951). *The psychology of intelligence.* New York: Norton.

Piaget, J. (1952). *The origins of intelligence in children.* New York: International University Press.

Piaget, J., & Inhelder, B. (1956). *The child's conception of space.* London: Routledge Kegan Paul.

Plotnik, R., & Mollenauer, S. (1986). *Introduction to psychology,* New York: Random House.

Queensland Human Rights Commission (2021). Race case studies. https://www.qhrc.qld.gov.au/resources/casestudies/race-case-studies

Ramachandran, V. S., & Hubbard, E. M. (2001). Synaesthesia: A window into perception, thought and language. *Journal of consciousness studies, 8*(12), 3–34.

Ramachandran, V. S., & Hubbard, E. M. (2003). The phenomenology of synaesthesia. *Journal of consciousness studies, 10*(8), 49–57.

Ramsey, N. F. (2012). Signals reflecting brain metabolic activity. In J. Wolpaw & E. Winter Wolpaw (Eds). *Brain–computer interfaces: principles and practice.* Online edn, Oxford Academic. doi: 10.1093/acprof:oso/9780195388855.003.0004

Raven, B. H., & French, J. R. P. Jr (1958). Legitimate power, coercive power and observability in social influence. *Sociometry, 21,* 83–97.

Rich, A. N., Bradshaw, J. L., & Mattingley, J. B. (2005). A systematic, large-scale study of synaesthesia: Implications for the role of early experience in lexical-colour associations. *Cognition, 98*(1), 53–84.

Riehm, K. E., Feder, K. A., Tormohlen, K. N., Crum, R. M., Young, A. S., Green, K. M., Pacek, L. R., La Flair, L. N., & Mojtabai, R. (2019). Associations between time spent using social media and internalizing and externalizing problems among US youth. *JAMA psychiatry, 76*(12), 1266–1273.

Robson, D. (2009). Do we all have the capacity for synaesthesia? *New scientist, 201*(2695), 13.

Rosenbloom, M. H., Schmahmann, J. D., & Price, B. H. (2012). The functional neuroanatomy of decision-making. *The journal of neuropsychiatry and clinical neurosciences, 24*(3), 266–277. https://doi.org/10.1176/appi.neuropsych.11060139

Rossi, S., Lanoë, C., Poirel, N., Pineau, A., Houdé, O., & Lubin, A. (2015). When i met my brain: participating in a neuroimaging study influences children's naive mind-brain conceptions. *Trends in neuroscience and education,* 4, 92–7. doi: 10.1016/j. tine.2015.07.001

Rousay, V. (2021, January 21). Bottom-up processing. *Simply Psychology.* https://www.simplypsychology.org/bottom-up-processing.html

Rozin, P. (1990). The importance of social factors in understanding the acquisition of food habits. In E. D. Capaldi & T. L. Powley (Eds.), *Taste, experience, and feeding.* Washington, DC: American Psychological Association.

Ruhr-University Bochum (2019, December 9). How playing the drums changes the brain: Many years of playing the instrument leave clear traces. *ScienceDaily.* www.sciencedaily.com/releases/2019/12/191209110513.htm

Rutgers University (2014, May 27). Learning early in life may help keep brain cells alive: Brain cells survive in young who master a task, *ScienceDaily.* https://www.sciencedaily.com/releases/2014/05/140527154750.htm

Rutter, M. (1995). Clinical implications of attachment concepts: Retrospect and prospect. *Journal of child psychiatry and psychology, 36,* 549–571.

Ryan P. (2021, October 9). Weideman CTE diagnosis should prompt sport to ask hard questions: concussion expert. *The Age.* https://www.theage.com.au/sport/afl/murray-weideman-cte-diagnosis-should-prompt-sport-to-ask-hard-questions-concussion-expert-20211009-p58ymm.html

Ryan, L., Debenham, J., Pascoe, B., Smith, R., Owen, C., Richards, J., Gilbert, S., Anders, R., Usher, K., Price, D., Newley, J., Brown, M., Le, L. H., & Fairbairn, H. (2022). *Colonial frontier massacres in Australia 1788–1930.* Newcastle: University of Newcastle. http://hdl.handle.net/1959.13/1340762. Accessed 16 March 2022.

Saberi Moghadam, S., Samsami Khodadad, F., & Khazaeinezhad, V. (2019). An algorithmic model of decision making in the human brain. *Basic and clinical neuroscience, 10*(5), 443–449. doi: 10.32598/bcn.9.10.395

Saalmann, Y. B., & Kastner, S. (2011). Cognitive and perceptual functions of the visual thalamus. *Neuron, 71*(2), 209–223. doi:10.1016/j.neuron.2011.06.027

Saarni, C. (2000). Emotional competence: A developmental perspective. In R. Bar-On, & J. D. A. Parker (Eds.), *The handbook of emotional intelligence: Theory, development, assessment, and application at home, school, and in the workplace* (pp. 68–91). San Francisco, CA: Jossey-Bass.

Sabbatini, R. M. E. (1997, March). Phrenology: The history of brain localization. *Brain & Mind.* http://www.cerebromente.org.br/n01/frenolog/frenloc.htm

Sacks, O. (1985). *The Man Who Mistook His Wife For a Hat and other clinical tales.* UK: Gerald Duckworth.

Saeri, A. K., Cruwys, T., Barlow, F. K., Stronge, S., & Sibley, C. G. (2018). Social connectedness improves public mental health: Investigating bidirectional relationships in the New Zealand attitudes and values survey. *The Australian and New Zealand journal of psychiatry, 52*(4), 365–374.

Salmon, M., Doery, K., Dance, P., Chapman, J., Gilbert, R., Williams, R., & Lovett, R. (2018). Defining the indefinable: Descriptors of Aboriginal and Torres Strait Islander Peoples' cultures and their links to health and wellbeing. https://openresearch-repository.anu.edu.au/bitstream/1885/148406/8/ Defining_the_Indefinable_WEB2_FINAL.pdf

Santrock, J. W. (1997). *Life-span development* (6th ed.). USA: McGraw-Hill.

Sarwar, A., & Emmady, P. D. (2021, updated August 30). Spatial neglect. In: *StatPearls.* Treasure Island (FL): StatPearls Publishing. https://www.ncbi.nlm.nih.gov/books/NBK562184/

Schwartz, M. F., Linebarger, M. C. & Saffran, E. M. (1985). The status of the syntactic deficit theory of agrammatism. In M. L. Kean (Ed.), *Agrammatism* (pp. 83–124). Academic Press. doi: 10.1016/B978-0-12-402830-2.50008-0B978-0-12-402830-2.50008-0

Schmahmann, J. D., & Sherman, J. C. (1998). The cerebellar cognitive affective syndrome. *Brain, 121,* 561–579. doi: 10.1093/brain/121.4.561

Schneps, M. H., Brockmole, J. R., Sonnert, G., & Pomplun, M. (2012). History of reading struggles linked to enhanced learning in low spatial frequency scenes. *PLOS ONE*, *7*(4), e35724.

Segall, M. H., Campbell, D. T., & Herskovits, M. J. (1966). *The influence of culture on visual perception.* USA: Bobbs-Merrill.

Shay, M., & Sarra, G. (2021). Locating the voices of Indigenous young people on identity in Australia: an Indigenist analysis. *Diaspora, Indigenous, and minority education, 15*(3), 166–179.

Sherif, M., Harvey, O. J., White, B. J., Hood, W. R., & Sherif, C. W. (1961). *Intergroup conflict and cooperation: The Robbers Cave experiment.* Norman: University Book Exchange.

Sherman, S. M., Cheng, Y. P., Fingerman, K. L., & Schnyer, D. M. (2016). Social support, stress and the aging brain. *Social cognitive and affective neuroscience, 11*(7), 1050–1058. doi: 10.1093/scan/nsv071

Shweder, R. A. (1991). *Thinking through cultures: Expeditions in cultural psychology.* Harvard University Press.

Sieber, J., & Ziegler, R. (2019). Group polarization revisited: A processing effort account. *Personality & social psychology bulletin, 45*, 1482–1498.

Siipola, E. M. (1935). A group study of some effects of preparatory set. *Psychological monographs, 46*(6), 27–38. doi: 10.1037/h0093376

Simons, D. J., & Chabris, C. F. (1999). Gorillas in our midst: Sustained inattentional blindness for dynamic events. Perception, 28, 1059-1074. Figure provided by Daniel Simons, (www.dansimons.com)

Singelis, T. M., Triandis, H. C., Bhawuk, D. P. S., & Gelfand, M. J. (1995). Horizontal and vertical dimensions of individualism and collectivism: A theoretical and measurement refinement. *Cross-cultural research, 29*(3), 240–275.

Singer, J. (n.d.). Reflections on neurodiversity. https://neurodiversity2.blogspot.com/p/what.html

Smith, L. (2019, October 2). My autism journey: how I learned to stop trying to fit in. *Aeon.* https://aeon.co/ideas/my-autism-journey-how-i-learned-to-stop-trying-to-fit-in

Society for Neuroscience. (2012). What does it mean to have synesthesia? *BrainFacts.org.* http://www.brainfacts.org/About-Neuroscience/Ask-an-Expert/Articles/2012/Synesthesia

Solomon, A. (2008, June 2). The autism rights movement. *New York Magazine.* http://nymag.com/news/features/47225/. Accessed September 21, 2022.

Sowell, E. R., Thompson, P. M., Holmes, C. J., Jernigan T. L., & Toga, A. W. (1999). In vivo evidence for post-adolescent brain maturation in frontal and striatal regions. *Nature neuroscience, 2*(10), 859–61. doi: 10.1038/13154. PMID: 10491602

Spear, L. P. (2010). *The behavioral neuroscience of adolescence.* New York: W. W. Norton & Company.

Springer, S. P., & Deutsch, G. (1998). *Left brain, right brain.* New York: Freeman

Srna S., Schrift, R. Y., & Zauberman, G. (2018). The illusion of multitasking and its positive effect on performance. *Psychological Science, 29*(12). doi:10.1177/0956797618801013

Stavick, K. (2022, May 21). 'My brain is different, not broken': One Mat-Su Central student speaks about her journey during graduation. *Frontiersman.com.* https://www.frontiersman.com/news/my-brain-is-different-not-broken-one-mat-su-central-student-speaks-about-her-journey/article_2a50d0f0-d89d-11ec-9dfb-df4875208ad3.html

Stern, R. A., Daneshvar, D. H., Baugh, C. M., Seichepine, D. R., Montenigro, P. H., Riley, D. O., Fritts, N. G., Stamm, J. M., Robbins, C. A., McHale, L., Simkin, I., Stein, T. D., Alvarez, V. E., Goldstein, L. E., Budson, A. E., Kowall, N. W., Nowinski, C. J., Cantu, R. C., & McKee, A. C. (2013). Clinical presentation of chronic traumatic encephalopathy. *Neurology, 81*(13), 1122–1129. doi: 10.1212/WNL.0b013e3182a55f7f

Stice, E., & Shaw, H. (1994). Adverse effects of the media portrayed thin-ideal on women and linkages to bulimic symptomatology. *Journal of social and clinical psychology, 13*(3), 288–308.

Stolier, R. M., & Freeman, J. B. (2016). The neuroscience of social vision. In J. Absher & J. Cloutier (Eds.), *Neuroimaging personality, social cognition, and character* (pp. 139–157). Academic Press.

Stoner, J. (1967). Risky and cautious shifts in group decisions: The influence of widely held values, Journal of Experimental Social Psychology, 4(4), 442-459. Web link: https://dspace.mit.edu/bitstream/handle/1721.1/48923/riskycautiousshi00ston.pdf

Stoodley, C. J., Valera, E. M., & Schmahmann, J. D. (2012). Functional topography of the cerebellum for motor and cognitive tasks: an fMRI study. *NeuroImage, 59*, 1560–1570. doi: 10.1016/j.neuroimage.2011.08.065

Taylor & Francis Group. (2018). Checking phones in lectures can cost students half a grade in exams.' *ScienceDaily.* www.sciencedaily.com/releases/2018/07/180731104224.htm, retrieved 1 January, 2022.

Teixeira-Machado, L., Arida, R., & de Jesus Mari, J. (2019). Dance for neuroplasticity: A descriptive systematic review. *Neuroscience & Biobehavioral Reviews, 96*, 232–240. doi: 10.1016/j.neubiorev.2018.12.010

The Times. (2009, December 23). Kim Peek: savant who was the inspiration for the film *Rain Man.*

The Tonight Show Starring Jimmy Fallon. (2021, August 9). *Billie Eilish talks happier than ever, directing music videos and her synesthesia | The Tonight Show* [video]. YouTube. https://www.youtube.com/watch?v=bRfgF_tXsGE

Torelli, C., & Stoner, J. (2015). Managing cultural equity: A theoretical framework for building iconic brands in globalized markets. *Review of marketing research, 12*, 83–120.

Treffert, D. A., & Christensen, D. D. (2005, December). Inside the Mind of a Savant. *Scientific American.* https://www.scientificamerican.com/article/inside-themind-of-a-sava-2005-12/

Triandis, H. C. (2001). Individualism-collectivism and personality. *Journal of personality, 69*(6), 907–924.

Triandis, H. (2003). The future of workforce diversity in international organisations: A commentary. *Applied psychology, 52*, 486–495.

Twilley, N. (2015, October 26). Accounting for taste. *The New Yorker.* https://www.newyorker.com/magazine/2015/11/02/accounting-for-taste

University of Haifa. (2014, September 22). Food memory: Discovery shows how we remember taste experiences. *ScienceDaily*.

Uono, S., & Hietanen, J. K. (2015). Eye contact perception in the West and East: a cross-cultural study. *PLOS ONE, 10*(2), e0118094. doi: 10.1371/journal.pone.0118094

Usher, K., Jackson, D., Walker, R., Durkin, J., Smallwood, R., Robinson, M., Sampson, N., Adams, I. Porter, C., & Marriott, R. (2021, March 31). Indigenous resilience in Australia: A scoping review using a reflective decolonizing collective dialogue. *Frontiers in public health*, 9:630601. doi: 10.3389/fpubh.2021.630601

Valkenburg, P. M., & Peter, J. (2009). Social consequences of the internet for adolescents: A decade of research. *Current directions in psychological science, 18*(1), 1–5.

VandenBos, G. R. (2007). *APA dictionary of psychology*. American Psychological Association.

Vaughan, G. M., & Hogg, M. A. (Eds.) (1998). *Introduction to social psychology*. Sydney: Prentice Hall.

VCAA (Victorian Curriculum and Assessment Authority) (2022). *Victorian Certificate of Education Psychology Study design, 2023–2027*. https://www.vcaa.vic.edu.au/Documents/vce/psychology/2023PsychologySD.docx

Vogel, E. A., Rose, J. P., Roberts, L. R., & Eckles, K. (2014). Social comparison, social media, and self-esteem. *Psychology of popular media culture, 3*(4), 206–222.

von Károlyi, C., Winner, E., Gray, W., & Sherman, G. F. (2003). Dyslexia linked to talent: Global visual-spatial ability. *Brain and language, 85*(3), 427–431.

Wang, S. S. (2014, March 27). How autism can help you land a job. *Wall Street Journal*. http://www.wsj.com/articles/SB10001424052702304418404579465561364868556. Accessed September 21 2022.

Ward, J., & Mattingley, J. B. (2006). Synesthesia: An overview of contemporary findings and controversies. *Cortex, 42*, 129–136.

Watson, D., Clark, L. A., & Tellegen, A. (1988). Development and validation of brief measures of positive and negative affect: the PANAS scales. *Journal of personality and social psychology, 54*, 1063–1070. doi: 10.1037//0022-3514.54.6.1063

Weber B. (2009, December 27). Kim Peek, inspiration for 'rain man', dies at 58. *The New York Times*. https://www.nytimes.com/2009/12/27/us/27peek.html

Weinberg, R. A. (1989). Intelligence and IQ. *American psychologist, 44*(2), 98–104.

Weinstein, Y., Nunes, L., & Karpicke, J. (2016). On the placement of practice questions during study. *Journal of experimental psychology applied, 22*(1), 72–84. doi 10.1037/xap0000071

Weir, K. (2014). The lasting impact of neglect. *Monitor on psychology, 45*(6). https://www.apa.org/monitor/2014/06/neglect

Westerman, T. (2021). Culture-bound syndromes in Aboriginal Australian populations. *Clinical psychologist, 25*(1), 19–35.

Wiklund-Hörnqvist, C., Stillesjö, S., Andersson, M., Jonsson, B. & Nyberg, L. (2022). Retrieval practice is effective regardless of self-reported need for cognition – behavioral and brain imaging evidence. *Frontiers in psychology. 12*. doi: 10.3389/fpsyg.2021.797395

Willis, R. H. (1963). Two dimensions of conformity–nonconformity. *Sociometry, 26*(4), 499–513.

Willis, R. H., & Hollander, E. P. (1964). An experimental study of three response modes in social influence situations. *The Journal of Abnormal and Social Psychology, 69*(2), 150–156.

World Health Organization. (2018). *The ICD-11 Classification of mental and behavioral disorders: Diagnostic criteria for research*. Geneva, Switzerland: World Health Organization.

Zaromb, F. M., & Roediger, H. L. (2010). The testing effect in free recall is associated with enhanced organizational processes. *Memory & cognition, 38*, 995–1008. doi: 10.3758/MC.38.8.995

Ziersch, A., Due, C., & Walsh, M. (2020). Discrimination: A health hazard for people from refugee and asylum-seeking backgrounds resettled in Australia. *BMC public health, 20*, 108.

Zimbardo, P. G., Haney, C., & Banks, W. C. (1973, April 8). *A Pirandellian prison. New York Times Magazine, 8*, 38–60.

Zimbardo, P., Haney, C., Banks, W. C., & Jaffe, D. (1971). *The Stanford Prison Experiment: A simulation study of the psychology of imprisonment*. Zimbardo, Incorporated.

Zubrick, S. R., Shepherd, C. C., Dudgeon, P., Gee, G., Paradies, Y., Scrine, C., & Walker, R. (2014). Social determinants of social and emotional wellbeing. *Working together: Aboriginal and Torres Strait Islander mental health and wellbeing principles and practice, 2*, 93–112.

Index